I0896283

Digital Citizenship and Building a Responsible Online Presence

Hakikur Rahman
Presidency University, Bangladesh

IGI Global
Scientific Publishing
Publishing Tomorrow's Research Today

Vice President of Editorial	Melissa Wagner
Managing Editor of Acquisitions	Mikaela Felty
Managing Editor of Book Development	Jocelynn Hessler
Production Manager	Mike Brehm
Cover Design	Phillip Shickler

Published in the United States of America by
IGI Global Scientific Publishing
701 East Chocolate Avenue
Hershey, PA, 17033, USA
Tel: 717-533-8845
Fax: 717-533-8661
Website: https://www.igi-global.com E-mail: cust@igi-global.com

Copyright © 2025 by IGI Global Scientific Publishing. All rights reserved. No part of this publication may be reproduced, stored or distributed in any form or by any means, electronic or mechanical, including photocopying, without written permission from the publisher.
Product or company names used in this set are for identification purposes only. Inclusion of the names of the products or companies does not indicate a claim of ownership by IGI Global Scientific Publishing of the trademark or registered trademark.

Library of Congress Cataloging-in-Publication Data

Names: Rahman, Hakikur, 1957- editor.
Title: Digital citizenship and building a responsible online presence / edited by Hakikur Rahman.
Description: Hershey, PA : IGI Global Scientific Publishing, [2025] | Includes bibliographical references and index. | Summary: "Through exploring subjects like digital literacy, privacy, online conduct, and worldwide viewpoints, the book aims to enable people to become involved, knowledgeable, and responsible members of the online community. This book offers readers helpful advice, case studies, and stimulating conversations to help them accept digital citizenship as a necessary part of contemporary life"-- Provided by publisher.
Identifiers: LCCN 2024054408 (print) | LCCN 2024054409 (ebook) | ISBN 9798369366752 (hardcover) | ISBN 9798369366769 (paperback) | ISBN 9798369366776 (ebook)
Subjects: LCSH: Computer literacy. | Information technology--Moral and ethical aspects. | Online etiquette. | Citizenship.
Classification: LCC QA76.9.C64 D5185 2025 (print) | LCC QA76.9.C64 (ebook) | DDC 004--dc23/eng/20241205
LC record available at https://lccn.loc.gov/2024054408
LC ebook record available at https://lccn.loc.gov/2024054409

British Cataloguing in Publication Data
A Cataloguing in Publication record for this book is available from the British Library.

ll work contributed to this book is new, previously-unpublished material.
he views expressed in this book are those of the authors, but not necessarily of the publisher.
This book contains information sourced from authentic and highly regarded references, with reasonable efforts made to ensure the reliability of the data and information presented. The authors, editors, and publisher believe the information in this book to be accurate and true as of the date of publication. Every effort has been made to trace and credit the copyright holders of all materials included. However, the authors, editors, and publisher cannot assume responsibility for the validity of all materials or the consequences of their use. Should any copyright material be found unacknowledged, please inform the publisher so that corrections may be made in future reprints.

Table of Contents

Detailed Table of Contents

It's critical to grasp digital citizenship in the connected world of today. The term "digital citizenship" describes the ethical and responsible use of technology, especially when it comes to the internet. It includes a variety of actions, such as protecting personal data and honoring the privacy of others. Digital citizenship encourages us to be responsible, involved, and active participants in the online community, just as we endeavor to be good citizens in our offline communities. We must live our digital lives as examples of respect, honesty, and responsibility if we want to responsibly traverse the digital realm. This chapter delves into the intricacies of managing one's digital footprint. It explores how online actions can shape personal and professional identities and the long-term implications of digital interactions. The chapter offers strategies for cultivating a positive online reputation, understanding the permanence of digital records, and using social media platforms to enhance personal branding.

In the contemporary digital world, the development of social capital and the advancement of digital citizenship are pivotal to fostering inclusive, supportive, and responsible online communities. This chapter discusses the importance of social capital in relation to the advancement of digital citizenship and explores how digital platforms and tools facilitate the formation of social bonds and civic engagement through innovative methods to create an inclusive environment. It examines the dual facets of digital interactions – enhancing community ties and promoting ethical behavior online. Through the data analysis on the current trend of social capital development, this chapter provides insights into the mechanisms by which digital environments can enhance social connectivity and civic responsibility, the quality of digital citizenship and ultimately contributing to a more cohesive society to promote the development of social capital.

The following chapter draws on qualitative data to look at two aspects of computing education: the digital economy and digital literacy. Finally, the chapter compares these to the concept of digital citizenship, considering how 'digital citizenship' can be used to draw the other two together and give young people a deeper understanding of having power and agency in the digital world. The data presented in this chapter is based on fieldwork conducted as part of the author's PhD. While the original fieldwork was conducted in 2016 in a discipline where relatively limited qualitative fieldwork is conducted, it adds valuable insight and perspective. The chapter consists of a brief review of the concept of digital economy and digital literacy before using the data to look at how young people viewed both concepts in the context of what they were being taught as part of 'computing'. The data from the field works followed by a discussion of the findings and takeaways from the data before giving a brief introduction to 'digital citizenship'.

This chapter examines digital citizenship, focusing on digital literacy, ethical behavior, privacy, and security in our interconnected world. It emphasizes the need for individuals to develop responsible online practices to protect their personal information from evolving cyber threats. Key themes include the importance of digital literacy for effective engagement with technology and understanding the ethical implications of online actions. The chapter discusses strategies for educators to instill digital responsibility in youth, ensuring they navigate the digital landscape confidently. It highlights the significance of safeguarding personal information and utilizing privacy settings effectively to mitigate risks like identity theft and cyberbullying. By promoting best practices for online security, the chapter aims to empower individuals to protect their digital identities. Ultimately, it provides a comprehensive overview of digital citizenship, advocating for a proactive approach to harnessing technology's benefits while safeguarding personal rights in the digital realm.

This chapter explores the ethical implications of digital interactions and challenges readers to critically examine their own online conduct as well as the wider effects of their decisions. In a time when social and personal values are shaped by digital contacts, this chapter discusses important ethical issues that come up in online settings, including privacy, respect, accountability, and openness. It provides a combination of theoretical understandings and useful recommendations to assist readers in responsibly and mindfully navigating the intricacies of the digital world, drawing on well-established ethical theories and real-life situations. The chapter promotes digital citizenship that values empathy, inclusion, and integrity and highlights the significance of comprehending how routine digital behaviors contribute to broader ethical frameworks and societal norms.

The rise of the internet and digital platforms has revolutionized the production and sharing of information, allowing individuals to express their opinions quickly and effortlessly. However, the ease of creating

In today's connected world, being a responsible digital citizen is extremely important. With our growing dependence on digital platforms for communication, business, and personal use, it is essential to practice healthy cyber hygiene habits to navigate the digital space effectively. This study highlights three key areas for maintaining basic online hygiene: using secure Wi-Fi networks, avoiding unknown links and attachments, and creating strong passwords. Based on a survey of 157 participants, the research emphasises that coping appraisal is vital in encouraging healthy cyber hygiene behaviours. Offering practical and easily doable advice through educational programs can foster more effective and sustainable online hygiene practices. In an increasingly unstable digital landscape, practicing good cyber hygiene is critical to mitigating risks of cybercrimes and their potential impacts. As digital citizens, we must stay informed and adopt safe online behaviours to navigate the digital world securely.

The emergence of the digital era has brought about a fundamental change in how information is generated, distributed, and consumed. Artificial intelligence (AI), previously celebrated as a champion of pursuing truth, now assumes a dual function in this intricate environment. This chapter examines the complicated relationship between AI and media manipulation, specifically focusing on its capacity to generate convincing deepfakes that control public sentiment and disseminate false information via social media bots. Therefore, a comprehensive literature review was undertaken by gathering many publications to enhance understanding and obtain clarification on the subject matter. To cultivate a knowledgeable and resilient information environment, it is crucial to provide individuals with the necessary skills to assess sources critically, recognize content generated by artificial intelligence, and acknowledge their own biases. This chapter intends to illuminate AI systems in moderating misinformation and determining the most effective strategy for addressing this significant issue.

This chapter explores Gen Z's authenticity code regarding influencer-driven social media advertising. In a context where social media is cluttered with sponsored content, Gen Z seeks for authenticity, contributing to an influencer marketing landscape where brands must crack a unique formula to engage this tech-savvy and socially conscious generation. By analyzing the data gathered from an online survey with a non-probabilistic sample of 105 Gen Z members, this research reveals details regarding transparency, influencer partnerships, and the delicate balance between creativity and authenticity. Our findings show that Gen Z prefers real content instead of amazing but hollow messages, values simplicity instead of megalomaniac advertising productions, prefers the truth over spectacularity, prefers to understand the

Preface

In today's interconnected world, understanding digital citizenship has become paramount. As technology continues to shape our lives, the way we interact online carries profound implications for individuals, communities, and societies. Digital citizenship refers to the responsible and ethical use of technology, particularly in the online sphere. It encompasses a range of behaviors, from respecting others' privacy and intellectual property to safeguarding personal information and engaging in constructive discourse. Just as we strive to be good citizens in our physical communities, digital citizenship urges us to be active, engaged, and accountable members of the online world.

The rapid growth of digital platforms has given rise to unprecedented opportunities for learning, communication, and collaboration. However, it has also introduced challenges, including misinformation, cyberbullying, digital security threats, and ethical dilemmas surrounding data privacy. Navigating these challenges requires a comprehensive understanding of digital literacy, online etiquette, and the legal and ethical responsibilities that come with an increasingly digital existence.

In light of our world's growing interconnectedness, the objective of this book, *Digital Citizenship and Building a Responsible Online Presence*, is to offer a thorough examination of digital citizenship. This volume aims to equip readers with the knowledge, skills, and ethical perspectives necessary to navigate the digital landscape responsibly. By exploring topics such as digital literacy, online privacy, cyber ethics, responsible social media usage, and global perspectives on digital engagement, the book empowers individuals to become informed, engaged, and conscientious digital citizens.

This book brings together contributions from leading experts, scholars, and practitioners in the field of digital citizenship, offering a multidimensional perspective on the subject. Through case studies, practical insights, and thought-provoking discussions, readers will gain a deeper understanding of how to cultivate a responsible online presence. Whether you are an educator, student, policymaker, or general internet user, this book provides valuable guidance on how to engage with digital spaces ethically and effectively.

CHAPTER OVERVIEW

Chapter 1

This chapter introduces the concept of digital citizenship, emphasizing ethical and responsible technology use, particularly in the online environment. It highlights the significance of safeguarding personal data and respecting others' privacy. The chapter explores the long-term effects of online actions, offering strategies to build a positive online reputation and use social media for personal branding. It encourages responsible engagement in the digital world to create a respectful, honest, and secure online presence.

Chapter 2

Exploring the connection between social capital and digital citizenship, this chapter discusses how digital platforms foster social bonds and civic engagement. It focuses on creating inclusive and supportive online communities, enhancing community ties, and promoting ethical behavior. The chapter delves into the current trends in social capital development and its role in building cohesive, responsible digital societies.

Chapter 3

Focusing on the intersection of digital economy and literacy, this chapter investigates how digital citizenship integrates these concepts to empower youth in navigating the digital world. It presents qualitative data on how young people understand digital economy and literacy in the context of computing education, ultimately linking them to the broader concept of digital citizenship, which provides them with agency in the online realm.

Chapter 4

This chapter examines digital literacy, emphasizing the ethical use of technology, privacy, and security. It highlights how individuals can protect their personal information from cyber threats, focusing on educators' roles in instilling responsible online practices in youth. The chapter provides strategies for safeguarding digital identities and fostering responsible online engagement, empowering individuals to navigate the digital landscape with confidence.

Chapter 5

This chapter delves into the ethical challenges posed by digital interactions. It explores privacy, respect, accountability, and openness, offering both theoretical insights and practical recommendations for responsible online behavior. It advocates for a digital citizenship framework that values empathy, integrity, and inclusion, while examining how digital behaviors contribute to societal norms and ethical standards.

Chapter 6

With the rise of misinformation online, this chapter provides tools for evaluating the credibility and relevance of digital content. It outlines five criteria for assessing credibility, three steps for evaluating relevance, and six strategies to enhance the evaluation process, fostering critical thinking and improving digital literacy for managing the complexities of online information effectively.

Chapter 7

This chapter explores digital literacy in the context of governance, using Kazakhstan's Astana IT University as a case study. It examines how digital literacy supports people-centered governance, highlighting the importance of individual participation in decision-making processes. The research aims to inform policy and practice, enhancing digital literacy skills to foster inclusive governance in higher education institutions.

Chapter 8

In the evolving digital landscape, this chapter explores the concept of fair use in digital content. It examines how fair use enables innovation and public discourse while protecting intellectual property rights. The chapter addresses the challenges posed by emerging technologies like AI and social media and advocates for a balanced approach that protects creators' rights while fostering knowledge exchange.

Chapter 9

This chapter emphasizes the importance of empowering young people with digital citizenship skills. It focuses on teaching principles of responsibility, ethics, and safety, helping learners critically evaluate online content, engage respectfully, and protect their privacy. It also highlights the role of integrating the United Nations Sustainable Development Goals into digital citizenship education, fostering global awareness and engagement.

Chapter 10

Exploring digital citizenship from an Indian perspective, this chapter addresses the challenges marginalized communities face on social media, including gender, ethnic, and cultural underrepresentation. It examines the societal impact of these exclusions and the need for diversity and inclusion in digital platforms to promote equitable representation in online discourse.

Chapter 11

This chapter focuses on the integration of green marketing within the healthcare sector, emphasizing how digital platforms can promote sustainability. It discusses how healthcare organizations can adopt environmentally conscious marketing strategies, build trust with eco-conscious consumers, and advocate for sustainability, linking it to the broader concept of digital citizenship in promoting responsible practices.

Chapter 12

In today's digital world, practicing cyber hygiene is crucial. This chapter presents essential practices for maintaining online safety, such as using secure networks, avoiding unknown links, and creating strong passwords. Based on a survey, the chapter emphasizes the role of coping appraisal in encouraging healthy cyber hygiene habits, aiming to reduce the risks of cybercrimes and enhance overall digital security.

Chapter 13

This chapter examines the dual role of AI in both promoting truth and spreading misinformation. It focuses on AI-generated deepfakes and social media bots that influence public opinion and manipulate digital content. The chapter advocates for developing critical thinking skills to assess sources and recognize AI-generated content, promoting a more resilient and informed digital environment.

Chapter 14

Focusing on Gen Z's expectations of authenticity in influencer-driven marketing, this chapter analyzes how brands can engage this generation through transparent, creative, and original content. It presents findings from a survey of Gen Z participants, revealing their preference for authentic content over scripted advertisements and exploring the delicate balance between creativity and transparency in social media marketing.

As we move forward in the digital age, fostering a culture of responsibility, empathy, and awareness in the online world is crucial. It is our hope that this book serves as a valuable resource in helping individuals embrace digital citizenship as an essential component of modern life. By promoting digital responsibility, we can contribute to a safer, more inclusive, and more ethical online environment for current and future generations.

Hakikur Rahman
Presidency University, Bangladesh

Chapter 1
Navigating the Complexities of Online Identity and Reputation

Hakikur Rahman
https://orcid.org/0000-0002-2132-1298
Presidency University, Bangladesh

ABSTRACT

It's critical to grasp digital citizenship in the connected world of today. The term "digital citizenship" describes the ethical and responsible use of technology, especially when it comes to the internet. It includes a variety of actions, such as protecting personal data and honoring the privacy of others. Digital citizenship encourages us to be responsible, involved, and active participants in the online community, just as we endeavor to be good citizens in our offline communities. We must live our digital lives as examples of respect, honesty, and responsibility if we want to responsibly traverse the digital realm. This chapter delves into the intricacies of managing one's digital footprint. It explores how online actions can shape personal and professional identities and the long-term implications of digital interactions. The chapter offers strategies for cultivating a positive online reputation, understanding the permanence of digital records, and using social media platforms to enhance personal branding.

INTRODUCTION

Overview of digital identity and its importance

The general definition of digital identity is the depiction of human identities in a datafied, machine-readable format. This process does not have a single, unified function like that of a digital artifact. According to Nyst et al. (2016, pp. 8–9), digital identity is associated with systems that allow for individual identification as well as authentication systems that control access rights and grant permission to carry out pre-specified tasks or obtain pre-arranged access to services. The three roles of identity, authentication, and authorization are all carried out digitally, according to the authors. This indicates that none of the three aforementioned functions can be matched by an offline process (Giannopoulou,

DOI: 10.4018/979-8-3693-6675-2.ch001

Copyright ©2025, IGI Global Scientific Publishing. Copying or distributing in print or electronic forms without written permission of IGI Global Scientific Publishing is prohibited.

2023). This study discusses the implementation of digital identity through self-sovereign technological infrastructures and associated philosophies, based on the definition of digital identity presented above.

Digital identity has become essential to both personal and professional life in an increasingly linked society. Digital identities include social media profiles, online interactions, and digital footprints. Digital identities are defined as an individual's online representation. This complex identity is a major factor in shaping how people see us in both virtual and real-world settings, in addition to reflecting who we are. Digital identity is significant because of how it affects connections, opportunities, and reputation. While a disjointed or unfavorable digital identity can result in lost opportunities and possible harm, a well-manicured online presence can increase credibility and build trust. Understanding and controlling our digital identity becomes essential as we negotiate the complexity of the online environment. This is necessary for responsible online citizenship as well as for creating a good, genuine presence that is consistent with our goals and beliefs. Figure-1 shows a conceptual view of digital identity.

Figure 1. Digital identity

A conceptual illustration based on digital identity. It visualizes the interplay between online interactions, opportunities, and risks, as well as the complexity of navigating the digital world.

Moreover, digital identity is a multifaceted concept that has been defined and interpreted differently across disciplines such as information technology, sociology, and law. At its core, digital identity refers to the collection of data that uniquely identifies an individual in a digital context. According to Cameron (2005), digital identity can be conceptualized as a set of claims made by an individual or entity about themselves, often validated within specific systems. This perspective emphasizes the contextual and system-dependent nature of identity, where the attributes or claims are recognized within the boundaries of the platform or service.

From a sociological perspective, digital identity extends beyond technical attributes to encompass the ways individuals present themselves and interact with others in online environments. Turkle (1995) argues that online identities can be fluid and multifaceted, allowing individuals to explore and construct versions of themselves that may differ from their offline personas. This aligns with Goffman's (1959) dramaturgical approach to identity, where individuals manage impressions depending on the audience and context—a principle equally applicable in digital interactions.

Frameworks for understanding digital identity also vary. The National Institute of Standards and Technology (NIST, 2017) provides a framework that emphasizes authentication and assurance levels, focusing on how identity attributes are verified to ensure security and trust in digital transactions. Meanwhile, the European Union's eIDAS regulation (EU, 2014) offers a legal framework for electronic identification and trust services, aiming to standardize and secure digital identity across member states. These frameworks highlight the interplay between technology, governance, and user-centric approaches to digital identity (Rahman, 2009).

The evolving definitions and frameworks reflect the complexities of digital identity in modern society, where the balance between security, privacy, and user autonomy is constantly negotiated. As technology continues to shape digital interactions, a comprehensive understanding of digital identity requires integrating technical, social, and legal dimensions.

This chapter explores the complexities of controlling one's digital footprint, which is defined as all the digital footprints we leave behind as a result of our interactions, creations of material, and involvement on different platforms. It investigates how these virtual behaviors—from sharing material and making comments on social media to networking professionally and posting on social media—can significantly mold one's identity on the professional and personal levels. Digital interactions have far-reaching effects that go well beyond the here and now. They can affect relationships, hiring decisions, and even a person's reputation in the community. In a time when information may be permanently saved and spread quickly, it is critical to comprehend the long-term effects of our online conduct.

This study emphasizes the value of deliberate interaction and intentional content sharing while providing practical methods for building a strong online reputation. It also emphasizes the enduring nature of digital records, serving as a helpful reminder to readers that it may be difficult to completely remove something that has been uploaded online. In addition, we will look at how to use social media platforms to further develop personal branding and make sure that the online persona is consistent with both personal and professional objectives. Readers will acquire the skills required to responsibly and confidently negotiate the complexities of their online presence after reading this thorough analysis. It is a thorough study of literatures ranging from classics and latest researches depending on the importance of their contributions in the aspect of digital identity and its intricacies.

A society's ability to identify its members is a prerequisite for any economy to operate. Relationship building would be impossible due to the lack of reliable methods for identifying natural persons. The emergence of the Internet has made it necessary to implement suitable methods of identification in the digital sphere. This gave rise to the concept of "digital identity," which has developed over time from straightforward digital copies of physical forms of identification (like scanning an ID card) to sophisticated cryptography-based forms of identification (like electronic signatures). A new framework for expressing digital identity has been made possible by recent technological advancements, which include the development of blockchain technology, improved smartphone designs, and the growth of cryptography. The user is at the core of this framework. The goal of this framework is to replace the existing "centralized" identification paradigm with a "decentralized" identity model. New concepts are so introduced, including "decentralized" digital identity and even a kind of it called self-sovereign identity (SSI). These ideas are explained in this chapter along with possible applications within the current regulatory framework in Europe and the global aspect (Capisizu, 2020).

Any piece of data that uniquely identifies an account at a networked service—typically an alphanumeric character string, but possibly a biometric like a fingerprint, voiceprint, or retinal scan—is referred to as a digital identifier. The email address is among the most widely used and flexible types of digital identity. Email addresses are useful identifiers because they allow communication between accounts and are globally unique, which is a need for their basic function (Mueller, et. al., 2006).

The speed at which digital identity solutions are being designed makes it imperative to develop reliable tools that are integrated into matching identity infrastructures. The key lies in controlling digital identity and the infrastructures that follow. Digital identity infrastructures are systems that create, manage, and market certain aspects of user identities. In order to mediate these identities, identity providers that exercise power and control over them (I) form these infrastructures through state actors and private commercial actors acting as identity providers; (II) construct identifiers with or without the direct control or intervention of the referred user; and (III) mediate these identities through technological design decisions (Giannopoulou, 2023).

It is commonly acknowledged that there is an issue with a globally and universally trusted digital identity system, or rather that there is one that does not exist. A strong body of knowledge about cryptographic protocols, different architectural styles, and operational but disconnected infrastructures has been accumulated during decades of research. Although achieving a universal internet-based and user-centric digital identification solution may be the ultimate objective, it is extremely unlikely that one special solution architecture will control the worldwide market (Bazarhanova, & Smolander, 2020).

Furthermore, users of the Internet typically need to log in in order to access services that require them to log in. In fact, entering a password and then an identity online has become a common practice. When using an online bank account to pay bills or transfer money, accessing an email account, booking an airline ticket, starting a computer at work, enrolling in an online course, or bidding in an online auction, one must keep in mind the special code linked to that particular service. They will most likely need to provide a form of identification and a password in order to make any kind of online purchase. To view the articles or leave comments, one must first register and receive an identity assignment from numerous free online newspapers and blogs (Mueller, et. al., 2006).

Thus, in order to build an online persona, an individual with an active online life will register with dozens of websites. They will divulge a great deal of personal information to service providers during the registration process, ranging from as little as their name and email address to as much as their home address, office number, national ID card number, gender, age, and credit card details. A person like this will most likely amass over a dozen of these virtual personas. They could find it difficult to recall which one to use or which password goes with which one. They will have to go through the process of presenting and validating these online identities multiple times a day, or hundreds of times a month, after they have registered and created them. Therefore, identity management plays a crucial role in the online economy for both users and the service providers who cater to them. Identity is actually the key to both e-commerce and a variety of other non-commercial communication and trade channels. Figure-2 illustrates the complexities of sharing data and information online.

Figure 2. A conceptual diagram illustrating online identity management.

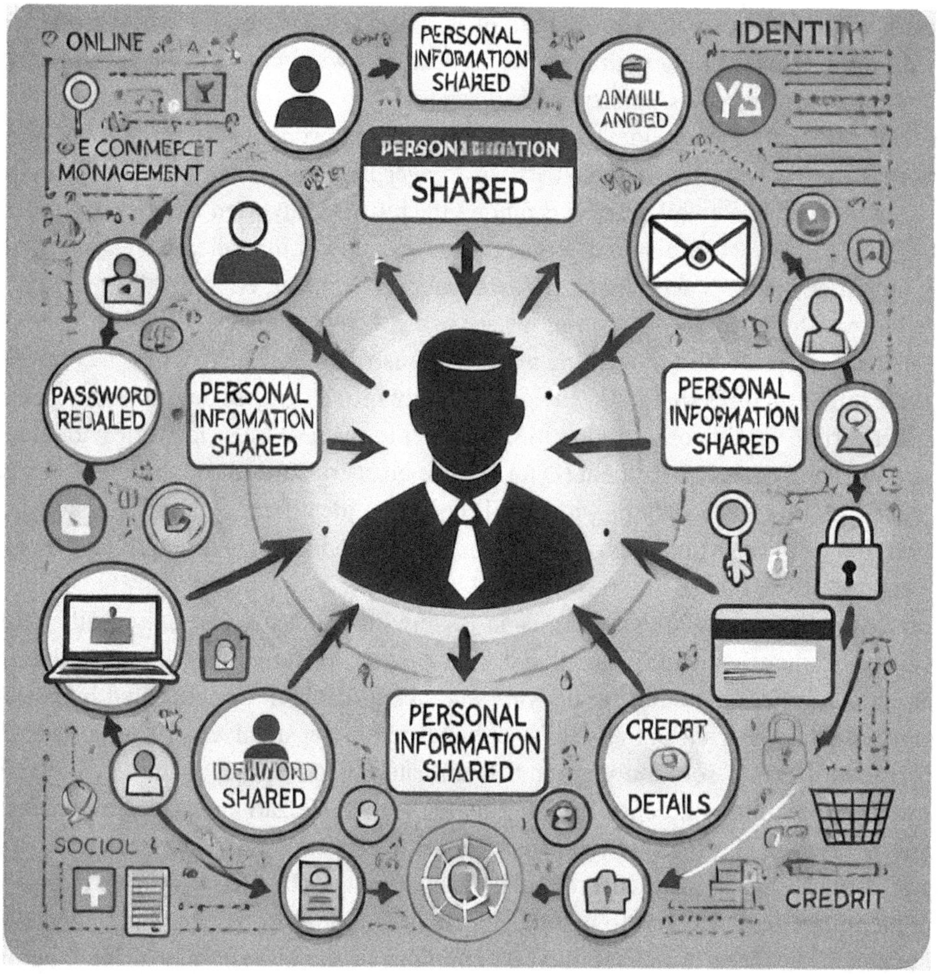

The central figure represents a user, surrounded by various icons symbolizing online personas, including social media, e-commerce, and banking. Arrows connect the user to these personas with labels such as "Personal Information Shared" (e.g., name, email, address, credit card details). The diagram highlights challenges like "Password Recall" and "Identity Validation," represented by icons of a key, a lock, and a checklist. At the bottom, the importance of "Identity Management" is emphasized with icons for e-commerce, communication, and trade.

To cite an example as the starter, this study mentions about a case of French Internet users.

They used 16.4 digital IDs on average in 2013, up from 12.2 in 2009, according to records on "digital trust" from ACSEL, the French Digital Economy Association. When we talk about digital ID, we're talking about the login and password information that's required to open an account as well as any digital footprints that people leave behind from using technology (such as IP addresses, images, and purchase categories). Laurent, et. al. (2015) find that Digital IDs are utilized by 90% of Internet users for e-banking, 7% for social networks, and 8% for e-administration and e-commerce services. According to the same ACSEL research, which assesses Internet users' perceptions of the hazards associated with their digital

ID, 42% of them worry about ID problems (up from 34% in 2009) and 40% worry about data misuse. It's critical to understand that, on average, 400,000 identity thefts occur in France each year. The "breaking point with regard to trust" for an Internet user, according to ACSEL, is "when the risks perceived by the user are considered to be greater than the use value of a digital service."

The concept of online reputation and its impact on individuals and societies

These days, thanks to the widespread use of the Internet, smartphones, tablets, and PCs, customers may express their opinions, wishes, and grievances on the goods and services they use. A growing number of people read, watch, and disseminate content that could harm companies' reputations and result in irreversible or irreparable harm. A business would need to have an online presence, keep an eye on any shared content about the brand, and respond to inquiries and issues from clients in order to obtain a competitive edge. It becomes imperative for businesses to recognize the significance of their reputation in both the real and virtual worlds, to respect consumer feedback, complaints, and ideas, and to actively work toward building a strong reputation (Şirzad, 2022).

At the same time, advances in information and communication technology have made it necessary for organizations to understand the value of their reputation both online and offline. Businesses with a strong reputation tend to draw in more investors and clients. Additionally, companies with a strong reputation for themselves are better at handling crises. One of the main purposes of public relations techniques is reputation management. Managing the relationship between businesses and their audiences and building and sustaining a solid reputation in the real and virtual worlds are two of the key goals of public relations strategies.

Researches have recently shown how the internet and other technologies are altering the dynamics of communication. Young people are the primary consumers of social media, which has grown to be a significant part of our life (Kemp, 2017; Marengo et al., 2018). Our daily usage of social media has affected nearly everything in our lives, from dating to parenting (Perrin, 2015). Studies have indicated that social media and the internet are significant venues for establishing one's reputation. Youth can build their social identities, maintain contacts, and exchange knowledge thanks to social media (Badenes-Ribera et al., 2019; Amin, & Khan, 2021).

Social networking services like Facebook and Instagram made it easier for users to share information online and have since become a vital part of many people's everyday lives in our digitally advanced and highly interconnected world. As a result, there is a noticeable rise in the quantity of personal data that is readily available online, and at the same time, it is simple to locate personal data online. The availability and simple retrieval of publicly available personal data makes it easy to create thorough web profiles and makes it easier to derive implicit information for a variety of uses. Such data creates an online persona and is utilized to assess an individual. Despite initial research indicating that people value their online reputation, people make minimal effort to control it (Yang, 2015). Figure-3 shows the intricate connectivity among various social networks.

Figure 3. A conceptual illustration showing the influence of social networking services like Facebook and Instagram on online personas.

At the center is a silhouette of a person surrounded by interconnected icons representing social media platforms, personal data (e.g., photos, names, and dates), and tools like magnifying glasses and graphs symbolizing data analysis. Arrows indicate data flow from social media to the creation of web profiles and the derivation of implicit information. A highlighted overlay emphasizes the concept of "online reputation," accompanied by a caution symbol.

Cultural and regional contexts play a significant role in shaping the ways digital identity is defined, managed, and perceived. In Western countries, particularly in North America and parts of Europe, digital identity is often centered on individual autonomy and privacy. These regions typically emphasize user-centric identity frameworks, such as self-sovereign identity (SSI), which allow individuals to control their digital identity without relying on centralized intermediaries (Allen et al., 2016). Such frameworks align with broader societal values around personal freedom and data ownership, as seen in the European Union's General Data Protection Regulation (GDPR), which grants individuals extensive rights over their personal data (European Parliament, 2016).

In contrast, collectivist cultures in regions such as East Asia often take a more centralized and community-oriented approach to digital identity. For instance, China has implemented a national digital identity system tightly integrated with its social credit system, where an individual's digital identity is linked to their financial, legal, and social behavior (Meissner, 2017). This approach reflects a cultural emphasis on societal harmony and collective accountability, though it raises concerns about surveillance and privacy.

In the Global South, including regions like Africa and South Asia, digital identity initiatives are often driven by development goals. Programs such as India's Aadhaar—a biometric-based digital identity system—aim to facilitate access to government services, financial inclusion, and social welfare (Banerjee, 2020). While such systems have been praised for their transformative potential, they have also sparked debates around data security, exclusion, and the risks of centralizing sensitive information in resource-constrained settings.

These varying approaches to digital identity reflect the interplay between cultural values, governance structures, and technological capabilities. While some regions prioritize individual privacy, others emphasize collective benefits or developmental objectives. As global digital interactions increase, reconciling these divergent approaches will be critical for creating inclusive and equitable digital identity systems (Rahman, 2021).

Thus the study sets forth to clarify issues related to the concept and complexities of digital footprint or online presence with reputation and try to find out relevant solutions through an in-depth literature vertical review. Firstly, it explores the understanding of digital footprints and the permanence of online actions.

SECTION 1: UNDERSTANDING DIGITAL FOOTPRINTS

Definition and components of a digital footprint

Digital footprint is the information about a particular person that exists on the internet as a result of their online activity (Google definition). A person's distinct collection of traceable digital behaviors, contributions, interactions, and activities that appear on the Internet or on digital devices is referred to as their "digital footprint" or "digital shadow." Digital footprints fall into one of two categories: passive or active.

Any information pertaining to an online activity is referred to as a digital footprint. The user leaves digital traces when they interact with various digital platforms, like web browsers, YouTube, photo-tagging, checked-in locations, and record purchases. Yes, using social media in whatever form contributes to one's digital footprint. The purpose of this study was to categorize the many kinds of digital footprint information that are utilized to forecast psychographic traits and human behavior. 48 studies were systematically analyzed by Nawi, et. al. (2020) to determine which type of digital footprint was considered in the continuing research. The findings demonstrate that digital footprints contain a variety of data kinds, including geographic, time-series, event, structured, unstructured, network, and connected data. To sum up, utilizing digital footprint data is a relatively novel approach to conducting research on psychographic and human behavior prediction. Utilizing digital footprint data offers a fantastic chance to enhance understanding of human behavior

In this aspect, a large portion of our everyday lives, both personal and professional, are now mediated by digital technology. This implies that, oftentimes even without realizing it, we are interacting with information systems all the time. This conversation is logged for the most part, if not all of it, and is frequently kept for a long time. It is possible to analyze this enormous digital footprint to learn more about a person's behavior, needs, and preferences, as well as to forecast their future activities. With this understanding, a new breed of interactive systems might be created, one in which computers would help human creativity and sense-making by doing what they do best, which is data processing and statistical modeling (Sjöberg, et. al., 2017).

How digital footprints are created and tracked

Digital footprints are distinct data trails, either purposefully or accidentally, that users of the Internet leave behind. It is crucial to share knowledge about digital footprints since data uploaded online is never retracted, even after it has been erased.

Digital traces are comparable to physical traces that are specific to each individual. Inadvertently, passive digital traces are created. With the advent of online research on the Internet, search histories and website visits were limited to the computer's IP address rather than being retained on a card catalog or local area network (LAN). Owners of personal computers are now required to register their machines with their IP address and name before they can use them. Consequently, when a user connects to the Internet and visits a website, a passive digital footprint is left behind; the server records the IP address, which indicates the Internet service provider and the computer's location (McDermot, 2018).

A greater percentage of human endeavors, including social connections and entertainment, are being mediated by digital services and technologies in recent times. The social sciences are undergoing a paradigm shift due to the records of those activities, or "big social data." These studies are going from small-scale, usually involving questionnaires or lab-based observations and experiments, to large-scale, where researchers watch the behavior of thousands or millions of people and look for statistical regularities and underlying principles (Lambiotte, & Kosinski, 2014). These studies give previously unheard-of empirical observations on a large scale, which has the potential to significantly advance our knowledge of people and social systems.

According to Blue, et. al. (2018) the distinction between people's online and offline lives is getting more and more hazy as people spend more time communicating on the internet. Digital footprints are the records and maps of the actions that validate an individual's online identity. This automated trail records the routines, passions, happenings, connections, and correspondences that are entwined with a person's bodily existence.

Hence, digital footprints are the collected data resulting from an individual's online activity and digitally detectable conduct. Not only may Internet use have positive or negative short- or long-term effects on an individual's life, but so can an individual's digital footprint. Users' online actions are important, but what is said about them online has an impact as well. Similar to physical footprints, digital footprints do not provide a whole image of a person, but they can enable a range of accurate and varied judgments (Micheli, et. al., 2018). Figure-4 shows a conceptual illustration of digital footprint.

Figure 4. A conceptual diagram illustrating digital footprints.

At the center is an oval labeled "Digital Footprints," surrounded by five connected elements: (1) "Collected Data from Online Activity," represented by icons of a computer, smartphone, and browsing history; (2) "Positive/Negative Effects on Life," depicted by a balance scale; (3) "Online Actions," symbolized by a cursor clicking on a webpage; (4) "Impact of What is Said Online," illustrated by chat bubbles and a megaphone; and (5) "Judgments Based on Digital Footprints," shown with a magnifying glass analyzing footprints. The design features soft blue and green tones with a subtle background of binary code and cloud icons.

People can showcase themselves in a variety of ways on the Internet, ranging from malicious identity fraud to basic self-enhancement. When it comes to choosing people to date, work with, or socialize with, we frequently base our decisions on our impressions of them from the Internet. Hinds & Joinson (2019) observe that researchers investigating personality perception have recently shifted their focus to social media and digital device analysis to see whether an individual's digital footprint might disclose aspects of their identity. Concurrently, developments in "big data" analytics have shown that characteristics of persons can be predicted by computer algorithms based on their digital footprints.

The permanence of online actions

A lot of people take it for granted that texts found online will always be there. While there is some truth to the premise of permanence, this article argues that it is not always the default mode that we think it is in the digital age. This article aims to shed light on the fractures and holes in the façade of permanence by analyzing different digital media. These holes show themselves as three main things: obscurity, insecurity and deletion, and incompatibility and obsolescence. These digital losses and gaps are concerning to an increasing degree as more composition professors adopt increasingly digital pedagogies (Burton, 2015).

Prior to the internet, people might suffer reputational harm if their personal information was made public. However, because digital spaces substantially increase the circulation and longevity of information, it can become practically impossible to forget about past transgressions and move past them (Park, 2022).

However, the reach and permanence of information in the digital age reflect how an individual can completely lose control over everything from their own "likeness" to deeply personal information through the information circulating on the internet. Warren and Brandeis believed that information published by an individual could no longer be granted privacy protections. Furthermore, the lack of confidentiality for information that a person chooses to share with others highlights an unanticipated barrier to the enduring nature of news and information in the modern era.

Hence, the next section dedicates to the ambiance of managing online presence through finding strategies for maintaining privacy and security, tools for monitoring and controlling digital footprints and shows two case studies of effective online presence management.

SECTION 2: MANAGING ONLINE PRESENCE

Strategies for maintaining privacy and security

The "right to be let alone" or privacy is highly valued in Western countries, as stated in an 1890 Harvard Law Review article by Warren and Brandeis. Since then, the right of citizens to control their personal information has been incorporated into the scope of privacy rights as established by US and European courts. According to Warren and Brandeis's (1890) Harvard Law Review article, Western societies consider privacy as a fundamental human right, or the "right to be let alone," even though these rights do not qualify as legally recognized privileges like doctor-patient privileges. Since then, the right of citizens to control their personal information has been incorporated into the scope of privacy rights as established by US and European courts. These rights are protected by American common and statute law, even though they do not equal officially recognized legal privileges (such as doctor-patient privileges) (Quan-Haase, & Ho, 2020).

Similar privacy and security concerns on private data online are not only treated importantly in the US and European countries, but also emerging as a concerned entity in the developing and less privileged countries. They ranges from simple online presence to participation in various social media.

An ecosystem with multiple entities is a social network. Users, the Open Storage Network (OSN) service provider, third-party apps, and advertising are a few examples of these entities. However, users—who receive a variety of social networking services—and OSN providers—who offer those social networking services—are the main players in this ecosystem. Users and OSN service providers alike suffer greatly as a result of the privacy and security issues.

For millions of Internet users, online social networks, or OSNs, have become a commonplace cultural phenomena. Through the integration of user-generated profiles and communication tools that allow users to maintain a virtual constant connection, OSNs capitalize on users' actual social connections and further merge their virtual and physical lives. According to a study made by Kayes & Iamnitchi (2017), Facebook is the third most popular website on the Internet as of 2017 with 1.94 billion monthly active users. More than 313 million people use the social microblogging site Twitter each month, and they post messages in more than 40 languages. Figure-5 shows a conceptual diagram of the online social network (OSN) ecosystem.

Figure 5. Figure shows a conceptual diagram of the online social network (OSN) ecosystem.

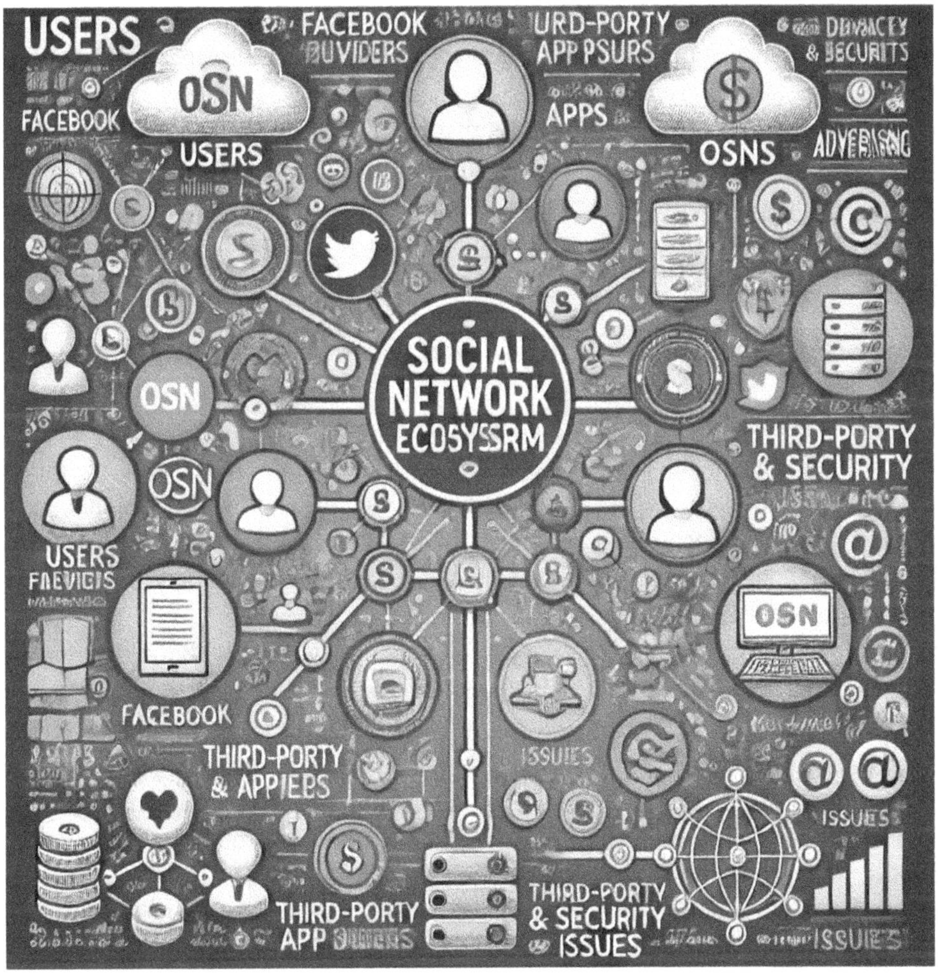

At the center is a node labeled "Social Network Ecosystem," with four connected elements: (1) "Users," depicted by icons of people and a smartphone, representing recipients of social networking services; (2) "OSN Providers," represented by a server and cloud icon, symbolizing service providers;

(3) “Third-Party Apps & Advertising,” shown with app icons and a dollar symbol; and (4) “Privacy & Security Issues,” visualized with a shield, lock, and warning signs. Additional annotations include user statistics like “1.94B Facebook users” and “313M Twitter users.” The diagram employs interconnected lines and a modern, tech-inspired aesthetic with blue and gray tones.

More specifically, if we argue about the mostly utilized version of e-commerce, one of the main barriers to consumers engaging in electronic commerce (e-commerce) transactions—which necessitate the disclosure of personal information such as date of birth, social security number, home phone number, credit card information, and so forth—is their concern for privacy. Thus, safeguarding the privacy of customers is crucial to the success of online shopping. However, Gurung, & Raja, (2016) mention that in order to better understand customer preferences, e-commerce also requires obtaining consumer information. As a result, gathering the customer data needed to increase sales and profitability presents a difficult problem for the managers. Managers must take care to protect their customers' privacy; otherwise, there may be negative public reaction.

Therefore, people's anxiety about the possibility of information misuse is reflected in their online privacy worries, which underscores the need to protect their personal information and online activities from prying eyes. The study of privacy concerns is important because people who don't care about their privacy could unintentionally give up information that could be used for fraud or identity theft. On the other hand, excessive levels of worry could prevent people from fully engaging in digital activities and from utilizing social and informational opportunities. Users have expressed concerns about identity theft, fraud, and online bullying, the misuse of personal information posted online, distrust for online banking, and worries about smart homes and medical/monitoring technologies (Quan-Haase, & Ho, 2020).

Tools for monitoring and controlling digital footprints

Since the internet became such a widely used tool for communication, concerns about digital privacy have grown. People are increasingly disclosing very personal information online, often unintentionally, due to the widespread use of computers and smartphones, the expansion of communication channels, and the explosion of digital information. Teens and young adults in general spend a lot of time on Facebook and MySpace, among other social media websites. Users may not always have their own or their children's best interests at heart when posting certain material.

With the emergence of Web 2.0 and the growth of social networking (Facebook, Twitter, etc.) over the past few years, I have attempted to raise awareness about the potential of these technologies for education as well as the crucial responsibility of managing digital footprints, or the records of an individual's online activities. Professionally speaking, the two have to work together. Strong search engines like Google and identity aggregation businesses like CVgadget, Spokeo, and Pipl are creating sophisticated tools that enable people to profile us when we reveal more of ourselves online and others expose us. Opportunities for relationships, work, education, and other areas are all impacted by this convergence of content and aggregation capabilities, some in positive ways and some negatively. Trackur, ReputationDefender, and Lookuppage are examples of how online reputation management is becoming a profitable industry in reaction to this. (Hengstler, 2011; Moore, 2012).

In a similar study, Nithiananthan (2021) mentions that digital technology is used in the quantified self-movement to gather data on a person's life in order to enhance human performance, self-awareness, and decision-making. Self-awareness and self-knowledge are desirable traits for people to have in order to enhance their performance and results. In order to measure every facet of our everyday lives and increase life efficiency, technology nowadays logs nearly everything. People who use digital services leave behind digital traces of their activity, which present an opportunity to learn more about who they are, what interests them, and how they behave.

Hence, the Internet is no longer private these days. To provide targeted advertisements or customized content, web service providers are attempting to gather every bit of user data for analytics or advertising purposes. Hackers with bad intentions have also found it easier to obtain consumer data. As a result of publicly accessible data, such as personally identifiable information (PII) or compromised account credentials, attackers have more ways to access user accounts and data. As a result, users need to be mindful of the personal information they share and keep a close eye on account security.

The Digital Footprint and Breach Check (DFBC) Reconnaissance Tool, which was created to streamline and expedite the process of obtaining user data that is publicly available and verifying the status of account breaches, is presented in this paper. Because it both a Command Line Interface (CLI) and a Graphical User Interface (GUI), customers with varying levels of experience can use it to better understand the dangers associated with their data and accounts. The tool's primary functions include email breach and social network data collection on Facebook and Twitter. According to the results, 60% of respondents highly agree, and 40% of respondents believe that the information obtained with this suggested method is thorough (Ng, et. al., 2021).

Sjöberg, et. al. (2017) have developed a PDS system implementation, the Digital Me (DiMe) platform. DiMe is a personal data storage system that gathers a person's digital footprint from mobile devices. Its main goal is to make it possible for various information processing and machine learning applications to function inside the user-controlled private data repository. To enable efficient management of vast amounts of personal data, interactive manipulation and automated analysis must work together. DiMe supports both interaction and automated modeling at multiple points throughout the system. DiMe was created primarily for knowledge work applications, but it can also be used in other contexts involving the administration of personal data, including e-commerce, home automation, well-being, personal training, and education. The DiMe platform can be obtained from https://reknow.fi/dime/ and is accessible as free and open source software.

Case studies of effective online presence management

Business websites can now enhance their communication quality with the use of online interactivity. Therefore, it's critical that web designers comprehend the idea and how it impacts the caliber of their work. The five interaction factors proposed by Ha and James (playfulness, connectivity, reciprocal communication, information collecting, and choice) and their association with design excellence were validated empirically in this study. The results indicated that the qualities of reciprocal communication, fun, and connectivity are significant indicators of the quality of a website. Although site engineering and user tracking benefit greatly from information gathering tools, the companies in our survey have not yet completely realized these technologies' tremendous potential. Despite providing consumers with customization and freedom, the choice dimension did not significantly predict the quality of websites. The outcomes emphasized how interactive elements are viable in an online setting (Chen, & Yen, 2004).

In a classic case study conducted by Dow (2008) find that according to students' opinions, there is a conflict between what they want from traditional classroom experiences and what they get from taking classes online. Online learning satisfaction is predicted by social presence. The need for a deeper understanding of social presence in online learning contexts is shown by this grounded theory case study of graduate students studying library and information science. The findings show that efficient dialogue, well-structured interactions, usability, and transparency in computer-mediated interactions are characteristics that influence online social context, online communication, and online interactivity. According to the study's student voices, online courses can make it difficult for students to maintain person-to-person awareness. They also highlight the necessity for a blend of in-person and online education to improve learning outcomes and increase student satisfaction.

Here are several real-life case studies that illustrate effective online identity and reputation management:

1. Repairing a Hedge Fund's Online Reputation

A New York hedge fund experienced significant reputational damage due to a negative post by a disgruntled ex-investor on an anonymous review site. This post quickly rose to prominence in search engine results, adversely affecting the firm's business. Through strategic online reputation management (ORM) efforts, the negative link was suppressed within approximately four months, restoring the firm's online image (Recover Reputation, n.d.).

2. Rebranding an Entertainment CEO's Online Presence

An entertainment industry CEO aimed to revitalize their online reputation to align with a new professional role. By implementing comprehensive ORM strategies, including content creation and search engine optimization, the individual's online presence was successfully reshaped to reflect their current position and aspirations (Recover Reputation, n.d.).

3. Yelp's Legal Action Against Deceptive Reputation Management Practices

In 2024, Yelp pursued legal action against ReviewVio, a reputation management company operating as Dandy, for falsely advertising its ability to remove negative reviews from Yelp's platform. Yelp alleged that such practices misled businesses and consumers, undermining the integrity of online reviews. The court allowed Yelp to proceed with claims of trademark infringement and unfair competition, highlighting the importance of ethical practices in online reputation management (Reuters, 2024).

4. Enhancing Online Profiles for Businesses and Individuals

Reputation X documented various case studies where they improved online profiles for businesses and individuals. Using targeted reputation management strategies, they addressed specific challenges, resulting in enhanced online images for their clients (Reputation X, n.d.).

5. Online Reputation Management for Brands and Executives

ReputationManagement.com offers case studies that showcase their work with global brands, CEOs, and executives. These studies highlight the strategies employed to manage and improve the online reputations of high-profile clients (ReputationManagement.com, n.d.).

Now we talk about the personal branding in the digital age by keeping eyes on the role of social media in shaping public perception, efforts on building a positive online persona and learn on how to leverage digital platforms for career advancement.

SECTION 3: PERSONAL BRANDING IN THE DIGITAL AGE

The role of social media in shaping public perception

Social media offers a forum for the exchange of beliefs, knowledge, and positions. It can also be used to voice concerns and pose inquiries about new developments. Businesses frequently use social media platforms to promote their offerings, counting on users to re-post and distribute pertinent content (Silvia, 2019), which eventually boosts revenue (Etim et al., 2018). Social media analytics, including as engagement, conversion, response rate, and reach, can demonstrate the actual outcomes of advertised initiatives. Given the analytical advantages, the supposition is that policymakers can utilize public health awareness to encourage protective behaviors if companies believe social media to be the most effective medium for promoting their goals. Numerous nations concentrated on this strategy.

For example, social media campaigns promoting health hygiene, social isolation, and stay-at-home advice were started in Sweden. The information was distributed in Arabic, English, Russian, and the official language of Sweden. Even while measurements like reach are traceable, it is difficult to determine whether or not the person followed the instructions, whether it was because of government-sponsored campaigns and banners, or whether it was because of the influence of all the news and conversations on social media (Tsoy, et. al., 2021).

In the current digital era, social media's impact on public opinion about court matters has become a crucial aspect of the justice delivery system, especially in the context of India. This article delves at the different ways that social media shapes public perceptions of court cases and how it affects the justice system. As the world has grown more connected, social media platforms have developed into crucial tools for public discourse, mobilization, and information dissemination. Kumar, & Vyas, (2023) explore the profound effects of social media on the legal landscape, highlighting both the advantages and disadvantages of the platform. The remarkable reach and immediacy that social media offers make it one of the main advantages for disseminating information regarding court matters. They find that participation in court proceedings promotes transparency and accountability within the legal system. Social media may also organize public opinion and activity in a way that leads to significant change and legal reform.

Regarding the role of interpretation, the media ought to offer impartial and morally sound information. In addition to giving users access to already-existing material, new media facilitates the creation of new knowledge by allowing people to share, debate, and disseminate information on viruses. Social media gives anyone a forum to express their opinions, worries, and inquiries. Social media serves as a forum for conversation, in contrast to traditional media, which primarily serves as an informational tool. From an old study made by Johnson & Kaye's (2004) it reveals that blogs have greater trust than traditional media, such as television. The notion that the writer, who is the information source, has firsthand familiarity with the subject matter of the work lends credibility to it.

Furthermore, the material on social media can be edited, shared, and commented upon by other users, enabling the reliability of the data to be confirmed. That could, however, also have unfavorable effects. Inaccurate information and biased media, particularly social media, can worsen prejudicially based problems by exacerbating the situation or perpetuating a false perception of the facts (Cho et al., 2020).

Building a positive online persona

A digital persona is a component of a person's identity that has been expanded into the online space. It is equivalent to a digital unconscious that organizes a split digital self. In order to make room for new approaches to identity management and comprehension, it has to have its personal, social, institutional, legal, scientific, and technical components reexamined. But the fragmented nature of scientific study prevents an interdisciplinary explanation of the fate of the digital personae. This is demonstrated by the gaps in the legal frameworks, the propensity to create fragmented management tools, and the absence of a comprehensive framework that exists today.

When it comes to the opportunities and issues surrounding the management of digital personas in this environment, society, experts, institutions, and groups are still in the pre-conscious, or fragile unconscious, stage. This article's goal is to provide a preliminary list of comprehensive features that influence how individuals and society perceive their digital identities and selfhood, as well as to elicit thought on how our digital personae will be managed on a personal, societal, and institutional level in the future. The goal of this ongoing research is to help define the digital persona and create still-developing models and typologies of digital personhood (de Kerckhove, & De Almeida, 2013).

In terms of creating Academic persona, as part of the Times Higher Education World University Rankings, Thomson Reuters, one of the world's biggest publishers, has collaborated on the "Global Institutional Profiles Project." In order to prove that its data are the most trustworthy for the industry and useful for those 'using' the industry, such as governments, businesses, and students who want to know where best practices can be found, Thomson Reuters is attempting to rank institutions and disciplines within institutions. This pilot study sheds light on how difficult it is to create and keep up an intellectual online presence.

Marshall, et. al. (2017) find that providing participants with access to and knowledge about a variety of potential platforms that can be utilized in the building of an academic identity was one of the pilot project's main goals. The academics that took part in this pilot study were individuals who were active on various social networking and/or professional platforms.

Before Facebook, Twitter, and the majority of other digital media platforms became commonplace in our online life. Jay Bolter (2000) predicted that online activities will change the way that people interact with the internet. A "networked self" is "displacing Cartesian printed self as a cultural paradigm," he said, highlighting how we comprehend and create identity (Bolter, 2000, p. 26). The twenty-first century has seen a boom in scholarly research on the relationship between public digital identities and technology, in addition to the growth and widespread popularization of venues for their creation.

These methods have mostly ignored the emergence of the non-human as a component of social networks and the relationships between people and their networks of other people, frequently missing the wider implications of what personas are and might be. In this introduction, we aim to both map out the existing literature on subjectivity and the public self-presentation through networked technologies and add to these growing accounts by offering a succinct synopsis of the five key aspects of an online identity.

The five dimensions of persona that we identify and explain below are public, mediatized, performative, collective, and having purposeful value. These dimensions are primary, while we accept that they are not all-inclusive or full. They are:

- The Public Dimension of Online Persona
- Mediatised Dimension of Persona
- Performative Dimension of Persona
- Collective Dimension of Persona
- Value Dimension of Persona (Moore, et.al., 2017).

Leveraging digital platforms for career advancement

A crucial sector of the economy, work and employment are changing significantly as a result of increased Internet usage and connectivity as well as the emergence of digital platforms. A new and rapidly expanding method of working is called crowdwork. It is accompanied by a constant dispute between opposing viewpoints over how it affects workers. Nevertheless, both viewpoints take a static approach that ignores how crowdworkers advance in their careers over time. In order to further this discussion, we use a dynamic perspective on crowdwork in this study to examine the temporal trajectory of workers based on their own perspectives.

A study made by Idowu, & Elbanna, (2022) reveals that via extensive qualitative data and inductive analysis, that crowdworkers create what may be thought of as a path toward career growth. It delineates five phases in this professional trajectory and the endeavors of laborers to shape their work obligations and job-related assets in order to fashion a future for themselves. The conversation demonstrates the useful understanding that this method offers to both theory and practice. Figure-6 gives a conceptual flowchart illustrating the professional trajectory of crowdworkers, as described in Idowu & Elbanna (2022).

Figure 6. Flowchart

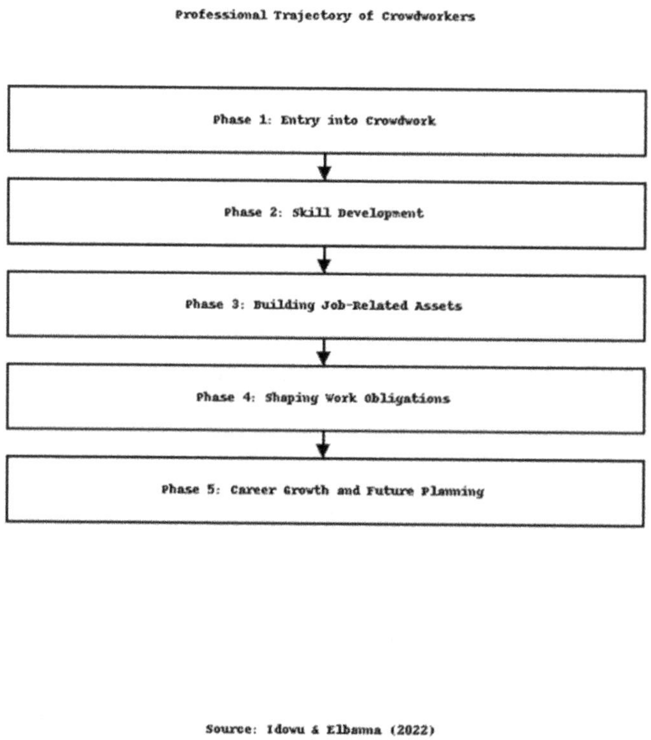

The flowchart consists of five phases arranged sequentially from top to bottom: 1. Entry into Crowdwork, 2. Skill Development, 3. Building Job-Related Assets, 4. Shaping Work Obligations, and 5. Career Growth and Future Planning. Arrows connect each phase, showing the progression from one to the next. The diagram emphasizes the efforts of laborers to shape their obligations and assets to build a future.

Retrospectively, information technology is becoming a necessary component in many disciplines due to its widespread use in modern life. In the current globalized and rapidly developing technological era, the role of information technology in the progress of human resource jobs is becoming more and more important. The way that organizations manage their human resources has changed significantly as a result of the effects that information technology has had on the corporate environment.

The process of gathering data necessitates a meticulous approach to data observation and documentation. To extract meaningful conclusions, data is then subjected to analytical techniques including data reduction, visualisation, and inference. The results of the study suggest that there is a strong correlation between the use of information technology and the advancement of human resources' professional careers. Additionally, the results show that information technology makes knowledge easy to discover, opens doors for online learning, promotes professional networking, and gives tools to increase efficiency.

Additionally, the results show that information technology makes knowledge easy to discover, opens doors for online learning, promotes professional networking, and gives tools to increase efficiency. This makes it easier for HR departments to grow their professional networks, add new skills to their repertoire, and improve current job opportunities (Prastyaningtyas, et. al., 2023).

Thereafter, we like to discuss about the consequences of digital actions through legal implications of online behavior, the right to be forgotten and digital erasure to save the entities with a few real-life examples of online reputation affecting personal and professional lives.

SECTION 4: THE CONSEQUENCES OF DIGITAL ACTIONS

Legal implications of online behavior

Along these contexts Harris, B., & Birnbaum, R. (2015) thoroughly examine the advantages of internet counseling as well as its ethical and legal ramifications. They cover topics like technology, asynchronous communication, accessibility, anonymity, online security, informed consent, and the difficulties associated with regulation, liability, and licensing in the field of online counseling. According to the research, internet counseling is crucial in offering a substitute service delivery model that gets around some of the obstacles that traditional face-to-face counseling services have. The authors do draw the conclusion that there is still a dearth of empirical data and a number of open questions regarding the suitability of online counseling for a range of demographics and the efficacy of online mental health therapies.

In another research Bailey, R. A. (2011) finds that a 22-year-old woman who has never had any serious medical history is experiencing burning when she urinates. As a busy, tech-savvy college student, she chooses not to wait to see a doctor in person and instead has an online consultation. She uses a credit card to pay $45 for an online medical consultation, logs into the website, and fills out a form outlining her medical history and describing her symptoms and concerns at hand. That day, the on-call online physician, whom she had never met before, looks over her records, emails her, and determines that she has a urinary tract infection (UTI).

The doctor could not see the patient because they were only communicating via email. He gives her an antibiotic prescription, which he electronically submits to a nearby drugstore. The research particularly places a question here, as "What ethical and legal ramifications result from this interaction?"

Customers are using the internet more and more to get information to help them decide what to buy when it comes to travel-related products. They frequently depend on advice about where to go from a variety of sources, including independent experts, sellers, and more and more other customers. User-generated content (UGC) is a new breed of online content that presents a variety of legal and social concerns to content creators and consumers, particularly with regard to issues like defamation, misrepresentation, and social shame. A study conducted by Burgess, et. al. (2011) presents research that looked at how reliable these various sources of online travel information were.

The study included a poll of 12,000 Australian travelers, and the findings are consistent with the idea that people's degree of trust in various internet travel sources varies. The majority of respondents "tended to agree" that they trusted information from commercial operators, travel agents, and other websites as well as comments left by other travelers. However, material found on tourism websites run by state governments received the highest degree of trust. These findings imply that comments about travel made online are more trustworthy when they are found on specialized travel websites as opposed to more general social networking sites.

The right to be forgotten and digital erasure

The capacity to remove out-of-date personal data has come under scrutiny as online privacy becomes a more pressing problem. As a result, despite scant public opinion research, numerous state and federal legislation modeled after the European "Right to be Forgotten" have been proposed in the United States. Public opinion is a crucial factor to take into account when considering the possibility of a right to be forgotten in the US because legislators and judges frequently take it into account when determining what laws to enact and how to interpret those laws.

To find out further, Bode, L., & Jones, M. L. (2018) examine survey data obtained from Amazon's Mechanical Turk (N = 1,380) to ascertain expected levels of support and factors that indicate popular support for such a bill. They discover that women, people who show the most faith in Google, and those who are especially worried about internet privacy are more likely to favor such a rule. People who believe that the government is overly expansive are less inclined to back such laws. This points to a convoluted path ahead for the United States in the context of a Right to be Forgotten, with support coalitions spanning conventional political divides.

Moreover, the right to oblivion, often known as the right to be forgotten, is currently the subject of heated discussion in high-level circles. Legislators in the European Union have been debating the necessity of this kind of right in the digital age for months. Officials from the Council of Europe have also voiced their concerns, while politicians from various countries have also raised their voices, data protection authorities, human rights organizations, academics, and experts from a variety of backgrounds have joined the march.

The right of natural persons to have their personal data removed after a predetermined amount of time is in question.

According to two distinct perspectives—the criminal history and the data protection laws—this has already been partially acknowledged as a right (see below, points 4.1 and 4.2). However, the growth of information and communication technology (ICT) has been influencing the need to reconsider the expansion of that right's purview. In this field, technological advancement has had a significant impact. New checks and balances between the unrestricted flow of information and personal autonomy are necessary in the age of the Internet. The right to be forgotten nowadays is all about striking this balance (De Terwangne, 2014).

Following the recent European Court of Justice decision in Google Spain v. Spanish Data Protection Agency (ECJ C-131/12), the "right to be forgotten" has garnered global media attention once more. When the European Commission first suggested a right to be forgotten in 2012, there was a lot of public interest in the idea and much discussion around it. Individuals should therefore have the ability to remove personal information about them under specific circumstances.

More recently, the idea appears to be nearing completion in light of the European Parliament's adoption of the LIBE Committee's changes on March 14, 2014. It is now known as a "right of erasure," even though it is mostly unchanged from the previously distributed iterations.

Researching in this aspect, Tamò, A., & George, D. (2014) contend that, despite its catchy name, the right to be forgotten is actually a general word that unifies two existing legal provisions: the significant right of erasure derived from data protection, and the rather procedural right of oblivion.

Real-life examples of online reputation affecting personal and professional lives

Over the past ten years, online review sites have grown in importance as a source of consumer information. Review websites for professionals, such as physicians and attorneys, are starting to catch up by following the example set by eateries and lodging establishments. According to a 2014 survey, potential clients actively look for online reviews of attorneys and are willing to go further to work with an attorney who has received good feedback. However, the media is increasingly covering examples of people and companies who have handled internet reviews in remarkably harmful ways, coinciding with the rise in significance of reputational data. Figure-7 shows a hierarchical flowchart titled "The Growing Role of Online Review Sites."

Figure 7. Diagram

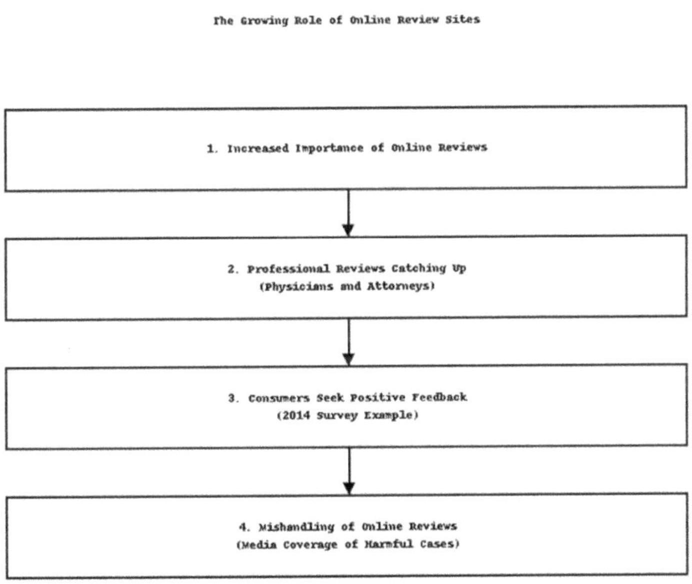

The diagram consists of four connected boxes: 1. "Increased Importance of Online Reviews" at the top. 2. "Professional Reviews Catching Up (Physicians and Attorneys)" in the second layer. 3. "Consumers Seek Positive Feedback (2014 Survey Example)" in the third layer. 4. "Mishandling of Online Reviews (Media Coverage of Harmful Cases)" at the bottom. Each box is connected with downward arrows, showing the progression. At the bottom, the source is noted as "Based on the provided text."

Lawyers are not above the law; as recent disciplinary cases and ethics opinions have shown, attorneys may also feel inclined to lash out at former clients in a way that goes beyond their duties of loyalty and secrecy in response to unfavorable assessments (Robertson, 2016).

The phenomena of voluntary self-disclosure on social networking sites is examined by the author in this article. Self-disclosure is essential to the success of social networking sites because it allows member interactions, service adjustments, targeted marketing, and the creation of digital content. And yet, in spite of their desire to socialize inside the increasingly popular virtual community, members are frequently reluctant to divulge personal information. We created a brief study methodology that identifies important facilitators and barriers to member self-disclosure by consulting pertinent literature.

These factors include computing settings (e.g., perceived Internet risk), networking service attributes (e.g., perceived critical mass), and personality features (e.g., extroversion and privacy value). In a study conducted by Chen, R. (2013) 222 social networking site users' survey responses were used to validate the research model, and the analysis's findings supported the links that had been theorized. The current study adds to our understanding of the ways in which various elements influence members' self-disclosure and provides guidance to networking service providers on how to more effectively encourage members to share personal information online.

Implications of artificial intelligence and machine learning on digital identity management

Artificial intelligence (AI) and machine learning (ML) are revolutionizing digital identity management, bringing both opportunities and challenges. AI and ML technologies have enabled more sophisticated methods of identity verification, fraud detection, and personalized user experiences. For instance, biometric systems powered by AI, such as facial recognition and fingerprint scanning, are increasingly used to verify identities in real-time. These systems offer higher accuracy and efficiency compared to traditional methods, reducing the risk of identity fraud (Jain et al., 2016). Additionally, ML algorithms can analyze vast amounts of data to detect anomalies in user behavior, allowing systems to identify and mitigate identity theft or account compromise proactively (Rahman et al., 2019).

However, the integration of AI and ML into digital identity management raises significant ethical and technical concerns. One critical issue is algorithmic bias, where AI systems may inadvertently discriminate against certain demographic groups due to biases present in the training data. For example, studies have shown that facial recognition systems often perform less accurately for individuals with darker skin tones or those from underrepresented groups (Buolamwini & Gebru, 2018). Such biases can exacerbate digital exclusion, particularly in regions where access to technology is already limited.

Privacy concerns also emerge when AI and ML systems are used to aggregate and analyze personal data for identity management. The ability of ML algorithms to infer sensitive information from seemingly unrelated data poses risks to user privacy. For instance, predictive analytics could inadvertently reveal personal characteristics such as sexual orientation, political preferences, or health conditions without user consent (Zuboff, 2019). This has led to calls for stricter data governance frameworks and transparency in AI systems to ensure that user privacy is protected.

Moreover, the reliance on AI and ML for identity management raises questions about accountability and trust. If an AI-driven identity verification system denies access to a legitimate user or misidentifies an individual, it is often unclear who should be held accountable—the developers, the operators, or the technology itself. This "black-box" nature of AI systems complicates efforts to build trust in digital identity solutions, necessitating the development of explainable AI (XAI) models that make the decision-making process more transparent (Arrieta et al., 2020).

Despite these challenges, the potential of AI and ML to enhance digital identity management is undeniable. By leveraging these technologies responsibly, organizations can create more secure, inclusive, and user-friendly identity systems. However, achieving this balance requires robust regulatory frameworks, continuous monitoring for biases, and ethical considerations in system design. As AI continues to evolve, its role in shaping digital identity will be pivotal in defining the future of digital citizenship.

To comprehend the theme of this title, now the emphasis is given on educating for a responsible online presence through a form of curriculum development for digital citizenship education, learning teaching strategies for different online age groups and at the same time discussing about the role of parents, educators, and policymakers in guiding online behavior.

SECTION 5: EDUCATING FOR A RESPONSIBLE ONLINE PRESENCE

Curriculum development for digital citizenship education

The rapid advancement of technology, especially the widespread availability of the Internet and the emergence of social media in the 21st century, offers never-before-seen chances for connection and empowerment to people and communities. The idea of citizenship and how it is practiced have also been tested and altered by it. Particularly, the Internet's explosive growth has revolutionized people's civic and sociopolitical engagement, fostering a more self-actualized form of citizenship, and having a direct impact on the processes of social and political decision-making as well as democratization.

In this context, Bennett, W. L. (2008) gives a brief overview of how the pervasive influence of technology, especially globalization and ubiquitous connectedness through social media, is changing the concept of "citizenship" in the twenty-first century. He emphasizes that traditionally, the concept of "citizenship" was inextricably linked to the geopolitical realities of specific nation states. He also talk about how these developments affect citizenship education, including how to define digital citizenship, what makes a curriculum for it, and how difficult it is to put into practice.

Therefore, education authorities must adopt a coordinated and comprehensive approach to digital citizenship education and integrate it into school curricula to ensure that it is implemented effectively, even in the face of global efforts to address issues regarding the role of education for the development of digital citizenship.

Furthermore, rather than empowering children through education or the acquisition of competencies to enable them to actively participate in digital society, the Council of Europe's actions concerning children's digital lives over the past ten years have primarily focused on their safety and protection in the digital environment. The Council of Europe's member states are guided by a number of legally enforceable instruments when it comes to safeguarding children in the digital age. Additionally, the European Court of Human Rights has established precedents on ICTs and human rights (Frau-Meigs, et. al., 2017).

Başarmak, et. al (2019) did a research to examine secondary education curriculum from the perspective of digital citizenship subdimensions. The data source was the secondary school curriculum for the first, second, third, and fourth grades in high school for the 2017–2018 academic year. Computer Science; English; Geography; Democracy and Human Rights; Religion and Ethics; Philosophy; French; Visual Arts; Music; Medical Knowledge and Traffic; Revolution History and Atatürkism; History; Turkish Language and Literature; and Contemporary Turkish and World History are among the curricula that are examined in this scope. Data for the study were gathered through document examination, and qualitative research methodologies were used throughout.

In their research "Digital Citizenship Scale for the Young,", using Kuş, Güneş, Başarmak, and Yakar's (2017) the induction analysis approach was employed to analyze the data. They mentioned that according to their current survey, the courses with the greatest number of curriculum references to digital citizenship are "computer science" and "democracy and human rights." Though digital skills are extensively covered in computer science curricula, there is less emphasis on ethical and critical thinking abilities as well as rights and obligations in the digital world? However, there is very little content on digital citizenship in the other courses' curricula.

Teaching strategies for different online age groups

Two researchers (Simonds, T. A., & Brock, B. L., 2014) investigated the online learning preferences of students in their study. The research paradigm they employed was the scholarship of teaching and learning process, and they included an online survey and virtual focus groups into the online courses they were instructing. Following data collection, the researchers tested their hypothesis—that there was a relationship between various student variables and student preferences for particular kinds of online learning activities—by performing multiple logistic regression analyses. A statistically significant correlation was found between the age of the students and their preferences for particular kinds of online learning activities, according to the data analysis results.

They (Simonds, T. A., & Brock, B. L., 2014) also found that younger students tended to choose more participatory learning tactics, but older students in the study showed a far stronger preference for videos of the professor lecturing. Older students' focus group remarks shed light on some of the reasons why they thought watching video lectures was beneficial for their learning, while younger students' responses show how they learn best in online courses. Based on the study's findings, the researchers make recommendations for online instructors and explain why the scholarship of teaching and learning research process may be particularly beneficial to them in their endeavors for both teaching and research.

However, the employment of Internet-based technology in education has been one of the primary trends since the late 20th century. The Internet is becoming a common educational instrument, and learners of all ages and genders are increasingly choosing to receive their education online. Nonetheless, compared to traditional classroom learning, online learning is thought to be more difficult. Since online learning requires the use of computer technology as a delivery method, students who are not accustomed to utilizing computers may find it challenging. For instance, female students typically have lower computer self-efficacy or are less familiar with computers than male students, and older students typically have less computer competence than younger students (Chyung, 2007). Figure-8 shows a hierarchical flowchart titled "Internet-Based Technology in Education."

Figure 8. Diagram

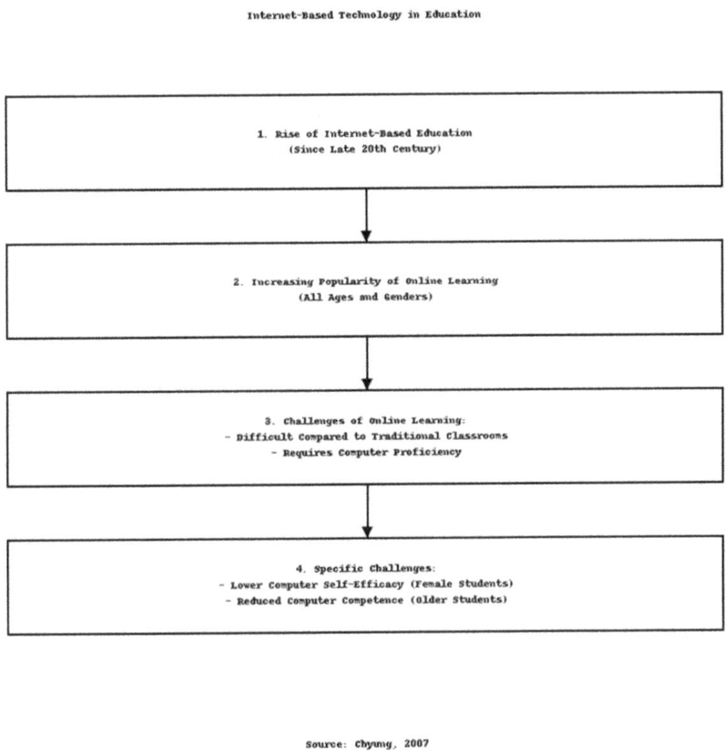

The diagram consists of four connected boxes: 1. "Rise of Internet-Based Education (Since Late 20th Century)" 2. "Increasing Popularity of Online Learning (All Ages and Genders)" 3. "Challenges of Online Learning: - Difficult Compared to Traditional Classrooms - Requires Computer Proficiency" 4. "Specific Challenges: - Lower Computer Self-Efficacy (Female Students) - Reduced Computer Competence (Older Students)." Arrows connect each box sequentially, illustrating the progression from the rise of internet-based education to specific challenges faced by different demographics.

In a mixed-method study (Ke, F., & Kwak, D., 2013) investigated the potential for consistent learning across age and ethnicity groups in terms of perception, involvement, and learning satisfaction in online learning interactions. Students registered in 28 online courses provided data through interviews, online interaction transcripts used for content analysis, and learning-experience survey responses used for structural equation modeling. Students who are non-traditional in age or who identify as minorities did not show any discernible benefit or disadvantage in terms of the quantity and quality of online interaction involvement, according to the interaction-transcript study. The survey results' structural equation modeling, however, showed that minority status of the students was linked to a more positive assessment of the learner-teacher connection but a lower level of satisfaction with web-based distance learning.

The role of parents, educators, and policymakers in guiding online behavior

To find out the role of parents, educators, and policymakers in guiding online behavior Youn, S. (2008) in his study looks at how parents' influence affects teenagers' attitudes on privacy protection. According to survey results, youths who have high levels of concept-oriented family communication also have a tendency to participate in conversation mediation, which has an impact on their privacy concern. Teens with excellent socio-oriented communication skills, on the other hand, typically have more family norms and use the Internet alongside their parents. While co-surfing mediation is related to teens' level of privacy worry, rulemaking mediation is not directly related to their level of privacy concern. This study also reveals that teenagers' attitudes regarding privacy protection measures are explained by parental mediation and their degree of concern.

However, Banić & Orehovački (2024) find that parental involvement in guiding their children's internet usage has become increasingly important in today's digital world. This study explores the changing patterns of parental supervision over their kids' online use. Parents have a critical role in monitoring and protecting their children's online experiences as the digital world has a greater and greater impact on young people's lives. In order to investigate the tactics parents use, the content they limit, the guidelines they set, the possibility of exposure to unsuitable content, and the influence of parents' computer literacy on their kids' online safety, the study tackles important research problems. A thematic topic added to the research broadens the analysis by including ideas from studies that don't directly address the main concerns but yet provide important context and understanding to the field of digital parenting.

Building on this, a movement towards more proactive parental involvement is highlighted by the results of a systematic literature review that was carried out in compliance with the Preferred Reporting Items for Systematic Reviews and Meta-Analyses (PRISMA) criteria. These results, which draw from 49 studies across 11 databases, highlight the most recent developments in parental mediation techniques and trends. In addition to the widely used restrictive mediation techniques, active mediation strategies—which entail constructive dialogue and beneficial engagement with online content—are becoming more and more well-known. In an effort to protect their children's safety and privacy, parents are deliberately banning some types of online content.

The term "digital parenting" describes techniques for raising children that optimize the positive effects and reduce any possible hazards associated with their use of digital media and online places. For many caretakers, weighing the benefits and drawbacks of early digital usage can be difficult.

Choy, et. al. (2024) made a scoping review, which draws from 40 studies published in international peer-reviewed journals between 2010 and 2023, synthesizes the evidence addressing digital parenting techniques and their impact on children's digital use and development. This scoping review produced four themes.

- First, there were disparities in the ways that parents saw early internet use: as "educational aids," as "distractions," and as a result of cultural differences.
- Second, digital parenting techniques—parental modeling, parenting style, parental mediation, and the aim of children's digital use—had an impact on kids' digital use.
- Thirdly, there was a correlation found between the different outcomes of digital parenting and the amount of time kids spent on the internet. The outcomes included the kids' digital literacy, parent-child interactions, language and social-emotional development, behavioral problems, and emergent literacy.

- Fourthly, the ages of the children, the kinds of digital resources available, and parental and family-related factors (such as gender, socioeconomic level, ethnicity, family structure, religion, and parents' digital competence) all had an impact. According to the review, future studies should focus on parental digital literacy training programs and use monitoring tools to more accurately gauge their children's digital usage.

Addressing the need for user education on digital footprint

In an increasingly digital world, understanding one's digital footprint—the trail of data created through online interactions—has become crucial. A digital footprint can include anything from social media posts and online purchases to search engine queries and geolocation data. While a digital footprint can facilitate convenience and personalization, it also poses risks such as privacy breaches, reputational harm, and identity theft. Yet, many users remain unaware of how their online actions contribute to their digital footprint and the long-term implications of these actions. This highlights the pressing need for user education in this domain (Madden et al., 2013).

Education on digital footprints is essential for fostering digital literacy and empowering users to make informed decisions about their online presence. Research has shown that individuals who understand how their data is collected, shared, and monetized are more likely to take proactive measures to protect their privacy (Livingstone et al., 2018). For instance, users educated on digital footprints are more inclined to adjust privacy settings, use secure passwords, and think critically before sharing personal information online. Such practices not only safeguard individual users but also contribute to building a more secure and responsible online environment.

Moreover, user education must extend beyond basic technical knowledge to address the ethical and reputational dimensions of online behavior. A poorly managed digital footprint can have lasting consequences, especially for young people who may not fully grasp the permanence of online actions. Inappropriate or controversial posts made during adolescence, for example, can resurface years later, affecting professional or personal opportunities (Peterson & Johnston, 2015). Educational programs must therefore emphasize the importance of curating a positive digital identity, teaching users to think critically about how their online actions may impact their long-term reputation.

Governments, educational institutions, and technology companies all have roles to play in promoting awareness and providing resources for digital footprint education. Initiatives such as eSafety programs and online tutorials can help users of all ages understand the implications of their digital actions. Additionally, incorporating digital literacy into school curricula can equip students with the skills necessary to navigate the complexities of digital identity and reputation management. For instance, schools in countries like Finland and Estonia have successfully integrated digital literacy, including digital footprint awareness, into their education systems, setting an example for global adoption (OECD, 2021).

Ultimately, empowering users with the knowledge and tools to manage their digital footprints is essential for cultivating responsible online citizenship. As digital interactions increasingly permeate every aspect of modern life, understanding and managing one's digital footprint is no longer optional but a vital skill for navigating the complexities of the online world.

Detailing out on the concept of digital footprint, its perception in the digital arena and various aspects on privacies, concerns and realization of digital footprint in our daily livelihood, we like to draw our conclusion.

CONCLUSION

Keeping an eye on one's online identity and reputation is essential for both professional and personal development in the linked world of today. This chapter has looked at the significant influence that digital behaviors can have on how people are viewed in different contexts. The tactics addressed underscore the significance of exercising intention and diligence in all online interactions, ranging from comprehending the enduring nature of digital records to harnessing social media for personal branding purposes. People can improve their reputations, gain trust, and generate possibilities that support their long-term objectives by developing a positive digital footprint.

In the future, as technology develops and social standards shift, the field of online identity management will also undergo changes. In order to protect their reputations, keep up with new developments, and adopt ethical online conduct, digital citizens must continue to be flexible and proactive. The values of integrity, accountability, and respect must guide us as we negotiate the intricate world of digital technology. By doing this, we support a more moral and reliable online community in addition to safeguarding our own identities. By adhering to these principles, we may navigate the digital terrain in an ethical and successful manner, building a solid and long-lasting online presence. Thereby what we have found to endeavor the future outlook on digital identity and reputation management.

Future outlook on digital identity and reputation management

In addition to using the professional services that identity management helps them with, more and more Internet users are interacting with one another and using services that other users have built. Security needs and partner trust are major concerns in dealings with professional services, but they are even more so when interacting with strangers. A user may or may not get their expectations met regarding the behavior of their interaction partners. Future perceptions of interaction partners who live up to these standards will reward those who do, while those who don't look to be reliable.

Steinbrecher, et. al. (2011) found in their research that most users modify their behavior based on past interaction partners' reliability in subsequent encounters. Although professional services make an effort to adhere to regulatory requirements and quality standards, it is frequently unrealistic to expect individuals to provide such a degree of professional service. However, identity management as it exists now does not help with or even address this problem.

This study draws attention to the issue of digital identity, or cross-set unique identifying tokens, which arises when social software is used in different dimensions. However, Jennings, B., & Finkelstein, A. (2009) find that a single source for digital identities is required when social software—blogs, wikis, and other Web 2.0 tools—begins to add value to the organization. On the basis of this, notions of trust and reputation can be used in the context of bigger workflow processes and human agents interacting with such software. It is possible to link human agent behaviors and data generation via a digital identity across diverse data sources by analyzing human agent activity within pre-existing data sets and provide a method for contributing new data. With this improved understanding of human agents, new business procedures could be developed.

The book chapter examines the relationship between technologies 4.0 and branding strategies by emphasizing the importance of stakeholders in the innovative context. It focuses on corporate branding and digital transformation, highlighting the potential and difficulties associated with managing one's identity and reputation. The study retraces the pivotal role that consumers play in marketing and demonstrates how methods that placed the product at the center have evolved to take the target market into account more and more. It focuses on how Internet technology is changing society and how professional communication has evolved as a result of its application in various sectors (Cuomo, & Foroudi, 2023). Furthermore, the study also looks at how brand equity is affected by social media marketing initiatives and activities in terms of digital footprint.

Ethical Implications of Data Usage and User Consent

The ethical implications of data usage and user consent have become a central concern in the digital age, as vast amounts of personal information are collected, stored, and analyzed by online platforms and organizations. At its core, the principle of user consent serves as a cornerstone for respecting individual autonomy and privacy. However, the implementation of this principle often falls short due to opaque data collection practices and complex terms of service agreements that are difficult for users to understand. Studies have shown that most users do not read or fully comprehend privacy policies, leading to uninformed consent and potential misuse of their data (Obar & Oeldorf-Hirsch, 2020). This raises significant ethical concerns about whether such consent can truly be considered valid. Figure-9 illustrates a conceptual diagram on ethical implications of data usage.

Figure 9. Diagram

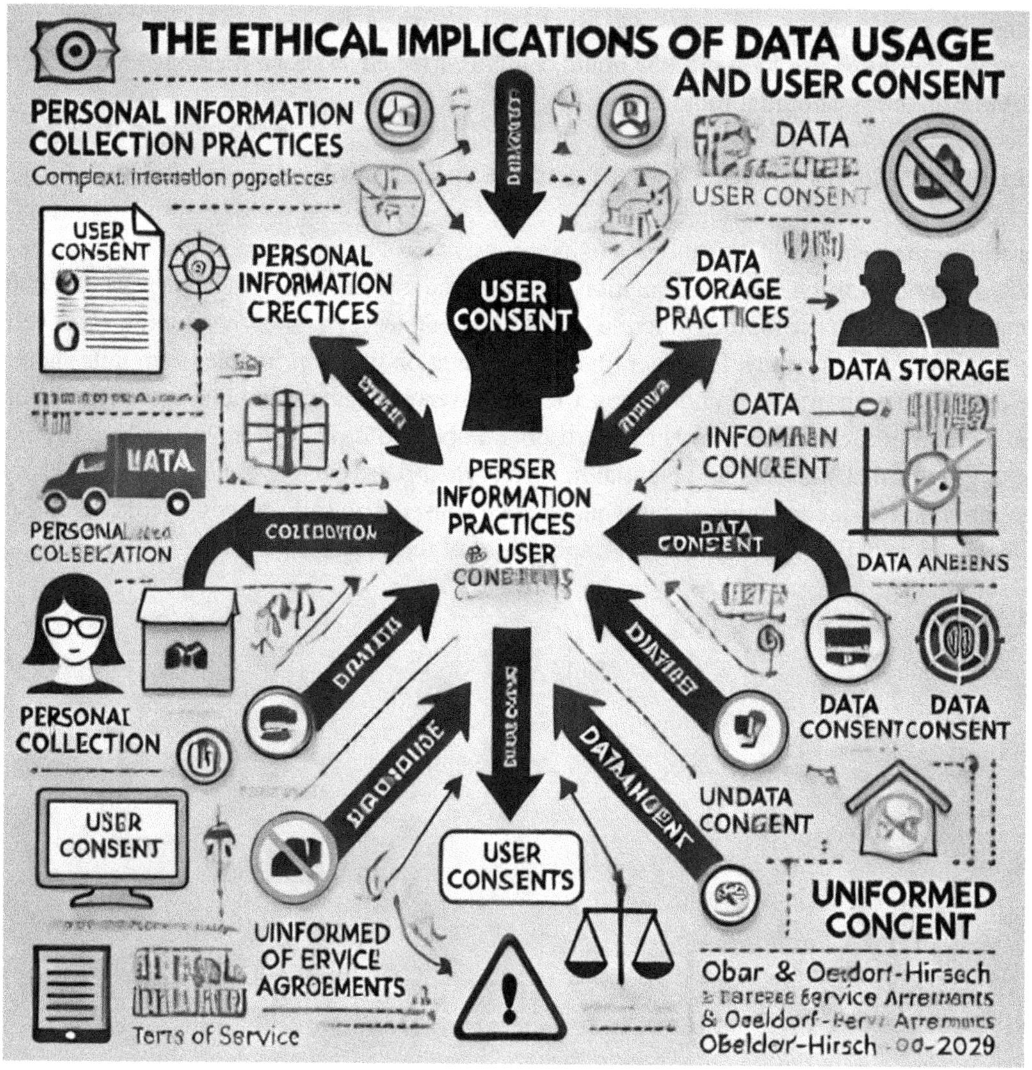

Shows a conceptual diagram illustrating the ethical implications of data usage and user consent.

A key ethical challenge lies in balancing business interests with user rights. While organizations leverage user data to improve services, personalize experiences, and drive revenue through targeted advertising, these practices often involve collecting more data than is strictly necessary. This over-collection of data, coupled with the lack of transparency in how it is used, risks eroding user trust and violating their right to privacy (Zuboff, 2019). For instance, scandals like the Cambridge Analytica incident highlight how data can be weaponized for purposes beyond the user's understanding or consent, such as manipulating political opinions or influencing elections (Cadwalladr & Graham-Harrison, 2018).

Furthermore, the ethical implications extend to the broader societal impact of data usage. Without robust safeguards, marginalized groups can face disproportionate risks of harm, such as surveillance, discrimination, or exclusion. For example, biased algorithms trained on unrepresentative data can perpetuate inequalities in areas like hiring, lending, or law enforcement (Noble, 2018). This underscores the importance of ethical data governance that prioritizes fairness, accountability, and inclusivity.

Addressing these ethical concerns requires rethinking the framework for user consent. Emerging concepts like "dynamic consent" offer a more user-centric approach, allowing individuals to provide, revoke, or modify their consent in real-time as their understanding of data practices evolves (Kaye et al., 2015). Additionally, regulatory frameworks such as the General Data Protection Regulation (GDPR) emphasize transparency, data minimization, and user empowerment, setting a global benchmark for ethical data practices. However, achieving true ethical data usage demands not only compliance with regulations but also a cultural shift among organizations toward prioritizing user rights and trust over profit.

Final thoughts on the importance of a responsible online presence

It has been observed that teenagers in particular do not believe that responsible and ethical Internet use is necessary, and irresponsible and offensive online behavior is starting to have major repercussions. In this aspect, instructors and teacher-librarians have a responsibility to help students make responsible decisions when interacting in the digital world by educating them about the potentially disastrous consequences of careless, improper, or malicious online behavior (Oxley, 2011).

It has also been observed that student participation is a requirement for course design that maximizes student learning. In this context, Stearns, S. A. (2017) looked at one strategy for increasing student involvement in online forums. How to persuade students to read the bulk of their student-colleagues' comments in online discussions—rather than just a minimum number of them—was particularly crucial. The findings show that the theme analysis approach used is successful in raising students' reading rates. The students' assertions that using this learning tool prompted them to think critically and that much of their learning was based on social constructivist theory are maybe even more intriguing. Students also mentioned how much they enjoyed the learning experience.

Since there has been a considerable rise in cybercrime, the importance of ethics becomes even more crucial in recent days. The threat of unethical behavior in cyberspace must be stopped immediately, and ethical standards must be included to a secure digital environment. Users need to grasp how current societal norms and ethics might be expanded in response to the challenges given by ethical dilemmas in cyberspace in order to minimize the spread of cybercrimes and unethical concerns online. Therefore, it is now more important than ever to comprehend the significance of cyber ethics in day-to-day living (Santhosh, & Thiyagu, 2024). Cyber ethics is actually about social responsibility in cyberspace and aims to instill in people the knowledge of responsible behavior when utilizing the online environment. It involves the rules of social engagement and accountability in cyberspace.

To strengthen the security aspects, potentially disruptive identity-related issues have surfaced in recent years, such self-sovereign identity (SSI) for sovereign data control or digital twin technology for product lifecycle management. Ante, et. al. (2022) find out through a bibliometric analysis of 1,395 peer-reviewed articles and associated 44,412 references, uncovered research streams and emerging trends in academic research on digital identity. Using co-citation analysis, they identify six different research streams and how they relate to each other. The six areas of research are as follows:

i) attribute-based encryption schemes;
ii) distributed networks and user privacy;
iii) user authentication in wireless sensor networks;

iv) attribute-based encryption schemes;
v) secure data exchange in the Internet of Things; and
vi) blockchain and smart contracts for secure data management.

What we conclude is that the literature on digital identity management systems (IdM) is extensive, and the non-technical requirements and technology components of the solutions vary. Eventually, though, identity exchange will be required across domains and even borders, necessitating flexible structures and interoperable solutions. An overview of recent research on digital identity management is what this study attempts to provide. In similar context, Bazarhanova & Smolander (2020) derive the underlying non-technical assumptions from a thorough literature study of digital identification solution architectures. In both cases the results demonstrate that preconceptions about organizations, businesses, trust, and human users may all inform solution designs. Specifically, building confidence and cooperation between participating organizations; the ability of human users to preserve confidential cryptographic content; or the presumption that win-win business models are readily discernible.

REFERENCES

Allen, C., Berg, C., & Lane, A. M. (2016). *The Path to Self-Sovereign Identity*. Retrieved December 22, 2025 from https://ssimeetup.org/

Amin, F., & Khan, M. F. (2021). Online reputation and stress: Discovering the dark side of social media. *FIIB Business Review*, *10*(2), 181–192. DOI: 10.1177/2319714520948171

Ante, L., Fischer, C., & Strehle, E. (2022). A bibliometric review of research on digital identity: Research streams, influential works and future research paths. *Journal of Manufacturing Systems*, *62*, 523–538. DOI: 10.1016/j.jmsy.2022.01.005

Arrieta, A. B., Díaz-Rodríguez, N., Del Ser, J., Bennetot, A., Tabik, S., Barbado, A., & Herrera, F. (2020). Explainable artificial intelligence (XAI): Concepts, taxonomies, opportunities, and challenges toward responsible AI. *Information Fusion*, *58*, 82–115. DOI: 10.1016/j.inffus.2019.12.012

Badenes-Ribera, L., Fabris, M. A., Gastaldi, F. G. M., Prino, L. E., & Longobardi, C. (2019). Parent and peer attachment as predictors of Facebook addiction symptoms in different developmental stages (early adolescents and adolescents). *Addictive Behaviors*, *95*, 226–232. DOI: 10.1016/j.addbeh.2019.05.009 PMID: 31103243

Bailey, R. A. (2011). The legal, financial, and ethical implications of online medical consultations. *J. Tech. L. & Pol'y*, *16*, 53.

Banerjee, S. (2020). Aadhaar: Digital identity as a developmental tool in India. *Information Systems Frontiers*, *22*(3), 557–569. DOI: 10.1007/s10796-019-09931-5

Banić, L., & Orehovački, T. (2024). A Comparison of Parenting Strategies in a Digital Environment: A Systematic Literature Review. *Multimodal Technologies and Interaction*, *8*(4), 32. DOI: 10.3390/mti8040032

Başarmak, U., Yakar, H., Güneş, E., & Kuş, Z. (2019). Analysis of digital citizenship subject contents of secondary education curricula. *Turkish Online Journal of Qualitative Inquiry*, *10*(1), 26–51. DOI: 10.17569/tojqi.438333

Bazarhanova, A., & Smolander, K. (2020). The review of non-technical assumptions in digital identity architectures.

Bennett, W. L. (2008). Changing citizenship in the digital age. In Bennett, W. L. (Ed.), *Civic life online: Learning how digital media can engage youth*. The MIT Press.

Blue, J., Condell, J., & Lunney, T. (2018, July). Digital footprints: your unique identity. In *Proceedings of the 32nd International BCS Human Computer Interaction Conference*. BCS Learning & Development.

Bode, L., & Jones, M. L. (2018). Do Americans want a right to be forgotten? Estimating public support for digital erasure legislation. *Policy and Internet*, *10*(3), 244–263. DOI: 10.1002/poi3.174

Bolter, J. (2000). Identity. In Swiss, T. (Ed.), *Unspun: Key Concepts for Understanding the World Wide Web* (pp. 17–29). New York University Press.

Buolamwini, J., & Gebru, T. (2018). Gender shades: Intersectional accuracy disparities in commercial gender classification. *Proceedings of the 1st Conference on Fairness, Accountability, and Transparency*, 81, 77–91. Retrieved December 24, 2024 from https://proceedings.mlr.press/v81/buolamwini18a.html

Burgess, S., Sellitto, C., Cox, C., & Buultjens, J. (2011). Trust perceptions of online travel information by different content creators: Some social and legal implications. *Information Systems Frontiers*, *13*(2), 221–235. DOI: 10.1007/s10796-009-9192-x

Burton, K. L. (2015). Eternal or ephemera? The myth of permanence in online writing. *Computers and Composition*, *38*, 57–67. DOI: 10.1016/j.compcom.2015.09.010

Cadwalladr, C., & Graham-Harrison, E. (2018). Revealed: 50 million Facebook profiles harvested for Cambridge Analytica in major data breach. *The Guardian*. Retrieved December 15, 2024 from https://www.theguardian.com/

Cameron, K. (2005). *The Laws of Identity*. Retrieved December 24, 2025 from https://www.identityblog.com/?p=354

Capisizu, L. A. (2020). Digital identity. *Conferin a Interna ională de Drept, Studii Europene i Rela ii Interna ionale, 8*(VIII), 256-263.

Chen, K., & Yen, D. C. (2004). Improving the quality of online presence through interactivity. *Information & Management*, *42*(1), 217–226. DOI: 10.1016/j.im.2004.01.005

Chen, R. (2013). Living a private life in public social networks: An exploration of member self-disclosure. *Decision Support Systems*, *55*(3), 661–668. DOI: 10.1016/j.dss.2012.12.003

Cho, H., Li, W., Cannon, J., Lopez, R., & Song, C. (2020). Testing three explanations for stigmatization of people of Asian descent during COVID-19: Maladaptive coping, biased media use, or racial prejudice? *Ethnicity & Health*, *0*(0), 1–16. PMID: 33059486

Choy, Y. N., Lau, E. Y. H., & Wu, D. (2024). Digital parenting and its impact on early childhood development: A scoping review. *Education and Information Technologies*, *29*(16), 1–41. DOI: 10.1007/s10639-024-12643-w

Chyung, S. Y. Y. (2007). Age and gender differences in online behavior, self-efficacy, and academic performance. *Quarterly Review of Distance Education*, *8*(3).

Cuomo, M. T., & Foroudi, P. (2023). Digital transformation: Corporate branding, identity, and reputation. In *Digital Transformation and Corporate Branding* (pp. 1–6). Routledge. DOI: 10.4324/9781003263784-1

de Kerckhove, D., & De Almeida, C. M. (2013). What is a digital persona? *Technoetic Arts : a Journal of Speculative Research*, *11*(3), 277–287. DOI: 10.1386/tear.11.3.277_1

de Terwangne, C. (2014). The right to be forgotten and informational autonomy in the digital environment. In *The ethics of memory in a digital age: Interrogating the right to be forgotten* (pp. 82–101). Palgrave Macmillan UK.

Dow, M. J. (2008). Implications of social presence for online learning: A case study of MLS students. *Journal of Education for Library and Information Science*, •••, 231–242.

Etim, A. E., Uzonna, I., & Worgu Steve, C. (2018). Social media usage and firm performance: Reflections from the Nigerian telecommunication sector. *International Journal of Management Science and Business Administration, 4*(6), 7–16. DOI: 10.18775/ijmsba.1849-5664-5419.2014.46.1001

European Parliament. (2016). *General Data Protection Regulation (GDPR).* Retrieved December 20, 2024 from https://eur-lex.europa.eu/legal-content/EN/TXT/?uri=CELEX%3A32016R0679

European Union (EU). (2014). *Regulation (EU) No 910/2014 on electronic identification and trust services for electronic transactions in the internal market (eIDAS).* Retrieved December 20, 2025 from https://eur-lex.europa.eu

Ferdinand, P. (2000). The internet, democracy and democratization. *Democratization, 7*(1), 1–17. DOI: 10.1080/13510340008403642

Frau-Meigs, D., O'Neill, B., Soriani, A., & Tomé, V. (2017). Digital citizenship education: Volume 1: Overview and new perspectives.

Giannopoulou, A. (2023). Digital identity infrastructures: A critical approach of self-sovereign identity. *Digital Society : Ethics, Socio-Legal and Governance of Digital Technology, 2*(2), 18. DOI: 10.1007/s44206-023-00049-z PMID: 37200582

Goffman, E. (1959). *The Presentation of Self in Everyday Life.* Doubleday.

Gurung, A., & Raja, M. K. (2016). Online privacy and security concerns of consumers. *Information and Computer Security, 24*(4), 348–371. DOI: 10.1108/ICS-05-2015-0020

Harris, B., & Birnbaum, R. (2015). Ethical and legal implications on the use of technology in counselling. *Clinical Social Work Journal, 43*(2), 133–141. DOI: 10.1007/s10615-014-0515-0

Hengstler, J. (2017). Managing your digital footprint: Ostriches v. eagles. Education for a digital world, 2(1), 89-139.

Hinds, J., & Joinson, A. (2019). Human and computer personality prediction from digital footprints. *Current Directions in Psychological Science, 28*(2), 204–211. DOI: 10.1177/0963721419827849

Idowu, A., & Elbanna, A. (2022). Digital platforms of work and the crafting of career path: The crowdworkers' perspective. *Information Systems Frontiers, 24*(2), 441–457. DOI: 10.1007/s10796-020-10036-1

Jain, A. K., Ross, A., & Nandakumar, K. (2016). *Introduction to Biometrics.* Springer., DOI: 10.1007/978-0-387-77326-1

Jennings, B., & Finkelstein, A. (2009). Digital identity and reputation in the context of a bounded social ecosystem. In *Business Process Management Workshops: BPM 2008 International Workshops, Milano, Italy, September 1-4, 2008. Revised Papers 6* (pp. 687-697). Springer Berlin Heidelberg. DOI: 10.1007/978-3-642-00328-8_69

Johnson, T. J., & Kaye, B. K. (2004). Wag the blog: How reliance on traditional media and the Internet influence credibility perceptions of Weblogs among blog users. *Journalism & Mass Communication Quarterly, 81*(3), 622–642. DOI: 10.1177/107769900408100310

Kaye, J., Whitley, E. A., Lund, D., Morrison, M., Teare, H., & Melham, K. (2015). Dynamic consent: A patient interface for twenty-first-century research networks. *European Journal of Human Genetics : EJHG, 23*(2), 141–146. DOI: 10.1038/ejhg.2014.71 PMID: 24801761

Kayes, I., & Iamnitchi, A. (2017). Privacy and security in online social networks: A survey. *Online Social Networks and Media, 3*, 1–21. DOI: 10.1016/j.osnem.2017.09.001

Ke, F., & Kwak, D. (2013). Online learning across ethnicity and age: A study on learning interaction participation, perception, and learning satisfaction. *Computers & Education, 61*, 43–51. DOI: 10.1016/j.compedu.2012.09.003

Kemp, S. (2017). *Digital in 2017: Global Overview. A collection of internet, social media, and mobile data from around the world.* Wearesocial & Hootsuite.

Kumar, D., & Vyas, S. (2023). *The Role of Social Media in Shaping Public Perception of Legal Cases and Its Influence on the Justice Delivery System.* Rawat Prakashan.

Lambiotte, R., & Kosinski, M. (2014). Tracking the digital footprints of personality. *Proceedings of the IEEE, 102*(12), 1934–1939. DOI: 10.1109/JPROC.2014.2359054

Laurent, M., Denouël, J., Levallois-Barth, C., & Waelbroeck, P. (2015). Digital identity. In *Digital identity management* (pp. 1–45). Elsevier. DOI: 10.1016/B978-1-78548-004-1.50001-8

Law, N. W. Y., Chow, S. L., & Fu, K. W. (2018). Digital citizenship and social media: A curriculum perspective. *Second handbook of information technology in primary and secondary education.*

Livingstone, S., Stoilova, M., & Nandagiri, R. (2018). Children's data and privacy online: Growing up in a digital age. *London School of Economics and Political Science*. Retrieved December 18, 2024 from https://www.lse.ac.uk/

Madden, M., Lenhart, A., Cortesi, S., & Gasser, U. (2013). Teens, social media, and privacy. *Pew Research Center.* Retrieved December 21, 2024 from https://www.pewresearch.org/

Marengo, D., Longobardi, C., Fabris, M. A., & Settanni, M. (2018). Highly-visual social media and internalizing symptoms in adolescence: The mediating role of body image concerns. *Computers in Human Behavior, 82*, 63–69. DOI: 10.1016/j.chb.2018.01.003

Marshall, P. D., Kim, B., & Christopher, M. (2017). Academic persona: The construction of online reputation in the modern academy. In *The digital academic* (pp. 47–62). Routledge. DOI: 10.4324/9781315473611-4

McDermot, M. (2018). Digital footprints: Creation, implication, and higher education. *FDLA Journal, 3*(1), 11.

Meissner, M. (2017). China's social credit system: A big-data enabled approach to market regulation with broad implications for doing business in China. *Mercator Institute for China Studies (MERICS). Papers on China, 16*, 1–24. Retrieved December 20, 2024, from https://merics.org

Micheli, M., Lutz, C., & Büchi, M. (2018). Digital footprints: An emerging dimension of digital inequality. *Journal of Information. Communication and Ethics in Society, 16*(3), 242–251. DOI: 10.1108/JICES-02-2018-0014

Moore, C., Barbour, K., & Lee, K. (2017). Five dimensions of online persona. *Persona Studies, 3*(1), 1–11. DOI: 10.21153/ps2017vol3no1art658

Moore, S. C. (2012). Digital Footprints on the Internet. *The International Journal of Childbirth Education, 27*(3).

Mueller, M. L., Park, Y., Lee, J., & Kim, T. Y. (2006). Digital identity: How users value the attributes of online identifiers. *Information Economics and Policy, 18*(4), 405–422. DOI: 10.1016/j.infoecopol.2006.04.002

National Institute of Standards and Technology (NIST). (2017). *Digital Identity Guidelines.* Special Publication 800-63-3. Retrieved from DOI: 10.6028/NIST.SP.800-63-3

Nawi, A., Hussin, Z., Ren, C. C., Norsaidi, N. S., & Mohd Pozi, M. S. (2020). Identifying the types of digital footprint data used to predict psychographic and human behaviour. In *Digital Libraries at Times of Massive Societal Transition: 22nd International Conference on Asia-Pacific Digital Libraries, ICADL 2020, Kyoto, Japan, November 30–December 1, 2020* [Springer International Publishing.]. *Proceedings, 22*, 287–296.

Ng, C. K., Yusof, Y., & Ab Aziz, N. S. N. (2021, December). DFBC Recon Tool: Digital Footprint and Breach Check Reconnaissance Tool. In *2021 14th International Conference on Developments in eSystems Engineering (DeSE)* (pp. 526-530). IEEE.

Nithiananthan, N. (2021). Identify User behavior on digital devices by using digital footprints (Doctoral dissertation).

Noble, S. U. (2018). *Algorithms of oppression: How search engines reinforce racism.* NYU Press. DOI: 10.18574/nyu/9781479833641.001.0001

Nyst, C., Makin, P., Pannifer, S., & Whitley, E. (2016). *Digital identity: Issue analysis: executive summary.* Consult Hyperion.

Obar, J. A., & Oeldorf-Hirsch, A. (2020). The biggest lie on the internet: Ignoring the privacy policies and terms of service policies of social networking services. *Information Communication and Society, 23*(1), 128–147. DOI: 10.1080/1369118X.2018.1486870

OECD. (2021). *21st-century readers: Developing literacy skills in a digital world.* Organisation for Economic Co-operation and Development., DOI: 10.1787/a83d84cb-

Oxley, C. (2011). Digital citizenship: Developing an ethical and responsible online culture. *Access : the Newsmagazine of the American Dental Hygienists' Association, 25*(3), 5–9.

Parate, S., Josyula, H. P., & Reddi, L. T. (2023). Digital identity verification: Transforming KYC processes in banking through advanced technology and enhanced security measures. *International Research Journal of Modernization in Engineering Technology and Science, 5*(9), 128–137.

Park, R. (2022). Forgiving Without Forgetting? Privacy in an Age of Digital Permanence.

Perrin, A. (2015). *Social media usage: 2005–2015.* Pew Research Trust.

Peterson, H. L., & Johnston, M. A. (2015). Digital footprints: Creating responsible digital citizens. *Journal of Digital and Media Literacy*, *3*(2), 101–118.

Prastyaningtyas, E. W., Ausat, A. M. A., Muhamad, L. F., Wanof, M. I., & Suherlan, S. (2023). The Role of Information Technology in Improving Human Resources Career Development. *Jurnal Teknologi Dan Sistem Informasi Bisnis*, *5*(3), 266–275. DOI: 10.47233/jteksis.v5i3.870

Quan-Haase, A., & Ho, D. (2020). Online privacy concerns and privacy protection strategies among older adults in East York, Canada. *Journal of the Association for Information Science and Technology*, *71*(9), 1089–1102. DOI: 10.1002/asi.24364

Rahman, H. (2009, November). Framework of e-governance at the local government level. In *Proceedings of the 3rd international conference on Theory and practice of electronic governance* (pp. 372-374). DOI: 10.1145/1693042.1693121

Rahman, H. (2021). Framework of E-governance at the grass roots level. In *Ubiquitous Technologies for Human Development and Knowledge Management* (pp. 78–114). IGI Global. DOI: 10.4018/978-1-7998-7844-5.ch005

Rahman, M. S., Liu, X., & Hossain, M. A. (2019). Anomaly detection in online social networks using machine learning algorithms. *Social Network Analysis and Mining*, *9*(1), 1–16. DOI: 10.1007/s13278-019-0566-6

Recover Reputation. (n.d.). *5 online reputation management case studies*. Retrieved December 23, 2024 from https://www.recoverreputation.com/5-online-reputation-management-case-studies

Reputation, X. (n.d.). *Case studies*. Retrieved December 23, 2024 from https://blog.reputationx.com/case-studies

ReputationManagement.com. (n.d.). *Case studies*. Retrieved December 21, 2024 from https://www.reputationmanagement.com/case-studies

Reuters. (2024, June 7). *Yelp can sue reputation company for promising to suppress bad reviews*. Retrieved December 21, 2024 from https://www.reuters.com/legal/yelp-can-sue-reputation-company-promising-suppress-bad-reviews-2024-06-07

Robertson, C. B. (2016). Online reputation management in attorney regulation. *The Georgetown Journal of Legal Ethics*, *29*, 97.

Santhosh, T., & Thiyagu, K. (2024). Fostering Responsible Behavior Online-Relevance of Cyber Ethics Education. *Malaysian Online Journal of Educational Technology*, *12*(1), 32–38. DOI: 10.52380/mojet.2024.12.1.428

Silvia, S. (2019). The Importance of Social Media and Digital Marketing to Attract Millennials' Behavior as a Consumer. Marketing, 4(2).

Simonds, T. A., & Brock, B. L. (2014). Relationship between age, experience, and student preference for types of learning activities in online courses. *The Journal of Educators Online*, *11*(1), n1. DOI: 10.9743/JEO.2014.1.3

Şirzad, N. (2022). A review on online reputation management and online reputation components. *Doğuş Üniversitesi Dergisi*, *23*(1), 219–242.

Sjöberg, M., Chen, H. H., Floréen, P., Koskela, M., Kuikkaniemi, K., Lehtiniemi, T., & Peltonen, J. (2017). Digital me: Controlling and making sense of my digital footprint. In *Symbiotic Interaction: 5th International Workshop, Symbiotic 2016, Padua, Italy, September 29–30, 2016, Revised Selected Papers 5* (pp. 155-167). Springer International Publishing.

Stearns, S. A. (2017). Student Responsible Learning: Getting Students to Read Online Discussions. *College Teaching*, *65*(2), 69–78. DOI: 10.1080/87567555.2016.1244654

Steinbrecher, S., Pingel, F., & Juschka, A. (2011). Reputation management as an extension of future identity management. In *Digital Privacy: PRIME-Privacy and Identity Management for Europe* (pp. 557–568). Springer Berlin Heidelberg. DOI: 10.1007/978-3-642-19050-6_19

Tamò, A., & George, D. (2014). Oblivion, erasure and forgetting in the digital age. *J. Intell. Prop. Info. Tech. & Elec. Com. L.*, *5*, 71.

Tsoy, D., Tirasawasdichai, T., & Kurpayanidi, K. I. (2021). Role of social media in shaping public risk perception during COVID-19 pandemic: A theoretical review. *International Journal of Management Science and Business Administration*, *7*(2), 35–41. DOI: 10.18775/ijmsba.1849-5664-5419.2014.72.1005

Turkle, S. (1995). *Life on the Screen: Identity in the Age of the Internet*. Simon & Schuster.

Yang, S. (2015). Understanding personal online reputation management: A grounded theory study.

Youn, S. (2008). Parental influence and teens' attitude toward online privacy protection. *The Journal of Consumer Affairs*, *42*(3), 362–388. DOI: 10.1111/j.1745-6606.2008.00113.x

Zuboff, S. (2019). *The Age of Surveillance Capitalism: The Fight for a Human Future at the New Frontier of Power*. PublicAffairs.

Chapter 2
The Digital Horizon:
Opportunities for Developing Social Capital and Advancing Digital Citizenship

Youmei Liu
University of Houston, USA

ABSTRACT

In the contemporary digital world, the development of social capital and the advancement of digital citizenship are pivotal to fostering inclusive, supportive, and responsible online communities. This chapter discusses the importance of social capital in relation to the advancement of digital citizenship and explores how digital platforms and tools facilitate the formation of social bonds and civic engagement through innovative methods to create an inclusive environment. It examines the dual facets of digital interactions – enhancing community ties and promoting ethical behavior online. Through the data analysis on the current trend of social capital development, this chapter provides insights into the mechanisms by which digital environments can enhance social connectivity and civic responsibility, the quality of digital citizenship and ultimately contributing to a more cohesive society to promote the development of social capital.

INTRODUCTION

The Role of Emerging Technologies

The world is currently undergoing a significant and transformative period, facing social and economic challenges driven by the rapid advancement of digital technologies. The digital age has fundamentally changed how individuals connect, interact, and contribute to their communities. Emerging technologies such as artificial intelligence (AI) and blockchain are reshaping digital interactions, influencing trust-building, transparency, and community engagement in profound ways. AI-powered algorithms enable personalized online experiences, fostering deeper engagement and participation, while blockchain

DOI: 10.4018/979-8-3693-6675-2.ch002

Copyright ©2025, IGI Global Scientific Publishing. Copying or distributing in print or electronic forms without written permission of IGI Global Scientific Publishing is prohibited.

technology enhances transparency and security, reducing misinformation and strengthening trust in digital interactions.

Generative AI agents can be used to autonomously transact and manage digital assets on behalf of users, introducing a digital future where agents can mitigate risks, enhance automation, and authenticate content on the blockchain (Prove AI, 2024). "Blockchain technology has a decentralized framework that enhances trust by allowing users to fine-tune foundation models. This collaborative approach enables both proprietary and open-source models to learn from each other, ensuring checks and balances to prevent unexpected outcomes" (Prove AI, para 7). "One of the key benefits of decentralized identity is enhanced security. By distributing data across a network of nodes, blockchain technology ensures that there is no single point of failure. This setup significantly reduces the risk of data breaches and unauthorized access" (Anglen, 2024). Thus, these innovations provided efficient and effective technologies for promoting and developing global online community network.

As of July 2024, Northern Europe ranked first by the internet penetration rate, with over 97.5% of its population using the internet. Northern America followed, with 96.9%. Overall, the global average internet penetration rate was roughly 67.1% (Petrosyan, 2024). This widespread adoption, particularly within Industry 4.0, has been shown to positively impact productivity and foster sustainable growth (Haseeb et al., 2019). Virtual communities represent a modern extension of open societies, where traditional patterns of connection have evolved, and allowing individuals to engage and interact within digital environments (Stefanoudis et al., 2022). The proliferation of virtual communication methods has led to a growing number of interconnected individuals, facilitating both the preservation of existing relationships and the formation of new connections (Moore et al., 2021). This transformation has expanded the scope of social interaction, making it easier for individuals to maintain meaningful ties regardless of physical distance.

Digital Citizenship and Social Capital Development

"The emergence of digital technologies and a more global and digital society has brought about the need to develop and educate in Digital Citizenship, as well as to study how youth are taught to participate and learn citizenship in a digital age" (Peart et al, 2022, abstract). "In an era where technology and digital platforms are integral to our everyday lives, the role of digital citizenship is ever more critical. This is especially true in educational settings, where equipping students with digital citizenship skills is essential for their safe and responsible navigation of the digital landscape" (Misha, 2024, papa. 1). Digital citizenship is the responsible and respectful use of technology to engage online, find reliable sources, and protect and promote human rights (Mossberger et al, 2008, Hintz, 2019, Balkin, 2017, & Isin, 2020). The Congressional Research Service (CRS) introduced in the Senate the Digital Citizenship and Media Literacy Act on February 13, 2023. The bill is aimed at increasing media literacy education and helping students navigate disinformation, ultimately strengthening digital citizenship. "One of the fundamental components of the United Nations' sustainable development 2030 agenda is quality education. It aims to ensure inclusive and equitable quality education for all. Digital technologies have emerged as an essential tool to achieve this goal" (Haleem et al, 2022, abstract). The call for enhancing digital education is global wide and urgent for the sustainability of the human civilizations.

Digital technologies not only offer new avenues for communication but also present unique opportunities for building social capital and fostering digital citizenship. Social capital has been a significant area of interest for researchers since the late 19th century, with sociologists like Lester Ward, Émile Durkheim, and Jane Addams being recognized for their pioneering contributions to its foundational theories. As defined by the Oxford dictionary via Lexicon.com, social capital refers to "the networks of relationships among people who live and work in a particular society, enabling that society to function effectively" (n.d.). Over the years, scholars across disciplines such as economics, sociology, politics, and psychology have deepened their exploration of the core components of social capital, examining it in various contexts. These interdisciplinary efforts have broadened the scope of research and enhanced our understanding of how social capital impacts individuals, organizations, and societies. Bauernschuster et al. (2014) stated that there exists significant positive effects of the internet on some particular dimensions of social capital. Liu et al. (2016) provided a thorough meta-analysis of the positive relationship between social networking online and social capital.

Digital Inequality: Regional Disparities and the Digital Divide

However, the technological innovations have the potential to either bridge or exacerbate existing social and economic divides, depending on their accessibility and implementation. The digital divide remains a major barrier to the equitable development of social capital and digital citizenship. This divide manifests in multiple dimensions, including disparities in internet access, differences in digital literacy levels, and unequal opportunities for participation in online communities. "Over time, the digital divide will lead to increasingly wider skills gaps that prevent people from finding employment and companies from finding the talent they need" (Viasat, 2023, para, 7). Low-income and rural populations often face systemic obstacles in accessing high-speed internet, digital devices, and relevant digital skills training, limiting their ability to benefit from digital advancements. "Most offline households do not have the resources to pay market prices for broadband service even where it is available" (Pew, 2022, para. 6). Consequently, gaps in digital citizenship emerge, affecting individuals' ability to critically evaluate online content, engage in informed discourse, and navigate the responsibilities of digital participation. Addressing this divide is essential to ensuring that digital citizenship and social capital development are inclusive and accessible to all through appropriate and effective channels.

Social capital thrives on networks, trust, and reciprocity – all of which are significantly influenced by digital access. When regional disparities in digital access persist, they contribute to the weakened bonding social capital – families and close communities in digitally excluded areas struggle to maintain strong social ties due to communication barriers. Research indicates that those without reliable internet often experience weakened relationships, which diminishes community cohesion and support systems. These barriers can particularly impact participation in communal activities that facilitate bonding social capital (Sanders, & Scanlon, 2021). They also contributed to the limited bridging social capital – individuals cannot form diverse connections across cultural and economic groups, reducing opportunities for cross-community collaboration. A lack of access to online platforms limits individuals' ability to explore new perspectives and engage in meaningful dialogue with others outside their immediate networks (Sanders, & Scanlon, 2021). Consequently, this dynamic impedes the generation of collaborative efforts that could address shared challenges and foster social innovation (Wang, et al. 2021).

Regional disparities in digital access create significant challenges for the development of social capital by limiting economic mobility, educational opportunities, civic engagement, and social connectivity. Addressing these disparities through targeted policies, investment in infrastructure, and digital literacy programs is essential for creating an inclusive digital society where all individuals can participate fully in the opportunities offered by the digital world.

Therefore, this chapter explores the transformative potential of the digital sphere, focusing on both the opportunities and challenges it presents. It examines the concept of social capital in digital contexts, provides a comprehensive definition of digital citizenship, and highlights the importance of these elements in cultivating sustainable online communities. Alongside a review of current literature and research in the field, the chapter draws upon data from a 10-year longitudinal study conducted from 2013 to 2022 among college students at the University of Houston (UH). A total of 2,185 students participated in the study (68% female and 32% male). The age of participants ranged from 18 to 54, with the majority falling within the 18-24 and 25-34 age groups, which together comprised 93% of the total sample. According to a 2022 Statista report, the leading social media platforms in the U.S. have a similar age demographic, with Twitter users at 80% and Facebook users at 49% within this age range. Therefore, the study's participant profile closely mirrors the broader population of global social media users.

This longitudinal study offers valuable insights into the dynamics of trust and social capital on social media platforms, allowing for a deeper understanding of current trends in social capital development. These findings provide a strong foundation for exploring how educational institutions can integrate these trends into responsive curricula to advance the skills and awareness of digital citizens. Thus, the following three topics are discussed to support the theme of the chapter.

1. Understanding social capital in the digital age
2. Digital citizenship and social capital: responsibilities and rights
3. The role of education in fostering digital citizenship

Through examining these areas, the chapter will provide a holistic view of the evolving digital landscape and its implications for social capital and digital citizenship development.

UNDERSTANDING SOCIAL CAPITAL IN THE DIGITAL AGE

According to Britannica (n.d.), social capital can be understood through three key dimensions: interconnected networks of relationships (social ties), the levels of trust that characterize these networks, and the resources or benefits gained through these relationships. Regardless of the specific focus within social capital research, trust remains a critical component. It serves as the foundation for any form of social capital, fostering the relationships and cooperation necessary for effective societal functioning. As Cialdini and Fincham (1990) state that trust is a fundamental component of interpersonal relationships, emphasizing its central role in the development and sustainability of social capital.

Social capital – the networks of relationships, trust, and cooperation that facilitate collective action and community resilience – has found a new frontier in the digital realm. In this context, social capital extends beyond physical boundaries, enabling individuals to form connections, share knowledge, and collaborate across vast distances. The development of social capital in digital spaces is not merely an extension of people's offline lives but a transformation of how people build communities, solve problems, and support each other in an increasingly virtual world. Social capital goes beyond simply knowing other people or having friends. It consists of having relationships with people whom one can count on to provide benefits in the future. (Liu, et al, 2016). "Social capital theory has received considerable attention from diverse fields because it brings tangible and intangible benefits to individuals and organizations by helping atomized individuals interact through social networks and it promotes citizens' social participation, decision making, cooperation in the public good, and abilities for achievements" (Ruo-Fei, et al, 2022, para.7).

Bridging vs. Bonding Social Capital in Digital Spaces

Social capital is typically categorized into bonding social capital, which involves close-knit, strong ties within familiar groups, and bridging social capital, which facilitates connections across diverse networks (Putnam, 2000). Digital environments uniquely shape both forms, but not always equally. "Bonding social capital associates to the close and tight connections exist among the individuals or groups who share similar attitude, characteristics along with potential available info and resources" (Bin, et al., 2021, p. 27). Coleman (1988) found that ties in the bonding social capital support unity among each other and it is based on complex and dense relationship because of homogenous characteristics and the member's association with each other over a long period. Woolcock (2010) argued that bonding relates to connections and interactions between homogenous community members such as neighbors and family members and is rooted in particularized forms of trust and reciprocity. In the digital age, bonding social capital is strengthened through social media platforms, where individuals maintain and reinforce relationships with family, close friends, and known communities. The UH longitudinal study found that 80% of students used social media often (43%) and very often (37%) to maintain close relationships, as compared to only 26% who used it often (20%) and very often (6%) to meet new people. This suggests that social media is more effective in reinforcing existing strong ties rather than expanding social circles.

While bridging social capital is associated with a bridge or connections through which people from different demographics (race, ethnicity, class or religion) come closer and share similar socio-economic status (Putnam, 2001). Granovetter (1982) found that bridging social capital is related to loose connection and is referred to "weak tie." It is because of these weak ties people get the most useful information and perspectives without having emotional attachment or support (Bin, et al, 2021). Bridging social capital thrives on platforms like Reddit, Twitter (X), and LinkedIn, where users engage with broader communities and diverse perspectives. For example, Reddit provides spaces for knowledge-sharing across global interest groups, while LinkedIn fosters professional networking and mentorship opportunities. However, the potential for echo chambers, where individuals are exposed primarily to information, opinions, and perspectives that reinforce their existing beliefs, often due to algorithmic filtering, selective exposure, or social dynamics, can limit exposure to diverse viewpoints and can contribute to polarization and misinformation, and also limit true bridging social capital by reinforcing existing beliefs rather than exposing individuals to diverse viewpoints.

Understanding how different platforms contribute to bonding and bridging social capital is crucial for designing digital environments that promote both community cohesion and cross-network collaboration. The research study conducted by Ali, et al. in 2023 reveals that "Positive attitudes toward virtual society platforms are associated with increased social capital, particularly in terms of bridging and bonding" (Abstract).

Traditional Social Capital vs Digital Social Capital

"The digital age has changed the way people interact with each other, and the digital world has got the largest share of people's relationship with each other. In this context, putting a premium on social capital in the digital world seems necessary" (Jafari & Mohsttsmi, 2019, abstract). The use of digital technologies to enhance traditional public participation processes and to build a new form of social capital – digital social capital – is being embraced in a range of planning practices (Mandarano, 2010). Traditional social capital and digital social capital both refer to the networks of relationships, trust, and norms that enable individuals and communities to work together effectively. However, they differ in their characteristics, modes of interaction, and the contexts in which they are built and maintained. "Digital technology has so changed the way we establish and maintain relationships that 'friending' is now a regularly used verb" (Walden University, n.d. para.3).

Traditional Social Capital is deeply rooted in physical communities, where relationships are built through face-to-face interactions, long-term connections, and shared experiences. This form of social capital tends to be more stable and geographically bounded, emphasizing local community cohesion and shared cultural norms, and the interaction among individuals in these communities cultivates trust and reciprocity, enabling greater political engagement and social participation (Leyden, 2003). Digital Social Capital, on the other hand, is characterized by its global reach and a dynamic nature, facilitated by online platforms that connect diverse individuals across vast distances (Zhou, et al., 2022). It relies on both strong and weak ties, enabling broad networks of connections that can rapidly evolve in response to new technologies and trends. Digital social capital empowers individuals to create connections that transcend geographical boundaries, promoting collaboration and innovation (Ellison, et al., 2007). However, digital social capital also faces unique challenges, such as maintaining trust in less personal online environments and navigating the complexities of digital communication.

While digital platforms have made it easier than ever to form social connections across vast distances, these connections often come with inherent risks that can undermine trust, cooperation, and community cohesion. "It is now commonplace to say that digital technologies and data have profoundly reshaped the ways we live, interact, think and know the world" (Gangneux, 2019, para.1). "Today, the world faces a crucial and dramatic transition with social and economic challenges related to the advancement of digital technologies" (Zhou, et al., 2022, para. 2). These challenges stem from concerns about the potential negative effects of new technologies on social life (Ali et al., 2017), including disruptions to daily interactions, feelings of loneliness, increased rates of depression, and fears of losing social capital (Facer, 2011).

There are mixed findings regarding the impact of the internet on trust. Sabatini and Sarracino (2019) observed that participation in online networks led to a significant decline in all forms of trust in Italy. In China, Zhou et al. (2022) discovered that internet usage negatively affected political and generalized trust, although it did not impact trust in parents. Contrary to concerns that internet access might reduce social capital, Bauernschuster et al. (2014) found no evidence to support this claim. In fact, their research suggests that the internet has a positive impact on specific aspects of social capital. Similarly, Liu et al. (2016) conducted a comprehensive meta-analysis, concluding that online social networking enhances social capital by providing a platform to strengthen existing interpersonal relationships. Lu et al. (2020) found a positive effect of internet use on trust in governments based on the 2017 Netizen Social Consciousness Survey.

Both traditional and digital social capital play crucial roles in building and sustaining communities, but they operate in different contexts and environments. Traditional social capital remains vital for maintaining local community cohesion and face-to-face relationships, while digital social capital offers opportunities for global collaboration, diversity, and the rapid mobilization of collective action. Understanding the strengths and limitations of each form of social capital is essential for effectively leveraging them to build stronger, more connected, and resilient communities in both physical and digital spaces.

The Impact of Social Media Platforms on Trust and Social Capital

Social media is now a central part of daily life, revolutionizing how people communicate, share information, and interact with others. According to statista.com, social network user penetration in the United States in 2024 is 92.65% and it will continue to grow to 95.71% by 2028, estimated with growth of over 3% in four years. Meanwhile, the internet users of 300.06 million in 2024 will increase to 326.75 million in 2029 by 8.61%. As a result, social media serves as a crucial medium for expanding existing networks and developing social capital. Understanding how people perceive and trust in social media is crucial since social media platform is an indispensable part of people's interactions and life experiences, which are the key to the development of social capital.

In the UH longitudinal study, 100% students reported that they are familiar with the concept of social media platforms in 2022 versus 45% in 2013. Students' preferences of using social media platforms fluctuate throughout the 10-year's investigation. Kim and Park stated in their study "ease of use and feature relevance are important factors that influence social media use" (2021, page 444). When social media platform provides misinformation and low quality message, people tend to leave that platform. The "perceived credibility and trustworthiness are important factors that influence social media use" (Wang & Zhange, 2020, p.191). Another important factor is the commitment from social media platforms to protecting user's safety and security of private information. Twingate data report in 2020 found out that 45.9% of people were extremely or moderately concerned about their personal information on social media among generation X and millennials.

People use social media platforms for a variety of reasons, as each social media platform is unique and designed to serve specific purposes or cater to specific user needs. In the UH study, regarding social media tools adoption, the top five reported reasons from students in order of importance are: social engagement (20%), direct communication (12%), speed of feedback (10%), relationship building and making new friends (9%). Social engagement remain the first reason to adopt social media tools throughout ten-year period.

The UH study examined social capital through two specific questions. The first question focused on how social media tools were used to maintain close relationships with friends and family, thus preserving existing communities (bonding social capital). The second question investigated the use of social media platforms to meet new people and expand social circles, fostering new relationships (bridging social capital). Over a ten-year period, the findings revealed that, on average, 37% of students reported using social media very often to maintain existing relationships, 43% often, 14% rarely, and 6% not at all. In contrast, when it came to expanding their social networks, the trends shifted: only 6% reported very often, 20% often, 45% rarely, and 29% not at all using social media to meet new friends and extend their social circles.

The behavioral patterns observed in students' social media use indicate a stronger focus on maintaining existing relationships (bonding social capital) rather than actively seeking to expand social networks (bridging social capital). According to the data, while social media is effective in maintaining strong ties, it plays a less significant role in developing new connections or accessing more diverse networks, which are essential for enhancing bridging social capital (Liu, 2023). Consistent communication and interactions on these platforms help to sustain and reinforce existing relationships, leading to the accumulation of social capital through trust, reciprocity, and social support.

Regarding the level of trust in social media platforms, the data indicate that 1) students had increased caution in using social media platforms by almost 26% in ten-year span from 42.9% (strongly agree) in 2013 to 68.6% in 2022; 2) student trust in people through social media declined, with agreement dropping from 27.8% in 2013 to 17.9% in 2022, while disagreement rose from 27.3% to 47.6%, reflecting a 20% increase in distrust, and 3) students consistently expressed strong trust in close friends and family on social media, with an average of 50.7% strongly agreeing and 37.9% agreeing, while disagreement remained low at under 3.1%. The results of the study shows a strong bonding social capital and a challenging weak tie in bridging social capital.

The increased caution on internet is a very complicated issue and it could related to cyber security, such as identity fraud and theft, online bullying and harassment, misinformation and fake news, online scams and phishing, unfair news reports, lack of confidence in government, just to mention a few. In recent years, trust level is declining in different areas, such as in news media, and in government. A research study conducted by the University of Oxford in virtual roundtable discussions with small groups of journalists and publishers in 2021, stating that many expressed concern that digital platforms were at least partly to blame for declining levels of trust in news in many places around the world (Newman et al., 2022). Some worried that platforms enable bad-faith criticism of journalism to circulate more easily and insidiously while polluting the information environment with low quality substitutes for factual reporting (Mont'Alverne et al, 2022).

Trust in the media has also declined in the United States. In 2021, only 23% of Americans say they trust the media to report the news "fairly and accurately" (p. 2). According to Pew Research Center, "Trust in the government has declined in the United States. In 2021, only 22% of Americans say they trust the government to do what is right "just about always" or "most of the time." (p. 1). "Since 2007, the shares saying they can trust the government always or most of the time has not surpassed 30%" (Pew Research Center, 2022). Edelman, a global public relations and marketing consultancy firm, states that the trust crisis is a global phenomenon. In 2022, trust in institutions is at an all-time low in most countries around the world and the trust crisis is having a negative impact on society. It is leading to a decline in civic engagement, a rise in extremism, and a decrease in economic growth (Edelman, 2022). These data are very concerning. When social media became the vital platform for people to access various resources, their level of trust will seriously affect their attitudes towards using and sharing information.

While it is important to approach social media interactions with a degree of caution, there are undeniable benefits that social media platforms bring to people's life. Social media platforms removed communication barriers across the globe, regardless of geographical boundaries, serve as powerful tools for sharing information and raising awareness on various topics, facilitating the dissemination of information on social, cultural, political, and environmental issues. They allow individuals to connect with professionals in their field, build business networks, and foster collaborations. "Positive impacts of social media on youth include increased social connection, civic engagement, and self-expression" (Smith, et al., 2018, p.402). Social media can be used to build social capital by facilitating information sharing, network formation, and civic engagement (Clark & Barbour, 2021).

The development of social capital in the digital age presents a complex landscape of both opportunities and challenges. To address these challenges effectively, it is essential for netizens to embrace their roles as responsible digital citizens, ensuring their actions contribute positively to the development of social capital. This involves understanding both the responsibilities and rights associated with digital citizenship. By promoting ethical behavior, safeguarding privacy, encouraging inclusivity, and protecting the freedom of expression, individuals can strengthen the foundation of digital social capital. Understanding these responsibilities and rights is crucial for utilizing the strengths of the digital world to build connected, safe, and resilient online communities that thrive on mutual trust and collaboration.

DIGITAL CITIZENSHIP AND SOCIAL CAPITAL: RESPONSIBILITIES AND RIGHTS

Karen Mossberger, co-author of Digital Citizenship: *The Internet, Society, and Participation*, defines digital citizens as individuals who use the internet regularly and effectively (2008). "The idea of citizenship being digital, for example, carries within it the principles of scale, immediacy and information control that are all immanent in the idea of big data and which thus exert a change in understanding" (Pangrazio & Sefton-Green, 2021, p. 17). In this context, a digital citizen actively engages with information technology (IT) to participate in societal, political, and governmental activities in a global sphere. Digital citizenship can thus be defined as simply 'the right to participate in society online' (Mossberger, et al, 2008). Digital citizenship also refers to knowledge, skills, and values related to digital communication, commerce, literacy, security, rights, and etiquette (Ribble & Bailey, 2007). This definition emphasizes the role of digital literacy and online engagement in fostering meaningful participation in the digital age. Digital citizenship is the responsible and respectful use of technology to engage online, find reliable sources, and protect and promote human rights (Mossberger, 2008, Hintz, 2019, Balkin, 2017, & Isin, 2020). It teaches skills to communicate, collaborate, and act positively on any online platform (Hintz, 2019, Balkin, 2017). "Digital Citizenship is a way for business leaders, educators, governmental workers, and advocates to understand how we should use technology appropriately" (USIDHR, 2022, para. 1).

Digital citizenship encompasses a range of behaviors and norms that are deemed appropriate and beneficial within the digital space. These include understanding and respecting privacy rights, engaging in positive and respectful interactions online, and being aware of the ethical and legal implications of one's digital activities. "Digital citizenship is an increasingly vital concept in our interconnected digital age. It encompasses the norms of appropriate and responsible behavior when using technology and includes everything from online ethics and safety to privacy and managing one's digital footprint" (Lcom Team, 2024, para. 1). As digital platforms become central to personal, educational, and professional activities, the need for a comprehensive understanding of these responsibilities grows such as the characteristics and responsibilities of digital citizens, the digital right to everyone and the challenges to achieve digital citizenship, as well as enhancing digital community to promote the development of social capital.

The Interplay of Digital Citizenship and Social Capital Development

Digital citizenship plays a vital role in building social capital by facilitating connections, enhancing civic engagement and bridging the digital divide. In the interconnected world, the ability to engage responsibly, ethically, and effectively online is not just a personal asset but a societal necessity. A good digital citizen knows their rights to privacy and freedom of expression online, but also understands the responsibility to respect others' privacy and not spread harmful information (Lcom Team, 2024). Digital citizenship, the practice of understanding and fulfilling the rights and responsibilities in the digital environment, is the cornerstone of a healthy and thriving online community. "Digital citizenship refers to the appropriate behavior, attitudes, and skills required to engage responsibly and respectfully in the digital sphere. It involves being mindful of the consequences of our actions online, respecting the rights and privacy of others, and contributing positively to the digital community" (Abhishek, 2023, para. 2).

It ensures that the digital interactions contribute positively to the collective digital well-being, fostering a safe, inclusive, and respectful digital space for all. The relationship between digital citizenship and social capital development is intricate and multifaceted, grounded in organic, logical, and philosophical connections. Here are some key ways in which these two concepts are interrelated:

1. *Organic Relationship*: The Role of Digital Citizenship in Community Building

Digital Interactions as Building Blocks of Social Capital

At its core, social capital is built through interactions within a community, and digital citizenship provides the framework for these interactions in the online world. Digital social media platforms allow individuals to connect, collaborate, and communicate more easily, fostering stronger community ties (Education Uplifted, 2024). Responsible digital citizenship – through respectful communication, ethical behavior, and positive engagement can naturally fosters trust, reciprocity, and networks that are essential components of social capital. Van Dijk, (2005) argues that digital citizenship is essential for creating a knowledge society and fostering social capital.

While digital interactions can foster strong community ties, challenges such as online misinformation, echo chambers, and digital inequality can hinder the organic development of social capital. For example, the effect of echo chambers, where users engage primarily with like-minded individuals, exacerbates the problem of misinformation. This phenomenon reinforces existing biases and prevents exposure to diverse viewpoints, leading to polarization and fragmented communities (Chan, et al., 2022). Polarization helps to create and disseminate false and extremist information (Spohr, 2017). Without critical digital literacy, individuals may fall prey to divisive rhetoric and misinformation, undermining trust within online communities. Additionally, the digital divide and regional disparity, where access and competencies differ based on socio-economic and demographic factors can hinder the ability of some individuals to navigate online environments effectively, further exacerbating issues related to misinformation and trust (Wang, et al. 2021). Addressing these challenges requires proactive digital literacy education and robust platform moderation.

2. *Logical Relationship*: The Reciprocal Nature of Digital Citizenship and Social Capital

Social Capital Enhancing Digital Citizenship

The presence of strong social capital in digital spaces reinforces responsible digital citizenship. Strong social connections can provide emotional support and guidance, helping individuals to develop responsible digital behaviors. When individuals are part of a trusted network, they are more likely to act responsibly online, knowing that their actions have consequences within their community. Social capital can enable individuals to work together to address digital challenges and promote positive online experiences. Tufekci and Wilson (2012) highlights the role of social media in facilitating collective action and social change. Thus, social capital creates a supportive environment that encourages ethical behavior and mutual respect. Social capital has been employed to shed light on the challenges concerning civic engagement, the role of social capital in fostering civic ethics, and the potential economic benefits associated with social capital (Portes, 2014).

However, a logical extension of this argument is that digital spaces lacking accountability mechanisms may enable negative behaviors, such as trolling or cyberbullying, eroding social capital. "There is a disturbing growth in the number of cyberbullies and trolls because of global conditions and the increase of internet communication. Cyberbullying is a virtual form of real-life bullying, and trolling is a kind of online harassment" (Aydin, et al., abstract, 2021). Digital citizenship frameworks must therefore emphasize the importance of accountability, fostering ethical engagement through community reporting systems and responsible content moderation. Digital citizenship involves behaving safely, thinking critically, and participating responsibly and ethically in the digital world (Let's Talk Science, 2020).

3. *Philosophical Relationship*: The Ethical Imperative of Digital Citizenship for the Common Good

The Ethical Responsibility to Contribute to Social Capital

From a philosophical standpoint, digital citizenship carries an ethical responsibility to contribute to the common good, which aligns with the principles of social capital development. It encompasses appropriate and responsible behavior in technology use, highlighting safety, online ethics, and respect (Lcom Team, 2024). Digital citizens have a moral obligation to engage in ways that strengthen the community, promote equity, and support collective well-being. In communicating with digital members, "respecting others in the digital world is paramount. This includes refraining from cyberbullying, hate speech, or any form of online harassment. Always remember that there are real people with feelings on the other side of the screen. Treat them with the same courtesy and respect you would in person" (Cohen, 2023, para, 3). This ethical dimension of digital citizenship is essential for nurturing the social networks and trust that underpin social capital.

Failure to uphold ethical digital citizenship leads to significant societal challenges, including online harassment, misinformation proliferation, and digital exclusion (Zhong, et al. 2022, Lee & Rubin, 2021), for example, online harassment, misinformation proliferation and digital exclusion. Social media enables these actions within a convenient environment that attracts a wide audience. Repeated harm from cyberbullying marked by power imbalances can lead victims to display low self-esteem, anxiety, depression, and even suicidal ideation (Zhong, et al., 2022). A significant challenge in the digital age is the abundance of information available online, which poses the risk of information overload and the spread of misinformation (Hultenius, 2023). Digital exclusion equates to social exclusion, where a lack of information and digital technology negatively impacts personal, political, and economic capabilities (Sanders & Scanlon, 2021). Efforts should be made to bridge the digital divide and ensure that all individuals, regardless of their socioeconomic background, have equal opportunities to participate in the digital world. This includes promoting affordable internet access, providing digital literacy training, and supporting initiatives that increase accessibility for individuals with disabilities (Hultenius, 2023).

4. *Synergistic Relationship*: Digital Citizenship as a Catalyst for Expanding Social Capital

Expanding Social Networks beyond Physical Boundaries

Digital citizenship enables individuals to form connections across geographic, cultural, and socioeconomic boundaries, significantly expanding the potential for social capital development. Online platforms allow for the creation of global networks that transcend the limitations of physical communities, fostering a more interconnected world. Digital citizenship provides communities with the resources to mobilize and advocate for civic interests, thus augmenting social capital through localized network effects (Oscar. howell/Social Capital and Digital Citizenship, 2024).

Digital citizenship also facilitates collective action, where individuals can mobilize around shared goals and causes, such as the #MeToo Movement: a digital campaign launched in 2017 that connected individuals across the world, empowering survivors of sexual harassment and influencing policy reforms. This movement facilitated healing, empowerment, and solidarity for those who connected and found support through story-sharing (Kundu, 2024). Another powerful example is the Civic Tech in Disaster Relief: platforms like Ushahidi – a crisis preparedness platform, exemplifies how civic tech can enable global communities to mobilize resources during natural disasters, reinforcing linking social capital (Ushahidi, 2023). The platform has evolved into a global tool used in over 160 countries for disaster response, human rights protection, governance and accountability, and climate action since it was developed in 2007-2008 (Ushahidi, 2024). This collective action, made possible by digital communication and organization tools, enhances social capital by strengthening community ties and achieving common objectives that benefit the broader society. The use of digital tools encourages social and political engagement among users, enhancing the sense of community and collective problem-solving (You, et al., 2024).

5. *Foundational Relationship*: Digital Citizenship as the Foundation for Sustainable Social Capital

Sustainability of Social Capital through Responsible Digital Citizenship

The sustainability of social capital in digital environments depends on the consistent practice of responsible digital citizenship. Digital literacies enhance social advantages by facilitating interactions and fostering relationships online. This connectivity empowers individuals to engage in collaborative efforts, thereby accumulating social capital (Jaberian, 2023). As individuals continue to engage ethically, share knowledge, and support each other online, they contribute to a sustainable cycle of social capital development that benefits both current and future generations. The influence of social media and digital tools on public opinion and policy highlights the need for active participation, which directly contributes to the cultivation of social capital (Lozano-Díaz, & Fernández-Prados, 2020).

The relationship between digital citizenship and social capital development is organic, logical, and philosophical. Digital citizenship serves as both a foundation and a catalyst for social capital, while social capital reinforces and enhances the practice of digital citizenship. Together, they contribute to the creation of a more connected, ethical, and equitable digital society. Building on the interconnected relationship between digital citizenship and social capital, it is essential to explore the responsibilities and rights inherent to digital citizenship. As individuals engage in the digital world, their actions and behaviors not only shape personal networks but also influence the broader digital society.

Digital Citizenship and Ethics: Understanding Rights and Responsibilities

The rise of digital technologies necessitates a nuanced comprehension of digital citizenship, which, like traditional citizenship, is grounded in rights and responsibilities (Lcom Team, 2024, Contributors to Wikimedia projects, 2024). The responsibilities of digital citizens, such as ethical online behavior, critical thinking, and maintaining online safety, play a crucial role in fostering trust and cooperation, which are the building blocks of social capital. Equally important are the rights of digital citizens, including access to information, privacy protection, and freedom of expression, which empower individuals to participate actively and responsibly in digital communities.

Understanding these responsibilities and rights is fundamental to creating a more inclusive and ethical digital environment, where social capital thrives. As digital citizens, individuals have a set of responsibilities that ensure the digital world remains a positive, safe, and sustainable environment for everyone. These responsibilities include practicing digital etiquette, exercising critical thinking, ensuring online safety and security, and being mindful of the environmental impact of digital activities. Currently, digital rights are not a set of rights in and of themselves, but are related to other human rights, particularly freedom of expression and the right to privacy in online and digital environments (Hutt, 2015). Digital citizenship rights are essential for ensuring that individuals can participate fully and safely in the digital world. These rights include access to technology, privacy, freedom of expression, and intellectual property protection. Each of these components plays a critical role in shaping how we interact online and how we contribute to the digital society.

The integration of digital responsibilities and rights with social capital development is fundamental to fostering a healthy and inclusive digital environment. Digital responsibilities – such as practicing respectful communication, critically evaluating information, ensuring online safety, and minimizing environmental impact, are essential for building trust, collaboration, and cooperation within online communities. These responsibilities help cultivate bonding social capital by strengthening relationships within tight-knit groups and bridging social capital by connecting diverse individuals and fostering inclusivity.

Equally important are the rights associated with digital citizenship, including access to technology, privacy protection, freedom of expression, and intellectual property rights. These rights empower individuals to participate actively and responsibly in digital spaces, facilitating the development of bridging and linking social capital. By ensuring that all individuals have equal access to digital tools and the ability to engage freely and safely, the digital society becomes more equitable and collaborative, promoting the growth of networks that benefit both individuals and the broader community. Together, digital rights and responsibilities form the foundation for a well-functioning, ethical digital society where social capital can thrive, contributing to stronger, more connected, and more resilient communities, which rely on ethical behaviors in digital citizenship.

Ethical considerations in digital citizenship are central to ensuring that the digital environment remains a space where individuals can interact responsibly, safely, and equitably. Cultivating a sense of digital ethics encourages honesty and integrity (Lcom Team, 2024). Understanding the balance between rights and responsibilities, fostering inclusive communities, and recognizing the importance of accountability are crucial elements of ethical digital engagement. By respecting privacy, promoting inclusivity, engaging in truthful communication, and fostering respectful interactions, digital citizens can enhance both bonding and bridging social capital. Understanding how to navigate online spaces safely helps mitigate risks associated with inappropriate content or interactions (Lcom Team, 2024). These ethical behaviors help build trust, cooperation, and networks, creating a more connected and equitable digital society. The following lists three areas that cover the ethical considerations in digital citizenship.

1. *Balancing Rights and Responsibilities*

- Digital Rights and Their Corresponding Responsibilities

In the digital world, rights such as freedom of expression, privacy, and access to information are fundamental. However, these rights come with corresponding responsibilities that ensure the digital space is safe and respectful for everyone. Digital citizens are expected to recognize their rights to privacy and freedom of expression while also understanding that these rights come with responsibilities, such as reporting any unethical behavior encountered online (Let's Talk Science, 2024; Dahal, 2023.) For instance, while individuals have the right to express their opinions online, they also have the responsibility to do so in a way that does not harm others or spread misinformation. This balance between rights and responsibilities is essential for maintaining the integrity of digital platforms and protecting the rights of all users. Individuals should treat others with respect and dignity, mirroring the ethical standards expected in face-to-face interactions (Let's Talk Science, 2024).

- Ethical Use of Digital Platforms

Ethical digital citizenship involves using digital platforms in a way that respects the rights of others and adheres to ethical standards. This includes avoiding behaviors that infringe on others' rights, such as spreading false information, engaging in cyberbullying, or violating privacy. Digital citizens must also be aware of the impact their actions can have on the broader digital community and strive to contribute positively. For example, while sharing content is a right, it is also a responsibility to ensure that the content is accurate, not harmful, and respects intellectual property rights. "By promoting responsible and respectful online interactions, digital citizenship paves the way for more effective global partnerships and a better understanding of different cultures and perspectives" (Lcom Team, 2024, para.10).

- Avoiding Infringement on Others' Rights

Respecting the rights of others is a cornerstone of ethical digital citizenship. A key issue in digital rights discourse is the tendency to 'universalize' the human subject, when in reality human rights issues are locally situated (Livingstone & Third, 2017). "What is a digital right in some parts of the world may not be in others, yet individuals and their data can easily cross state and regional borders online" (Pangrazio & Sefton-Green, 2021, p. 19). This means being mindful not to infringe on others' privacy, not engaging in activities that could harm someone's reputation or well-being, and respecting intellectual property by not engaging in piracy or unauthorized sharing of content. Digital citizens must recognize that their actions online can have real-world consequences and that they have a responsibility to ensure that their digital behavior does not negatively impact others.

2. *The Role of Digital Citizens in Fostering Inclusive Communities*

- Promoting Inclusivity and Diversity in Online Spaces

Digital citizens have a vital role in creating and maintaining inclusive online communities. "Being inclusive within your communities on and offline fosters a sense of belonging, respect, and empowerment among users. When prioritized, it leads to more meaningful interactions, increased community engagement, and a greater sense of connection" (Couto, 2024, para. 3). Inclusivity involves ensuring that all individuals, regardless of their background, gender, race, ethnicity, religion, sexual orientation, or ability, feel welcome and valued in digital spaces. Promoting diversity means actively supporting and amplifying the voices of marginalized groups and ensuring that online discussions reflect a wide range of perspectives. This not only enriches the digital community but also strengthens social capital by fostering a sense of belonging and mutual respect.

- Ensuring Equity in Digital Participation

Equity in digital participation involves recognizing and addressing the barriers that may prevent some individuals from fully engaging in digital spaces like regional disparities and divide. These barriers can include lack of access to disability, technology, digital literacy gaps, language differences, or systemic biases that marginalize certain groups. Inaccessible online platforms can exclude individuals with disabilities (Rex, 2024). Digital citizens can promote equity by advocating for accessible technologies, supporting initiatives that bridge the digital divide, and being mindful of how their own behavior and the design of digital platforms may impact different user groups. Ensuring equitable participation helps to create a more just and fair digital environment.

- Amplifying Marginalized Voices

One of the key ways digital citizens can foster inclusivity is by actively amplifying the voices of marginalized communities. Actively seeking to amplify voices from underrepresented groups enhances community inclusivity (Rex, 2024). This can involve sharing content created by underrepresented groups, participating in discussions that highlight issues affecting these communities, and standing against online behaviors that silence or marginalize others. By doing so, digital citizens help to create a more balanced and representative digital discourse, where diverse perspectives are valued and heard. Encouraging empathy and understanding through active listening helps to enrich discussions and build relationships among members (Rex, 2024).

- Creating Safe and Respectful Spaces

Fostering inclusive communities also involves creating spaces where everyone feels safe to express their opinions without fear of harassment or discrimination. Community guidelines are instrumental in preserving the pertinence and advantage of discussions by setting clear expectations for behavior and create a safe space for all members to engage without fear of abuse or offensive conduct (Disco, 2024). Digital citizens can contribute to this by calling out inappropriate behavior, supporting victims of online harassment, and promoting respectful dialogue. Creating a culture of respect and support in digital spaces encourages more people to participate and share their perspectives, enriching the digital community as a whole.

3. *Accountability and Consequences*

- Accountability for Online Actions

"Digital citizenship is not just about civic responsibilities or self-responsibilization, but rather how the digital facilitates new forms of participation" (Pangrazio & Sefton-Green, 2021, p. 18). Accountability in the digital world means that individuals are responsible for their online actions and must be prepared to face the consequences of unethical behavior. Unlike in the physical world, where actions are often directly linked to consequences, the perceived anonymity of the internet can sometimes lead people to believe they can act without accountability. However, ethical digital citizenship requires recognizing that actions taken online—whether posting a comment, sharing content, or participating in a discussion—can have significant impacts on others and the broader digital community.

- Consequences of Unethical Behavior

Online harassment and doxxing are particularly harmful forms of unethical behavior that can have devastating effects on individuals. Harassment can take many forms, including bullying, threats, or sustained attacks against someone's character. Doxxing involves revealing personal information, such as home addresses or phone numbers, which can lead to real-world harm. Digital citizenship has become a useful cipher to challenge cyberbullying and inappropriate behavior (Ribble, et al, 2004) particularly with regard to children and young people. Digital citizens must be aware that engaging in or supporting these activities is not only unethical but also illegal in many jurisdictions. They must also recognize their responsibility to report such behavior and support those affected.

- Promoting a Culture of Accountability

Beyond individual actions, fostering a culture of accountability in digital spaces is crucial. This means that digital platforms, communities, and leaders must create and enforce clear guidelines for behavior, provide mechanisms for reporting and addressing unethical actions, and promote awareness of the importance of accountability. By holding themselves and others accountable, digital citizens help to ensure that digital spaces are safe, respectful, and conducive to positive interactions.

Ethical considerations in digital citizenship revolve around balancing rights with responsibilities, fostering inclusivity and equity, and ensuring accountability for online actions. Digital citizens have a responsibility to use digital platforms ethically, respecting the rights of others while promoting a positive and inclusive digital environment. Accountability for one's actions online is crucial for maintaining the integrity of digital spaces and protecting the well-being of all users. By embracing these ethical considerations, digital citizens contribute to a healthier, more respectful, and equitable digital world to promote the development of social capital.

THE ROLE OF EDUCATION IN FOSTERING DIGITAL CITIZENSHIP

The cultivation of digital citizenship is essential for developing effective education programs aimed at enhancing online participation and political efficacy, particularly among youth (You, 2024). Achieving ideal digital citizenship requires ongoing education and active participation from all stakeholders, including individuals, educational institutions, and governments. It is the educational institutions' responsibility to be proactive in creating an optimal environment to nurture future students to be model digital citizen to act intelligently to contribute to the betterment of humankind. "Digital acts involve interpreting multiple streams of local and global information, and, in the age of datafication, anticipating unknown consequences" (Pangrazio & Sefton-Green, 2021, p. 18).

In an era where digital literacy is as crucial as reading and writing, the role of education in fostering digital citizenship becomes paramount. In particular, there needs to be a focus on extending digital literacies across different contexts, so that critical, ethical, and technical mastery can be developed (Pangrazio, 2019). As digital technologies permeate every aspect of life, educational institutions are uniquely positioned to equip students not only with the technical skills required to navigate digital tools but also with the ethical and social understanding necessary to engage responsibly in digital environments. Digital literacies should be approached as an evolving set of skills and practices with implications for identity (Littlejohn et al., 2012). Education in digital literacy can shape informed, ethical, and engaged digital citizens, who are capable of contributing positively to their online communities, to create an inclusive online learning environment through a structured curriculum that integrates digital citizenship in which students learn to recognize the impact of their online actions on themselves and others, fostering a culture of respect and responsibility. Educational institutions should be in the forefront of developing strategies and initiatives and implement them to promote digital citizenship across all levels of education, not only tech-savvy but also ethically aware and actively engaged in building a positive digital future.

Educational Strategies for Digital Citizenship via Curriculum Integration

Integrating digital citizenship into the curriculum is essential for preparing students to navigate the digital world responsibly and effectively. "Digital literacy is perhaps foundational for digital citizenship and digital rights: individuals cannot participate or claim their digital rights if they are not 'literate' in the first place" (Pangrazio & Sefton-Green, 2021, p. 21). Integrating digital citizenship into the curriculum can be achieved through both targeted curriculum development focused on digital literacy and cross-curricular approaches that embed digital citizenship concepts across various subjects for preparing students to navigate the digital world responsibly. By developing specific courses or modules that address digital literacy, online ethics, and safe internet practices, and by embedding these concepts across various subjects, schools can create a comprehensive educational experience that equips students with the knowledge and skills needed to become informed and ethical digital citizens. This integrated approach ensures that digital citizenship is not just an add-on but a core component of students' education, preparing them for the challenges and opportunities of the digital age. The following Table 1 suggests some implementation ideas and examples.

Table 1. Integrating Digital Citizenship into Curriculum Development and Digital Literacy

Integration Aspect	Implementation Examples
Integrating Digital Citizenship into Existing Curricula	Digital citizenship can be effectively integrated into existing curricula. Schools can start by assessing their current curriculum to identify gaps where digital citizenship education is needed. From there, they can introduce dedicated lessons or units that address these topics, ensuring that students receive a well-rounded education that includes digital competency as a core component.
Developing Specific Courses or Modules	Schools can develop specific courses or modules dedicated to digital citizenship. These courses should cover essential topics such as: - Digital literacy, which includes understanding how to use digital tools and platforms, evaluating online information, and protecting personal data. - Online ethics should be another focal point, addressing the importance of respectful communication, understanding the ethical implications of online behavior, and recognizing the responsibilities that come with digital rights, including respecting intellectual property, avoiding cyberbullying, and understanding the importance of consent in digital interactions. - Safe internet practices, teach students how to protect their online privacy, recognize phishing attempts, cyber threats, using secure passwords, and understanding privacy settings, crucial components and navigate social media safely. These courses can be standalone offerings or integrated into broader technology or social studies programs, depending on the school's structure and resources. By embedding these topics into the curriculum, schools can help students develop the critical thinking skills needed to make informed and ethical decisions online.
Incorporating Real-World Scenarios or Practices	To make digital citizenship education more engaging and relevant, curriculum developers should incorporate real-world scenarios that students are likely to encounter online. - Case studies, simulations, and interactive activities can help students understand the practical implications of digital citizenship concepts. For example, a module on online ethics could include a scenario where students must navigate the ethical dilemmas of sharing content or responding to inappropriate comments on social media. - Virtual cross-cultural projects, encouraging students to share their experiences and voice their concerns that they encounter in the digital environment to help each other understand the complexity of the reality. These real-world applications help students see the relevance of digital citizenship education and encourage them to apply what they have learned in their daily life.

There are other cross-curricular approaches that can effectively integrate digital citizenship components into education, weaving digital citizenship concepts into various subjects, and allowing students to explore digital citizenship from multiple perspectives and in various contexts. Through integrating digital citizenship into the courses in both areas of social and natural sciences, educators can create a holistic understanding of responsible online behavior that is not confined to a single course but is instead reinforced throughout a student's education. The following are some ideas worth trying.

Social Studies and Digital Citizenship

In social studies, digital citizenship can be explored in the context of civic engagement, democracy, and global citizenship. Students can examine the role of social media in shaping public opinion, the impact of digital technologies on political participation, and the ethical responsibilities of citizens in the digital age in their own countries or regions. Lessons might include analyzing case studies of digital activism, discussing the implications of surveillance on privacy rights, or exploring how digital platforms can be used to promote social justice globally. Students will gain a deeper understanding of their role as informed and responsible digital citizens through connecting digital citizenship to real-world social and political issues,

Science and Digital Responsibility

In science classes, digital citizenship can be connected to topics such as the ethical use of technology, the environmental impact of digital devices, and the role of data in scientific research. Students might investigate how digital tools can be used to collect and analyze data, while also discussing the responsibilities of scientists and citizens to protect privacy and use data ethically. Lessons could include exploring the environmental consequences of e-waste and considering how digital innovations can be harnessed to address global challenges. Thus, students learn to approach technology with a sense of responsibility and awareness of its broader implications.

A cross-curricular approach ensures that digital citizenship is not treated as an isolated topic but is instead woven into the fabric of students' overall education. Students develop a holistic understanding of what it means to be a responsible digital citizen through learning digital citizenship concepts in various subjects and contexts. This approach also reinforces the interconnectedness of digital literacy, online ethics, and safe internet practices, helping students see how these concepts apply to all aspects of their lives.

Digital Divide and Social Equity: Challenges, Impacts, and Solutions

The digital divide refers to the gap between individuals and communities that have access to modern information and communication technologies (ICTs) and those that do not. This divide can lead to significant disparities in educational opportunities, economic advancement, and social inclusion. In the research study conducted by Feng and Tan found out that digital cultural and social capital provides a useful theoretical basis for understanding the underlying effects of the digital divide on student development (2024). Based on the study, there are two levels of divide identified, level 1 is associated with the availability of the digital equipment, and level 2 is related to the ability on how well students can use the digital equipment. The gap limits students' ability to engage with information and communication technology, and meaningfully participate in today's rapidly evolving society (Noakes et al. 2018). Addressing the digital divide is crucial for ensuring that all students have equitable access to the digital resources they need to succeed in the modern world. This includes focusing on equity in access to digital resources and supporting underserved communities through targeted initiatives and programs.

Equity in Access to Digital Resources

- Understanding the Digital Divide in Education

The digital divide manifests in various ways within the education system, particularly in the unequal access to digital devices, reliable internet connections, and educational technology tools. "The digital divide is the socioeconomic gap created between communities, cultures, or individuals with different levels of access to digital tools" (IEEE, 2024, para. 1). Students from low-income families, rural areas, and other marginalized communities are often at a disadvantage, lacking the necessary resources to fully participate in digital learning environments. This inequity can lead to gaps in academic performance, digital literacy, and future opportunities, perpetuating cycles of poverty and social exclusion. The disparities can lead to significant educational disadvantages, making it challenging for these students to achieve academic success and develop essential digital citizenship skills (Kloza, 2023).

To address the digital divide, education systems must prioritize equitable access to technology for all students. Education systems can address this issue by partnering with internet service providers to offer affordable or subsidized internet plans to families. Schools can implement one-to-one device programs, where each student is provided with a device for schoolwork, ensuring that no student is left behind due to a lack of technology. Additionally, partnerships with technology companies or non-profit organizations can help supply these devices at little or no cost to families in need. Beyond devices and internet access, students need access to a range of educational technology tools that support digital learning.

Supporting Underserved Communities

To effectively support underserved communities, it is essential to first identify the specific needs and challenges these communities face. Numerous initiatives and programs have been developed to support students from underserved communities. These include government-funded programs that provide subsidies for internet access and digital devices, as well as non-profit organizations that donate technology and offer digital literacy training. For example, programs like the U.S. Federal Communications Commission's (FCC) Emergency Broadband Benefit or the ConnectED initiative aim to provide affordable internet access and devices to low-income families. Similarly, organizations like One Laptop per Child and the World Computer Exchange work to distribute low-cost devices to students in developing countries. These initiatives help bridge the digital divide by ensuring that students from underserved communities receive the tools and resources they need to participate fully in digital learning.

Community-Based Digital Literacy Programs

- Community-based programs play a critical role in supporting digital citizenship education for underserved students. These programs often provide digital literacy training in community centers, libraries, or through outreach initiatives that target families in low-income neighborhoods. Programs like these can teach parents and students how to use digital devices, navigate the internet safely, and access educational resources online. Community-based programs can create a supportive environment for digital education.

Long-Term Strategies for Sustainability

Addressing the digital divide requires long-term strategies that ensure sustainable access to technology and digital education. This involves not only providing immediate support but also building the infrastructure and systems needed to maintain equitable access over time. Education systems should work towards integrating digital literacy into the core curriculum, advocating for policies that support digital inclusion, and continuously assessing and addressing the needs of underserved communities. By taking a sustainable approach, schools and communities can ensure that all students have the opportunity to thrive in the digital age.

Addressing the digital divide is a critical challenge for education systems, requiring a focus on equity in access to digital resources and targeted support for underserved communities. By ensuring that all students have access to the necessary technology, internet connectivity, and educational tools, and by developing initiatives and programs that support those who are most in need, schools can help close the gap and provide every student with the opportunity to succeed in the digital world. These efforts not only benefit individual students but also contribute to a more inclusive and equitable society, where everyone has the chance to participate fully in the digital age.

Building a Responsible Digital Society: Global Trends and Best Practices

Digital citizenship education is a global imperative, with countries and cultures around the world adopting different approaches to prepare their citizens for the challenges and opportunities of the digital age. Numerous international organizations and educational bodies are promoting and implementing digital citizenship education. UNESCO (2024) emphasizes equipping learners with skills to critically access, engage with, and utilize digital technologies effectively. The Digital Citizenship Education Project spearheaded by the Council of Europe (COE.INT) focuses on empowering students through democratic competences that support active participation in society. It addresses key issues such as online safety, respect for others, and the ethical use of digital resources. Their guidelines advocate for educational frameworks that foster critical thinking and ethical decision-making, enabling students to navigate digital environments productively (2023). In the United State, the California Department of Education (CDE) outlines digital citizenship education as critical for ensuring that students understand their rights and responsibilities in a digital context. The state promotes various resources and policies to guide schools in teaching these concepts effectively (Digital Citizenship, 2024). Understanding these global perspectives provides valuable insights into how digital citizenship education can be tailored to diverse contexts and how international collaboration can enhance the effectiveness of these efforts. Key areas to explore include comparative approaches to digital citizenship education and the potential for global collaboration and exchange programs.

1. Comparative Approaches to Digital Citizenship Education

- Diverse Educational Frameworks Across Countries

Different countries have developed unique frameworks for digital citizenship education, reflecting their cultural values, educational priorities, and technological contexts. For example, in countries like Finland and Estonia, which are known for their advanced education systems, digital citizenship education is integrated into the national curriculum from an early age. These countries emphasize digital literacy, critical thinking, and the ethical use of technology, with a strong focus on preparing students to be active and responsible digital citizens. In contrast, countries like Japan and South Korea, where there is a high level of technological adoption, focus heavily on cyber safety and the ethical implications of technology use, with robust programs aimed at protecting students from online threats such as cyberbullying and internet addiction.

- Success Stories and Innovative Practices

There are several successful models and innovative practices in digital citizenship education from around the world that other countries can learn from. For instance, Singapore has implemented a comprehensive digital literacy program known as "Cyber Wellness," which is embedded across the curriculum and focuses on helping students make informed decisions online, understand the impact of their digital footprint, and practice responsible online behavior. The program is supported by extensive resources for educators, students, and parents, and includes regular updates to keep pace with emerging digital trends. Another example is the Canadian "MediaSmarts" program, a national initiative that provides digital and media literacy resources to educators across the country. The program covers a wide range of topics, including privacy, online ethics, digital health, and the responsible use of social media. MediaSmarts also offers professional development for teachers to help them integrate digital citizenship concepts into their teaching effectively.

- Cultural Influences on Digital Citizenship Education

Cultural values and societal norms significantly influence how digital citizenship education is approached in different regions. For example, in collectivist cultures like those in East Asia, digital citizenship education often emphasizes the collective good, social harmony, and respect for authority. This is reflected in programs that focus on the responsible use of digital platforms to maintain social order and protect community well-being. In contrast, individualistic cultures, such as those in the United States and Western Europe, tend to emphasize personal responsibility, freedom of expression, and individual rights in their digital citizenship education programs.

These cultural differences highlight the need for digital citizenship education to be context-sensitive, taking into account the values and norms of the society in which it is implemented. However, there are also commonalities across cultures, such as the need for digital literacy, online safety, and ethical behavior, which can serve as a foundation for global standards in digital citizenship education.

- Challenges and Opportunities in Global Implementation

While there are many successful models of digital citizenship education, there are also challenges in implementing these programs globally. These challenges include disparities in access to technology, differences in educational infrastructure, and varying levels of digital literacy among teachers and students. However, these challenges also present opportunities for innovation and collaboration. For instance, countries with advanced digital citizenship programs can share their resources and expertise with those that are just beginning to develop their digital education frameworks. Additionally, international organizations, such as UNESCO and the International Society for Technology in Education (ISTE), can play a role in promoting global standards and best practices in digital citizenship education.

2. Global Collaboration and Exchange Programs

- Opportunities for International Collaboration

Global collaboration offers significant opportunities to enhance digital citizenship education by allowing students and educators to learn from diverse perspectives and experiences. International collaborative projects can involve schools from different countries working together on digital citizenship initiatives, such as joint research projects, online discussions on global digital issues, or collaborative content creation. These projects provide students with a broader understanding of digital responsibility and the impact of their actions on a global scale.

One example of international collaboration in digital citizenship education is the "eTwinning" program, which connects schools across Europe to collaborate on projects using digital tools. Through eTwinning, students engage in cross-cultural exchanges that foster mutual understanding and respect, while also developing their digital literacy and communication skills. This type of collaboration helps students see the relevance of digital citizenship in a global context and encourages them to consider the broader implications of their online behavior.

- Exchange Programs and Digital Citizenship

Exchange programs that focus on digital citizenship provide students with the opportunity to immerse themselves in different cultural and technological environments. These programs can involve physical exchanges, where students visit schools in other countries, or virtual exchanges, where students connect with peers from around the world through online platforms. These experiences allow students to learn about different approaches to digital citizenship, understand global digital challenges, and develop skills in cross-cultural communication and collaboration.

For example, the "Global Nomads Group" offers virtual exchange programs that connect students from different countries to discuss global issues, including digital citizenship. Through these programs, students gain insights into how digital issues are perceived and addressed in different parts of the world, and they develop a more nuanced understanding of their role as global digital citizens. Another example is the "AIESEC" youth exchange program, which includes opportunities for students to engage in digital projects that address global challenges such as digital literacy, online safety, and cyber ethics. These projects allow students to apply their digital skills in real-world contexts, while also contributing to the global community. Global collaboration and exchange programs help students broaden their understanding of digital responsibility by exposing them to diverse perspectives on digital citizenship.

- Fostering a Global Digital Citizenship Community

Global collaboration and exchange programs contribute to the development of a global digital citizenship community, where students, educators, and policymakers from different countries come together to share knowledge, resources, and best practices. This community fosters a sense of solidarity and shared purpose, as members work together to address common digital challenges and promote ethical digital behavior on a global scale.

For example, the "Global Educators Network" provides a platform for educators around the world to collaborate on digital citizenship education, share resources, and develop joint initiatives. Through this network, educators can learn from each other's experiences, adapt successful practices to their own contexts, and contribute to the global conversation on digital citizenship. By fostering a global digital citizenship community, these programs help build a more inclusive, equitable, and responsible digital world, where everyone has the opportunity to participate and thrive.

Global perspectives on digital citizenship education highlight the diversity of approaches taken by different countries and cultures, as well as the potential for international collaboration to enhance digital citizenship learning. Comparative approaches reveal successful models and innovative practices that can serve as inspiration for educators worldwide. Meanwhile, global collaboration and exchange programs provide students with valuable opportunities to broaden their understanding of digital responsibility and contribute to a global digital citizenship community. By embracing these global perspectives, educators can prepare students to navigate the complexities of the digital world with a sense of responsibility, ethical awareness, and global solidarity.

FUTURE TRENDS AND IMPLICATIONS

The approach to digital citizenship education varies widely across countries, reflecting different cultural values, educational priorities, and levels of technological integration. For instance, Finland integrates digital citizenship into all levels of education, emphasizing critical thinking, digital literacy, and civic engagement. "Finland is a world-renowned leader in education, and digital literacy is a key component of its curriculum" (Finland Education Hub. 2023). While countries like Japan and South Korea focus on the ethical use of technology and cyber safety, driven by their high levels of digital connectivity and concerns about internet addiction. Understanding these differences can provide valuable insights into how digital citizenship education can be adapted to different cultural and educational contexts.

Successful models of digital citizenship education can serve as benchmarks for other countries. Singapore's "Cyber Wellness Program" by Minister of Education (updated 2024), Canada's "MediaSmarts" by Canada's Center for Digital Media Literacy, and the UK's "Computing at School" (CAS) initiative offer comprehensive frameworks that integrate digital citizenship across the curriculum. These programs focus on empowering students with the skills needed to navigate the digital world safely and responsibly, emphasizing both technical proficiency and ethical behavior.

Cultural values significantly shape how digital citizenship is taught. In collectivist societies like those in East Asia, the emphasis is often on the collective good and social harmony, which influences how digital behavior is governed. In contrast, more individualistic societies prioritize personal responsibility and freedom of expression. Recognizing these cultural influences is crucial for developing digital citizenship education programs that resonate with local values while also promoting universal digital literacy and ethical standards.

Implementing digital citizenship education globally presents challenges, such as varying levels of access to technology, differences in educational infrastructure, and diverse cultural attitudes toward digital rights and responsibilities. However, these challenges also offer opportunities for innovation and collaboration. Countries with advanced digital education frameworks can share resources and expertise with those developing their programs, fostering a global exchange of best practices. The following are some ideas for collaboration to develop international digital citizenship programs for excellence.

Research Topics to Address These Ideas

- Research Topic 1: "Comparative Analysis of Digital Citizenship Education in Collectivist vs. Individualist Cultures: Implications for Curriculum Design and Implementation"

This research would explore how cultural values influence the design and implementation of digital citizenship education programs in different regions. It would compare collectivist cultures, which emphasize social harmony and community welfare, with individualist cultures, which prioritize personal freedom and responsibility. The study could identify best practices and propose adaptable frameworks for diverse educational contexts.

- Research Topic 2: "Assessing the Effectiveness of Digital Citizenship Education in Promoting Ethical Online Behavior: A Cross-National Study"

This research would investigate the effectiveness of digital citizenship education programs in promoting ethical online behavior across different countries. It would compare the outcomes of programs like Singapore's "Cyber Wellness Program", Canada's "MediaSmarts", and the UK's "Computing at School" initiative, identifying the key factors that contribute to their success and the challenges they face.

- Research Topic 3: "The Role of National Policies in Shaping Digital Citizenship Education: A Comparative Study of Policy Approaches in Developed and Developing Countries"

This study would examine how national policies influence the development and implementation of digital citizenship education programs. It would compare policy approaches in developed and developing countries, analyzing how different levels of technological infrastructure, educational priorities, and cultural attitudes shape these programs.

These expanded ideas and research topics provide a comprehensive framework for exploring global perspectives on digital citizenship education. By examining comparative approaches, fostering international collaboration, and understanding the global implications of digital responsibility, researchers can contribute to the development of more effective and inclusive digital citizenship education programs worldwide. These studies would not only advance academic knowledge but also provide practical insights for educators, policymakers, and international organizations working to prepare students for the digital age.

CONCLUSION

This chapter explore the transformative impact of digital technologies on social capital and digital citizenship within contemporary society. Through detailed examination of how digital platforms facilitate social bonding and civic engagement, these elements are crucial in fostering inclusive and supportive online communities. The digital age has transformed how we interact, learn, and build communities, making it more important than ever to understand and practice responsible digital citizenship.

Education plays a pivotal role in this endeavor. By integrating digital citizenship into curricula, embedding digital citizenship across various subjects and emphasizing the development of critical digital skills, educational institutions can prepare students to navigate and contribute positively to the digital world to bridge the digital divide and create a more inclusive digital environment. It is important to understand the global perspectives on digital citizenship education, comparing how different countries approach this critical issue and exploring the potential for international collaboration.

The synthesis of these key points reinforces the overarching theme of the chapter: that fostering social capital and advancing digital citizenship are not just desirable outcomes but essential imperatives in the digital age. As we continue to navigate a world increasingly shaped by digital technologies, the ability to engage responsibly, ethically, and inclusively online becomes a crucial skill for individuals and communities alike. Social capital, which encompasses the networks of relationships, trust, and reciprocity that bind communities together, is deeply intertwined with digital citizenship. By practicing responsible digital citizenship, we contribute to the development of social capital, creating stronger, more resilient communities that can thrive in the digital era.

REFERENCES

Abhishek, G. (2023). Digital citizenship: empowering responsible online engagement. https://www.linkedin.com/pulse/digital-citizenship-empowering-responsible-online-abhishek-guguloth-/

Adyin, A., Arda, B., Güneş, B., & Erbaş, O. (2021). Psychopathology of cyberbullying and internet trolling. *Journal of Experimental and Basic Medical Sciences.*, *2*(3), 380–391. DOI: 10.5606/jebms.2021.75680

AIESEC. (n.d.). Leadership is for every young person. https://aiesec.org/

Ali, H., Hussain, T., Ali, S., Jumani, N. M., Li, B., & Zhang, G. (2017). Coverage of educational issues in Pakistan by Daily Dawn and the news: September 2015-December 2015. *Science International (Lahore)*, *29*(5), 1069–1072.

Ali, S., Wang, D., Hussain, T., & Li, B. (2023). The impact of virtual society on social capital formation: A comparative analysis of Facebook and WhatsApp. *Sage Journals*https://journals.sagepub.com/doi/full/10.1177/21582440231210077

Anglen, J. (2024). AI & Blockchain: digital identity 2024. *Rapid Innovation.* https://www.rapidinnovation.io/post/ai-and-blockchain-fusion-advancing-digital-identity-in-2024

Balkin, J. (2017). The three laws of robotics in the age of big data. *Faculty Scholarship Series.*http://hdl.handle.net/20.500.13051/4697

Bauernschuster, S., Falck, O., & Woessmann, L. (2014). Surfing alone? The internet and social capital: Evidence from an unforeseeable technological mistake. *Journal of Public Economics*, *117*, 73–89. DOI: 10.1016/j.jpubeco.2014.05.007

Bin, Y., Khan, W., & Mastafa, E., Hadi, Rooth, S., & Khan, R. A. (2021). Social media: A source of bridging and bonding social capital. *Journal of Media and Communication Studies*, *1*(1), •••.

Britannica. (n. d.). Social capital. https://www.britannica.com/topic/social-capital

Chan, C. K., Zhao, M. M., & Lee, P. S. N. (2022). Determinants of escape from echo chambers: The predictive power of political orientation, social media use, and demographics. In *Global Media and China.* https://journals.sagepub.com/doi/10.1177/20594364221140820

Cialdini, R. B., & Fincham, R. L. (1990). Trust and the development of interpersonal relationships. *Psychological Bulletin*, •••, 1990.

Cohen, C. (2023). Becoming a good digital citizen: navigating the online world ethically. https://www.cyberwise.org/post/becoming-a-good-digital-citizen-navigating-the-online-world-ethically#:~:text=1.%20Respect,for%20Others

Coleman, J. (1988). Social capital in the creation of human capital. *American Journal of Sociology*, *94*, 95–120. DOI: 10.1086/228943

Computing at School (CAS). (n.d.). Community for computing teachers. https://www.computingatschool.org.uk/

Couto, R. (2024). How to build an inclusive online community (and why it matters). Groundwork. https://groundwork.space/building-an-inclusive-online-community/Cyber

CRS - The Congressional Research Service. (2023). S.394 - Digital Citizenship and Media Literacy Act. https://www.congress.gov/bill/118th-congress/senate-bill/394

Cyber Wellness Program (updated 2024). Practising cyber wellness. https://www.moe.gov.sg/education-in-sg/our-programmes/cyber-wellness

Dahal, N. (2023). Digital citizenship and digital ethics. *Advances in Social Networking and Online Communities*, 249–257. Advance online publication. DOI: 10.4018/978-1-6684-8934-5.ch014

Digital Citizenship. (2024). Education Technology (CA Department of Education). ca.gov. https://www.cde.ca.gov/ls/et/dc/index.asp

Disco. (2024). 10 effective community rules examples and how to implement them. https://www.disco.co/blog/10-effective-community-rules-examples-and-how-to-implement-them#:~:text=Notably%2C%20the%20language%20and%20content,members%2C%20including%20other%20community%20members

Edelman (2022). 2022 Edelman trust barometer. https://www.edelman.com/trust/2022-trust-barometer

Education Uplifted. (2024). Enhancing civic engagement through technology in education. https://educationuplifted.com/civic-engagement-through-technology-2/

Ellison, N. B., Steinfield, C., & Lampe, C. (2007, July 1). The benefits of Facebook "Friends:" social capital and college students' use of online social network sites. *Journal of Computer-Mediated Communication, 12*(4), 1143–1168. DOI: 10.1111/j.1083-6101.2007.00367.x

eTwinning – European Commission. (n.d.). European School Education Platform. https://school-education.ec.europa.eu/en/etwinning

Facer, K. (2011). *Learning futures: Education, technology and social change*. Taylor & Francis. DOI: 10.4324/9780203817308

FCC - the U.S. Federal Communications Commission. (n.d.) Emergency broadband benefit. https://www.fcc.gov/broadbandbenefit

Feng, S., & Tan, C. Y. (2024). Toward conceptual clarity for digital cultural and social capital in student learning: Insights from a systematic literature review. *Humanities & Social Sciences Communications, 11*(1), 68. DOI: 10.1057/s41599-023-02519-8

Finland Education Hub. (2023). Digital literacy in Finnish education: A model for the world. https://finlandeducationhub.com/digital-literacy-in-finnish-education-a-model-for-the-world/

Gangneux, J. (2019). Digital citizenship in a datafied society: By Arne Hintz, Lina Dencik and Karin Wahl-Jorgensen, Cambridge, UK, Polity Press, 2019,180 pp. *Information Communication and Society, 22*(14), 2211–2213. DOI: 10.1080/1369118X.2019.1635186

Global Educators Network. https://www.gentefl.org/

Global Nomads Group – USA. (Since 1998). Designed by global youth for global youth. https://gng.org/

Haleem, A., Javaid, M., Qadri, M. A., & Suman, R. (2022). Understanding the role of digital technologies in education: a review. *Sustainable Operations and Computers*. V. 3, 2022, Pages 275-285https:// DOI: 10.1016/j.susoc.2022.05.004

Haseeb, M., Hussain, H. I., Ślusarczyk, B., & Jermsittiparsert, K. (2019). Industry 4.0: A solution towards technology challenges of sustainable business performance. *Social Sciences (Basel, Switzerland)*, *8*(5), 154. DOI: 10.3390/socsci8050154

Hintz, A., Dencik, L., & Wahl-Jorgensen, K. (2019). Digital citizenship in a datafied society. *Polity Press, Cambridge*, UK, 2019

Hultenius, J. (2023). The challenges of digital citizenship. https://www.linkedin.com/pulse/challenges-digital-citizenship-jonas-hultenius-g2ccf

Hutt, R. (2015). What are your digital rights? https://www.weforum.org/agenda/2015/11/what-are-your-digital-rights-explainer/

IEEE. (2024). Consequences of the digital divide in education. https://ctu.ieee.org/consequences-of-the-digital-divide-in-education/

Isin, E. F. (2020). *Being digital citizens*. Evelyn Sharon Ruppert (2nd ed.). London, United Kingdom: Rowman & Littlefield Publishing Group. p. 17. ISBN 978-1-78661-447-6. OCLC 1133661431.

ISTE - International Society for Technology in Education. (n.d.). Dream big. Transfer teaching. Empower learners. https://iste.org/

Jafari, S. M., & Moharrani, R. (2019). Conceptualizing digital social capital. *International Seminar on New Topics in Business Management*. https://www.researchgate.net/publication/340262485_Conceptualizing_Digital_Social_Capital

Javerian, N. (2023). Digital literacy and online social capital. *Iranian Journal of Educational Sociology*, *6*(4), 246–255. DOI: 10.61186/ijes.6.4.246

Kim, J., & Park, J. (2021). The impact of ease of use and feature relevance on social media use: A cross-platform study. *Information Systems Research*, *32*(2), 436–454.

Kloza, B. (2023). Consequences of the digital divide in education - connecting the unconnected. *Connecting the Unconnected*. https://ctu.ieee.org/consequences-of-the-digital-divide-in-education/

Kundu, K. (2024). The #MeToo Movement : Investigating the lasting international impacts. https://hir.harvard.edu/metoo-movement-international/

Lcom Team. (2024). The importance of digital citizenship: a 2024 perspective. https://www.learning.com/blog/the-importance-of-digital-citizenship-a-2024-perspective/#:~:text=Digital%20citizenship%20is%20an%20increasingly,Read%20the%20full%20article%20here)

Lee, D., & Rubin, T. (2021). Digital citizenship instruction: challenges and opportunities. https://infobase.com/blog/digital-citizenship-instruction-challenges-and-opportunities/

Let's Talk Science. (2020). Digital citizenship & ethics. *Let's Talk Science*. https://letstalkscience.ca/educational-resources/backgrounders/digital-citizenship-ethics

Leyden, K. M. (2003). Social capital and the built environment - PubMed Central. In *American Journal of Public Health*. https://pmc.ncbi.nlm.nih.gov/articles/PMC1448008/

Littlejohn, A., Beetham, H., & McGill, L. (2012). Learning at the digital frontier: A review of digital literacies in theory and practice. *Journal of Computer Assisted Learning*, *28*(6), 547–556. DOI: 10.1111/j.1365-2729.2011.00474.x

Liu, D., Ainsworth, S. E., & Baumeister, R. F. (2016). A meta-analysis of social networking online and social capital. *Review of General Psychology*, *20*(4), 369–391. DOI: 10.1037/gpr0000091

Liu, Y. (2023). Investigating the dynamics of trust in social capital on social media platforms. In Hoda, N., & Arabia, S. (Eds.), *Social Capital in the Age of Online Networking: Genesis, Manifestations, and Implications*. IGI Global., DOI: 10.4018/978-1-6684-8953-6.ch005

Livingstone, S., & Third, A. (2017). Children and young people's rights in the digital age: An emerging agenda. *New Media & Society*, *19*(5), 657–670. DOI: 10.1177/1461444816686318

Lozano-Díaz, A., & Fernández-Prados, J. S. (2020). Educating digital citizens: An opportunity to critical and activist perspective of sustainable development goals. *Sustainability (Basel)*, *12*(18), 7260. Advance online publication. DOI: 10.3390/su12187260

Lu, H., Tong, P., & Zhu, R. (2020). Does internet use affect netizens' trust in government? Empirical evidence from China. *Social Indicators Research*, *149*(1), 167–185. DOI: 10.1007/s11205-019-02247-0

Mandarano, L., Meenar, M., & Steins, C. (2010). Building social capital in the digital age of civic engagement. *Journal of Planning Literature*, *25*(2), 123–135. DOI: 10.1177/0885412210394102

MediaSmarts – Canada's Center for Digital Media Literacy. (n.d.). Media literacy week. https://mediasmarts.ca/?https://mediasmarts.ca/break-fake&gad_source=1&gclid=CjwKCAjw68K4BhAuEiwAylp3klut5HIt4l5pFNp9Ph-f_Bl8kerdbDHL10_Aj_-9j-OKiTuuubDG_BoCoVEQAvD_BwE

Misha, A. (2024), The growing importance of digital citizenship. https://www.edsurge.com/news/2024-02-19-the-growing-importance-of-digital-citizenship

Mont'Alerne, C., Arguedas, A. R., Toff, B., Fletcher, R., & Nielsen, R. K. (2022). The trust gap: how and why news on digital platforms is viewed more skeptically versus news in general. https://reutersinstitute.politics.ox.ac.uk/trust-gap-how-and-why-news-digital-platforms-viewed-more-sceptically-versus-news-general#header--0

Moore, E. J., Price, D. L., Van Abel, K. M., Janus, J. R., Moore, E. T., Martin, E., Morris, J. M., & Alexander, A. E. (2021). Association of Virtual Surgical Planning with external incisions in complex maxillectomy reconstruction. *JAMA Otolaryngology-Head & Neck Surgery*, *147*(6), 526–531. DOI: 10.1001/jamaoto.2021.0251 PMID: 33792635

Mossberger, K., Tolbert, C. J., & McNeal, R. S. (2008). *Digital citizenship: the internet, society, and participation*. MIT Press., DOI: 10.7551/mitpress/7428.001.0001

Newman, N., Fletcher, R., Robertson, C. T., Eddy, K., & Nielsen, R. K. (2022). *Digital news report 2022*. Reuters Institute for the Study of Journalism.

Noakes, T., Walton, M., & Cronjé, J. (2018) Capital meets capabilities: negotiating cultural exclusion in participatory culture. Proceedings of the *International Conference on e-Learning*, ICEL

Oscar.howell/Social Capital and Digital Citizenship - Technologies and Politics of Control. (2024). harvard.edu. https://cyber.harvard.edu/is2011/Oscar.howell/Social_Capital_and_Digital_Citizenship

Oxford dictionary - Lexicon.com (n.d.) Social capital. https://www.lexicon.com/search?q=social+capital

Pangrazio, L. (2019). *Young People's Literacies in the Digital Age: Continuities, Conflicts and Contradictions*. Routledge.

Pangrazio, L., & Sefton-Green, J. (2021). Digital rights, digital citizenship and digital literacy: What's the Difference? *Journal of New Approaches in Educational Research, 10*(1), 15–27. DOI: 10.7821/naer.2021.1.616

DOI: 10.7821/naer.2021.1.616

Peart, M., Cubo-Delgado, S. C., & Gutiérrez-Esteban, P. (2022). Exploring the role of digital and socio-civic skills for promoting youth participation and digital citizenship. *European Journal of Educational Research, 1*(2), 697–709. DOI: 10.12973/eu-jer.11.2.697

Petrosyan, A. (2024). Internet penetration rate worldwide 2024, by region. https://www.statista.com/statistiques/269329/penetration-rate-of-the-internet-by-region/

Pew (2022). Broadband access still a challenge in rural affordable housing. https://www.pewtrusts.org/en/research-and-analysis/articles/2022/12/08/broadband-access-still-a-challenge-in-rural-affordable-housing

Pew Research Center. (2022). Public trust in government: 1958-2022. https://www.pewresearch.org/politics/2022/06/06/public-trust-in-government-1958-2022/

Portes, A. (2014). Downsides of social capital. *Proceedings of the National Academy of Sciences of the United States of America, 111*(52), 18407–18408. DOI: 10.1073/pnas.1421888112 PMID: 25535346

Prove, A. I. (2024). Blockchain & generative AI: Fueling innovation within digital economy. https://proveai.com/news/blockchain-and-generative-ai-fueling-innovation-within-the-digital-economy

Rex, B. (2024). Breaking down barriers: building inclusive online communities. *Meeds DAO*. https://medium.com/meeds-dao/breaking-down-barriers-building-inclusive-online-communities-78b4e10ae379

Ribble, M., & Bailey, G. (2007). *Digital citizenship in schools*. ISTE.

Ribble, M., Bailey, G., & Ross, T. (2004). Digital citizenship: Addressing appropriate technology behavior. *Learning and Leading with Technology, 32*(1), 7–13.

Ruo-Fei, S., Zeng, J. Y., & Jin, C. H. (2022). The role of consumer' social capital on ethical consumption and consumer happiness. *SAGE Open, 12*(2), 21582440221095026. Advance online publication. DOI: 10.1177/21582440221095026

Sabatini, F., & Sarracino, F. (2019). Online social networks and trust. https://ideas.repec.org/a/spr/soinre/v142y2019i1d10.1007_s11205-018-1887-2.html

Sanders, C. K., & Scanlon, E. (2021). The digital divide is a human rights issue: advancing social inclusion through social work advocacy. *Journal of Human Rights and Social Work*. https://pmc.ncbi.nlm.nih.gov/articles/PMC7973804/

Smith, J. A., Spears, A. L., & Ellefsen, A. M. (2018). The impact of social media on youth: A review of the literature. *Developmental Review*, •••, 2018.

Spohr, D. (2017). Fake news and ideological polarisation: Filter bubbles and selective exposure on social media. *Business Information Review*, *34*(3), 150–160. DOI: 10.1177/0266382117722446

Statista. (2024). Social media and politics in the United States. https://www.statista.com/study/40641/social-media-and-politics-in-the-united-states/

Stefanoudis, P. V., Biancani, L. M., Cambronero-Solano, S., Clark, M. R., Copley, J. T., Easton, E., Elmer, F., Haddock, S. H. D., Herrera, S., Iglesias, I. S., Quattrini, A. M., Sigwart, J., Yesson, C., & Glover, A. G. (2022). Correction to 'moving conferences online: Lessons learned from an international virtual meeting'. *Proceedings. Biological Sciences*, *289*(1971), 20211769. DOI: 10.1098/rspb.2022.0321 PMID: 35350862

Tufekci, Z., & Wilson, C. (2012). Social media and the decision to participate in political protest: Observations from Tahrir square. *Journal of Communication*, *62*(2), 363–379. DOI: 10.1111/j.1460-2466.2012.01629.x

Twingate. (2020). Privacy for a premium. https://www.twingate.com/research/privacy-for-a-premium-exploring-peoples-sentiments-on-paying-for-social-media

UNESCO. (2024). Global citizenship education in a digital age: teacher guidelines. *unesco.org*. https://www.unesco.org/en/articles/global-citizenship-education-digital-age-teacher-guidelines

Useful Websites, C. O. E. I. N. T. (2023). Democratic Schools for All - www.coe.int. Democratic Schools for All. https://www.coe.int/en/web/campaign-free-to-speak-safe-to-learn/useful-websites

Ushahidi (2023). Crowdsourcing solutions to empower communities. https://www.ushahidi.com/

Ushahidi (2024). Ushahidi's vision for AI and social impact. https://www.ushahidi.com/about/blog/empowering-communities-through-technology-ushahidi-s-vision-for-ai-and-social-impact/

USIDHR – US Institute of Diplomacy and Human Rights. (2022). What is digital citizenship and why is it important? https://usidhr.org/what-is-digital-citizenship-and-why-is-it-important

van Dijk, J. (2005) The deepening divide: inequality in the information society. https://www.researchgate.net/publication/292683691_The_Deepening_Divide_Inequality_in_the_Information_Society, DOI: DOI: 10.4135/9781452229812

Viasat. (2023). Why is addressing the digital divide important? https://news.viasat.com/blog/corporate/why-is-addressing-the-digital-divide-important

Walden University. (n.d.). How is digital technology changing our interactions with each other? https://www.waldenu.edu/online-bachelors-programs/bs-in-psychology/resource/how-is-digital-technology-changing-our-interactions-with-each-other#:~:text=Digital%20technology%20has%20so%20changed,now%20a%20regularly%20used%20verb

Wang, J., & Zhang, J. (2020). The impact of perceived credibility and trustworthiness on social media use: A cross-platform study. *Journal of Business Research*, *2022*(112), 186–195.

Wang, T., Guo, X., & Wu, T. (2021). Social capital and digital divide: implications for mobile health policy in developing countries. *Journal of Healthcare Engineering*. https://pmc.ncbi.nlm.nih.gov/articles/PMC7857903/

Wellness Program. (2024). Practicing Cyber Wellness. Ministry of Education, Singapore. https://www.moe.gov.sg/education-in-sg/our-programmes/cyber-wellness

Wikimedia (2024). Citizenship - Wikipedia. Wikimedia Foundation, Inc. https://en.wikipedia.org/wiki/Citizenship

You, S., Lee, J., & Lee, Y. (2024). Effect of young adults' political efficacy on digital citizenship: Mediating effect of online social capital. *Current Psychology (New Brunswick, N.J.)*, *43*(41), 32154–32161. Advance online publication. DOI: 10.1007/s12144-024-06766-z

Zhong, J., Qiu, J., Sun, M., Jin, X., Zhang, J., Guo, Y., Qiu, X., Xu, Y., Huang, J., & Zheng, Y. (2022). To be ethical and responsible digital citizens or not: a linguistic analysis of cyberbullying on social media. *Frontiers in Psychology*. https://pmc.ncbi.nlm.nih.gov/articles/PMC9100568/

Zhou, D., Li, Y., & Tirasawasdichai, T. (2022, October 14). Do online social interactions cultivate social capital? Evidence from a longitudinal study. *Frontiers in Psychology*, *13*, 989137. DOI: 10.3389/fpsyg.2022.989137 PMID: 36312069

ADDITIONAL READING

Burke, K., & Larmar, S. (2021). Acknowledging another face in the virtual crowd: Reimagining the online experience in higher education through an online pedagogy of care. *Journal of Further and Higher Education*, *45*(5), 601–615. DOI: 10.1080/0309877X.2020.1804536

Choi, D. H. (2024). The impact of social media use on expressive political participation: Focusing on the role of citizenship and political efficacy. *The Journal of the Korea Contents Association*, *24*(3), 274–284. DOI: 10.5392/JKCA.2024.24.03.274

de Zúñiga, H. G., Barnidge, M., & Scherman, A. (2017). Social media social capital, offline social capital, and citizenship: Exploring asymmetrical social capital effects. *Political Communication*, *34*(1), 44–68. DOI: 10.1080/10584609.2016.1227000

Grishchenko N. (2022). Social Capital, Digital Divide and Their Relationship with Internet Use: A Longitudinal Study. *Digital Divide and Their Relationship with Internet Use: A Longitudinal Study*. (February 27, 2022).

Hong, S., Kim, D., & Kim, H. (2023). A study on the effect of political efficacy of youth generation on social trust and government trust: Focusing on the mediating effect of social fairness. *Modern Society and Public Administration, 33*(1), 1–32.

Hong, S. H., & Kim, S. G. (2022). The effect of social capital on creating shared value for educational companies. *Journal of Industrial Distribution & Business, 13*, 33–43.

Jaberian, N. (2023). Digital literacy and online social capital. *Iranian Journal of Educational Sociology, 6*(4), 246–255. DOI: 10.61186/ijes.6.4.246

Nelson, J. L., Lewis, D. A., & Lei, R. (2017). Digital democracy in America: A look at civic engagement in an internet age. *Journalism & Mass Communication Quarterly, 94*(1), 318–334. DOI: 10.1177/1077699016681969

Ng, Y.-L. (2022). Exploring the association between use of conversational artificial intelligence and social capital: Survey evidence from Hong Kong. *New Media & Society*. Advance online publication. DOI: 10.1177/14614448221074047

van Zoonen, W., Treem, J. W., & Ter Hoeven, C. L. (2022). A tool and a tyrant: Social media and wellbeing in organizational contexts. *Current Opinion in Psychology, 45*, 1–11. DOI: 10.1016/j.copsyc.2022.101300 PMID: 35231788

Zimmerman, T. (2022). Social noise: The influence of observers on social media information behavior. *The Journal of Documentation, 78*(6), 1228–1248. DOI: 10.1108/JD-08-2021-0165

KEY TERMS AND DEFINITIONS

Civic Engagement: refers to the active participation of individuals in activities aimed at influencing public policies, governance, or community affairs to improve the quality of life in a community.

Civic Responsivity: refers to the ability and willingness of individuals, organizations, or governments to respond actively and effectively to the needs, concerns, and aspirations of the community.

Digital Citizenship: refers to the responsible and ethical use of technology by individuals who participate in society online.

Inclusive Environment: refers to a setting where individuals of all backgrounds, abilities, and identities feel valued, respected, and supported.

Online Community: refers to a group of individuals who interact with each other through digital platforms, sharing common interests, goals, or experiences.

Social Bonding: refers to the process of forming and strengthening emotional connections and relationships between individuals.

Social Capital: refers to the networks, relationships, norms, and trust that enable individuals and groups to work together effectively and achieve mutual benefits.

Social Media Platforms: refers to digital tools that enable users to create, share, and interact with content, including text, images, videos, and audio.

Chapter 3
How to Become a Digital Citizen:
Using the Digital Economy and Digital Literacy to Understand Digital Citizenship

Bea Wohl
https://orcid.org/0000-0002-5969-9210
University of the Arts, London, UK

ABSTRACT

The following chapter draws on qualitative data to look at two aspects of computing education: the digital economy and digital literacy. Finally, the chapter compares these to the concept of digital citizenship, considering how 'digital citizenship' can be used to draw the other two together and give young people a deeper understanding of having power and agency in the digital world. The data presented in this chapter is based on fieldwork conducted as part of the author's PhD. While the original fieldwork was conducted in 2016 in a discipline where relatively limited qualitative fieldwork is conducted, it adds valuable insight and perspective. The chapter consists of a brief review of the concept of digital economy and digital literacy before using the data to look at how young people viewed both concepts in the context of what they were being taught as part of 'computing'. The data from the field works followed by a discussion of the findings and takeaways from the data before giving a brief introduction to 'digital citizenship'.

INTRODUCTION

The following chapter uses qualitative data to examine two aspects of computing education: the digital economy and digital literacy. Finally, it compares these to the concept of digital citizenship, considering how digital citizenship can connect the other two and give young people a deeper understanding of having power and agency in the digital world.

The data presented in this chapter is based on fieldwork conducted as part of the author's PhD. While the original fieldwork was conducted in 2016 in a discipline where relatively limited qualitative fieldwork is conducted, it adds valuable insight and perspective.

DOI: 10.4018/979-8-3693-6675-2.ch003

Copyright ©2025, IGI Global Scientific Publishing. Copying or distributing in print or electronic forms without written permission of IGI Global Scientific Publishing is prohibited.

Context of the Initial Research.

The chapter consists of a brief review of the concept of digital economy and digital literacy before using the data to look at how young people viewed both concepts in the context of what they were being taught as part of 'computing'. The data from the field works followed by a discussion of the findings and takeaways from the data before giving a brief introduction to 'digital citizenship' and considering why it could be a way to prepare young people for their digital future better.

The original research was conducted in the summer and fall of 2016. In 2014, England introduced a new statutory curriculum which replaced the subject of ICT (which had mostly focused on using specific software word processing to graphic design) with the broader subject of 'Computing', which encompassed a range of skills from digital literacy to computer science in the hope of preparing young people for the digital economy and a broader digital world. This followed a period of political debates and multiple reports ('Shut down or restart' in 2012 and 'Nexgen', in 2011) calling for reform in how computing was taught across England. There was a general feeling that the pre-2014 ICT curriculum was not giving young people the skills they needed.

The doctoral research this chapter draws on examined three themes in English computing education: digital economy, digital literacy, and computational thinking. The initial research consisted of in-depth interviews with 56 young people, nine computing teachers, and three computing education experts to validate the findings.

While the original research, as a doctoral thesis, made a substantial contribution to the field of computing education, it is relatively inaccessible, and its impact is, therefore, limited.

The author intends to use this chapter to highlight key findings from that work and the contribution that qualitative research with young people can make.

Including the concept of 'digital citizenship' allows the research to be more forward-looking rather than just highlighting the historical challenges of teaching computing and digital citizenship. It also demonstrates how to build computing education into a subject that is engaging and relevant to young people. Based on the interviews in the pages ahead, digital citizenship sounds a lot like what young people would like to learn but don't have a term for.

THE DIGITAL ECONOMY?

Defining the Digital Economy

Instead of thinking about a 'Digital Economy', it may be more useful to consider 'digital economies', reflecting on how the concept has evolved over time. The timeline below (fig 1) illustrates how the idea of the digital economy, rather than being merely descriptive, was employed as an anticipatory rhetorical term, aimed at making predictions about what might be on the horizon. The data in the figure, therefore, rather than being current, illustrates how assumptions were made about the future in the past.

Prior to the year 2000, there was little academic literature that refers specifically or attempts to define the digital economy. Literature from this period uses terms including "Knowledge Economy", "New Economy", "Innovation Economy", and "Information Economy" (Oxley et al., 2008; Singh, 2004), all of which align to the digital economy.

Carlsson (2004) referred to the internet as a general-purpose technology that could be compared to the steam engine or to electrification; in time, the term "digital economy" may seem as anachronistic as "steam economy" or "electric economy". As such, when considering the impact on young people, it may be worth thinking of the concept as adapting and changing; the 'digital economy' young people in the early 2000s were preparing for is fundamentally different to the one young people today need to look forward to.

Defining the Digital Economy in Phases

There have been a number of technological developments (i.e. transistors, personal computers, networked computing, the World Wide Web, the mobile phone, big data, artificial intelligence), specifically in the area of computers and computing, which over time have had an effect on economic activity (Ayres & Williams, 2004; Brynjolfsson & McAfee, 2016; Carlsson, 2004; Friedman, 2017; Hamid & Khalid, 2016; Keen, 2015). The digital economy now impacts countries across the world, where the conduct of goods or services is in the form of numbers transferred over digital networks (Irkinovich, N. R., 2022). These developments have ranged from the first invention of the transistors in 1947 (Ayres & Williams, 2004) to more recent developments of artificial intelligence and its use in medicine, transportation, and other areas (Brynjolfsson & McAfee, 2016). This chapter does not aim to provide a comprehensive timeline of all developments related to the digital economy; rather, selected authors have indicated a number of pivotal points when it comes to how technology has impacted economics. Using these pivotal points as signposts of key changes, it is possible to divide conceptions of the digital economy into three periods.

This paper examines the digital economy historically and can be seen as three distinctive periods of time: 1) before 1994, 2) from 1994–2007, and 3) after 2007- 2020. Understanding the past digital economy in this way can explain why much of the digital economy literature can seem somewhat contradictory.

The Pre-Digital Economy: Before 1994.

The pre-digital economy refers to the period during which much of the foundation and infrastructure essential for the later stages of the digital economy developed. This can be traced back to when the US Census Bureau first utilised computing technology for non-military purposes around 1951 (Ayres & Williams, 2004). With the advent of the desktop computer in the early 1990s, computers gradually became commonplace in businesses and homes. Before 1994, technology had already significantly impacted economic activity, although this influence was primarily focused on enhancing efficiency rather than transforming thc paradigm of business operations. For instance, early applications included replacing paper filing systems with digitised databases and substituting filing clerks with computers (Ayers & Williams, 2004, p. 319).

Digital Economy 1.0: Console-Based Digital Economy 1994 – 2007.

Circa 1994, several major companies were established, including the online retailer Amazon, the first widely adopted search engine, Yahoo, and the first web browser, Netscape (Keen, 2015). By the time access to the Internet became broadly and easily available to everyone, the nature of the economy began to change fundamentally (Carlsson, 2004). The Internet browser Netscape allowed consumers and businesses to access information from across the globe, search engines like Yahoo enabled users to explore this vast amount of data, and the online retailer Amazon facilitated online purchases (initially restricted to Books). Ayres and Williams (2004) identify the period from the 1990s onward as the start of the convergence of computing technology and communications technology, paving the way for the companies that emerged in 1994.

The Internet has permeated every office, home, and school, bringing about a significant shift. The distinction between consumers and producers began to blur. Non-market transactions, merely the movement of information from one place to another (such as a web search), have created genuine economic value. This transition was not only vastly different from the ways value had been generated in the past, but it was also quite difficult to measure and, therefore, not often studied (Brynjolfsson & Saunders, 2010).

Digital Economy 2.0: The Mobile Digital Economy.

The next stage of the digital economy can be referred to as "digital economy 2.0" and began in 2007. This year marked a pivotal moment when several technological events occurred, including the launch of the Apple iPhone, the Android operating system, Amazon's Kindle, and the emergence of the first "big-data" analytics companies (Friedman, 2017). These companies fundamentally transformed the perception of technology's impact. For instance, the introduction of the iPhone and other brands of 'smart' phones meant that the Internet was no longer something accessed solely through a console with a screen and keyboard; it could now be reached from anywhere with a mobile phone connection. Mobile web technology, when combined with big-data analytics, meant that the Internet was not just about connecting people and facilitating information exchange but also about linking devices that transmit and receive vast amounts of data. The widespread usage of both the iPhone and the Android operating system led to the development of online services where software applications could be purchased and downloaded through application marketplaces, making the sale of digital 'apps' possible in ways that had previously been unimaginable.

Figure 1. Timeline of the Three Phases of the Digital Economy

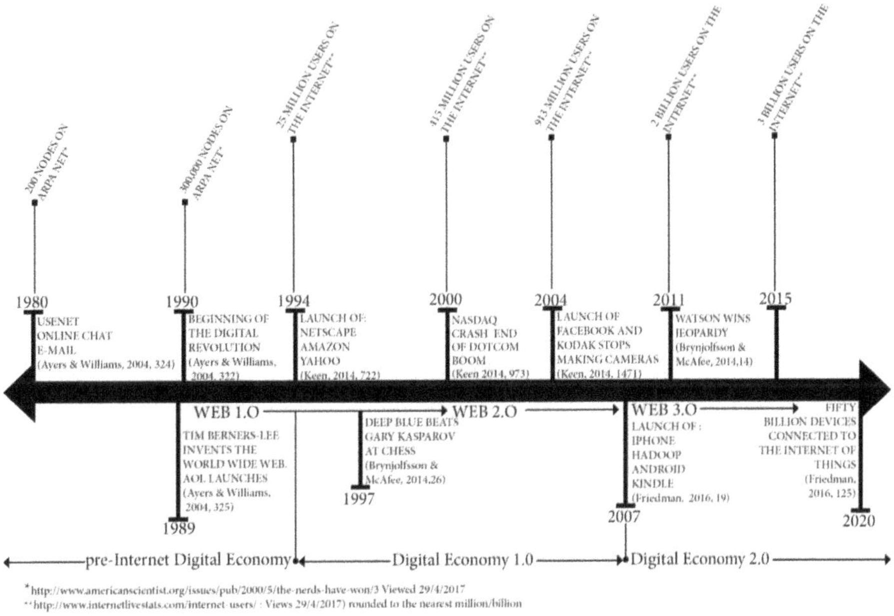

**http://www.americanscientist.org/issues/pub/2000/5/the-nerds-have-won/3. Viewed 29/4/2017*
***https://www.internetlivestats.com/internet-users/. Viewed 29/4/2017. Rounded to the nearest million/billion*

Three Phases of the Digital Economy.

Figure 1 presents an indicative timeline of the digital economy, with the terms "pre-digital economy," "digital economy 1.0," and "digital economy 2.0" serving as imposed classifications to refer to the evolution of literature and how perspectives on the digital economy have shifted over time. The points marked in 1994 and 2007 do not imply that significant technological advancements did not occur prior to these dates. Instead, they signify symbolic commercial or economic events that illustrate the technological progress reaching a broader consumer base.

The underlying purpose of using these three terms is not to undermine or invalidate any literature concerning the digital economy but rather to make it easier to understand this literature within the context it was written. The field of the digital economy suffers from being both a young field as an academic discourse and a fast-moving field as technology advances. This timeline makes it possible to provide a brief overview of the most relevant literature while at the same time not dwelling on literature which was written within in a different period of the digital economy.

The Digital Economy and Education

Understanding the phases of the digital economy can be very useful in computing education. It allows the question to be asked, "What sort of digital economy are young people being prepared for?"

Digital Economy 1.0.

In the 1998 ICT curriculum, young people learned to use software and computers as business tools but received little instruction on utilising the Internet or programming. These skills reflected the efficiencies of the pre-digital economy without leveraging the potential of digital networks, doing little to equip young people to fully capitalise on the Internet, the World Wide Web, or interconnected businesses.

Digital Economy 2.0.

This is the era of the digital economy when access to the Internet became possible almost anywhere and at any time for individuals using mobile devices. New types of products have started to emerge, and nearly all aspects of business and various sectors now rely on digital devices in some way. Once the digital economy 2.0 commenced, it is understandable that young people found it challenging to engage with what they were learning in ICT, as it represented a technological state that felt very distant from the world they experienced.

WHAT DOES IT MEAN TO BE DIGITALLY LITERATE?

The widespread use of the term "digital literacy" began in 1997 with a publication by Paul Gilster. This term is closely related to several other forms of "literacy" in the 21st century, including information literacy, computer literacy, network literacy, and Internet literacy.

The phrase "digital literacy" essentially refers to the ability to access and comprehend information using computers, mastering skills that go beyond just keystrokes. (Glister, 1997) (Bawden, 2008). The nature of information evolves in the digital age; to be digitally literate, one must navigate this new landscape effectively (Lanham, 1995). Digital literacy constitutes a new set of skills that encompass not only understanding but also attitudes towards information and technology (Bawden, 2001).

Digital literacy has changed what it means to be literate and, therefore, also designates a moral framework and a set of attitudes that sit alongside skills and knowledge (Bawden, 2008; Buckingham, 2008). It has become the underpinning ability to engage successfully in a digital environment (Martin, 2005).

Digital literacy is seen as a combination of what Aviram and Eshet-Akalai (2006) call "technical-procedural, cognitive and emotional social skills" —how one uses technological tools, how one thinks about those tools, and how one uses those tools to connect with other individuals specifically on an emotional level (Aviram & Eshet-Akalai, 2006).

When discussing digital literacy in the context of computing education, the Royal Society (2012) defined it as "the general ability to use computers." Chase and Laufenberg (2011) acknowledge that educators' understanding of digital literacy must constantly develop, just as their understanding of the future is also constantly developing.

Chase and Laufenberg (2011) propose that the essence of digital literacy lies in effectively leveraging technology to uphold the enduring values of education: inquiry, research, collaboration, presentation, and reflection. This viewpoint is similarly echoed in the Royal Society report (2012), which posits that digital literacy should be regarded as equivalent to the ability to read and write, emphasising the necessity of teaching these skills.

Research across several studies involving younger students shows that, irrespective of their ability to use technology or their backgrounds, all students lack skills in "evaluation" and "critical literacy" relating to digital literacy. Students must be explicitly taught how to understand, evaluate, and interpret information from digital sources (Eshet-Alkali & Amichai-Hamburger, 2004; Gui & Argentin, 2011; Ng, 2012).

The conundrum facing educators and students is that the term "digital literacy" is presented as the essential skill for navigating a digital environment. However, as the technology shaping this landscape evolves, must our understanding of digital literacy also be continuously developed, or are there key concepts that remain foundational?

CALLS FOR MORE DIGITAL LITERACY

Several reports over the years (Royal Society, 2017; Donoso et al., 2021; CECE, 2017) have highlighted the necessity of teaching digital literacy alongside computing. However, "there are rarely stand-alone curricula for teacher training in digital literacy. Thus, there is a risk that the subject is taught by teachers who lack the appropriate subject-matter knowledge." In other words, teachers are not receiving specific and dedicated training in digital literacy (CECE, 2017).

As the academic literature demonstrates, digital literacy should extend beyond young people being merely digitally skilled and should encompass a range of cognitive and social abilities: "Besides technical skills, experts attach great importance to non-technical skills such as critical thinking, information processing, the ethical use of digital technologies, protecting personal data and privacy, and managing one's digital identity" (Donoso et al 2021). The evidence gathered by the Royal Society suggests that teachers and young people in England (at least) view computing as "difficult" and only appropriate for the most high-achieving students (Royal Society, 2017). Furthermore, the subject is under pressure due to its limited allocation of time in school timetables and a lack of qualified teaching staff (Royal Society, 2017).

There remains a compelling argument that digital education encompasses both technical and non-technical elements, as young people will require both to participate in modern society (Donoso et al., 2021). Furthermore, to engage young people effectively, curricula must align with the realities of their lives (Donoso et al., 2021*)*.

DIGITAL LITERACY AND PREPARING FOR THE DIGITAL ECONOMY AND THE DIGITAL WORLD

While this chapter may appear to clearly distinguish between Digital Literacy and the Digital Economy, much of the more recent literature has brought these two themes together, specifically examining how digital literacy skills are key to preparing for a future in the Digital Economy—this is what can be referred to as something more akin to 'digital citizenship'.

The research found that young people had little sense of what being part of the digital economy might entail. It specifically noted that young people had little sense of how what they are learning is related to the world ahead of them.

This is echoed by Dinner and Polovinko (2018), who examined how young people were being prepared for the "fourth industrial revolution," and Brinda et al. (2018), who investigated the words high school students in Germany associate with the concept of the "Digital World." Their conclusion was that these associations related as much, if not more, to the media consumption of young people. Young people's understanding of the digital world is shaped by their experiences beyond the classroom. However, it can be challenging for schools to keep up with the ever-changing landscape, as the 'digital world' and 'digital economy' are constantly evolving.

However, there is no doubt that the digital economy affects all aspects of life, as stated in the OECD report from 2016, "Skills for a Digital World" (OECDa 2016). The report emphasises the critical importance of digital literacy skills, which are essential for "life-long learning" (OECDa 2016, 12). Integrating ICT across the curriculum can be one of the most effective means for young people to appreciate the relevance of digital skills. The OECD's findings indicate that while education must ensure all pupils possess "basic ICT skills", as these are essential for nearly all workplaces, they also assert that in terms of specialist skills, "Basic programming is no longer sufficient" (OECDb, 2016, 3).

There is a widening skills-based digital divide that is not immediately apparent, where those who engage successfully possess a high level of skills and access to resources, along with knowledge that many young people lack (Lombana-Bermudez et al. 2020). Although young people believe they understand how to interact online, they may miss the bigger picture. They are not always fully aware of the impact of sharing their data. While they recognise 'digital reputation' and the importance of not sharing personal information, they are less aware of how personal data is crucial to the business models of social media companies (Lombana-Bermudez et al. 2020, 13).

METHODOLOGY

The data presented in this chapter is based on in-depth group interviews conducted in 2016 with young people from three schools in the Northwest of England. The research included five group interviews at an initial school (School A) and a smaller number of interviews at two other schools (three in School B and one in School C). The group setting and open-ended questions elicit the best responses from young individuals (Maxwell, 2010; Cohen et al., 2011). The interviews were recorded, transcribed, and then analysed using a grounded theory-based approach to coding that involved open, axial, and selective coding to identify themes and key points from the data. The transcripts were checked against the recordings for accuracy and meaning (Bryman, 2004). The project was approved by the Lancaster University Ethics Committee, which oversaw the research during its execution.

For this project, pupils were typically interviewed in groups of six with the groups being chosen by the teacher. A total of 54 pupils took part in the interviews for this project, with about seven total hours of interviews with pupils.

School A: Total pupils interviewed 30, length of interview 20-50 minutes, all pupils in year 9.

- Interview 1: Year Nine, 6 pupils: 3 boys, 3 girls (43 minutes)
- Interview 2: Year Nine, 6 pupils: 3 boys, 3 girls (38 minutes)
- Interview 3: Year Nine, 6 pupils: 3 boys, 3 girls (20 minutes)

- Interview 4: Year Nine, 6 pupils: 3 boys, 3 girls (29 minutes)
- Interview 5: Year Nine, 6 pupils: 3 boys, 3 girls (54 minutes)

School B: Total pupils interviewed 18, length of interview 42-57 minutes, year 7 & 9 pupils.

- Interview 1: Year Nine, 5 pupils: 2 boys, 3 girls(45 minutes)
- Interview 2: Year Nine, 5 pupils: 3 boys, 2 girls (42 minutes)
- Interview 3: Year Seven, 7 pupils: 4 boys, 3 girls (57 minutes)

School C: Total pupils interviewed 6, length of interview 25 minutes, year 10 pupils.

- Interview 1: Year Ten, 6 pupils: 3 boys, 3 girls (25 minutes)

DATA

Data Regarding Digital Economy

This chapter presents the idea that the digital economy can be understood in three phases: the pre-Internet digital economy, digital economy 1.0, and digital economy 2.0, alongside a developing understanding of the potential nature of future digital economies.

While the students interviewed in 2016 were not considering the economy or their economic future in these terms, aspects of their experience closely follow this progression. They were aware of their grandparents' discomfort with Internet technology, as they belonged to a generation that grew up in the pre-digital economy. They observed their parents and teachers using tools, such as office-type software, which would be associated with digital economy 1.0, and they encountered daily the technologies of digital economy 2.0. Pupils expressed questions and concerns about the technologies linked to a future digital economy, alongside an ever-changing set of technologies. They felt they did not fully understand these new devices and had numerous concerns stemming from this lack of knowledge.

The People and Things of the Digital Economy: What is a Computer Scientist?

The young people displayed considerable confusion regarding what (or who) a computer scientist may be. While individuals like Bill Gates and Steve Jobs have undeniably had a profound impact on computing, many others from diverse backgrounds have also influenced the field. Most of the names mentioned by the young people were white men—ranging from the British TV presenter Dara O'Briain to the character Sheldon Cooper from "The Big Bang Theory." Several young individuals referred to physicists Stephen Hawking and Brian Cox; some confused Stephen Hawking with Steve Jobs. The young people appeared to have little understanding of what being a computer scientist involves, what type of person a computer scientist may be, or of role models who do not conform to the prevailing stereotype of a computer scientist.

Thinking about their own future and future employment

Young people identified a wide range of careers that they were interested in pursuing. While some of these were related to using computers, such as starting an online business (School A, Interview 1), many careers seemed far less related to technology (like becoming a foster carer (School A, Interview 5). However, the discussion about future jobs prompted the young people to consider how computers would affect a wider range of jobs, from being a veterinarian (School A, Interview 5) to being a beautician (School A, Interview 4). For example, the pupils who wanted to be beauticians thought they would use a computer to order products and communicate with clients (School A, Interview 4). Many of the pupils saw how IT/ICT skills could be used in a wide range of careers and that taking ICT at a higher level could be relevant for a range of careers (School A, Interview 4; School B, Interview 2; School B, Interview 3).

Making Choices About the Future.

This research suggests that computing education in the UK is impacting young people's choices. However, this does not necessarily mean that they feel they are more likely to choose a career in the digital economy. Instead, they are reflecting the skills they might need (in terms of computing) for a wide range of careers.

The young people seemed unsure if what they were learning about computing was either giving them skills for the future or introducing them to possibilities for the future (School A, Interview 5; School B, Interview 2; School C, Interview 1). However, when discussing computing broadly, they seemed to feel that it was important to their futures.

The workplaces these young people envisioned involved computers in various forms. For instance, some young individuals described the use of robots in manufacturing (School B, Interview 2). In Interview 1 at School C, the students mentioned the necessity of knowing the "basics," whether this knowledge applied to "most things in life" or to a specific career such as being a professional 'video gamer' (School A, Interview 2; School A, Interview 5; School C, Interview 1). However, they believed that the rapid advancement of technology meant that the computing classes could provide only basic skills (School A, Interview 5). In two of the interviews, the young people found it difficult to think of any activity that did not involve the use of computers (School A, Interview 4; School C, Interview 1).

While some felt that the curriculum was generally for life (School A, Interview 1), others felt that the main purpose of computing was specifically about having a future in computing (School A, Interview 2). For these young people, there was a question about computing: are the skills they are learning 'general skills' needed for a wide range of careers and life more widely in the same way that every child needs to learn maths, English, and science—or are the skills they are learning in computing for a specific future direction in the way skills they might be learning in design and technology or business studies.

Skills Needed in Future Employment.

The pupils did have a sense that the purpose of learning computing was the national skills shortage [in computer science] (School B, Interview 2). They were able to see the need to learn programming as not being connected to their needs as individuals but to the needs of the country. This argument left the pupils who were not interested in becoming programmers feeling less inclined to be engaged in the computing sessions as the purpose of the sessions was to prepare for a career which they had no intention of pursuing

(School A, Interview 1). While they felt they needed some skills for life, there were other aspects of computing that were more about preparing specifically for future employment (School A, Interview 5).

The feeling was that having a broad knowledge of computing would give them a wider range of options in the future (School B, Interview 3, School C, Interview 1). The young people could see that by learning computing from an early age, they would have more opportunities but could also see that it could change the nature of certain fields by increasing the amount of specifically gender diversity in STEM subjects, which tend to be male dominated. Specifically, for a number of young women interviewed, learning about computing had introduced them to new possibilities for their futures (School A, Interview 2; School B, Interview 1, School C, Interview).

Conclusion of data regarding the digital economy

Not all of the young people had thought about what career they wanted to pursue, and many of the perceptions they had about what skills they thought they would need were influenced more by the skills they saw in their parents than by what they thought the future digital world would be like. From the data, it seemed that they saw the need for skills like using specific software (such as office-type software). Where they thought that their careers would have a specific technical demand, they could see the connection to the skills they were learning in their computing lessons. For the students who saw themselves in a career that did not directly involve computers, they would rather be learning the general ICT skills rather than learning the more specific computing or Computer Science skills that have dominated much of the computing curriculum.

In the interviews, the young people were not thinking of the skills they were learning in terms of the digital economy. Though they were thinking about the skills they will need for the broader digital world, they did not to feel that they are being prepared for this and are left with many questions about how technology is changing society. Their sense of who is involved in the digital economy is limited and this may be affecting the extent to which they feel this is an industry that they might see themselves within.

In terms of the phases of the digital economy proposed earlier, the young people seem to be thinking about the skills they will need in terms of the skills that would be associated with digital economy 1.0. These are the skills that would be used on a computer at a fixed location but not leveraging newer developments such as big data, open data, or the Internet of Things.

While the young people appear to be aware, through their personal experiences, of the technologies associated with digital economy 2.0, this is the area where they have the most specific concerns and questions. Based on the data, the computing curriculum appears to be more rooted in digital economy 1.0, while the young people experience technology that is more like digital economy 2.0.

Data Regarding Digital Literacy

Digital Knowledge and Digital Skills—Cognitive, Critical, and Technical Skills Digital Literacy

One of the consistent themes across several interviews was that much of what young people valued learning, and desired to learn more about, relates to skills classified as office software, such as word processing, spreadsheets, and databases (School A, Interview 1; School A, Interview 2; School A, Interview 3; School A, Interview 4; School A, Interview 5; School B, Interview 1; School B, Interview 2).

While these skills are often regarded as less of a priority in computing education, the findings suggest that young people focus much more on concrete, applicable skills rather than broader concepts.

Thinking About Using Digital Tools.

The young people felt that they were learning computing to help them in future employment (School A, Interview 1; School A, Interview 5); however, others felt quite strongly that they were mainly learning it to explore a specific computing-based career (School A, Interview 2; School B, Interview 3).

There was a common opinion that if they learn about "this stuff" now, they will have the skills to keep up with technology; this is especially important as they observe their parents and grandparents "falling behind" (School A, Interview 5).

The pupils were aware of the ICT skills the adults in their lives are using every day, such as email, graphic design, and using the Internet (School A, Interview 2). This helps them imagine how they might use computing in the future. Similarly, because they use the Internet every day, they can see the value in learning how to make a webpage or an online CV (School A, Interview 1).

Learning Digital literacy - Value What They Have Learned in Regard to ICT.

When asked to think about the impact aspect of what they were learning, many of the young people focused on technical skills associated with using computers. Many of the things that the young people valued learning (and feel they would remember and use in the future) related to using office-type software, such as Microsoft (MS) Word, MS Excel, and using databases (School A, Interview 1; School A, Interview 2; School A, Interview 3; School A, Interview 4; School A, Interview 5; School B, Interview 1; School B, Interview 2). Many of the young people reported having learned this in primary [elementary] school(School A, Interview 1; School A, Interview 5; School B, Interview 1). The young people were aware of their parents using these skills; they could see the value in using these skills and felt they would remember and use them in the future.

The young people valued learning aspects of computing that related to their everyday lives, such as how search engines or computers in general worked, and even saw how the logic of computation was transferable (School C, Interview 1, School B, Interview 1).

Thinking Critically about Content and Information

Although critical thinking was not explicitly mentioned, the data highlights that young people engage with a variety of information sources (such as YouTube) but do not feel equipped to assess the validity of this content. These students recognised some of the threats that could pose problems for their computers at home, including viruses, malware, and hackers, yet felt they were not learning how to protect their devices (School A, Interviews 2; School A, Interviews 4). Several pupils disclosed during the interviews that they did not comprehend how viruses operate (School A, Interview 2) and were uncertain about how to prevent issues with their computers or how to address them once a problem arises.

They can find advice online from a range of sources but would like to hear that advice reinforced by what they are learning in school (School A, Interviews 4, 5). While the advice they find on sources like YouTube may not be wrong, it raises a number of questions and concerns about cyber security. The pupils seemed to have no forum in which to ask those questions.

Social-Emotional Literacy

One of the key aspects of digital literacy centres on the social-emotional skills that young people need to navigate their digital social lives and interact with others online. While the students did not use the term 'social-emotional literacy', the interviews revealed that many of these young individuals lead rich digital lives. Although they did not report learning how to navigate these environments in school, they had given considerable thought to how and why people behave the way they do online.

This section looks at data relating to how young people interact with others online, making friends online, how these young people engage with larger communities online, and putting their digital social lives in context.

Interacting with Others Online.

All group interviews with young people included discussions of online behaviour, both concerning others and the participants' own behaviour. There was an acceptance of some reasons why an online presentation may not fully align with an offline identity (School A, Interview 5; School B, Interview 3). The young people emphasised that there is significant pressure to look or behave in a certain way, and while it may not be easy to conform to this pressure in the physical world, in the digital realm, it is relatively simple to present oneself differently.

There was significant concern about being picked on. There was almost an allowance and acceptance of mean, trolling behaviour online (School A, Interview 5; School B, Interview 2; School B, Interview 3). Generally, people would say things online that they would not say in person, as there would be few or no consequences (School A, Interview 5; School B, Interview 1). However, while most young people believed there were few consequences for online behaviour, a smaller number actually thought that the potential consequences affecting online conduct the most were either social or through formal authority. Whereas, in face-to-face interactions, people might forget what was said, in an online context, "they can scroll back in conversation online" (School A, Interview 5; School B, Interview 1), providing proof that one had said what they did. There was a sense that it was not so much that people online intended to deceive each other; instead, it was more that the nature of the online environment altered someone's behaviour (School A, Interview 5).

Making Friends Online.

Several young people reported using the internet to maintain connections with friends and family world-wide, or with friends they had met in other contexts, such as a previous school (School A, Interview 2).

The pupils reported having connections with people in San Francisco, the Philippines, southern England, France, and Portugal (School A, Interview 1; School B, Interview 1). In some cases, these relationships involved maintaining contact with family, while in other cases, they were individuals whom the students had met through online activities. Similarly, other pupils mentioned using social media to engage with peers from the same school with whom they would not typically interact (School A, Interview 3). The pupils indicated that they benefited from these relationships, ranging from close friendships to practising language skills and maintaining family ties. They did not feel at risk from these connections; in fact, they believed they understood the precautions necessary and were very conscious of potential dangers (School A, Interview 2; School A, Interview 5).

Part of Something Bigger.

In addition to connecting with people around the globe, young individuals are also utilising digital networks to feel part of a larger community. While many young people did not indicate being part of any online groups, one interview revealed that they belonged to a "guild" or group within a game (School B, Interview 1). In a subsequent interview, it was noted that these groups were sufficient for young people to explore their interests and cultivate a sense of belonging.

For example, one young woman had discovered a way to pursue a relatively niche ambition of becoming an air traffic controller by taking part in an online simulation. She felt secure due to the infrastructure of the simulation, but she also sensed she was part of a greater task. It could be quite challenging for a teenage girl to connect with others who share her passion for aeroplanes, particularly control tower operations; however, this activity enables her to explore a potential career, hobby, and passion.

By contrast, another young woman who spoke of wanting to be politically active in later life highlighted that while being an observer may be her best way of engaging in online political activity, she may want to engage in a more interactive way as she gets older.

Digital Social Lives in Context.

These young people were also aware of the impact of social media and its history on other adults. They do not view social media as merely a sharing platform; instead, they see it as a means to create profiles and communicate with peers. The chat environment can be quite informal, serving as an extension of their face-to-face networks. They recognise that more public online performances may not necessarily reflect offline reality, as many digital networks provide opportunities to explore hobbies and make new friends.

DISCUSSION OF THE DATA COLLECTED

Digital economy

While computing education aims to prepare young people for direct entry into the digital economy as programmers, web developers, computer scientists, or network administrators, an important question arises regarding how well they are supported. Does computing education effectively enhance young people's understanding of the digital economy, enabling them to make informed decisions about their future? Previous research indicates that pupils' perceptions of how they will use computing and ICT skills are largely shaped by their school environments and the courses they undertake (Bradshaw, 2018).

What is the Digital Economy We Are Aiming For?

The young people involved in this research had little insight into how they would use technology in their future careers and a limited understanding of the individuals who have shaped the digital economy. This restricts the number of young people who can envision themselves as part of the digital economy in the future.

The young people involved in this study did not see computing as a particularly creative discipline. While some did express the desire to work in the games industry, this seemed to stem from their interest in computer games (as players) rather than any influence of education.

Understanding the Digital World Around Them

New technology is profoundly linked to society, and Lyon (1991) questions whether the social effects of Information Technologies in our 'new kind of society' are generally benign. Furthermore, if, as Carlson (2004, p. 262) suggests, the Internet and networked computer systems are "General Purpose Technologies" comparable to electricity or the steam engine, then society is potentially undergoing a significant transformation. The assumption, however, is that young people will adapt to technological developments with ease. However, the data indicates that the young individuals involved in this study do not feel they understand or can readily keep pace with the technological advancements they have witnessed even in their own lifetimes.

The young people are aware of the arguments about the changes in society but do not feel like they have influence over these changes.

If young people (as reported in this chapter) feel anxious, overwhelmed, and uninformed about new technology, they will struggle to apply it or use it as responsible, competent, confident, and creative users.

Computing education, if it is to engage and inform young people about the digital economy, must be engaged in emerging and new technologies, giving young people the tools they need to evaluate and make decisions about what technologies they use and how they use them. For example, young people should have an accurate understanding of privacy concerns with regard to new technologies such as smart homes, smart speaker technologies, or driverless cars (and other data gathering and 'smart' technologies).

It is important that this statement leaves room for young people to investigate and understand the affordances (Woollard, 2018) of different technologies. For example, are some technologies inherently unsafe? As a user of technologies, do you need to consider how those technologies could be abused or misused?

Understanding the Choices They Are Making

The choices these young people were making are rooted in their understanding of both the broad digital economy (the impact of digitisation across all sectors) and the narrow digital economy (sectors specifically dependent on digital technology). As Brynjolfsson and McAfee (2014) highlighted, while jobs like gardeners and cooks may not be directly at threat of digital replacements, many jobs will start to be complemented by digital technology (2014). Kids who want to become nurses may work alongside an artificial intelligence system to help with diagnoses (Brynjolfsson and McAfee, 2014). While nursing as a profession is not at risk of being replaced by automation, it requires a high level of technical skill and knowledge.

Current approaches to computing education tend to be more focused on the narrow needs of the digital economy rather than informing young people about the wide range of ways digital technology is used in a wide range of careers and job roles. In the same way that mathematics and science are not only relevant to those who will pursue careers in these fields, the premise of computing education could be (but is not currently) to prepare young people to live successful lives with a broad understanding of a digitised economy. In addition, the digital economy the young people are being prepared for is no longer the most relevant or current.

Computing Education and the Digital Economy

The young people involved in this study had a mixed understanding of the digital economy. While the computing curriculum has the potential to help young people better understand the digital economy, this does not seem to be happening currently. Computing lessons are more focused on specific concepts or skills rather than how those concepts or skills may be more broadly applied.

Digital literacy

The world in which these young people live, work, and play has been radically transformed by the advent of ubiquitous digital networks (Livingston & Sefton-Green, 2016). From the interviews examined in this thesis, the young people do not necessarily feel they possess or can demonstrate the skills required to navigate this altered world. Digital literacy can be regarded as an essential skill for fostering digital resilience and being equipped to navigate an online environment (Vissenberg, J., d'Haenens, L., & Livingstone, S. 2022).

The young people interviewed valued the ICT skills they had learned; they could see how using office-type software was key to success in the world of work. The subject content for computing leaves a wide scope in terms of the usage of specific software or skills.

With limited time given to a single subject that is intended to include such a wide range of topics, a degree of prioritisation must be made. From the general perspective of many of the students involved in this study, that prioritisation is not correct.

Critical Thinking and Media Literacy

Many of the young people involved in this study did not see how computing could be an avenue for creative endeavours. They could not see the creative aspects of their computing lessons and did not feel that computing was a particularly creative subject. Using the three-fold nature of digital literacy presented in this thesis, media literacy and critical thinking are essential parts of being digitally literate.

The students reported consuming a wide range of digital content and even raised questions about this content in the interviews. This is consistent with the findings of Livingston and Sefton-Green (2016). Glister (1997), in one of the first works on digital literacy, raised questions about the need to be able to contextualise content, warning that one of the problems of hypertext is that it is not always possible to judge the validity or usefulness of every link clicked. Based on the data collected, these very skills of contextualisation are missing from the current delivery of computing.

However, there seems to be relatively little support for the critical thinking, cognitive, and creative content creation aspects of digital literacy.

Complex Digital Social Lives, Require Digital Social Skills

The young people involved in this study reported having rich digital social lives, consistent with the findings of Livingstone and Sefton-Green (2016). While they were aware of and, in some instances, had negative experiences online, they also acknowledged the positive aspects of living digital lives. Whether it involved having friends and contacts around the globe, participating in events (such as flight simulations), or deepening their engagement in politics, these young individuals were capitalising on the social potential of interacting with others online. Although less visible diversity was overtly apparent in the lives of those who took part in this study, their digital social worlds extended globally, including locations like San Francisco, the Philippines, France, Portugal, and southern England. While digital socialising allowed young people to establish connections across the globe, this does not imply a sharp distinction between their online and offline relationships. Based on the interviews conducted, it appears that some young people have learned to harness the benefits of the rich social environment of the online world, while others are less aware of these advantages.

Even taking the most limited definition of digital literacy, there is little evidence from the interviews conducted that the young people felt they were learning anything that ensured that they were digitally literate, but rather having to build these skills independently.

Overarching discussion of the data collected

It is evident that all young people will be affected by the digital economy, and all students will require core digital literacy skills. The young people involved in this study already inhabit a 'digital world', so they do not perceive the computing curriculum as equipping them for an uncertain future. Instead, they acknowledge that their lives (and future careers) are being influenced by a broad array of digital tools. This does not imply that they lack questions regarding evolving technology; it appears their questions and concerns are both specific and informed. They understand that they need particular skills and knowledge to succeed. However, for reasons that merit further investigation in future research, the pupils seem to expect computing education to prepare them for utilising a wide variety of technologies; they wish to learn to be digitally literate but do not seem to feel they are acquiring these skills. More urgently, young people are leading rich digital social lives. Based on the literature reviewed in this chapter, these digital social skills are a critical component of digital literacy.

Digital Citizenship

While the study described in this chapter primarily focused on the digital economy and digital literacy, Digital Citizenship runs parallel to these concepts and has become increasingly prominent in the discussion of technology and education. The term has attracted increasing interest in the academic literature (Chen et al., 2021), but at the same time, it seems to have declined in teaching (Öztürk, 2021).

However, as this chapter has shown, young people are leading lives that are increasingly online and mediated by digital technology. They require a broad range of skills to navigate the complex landscape of both digital technologies and digital lives. The COVID-19 pandemic in 2020 specifically demonstrated that the use of digital tools is essential for all citizens to participate in education and work (Öztürk, 2021). While young people had been engaging in online socialisation long before the COVID-19 pandemic, the pandemic made it clear that there is now an expectation for social, educational, and even civic engagement to be conducted via digital platforms. Like the digital economy and digital literacy, this is an evolving field that demands greater attention from researchers, policymakers, and educators (Chen et al., 2021). Chen et al. (2021) go so far as to call for rigorous methodologies for defining and measuring digital citizenship.

With the increase in attention to computing education, there is a need to attend to how digital citizenship connects to related terms (as this chapter has highlighted), re-examining and defining how it relates to similar terms, Going being digital economy and digital literacy digital citizenship can include areas such as 'critical awareness' and 'digital activism' (Fernández-Prados, et al 2021). While these concepts came up in this study regarding digital literacy, areas such as 'activism' or even 'civic engagement' seem to stretch the term 'digital literacy' but do not quite fall under 'digital economy'.

Where digital citizenship has value is in its broadness of scope compared to the more 'skills' based understanding of 'digital literacy', which tends to be limited to a very specific and defined set of skills that are more difficult to update to the needs of young people (Jones and Mitchell, 2016). Equally, this allows for areas such as digital safety to go beyond focusing on negative behaviours such as bullying and rather evoke the opportunity to discuss positive behaviours online (Jones and Mitchell, 2016).

Citizenship refers to civic power and decision-making, engaging with a broader society beyond merely a set of skills or economic benefits, but rather focusing on social good and participation in social decision-making (Simsek and Simsek, 2013). While one may possess digital literacy skills or knowledge of the digital economy without the capacity to engage in a digital society (Simsek and Simsek, 2013), an emphasis on digital citizenship can encompass these concepts while empowering young people to harness the opportunities available to influence society, to be part of the digital community, and to fully embrace their roles as digital citizens.

CONCLUSION

Over the past decade, there has been an increased focus on computing education due to a sense that the needs of young people and the requirements of an emerging 'digital economy' have not been adequately addressed by what young people are learning in school.

While concerns have been raised about various industries struggling to find both the required skills and diversity, there is also a view that a range of digital literacies has become an increasingly essential skill set for optimally utilising the available technology. Young people exist in a world where digital technologies permeate their daily lives. Once they enter the workforce, regardless of the profession, computers (at the very least) will be crucial tools for success. The educational challenge has been how to equip young people for this reality.

Beyond economic activities, digital skills have become an essential part of civic engagement. Form filing taxes, registering to vote, and participating in local government meetings. Something the COVID-19 pandemic made abundantly clear.

The research reviewed in this chapter highlights the challenge that young people have not felt engaged with computing. The ideas surrounding the 'digital economy' are too abstract, and young people have little sense of what this might be. On the other hand, their digital literacy skills do not reflect their everyday lives, leaving them feeling unprepared for this ill-defined future.

Using the lens of Digital Citizenship, it becomes possible to go beyond what is covered in most computing classes but has the advantage of being forward-looking. Where digital literacy can be overly skills-focused, and the digital economy is overly abstract. This chapter suggests using the broader term 'digital citizenship' to help young people consider how they have agency and power in a digital world. Rather than focus on skills or careers, a focus on digital citizenship can focus on a set of behaviours and attitudes, under which skills can be taught with context and a range of futures can be imagined.

The key purpose of this study, however, is to highlight the role qualitative research can play in understanding young people's experiences. Computing education is just one area where research and policy-making greatly impact young people's lives. Yet, the challenges of conducting qualitative research with young people often limit the degree to which their voices are taken into account. This project demonstrates that by creating space for young people's voices in research, a deeper and more nuanced understanding of their experiences can be achieved.

Computing education, whether in the guise of a digital economy, digital literacy or digital citizenship, has the potential to be the blueprint onto which young people can map their future lives, seeing how they can use digital tools to be empowered. Over all, the purpose of teaching computing should be to prepare young people for their future and, at the same time, give them tools to make that future a better place.

REFERENCES

Aviram, A., & Eshet-Alkalai, Y. (2006). *Towards a theory of digital literacy: Three scenarios for the next steps.* http://www.eurodl.org/index.php?p=archives&year=2006&halfyear=1&article=223

Ayres, R., & Williams, E. (2004). The digital economy: Where do we stand? *Technological Forecasting and Social Change, 71*(4), 315–339. DOI: 10.1016/j.techfore.2003.11.001

Bawden, D. (2001). Information and digital literacies: A review of concepts. *The Journal of Documentation, 57*(2), 218–259. DOI: 10.1108/EUM0000000007083

Bawden, D. (2008). Origins and concepts of digital literacy. In Lankshear, C., & Knobel, M. (Eds.), *Digital literacies: Concepts, policies and practices* (1st ed., pp. 17–30). Peter Lang Publishing.

Bradshaw, P. (2018). Assessment of Computing and ICT at 16: What are the perceptions of learners? In Younie, S., & Bradshaw, P. (Eds.), *Debates in Computing and ICT Education* (1st ed., pp. 43–52). Routledge.

Brinda, T., Napierala, S., & Behler, G. (2018). What do secondary school students associate with the digital world? *Proceedings Of The 13Th Workshop In Primary And Secondary Computing Education, Postdam, Germany.* https://doi.org/ [Accessed 18 February 2021].DOI: 10.1145/3265757.3265763

Bryman, A. (2004). *Social research methods* (2nd ed.). Oxford University Press.

Brynjolfsson, E., & Kahin, B. (2002). *Understanding the digital economy.* The MIT Press.

Brynjolfsson, E., & McAfee, A. (2016). *The second machine age.* W. W. Norton.

Brynjolfsson, E., & Saunders, A. (2013). *Wired for innovation.* MIT Press.

Buckingham, D. (2008). Defining digital literacy - What do young people need to know about media? In Lankshear, C., & Knobel, M. (Eds.), *Digital literacies: Concepts, policies, and practices* (1st ed., pp. 73–90). Peter Lang Publishing.

Carlsson, B. (2004). The digital economy: What is new and what is not? *Structural Change and Economic Dynamics, 15*(3), 245–264. DOI: 10.1016/j.strueco.2004.02.001

Chase, Z., & Laufenberg, D. (2011). Embracing the squishiness of digital literacy. *Journal of Adolescent & Adult Literacy, 54*(7), 535–537. DOI: 10.1598/JAAL.54.7.7

Chen, L. L., Mirpuri, S., Rao, N., & Law, N. (2021). Conceptualization and measurement of digital citizenship across disciplines. *Educational Research Review, 33*, 100379. DOI: 10.1016/j.edurev.2021.100379

Cohen, L., Manion, L., & Morrison, K. (2011). *Research methods in education* (7th ed.). Routledge.

Dinner, I., & Polovinko, V. (2018). Dinner, Igor, and Vladimir *Polovinko. Features of Professional Self-determination and Professional Orientation of Young People in the Digital Economy.* In The 12th International Days of Statistics and Economics (pp. pp. 335-343). Prague. Retrieved 18 February 2021, from https://msed.vse.cz/msed_2018/article/204-Dinner-Igor-paper.pdf. [Accessed 18 February 2021].

Donoso, V., Pyzalski, J., Walter, N., Retzmann, N., Iwanicka, A., d'Haenens, L., & Bartkowiak, K. (2021). *Report on Interviews with Experts on Digital Skills in Schools and on the Labour Market*. Zenodo: European Commission. Retrieved from http://doi.org/ [Accessed 18 February 2021].DOI: 10.5281/zenodo.4274613

Eshet-Alkali, Y. (2004). 'Digital literacy: A conceptual framework for survival skills in the digital era. *Journal of Educational Multimedia and Hypermedia*, *13*(1), 93–106.

Eshet-Alkali, Y., & Amichai-Hamburger, Y. (2004). Experiments in Digital Literacy. *Cyberpsychology & Behavior*, *7*(4), 421–429. DOI: 10.1089/cpb.2004.7.421 PMID: 15331029

Fernández-Prados, J. S., Lozano-Díaz, A., & Ainz-Galende, A. (2021). Measuring Digital Citizenship: A Comparative Analysis. *Informatics (MDPI)*, *8*(1), 18. DOI: 10.3390/informatics8010018

Friedman, T. L. (2017). *'Thank you for being late: An optimist's guide to thriving in the age of acceleration*. Picador/Farrar Straus and Giroux.

Gilster, P. (1997). *Digital literacy*. John Wiley & Sons.

Gui, M., & Argentin, G. (2011). Digital skills of Internet natives: Different forms of digital literacy in a random sample of northern Italian high school students. *New Media & Society*, *13*(6), 963–980. DOI: 10.1177/1461444810389751

Hamid, N., & Khalid, F. (2016). Entrepreneurship and innovation in the digital economy. *The Lahore Journal Of Economics*, *21*(Special Edition), 273–312. DOI: 10.35536/lje.2016.v21.isp.a12

Irkinovich, N. R. (2022). THE DIGITAL ECONOMY TODAY. Academicia Globe. *Inderscience Research*, *3*(10), 198–203.

Jones, L. M., & Mitchell, K. J. (2016). Defining and measuring youth digital citizenship. New Media &amp. *New Media & Society*, *18*(9), 2063–2079. DOI: 10.1177/1461444815577797

Keen, A. (2015). *The Internet is not the answer*. Atlantic Books.

Koltay, T. (2011). The media and the literacies: Media literacy, information literacy, digital literacy. *Media Culture & Society*, *33*(2), 211–221. DOI: 10.1177/0163443710393382

Lanham, R. (1995). Digital literacy. *Scientific American*, *273*(3), 160–161.

Livingstone, I., & Hope, A. (2011). Next Gen [online]. Nesta., Available at https://www.nesta.org.uk/report/next-gen/, Retrieved May 28, 2019, from.

Livingstone, S., & Sefton-Green, J. (2016). *The class*. New York University Press.

Lombana-Bermudez, A., Cortesi, S., Fieseler, C., Gasser, U., Hasse, A., Newlands, G., & Wu, S. (2020). Youth and the Digital Economy: Exploring Youth Practices, Motivations, Skills, Pathways, and Value Creation. Berkman Klein Center for Internet & Society., Retrieved from https://cyber.harvard.edu/ publication/2020/youth-and-digital-economy, Retrieved February 18, 2021, from.

Lyon, D. (1991). The information society: Ideology or utopia. In Mackay, H., Young, M., & Beynon, J. (Eds.), *Understanding Technology in Education* (1st ed., pp. 93–108). The Falmer Press.

Martin, A. (2005). DigEuLit – A European framework for digital literacy: A progress report. *Journal Of Eliteracy*, *2*(2), 130–136.

Maxwell, J. (2010). Validity: How might you be wrong? In Luttrell, W. (Ed.), *Qualitative Educational Research* (1st ed., pp. 279–287). Routledge.

Ng, W. (2012). Can we teach digital natives digital literacy? *Computers & Education*, *59*(3), 1065–1078. DOI: 10.1016/j.compedu.2012.04.016

OECDa. (2016). Policy Brief on the Future of Work- Skills for a Digital World. OECD. Retrieved from https://www.oecd.org/employment/future-of-work.htm [Accessed 18 February 2021].

OECDb. (2016). Skills for a Digital World. OECD Publishing. htm., Retrieved February 18, 2021, from.

Oxley, L., Walker, P., Thorns, D., & Wang, H. (2008). The knowledge economy/society: the latest example of" Measurement without theory"?. Journal of Philosophical Economics, 2(Articles).

Öztürk, G. (2021). Digital citizenship and its teaching: A literature review. *Journal of Educational Technology & Online Learning*, *4*(1), 31–45.

Simsek, E., & Simsek, A. (2013). New Literacies for Digital Citizenship. *Contemporary Educational Technology*, *4*(2), 126–137. DOI: 10.30935/cedtech/6097

Singh, N. (2004). The digital economy: An overview. In Singh, N. (Ed.), *The Handbook of Information Security*.

The Committee on European Computing Education (CECE). (2017). Informatics Education in Europe: Are We All. In The Same Boat? Informatics Europe & ACM Europe. Retrieved from., Retrieved February 18, 2021, from, DOI: 10.1145/3106077

The Royal Society. (2012). *Shut down or restart? The way forward for computing in the UK schools.* [online] London: The Royal Society. Available at: https://royalsociety.org/topics-policy/projects/computing-in-schools/report/ [Accessed 28 May 2019].

The Royal Society. (2017). *After the Reboot: Computing Education in UK Schools*. London: Royal Society. Retrieved from http://royalsociety.org/computing-education [Accessed 18 February 2021].

United States Bureau of the Census. (2001). *Measuring the digital economy*.

Vissenberg, J., d'Haenens, L., & Livingstone, S. (2022). Digital literacy and online resilience as facilitators of young people's well-being? *European Psychologist*, *27*(2), 76–85. DOI: 10.1027/1016-9040/a000478

Woollard, J. (2018). Computer studies, information technology, ICT, and now computing. In Younie, S., & Bradshaw, P. (Eds.), *Debates in Computing and ICT Education* (1st ed., pp. 43–52). Routledge.

Chapter 4
Nurturing Responsible Digital Citizens in the Online World

Mohsen Mahmoudi-Dehaki
https://orcid.org/0000-0002-4962-6827
Department of English, Najafabad Branch, Islamic Azad University, Iran

Nasim Nasr-Esfahani
https://orcid.org/0000-0002-2388-6524
Department of English, Isfahan (Khorasgan) Branch, Islamic Azad University, Iran

ABSTRACT

This chapter examines digital citizenship, focusing on digital literacy, ethical behavior, privacy, and security in our interconnected world. It emphasizes the need for individuals to develop responsible online practices to protect their personal information from evolving cyber threats. Key themes include the importance of digital literacy for effective engagement with technology and understanding the ethical implications of online actions. The chapter discusses strategies for educators to instill digital responsibility in youth, ensuring they navigate the digital landscape confidently. It highlights the significance of safeguarding personal information and utilizing privacy settings effectively to mitigate risks like identity theft and cyberbullying. By promoting best practices for online security, the chapter aims to empower individuals to protect their digital identities. Ultimately, it provides a comprehensive overview of digital citizenship, advocating for a proactive approach to harnessing technology's benefits while safeguarding personal rights in the digital realm.

INTRODUCTION

The digital landscape presents remarkable opportunities and significant challenges in an increasingly interconnected world. As we navigate this complex online environment, the importance of digital literacy, ethical conduct, privacy, and security has never been more pronounced. This chapter addresses these critical concerns by exploring the multifaceted nature of digital citizenship, emphasizing the need for individuals to cultivate responsible online behaviors while safeguarding their personal information against a backdrop of evolving threats. At the heart of navigating the online world lies digital literacy, which encompasses the skills required to effectively and critically engage with digital technologies.

DOI: 10.4018/979-8-3693-6675-2.ch004

Copyright ©2025, IGI Global Scientific Publishing. Copying or distributing in print or electronic forms without written permission of IGI Global Scientific Publishing is prohibited.

Developing digital literacy is essential for personal empowerment and fostering ethical conduct in online interactions. As individuals increasingly rely on digital platforms for communication, education, and commerce, understanding the ethical implications of their online actions becomes imperative. This chapter will delve into the necessity of equipping individuals with the tools to navigate the complexities of the digital realm responsibly, ensuring they can contribute positively to the digital community.

The digital age has ushered in many privacy and security concerns that demand our attention. With the proliferation of personal data shared online, understanding how to protect this information is crucial. This chapter will highlight the importance of safeguarding personal information, recognizing the risks associated with oversharing, and utilizing privacy settings effectively. As cyber threats become sophisticated, individuals must be aware of potential dangers such as phishing, data breaches, and identity theft. By emphasizing best practices for online security, we aim to empower readers to take proactive measures to protect their digital identities. Education plays a pivotal role in shaping responsible digital citizens. This chapter will explore innovative strategies for educators and parents to instill values of digital responsibility in the next generation. By fostering a culture of digital citizenship, we can empower young individuals to navigate the online world with confidence and integrity. The discussion will include practical approaches for integrating digital citizenship education into curricula and community initiatives, ensuring that students have the knowledge and skills necessary to engage ethically with digital technologies.

As the current state of digital citizenship is examined, it is essential to consider the emerging trends and future perspectives shaping our online interactions. The impact of digital citizenship on societal norms and values will be explored, along with the evolving landscape of digital citizenship itself. By understanding these dynamics, individuals and communities can better prepare for the challenges and opportunities that lie ahead in the digital age. In summary, this chapter will provide a comprehensive overview of the essential components of digital citizenship, focusing on digital literacy, privacy, security, and ethical conduct. By addressing these concerns, we aim to equip individuals with the knowledge and skills necessary to navigate the online world responsibly, fostering a culture of digital responsibility that benefits both individuals and society. As we move forward, it is crucial to remain vigilant and proactive in our approach to digital citizenship, ensuring that we harness the power of technology while safeguarding our rights and well-being in the digital realm.

1. NAVIGATING THE ONLINE WORLD

1.1. Benefits of Being a Good Digital Citizen

Being a good digital citizen is increasingly essential in our interconnected world. Digital citizenship encompasses the responsible use of technology and the internet, fostering a positive online environment that enhances interactions and builds a supportive digital community (Mossberger et al., 2007). This part explores the benefits of being a good digital citizen, illustrating how responsible online behavior can significantly improve individual experiences and the broader digital landscape. Digital citizenship refers to the norms of appropriate and responsible behavior concerning technology use (Jones & Mitchell, 2016). It encompasses various aspects, including online safety, respect for others, and ethical behavior in digital interactions. An excellent digital citizen actively engages in online communities, respects the rights of others and contributes positively to the digital world. This concept is critical as our lives increasingly integrate with digital technologies, impacting how we communicate, learn, and participate in society.

One of the primary benefits of being a responsible digital citizen is improved online safety (Spears, 2013). Good digital citizens understand the importance of protecting their personal information and privacy. They are aware of online threats such as cyberbullying, identity theft, and scams, and they develop strategies to mitigate these risks. By practicing safe online habits, such as using strong passwords and understanding privacy settings, individuals can navigate the digital world more securely. This awareness protects them and contributes to a safer online environment for others. Responsible digital behavior fosters respectful and positive interactions in online spaces (Trier et al., 2023). Good digital citizens treat others with kindness and empathy, promoting a culture of respect and support. This positive engagement enhances communication and relationships, making online communities more welcoming and inclusive. By encouraging constructive discussions and standing against negativity, digital citizens help create a more harmonious digital landscape.

In an age of rampant misinformation, being a good digital citizen involves developing critical thinking and media literacy skills (Gonzalez-Mohino et al., 2023). Digital citizens learn to evaluate the credibility of sources, discern fact from fiction, and recognize bias in information. This skill set is essential for making informed decisions and participating meaningfully in discussions. By fostering these abilities, digital citizens contribute to a more informed and engaged society where individuals can navigate complex information landscapes effectively. A significant aspect of digital citizenship is understanding the long-lasting nature of one's online presence. Good digital citizens are mindful of the content they share and its potential impact on their future opportunities. By cultivating a positive digital footprint, individuals can enhance their online reputation, which is increasingly relevant in professional contexts. Employers often review candidates' online profiles, making it crucial to maintain a respectful and professional digital presence.

Digital citizenship promotes a mindset of lifelong learning and empowerment (Fabbri, 2019). The digital world offers countless personal and professional growth opportunities through online courses, webinars, and networking platforms. Good digital citizens actively seek these resources, enhancing their skills and knowledge. This commitment to continuous learning benefits individuals and contributes to a more knowledgeable and skilled workforce. Being a good digital citizen also involves active participation in civic life and social causes. Digital platforms enable individuals to advocate, raise awareness about important issues, and mobilize support for various causes. This engagement can lead to meaningful societal changes as digital citizens leverage their online presence to influence public discussions and hold institutions accountable. Individuals can contribute to a more equitable and just society by participating in online activism.

Digital citizenship promotes inclusivity and equal access to technology and digital resources (Tokovska et al., 2023). By advocating for digital inclusion, good digital citizens work to bridge the digital divide, ensuring everyone can participate in the digital world. This effort is crucial for fostering a more equitable society where all individuals can benefit from online resources and opportunities. The benefits of being a good digital citizen are multifaceted, impacting individual lives and the broader digital community. By embracing responsible online behavior, individuals can enhance their safety, foster positive interactions, develop critical thinking skills, and build a positive digital footprint. Moreover, good digital citizenship empowers individuals to engage in lifelong learning, participate in civic life, and advocate for inclusivity.

1.2. Developing Good Online Habits

Developing good online habits that promote healthy and productive interactions is crucial in today's digital age. As we increasingly rely on technology for communication, entertainment, and information, we must be mindful of our online behavior and its impact on ourselves and others. This part will provide practical tips for cultivating healthy online practices, focusing on respectful communication and critical thinking. When communicating online, it is easy to forget that real people are on the other side of the screen. Always strive to be respectful, kind, and considerate in your language, even if you disagree with someone's opinion. Please avoid profanity, insults, or aggressive language, which can escalate conflicts and hurt others.

When engaging in online discussions, understand others' perspectives before responding. Read posts carefully, ask clarifying questions, and avoid jumping to conclusions. Active listening demonstrates respect and can lead to more productive and meaningful conversations. Be cautious about sharing unverified information, significantly if it could damage someone's reputation. Before hitting "share," ask yourself if the information is factual, relevant, and necessary to share. It's best to refrain from spreading potentially harmful content if in doubt.

Respect others' privacy and boundaries by not sharing personal information without consent. Be mindful of what you post about yourself and others, and consider how it might affect them. Suppose someone asks you to remove content promptly and without argument. In the age of social media and fake news, verifying information before sharing it online is crucial. Cross-reference claims with reputable sources, look for signs of bias or misinformation, and be wary of sensational headlines or emotionally charged language. By sharing only verified information, you can help combat the spread of false or misleading content. We all have biases that can influence how we perceive and interpret information. Acknowledge your own biases and make an effort to consider different perspectives before forming an opinion. Be open to changing your mind if presented with compelling evidence contradicting your initial beliefs.

When encountering arguments or claims online, analyze them critically. Look for logical fallacies, such as hasty generalizations or ad hominem attacks. Consider the source's credibility, the strength of the evidence presented, and whether the conclusion follows logically from the premises. Engage in respectful discussions to challenge and refine your thinking. Actively seek out diverse perspectives and opinions on issues that interest you. Follow people and sources with different viewpoints and try to understand their origin. Exposure to diverse ideas can help you develop a more nuanced understanding of complex topics and avoid echo chambers. Good online habits are essential for fostering a healthier, more respectful, and more enlightened digital landscape. By practicing respectful communication and critical thinking, we can contribute to more productive and meaningful online interactions.

2. PRIVACY AND SECURITY

2.1. Guarding Your Digital Fortress

In today's digital landscape, where personal data is increasingly vulnerable to cyber threats, safeguarding one's online presence has become paramount. Cybersecurity breaches are not only frequent but can also have devastating consequences for individuals and organizations alike (Wilson, 2021). One of the most effective ways to protect personal information is through strong passwords and secure online practices. This essay delves into the significance of these measures and outlines best practices for enhancing digital security.

The rise of the internet and digital technologies has transformed how we communicate, shop, and manage our personal information. However, this convenience comes with risks. In 2022 alone, over 2,000 data breaches were reported, with a staggering 60% attributed to hacking incidents (Pengl & Li, 2022). These breaches can lead to identity theft, financial loss, and long-term reputational damage. As cybercriminals become sophisticated, individuals must proactively safeguard their digital identities.

Passwords serve as the first defense against unauthorized access to personal accounts. However, many people still use weak passwords that are easy to guess, such as birthdays or simple words. This practice significantly increases the risk of account compromise. To create effective passwords, individuals should follow these guidelines:

- Complexity: Use uppercase and lowercase letters, numbers, and special characters. For example, instead of "Password123," consider using "P@ssw0rd!23".
- Length: Aim for a minimum of 12 characters. Longer passwords are generally harder to crack.
- Uniqueness: Avoid reusing passwords across multiple accounts. If one account is compromised, others may be at risk.
- Password Managers: Utilize password management tools to generate and store complex passwords securely. These tools can help users maintain unique passwords for each account without the burden of memorization.

While strong passwords are crucial, they are not foolproof. Cybercriminals can employ various techniques, such as phishing or keylogging, to obtain passwords. Therefore, implementing Multi-Factor Authentication (MFA) (Kim & Hong, 2011) is essential. MFA adds a layer of security by requiring users to provide two or more verification factors to access their accounts. This could include a text message code, an email confirmation, or a biometric scan, such as a fingerprint. In addition to strong passwords and MFA, individuals should adopt secure online practices to protect their data further. Keeping software, operating systems, and applications up to date is vital for security. Developers frequently release updates to patch vulnerabilities that cybercriminals could exploit. Regularly updating devices ensure users have the latest security features and protections against known threats.

Phishing scams (James, 2005) are a standard method cybercriminals use to trick individuals into revealing sensitive information. Users should be cautious of unsolicited emails or messages that request personal information or prompt them to click on suspicious links. Always verify the sender's identity and look for signs of phishing, such as poor grammar or mismatched email addresses. Public Wi-Fi networks can expose individuals to significant risks, as these networks often lack adequate security measures. Using a secure, private connection is advisable when accessing sensitive information, such as online banking. If necessary, consider using a Virtual Private Network (VPN) to encrypt internet traffic and protect data from potential eavesdroppers.

Oversharing on social media can inadvertently expose individuals to cyber threats. Users should review their privacy settings and limit the amount of personal information shared publicly. Avoid posting sensitive details like birthdays, addresses, or travel plans; malicious actors could exploit that. Despite the best security measures, breaches can still occur. Therefore, having a robust data backup strategy is essential. Regularly backing up important files ensures that individuals can recover their information without significant disruption in a cyberattack or data loss. Cloud storage solutions and external hard drives are practical options for maintaining backups.

In an era of pervasive cyber threats, protecting personal data is more critical than ever. Strong passwords, multi-factor authentication, and secure online practices form the foundation of a robust cybersecurity strategy. By adopting these measures, individuals can significantly reduce their risk of falling victim to cyberattacks and safeguard their digital lives. As technology evolves, remaining vigilant and proactive in cybersecurity will be essential in defending against emerging threats.

2.2. Embracing Privacy Settings

Understanding and utilizing privacy settings is crucial in today's digital landscape, where oversharing can lead to significant risks (Shabahang et al., 2024). As our lives become increasingly intertwined with online platforms, the potential for misuse of personal information grows. This discussion delves into the risks associated with oversharing online and emphasizes the importance of embracing privacy settings. One of the most pressing risks of oversharing personal information online is identity theft (Guedes et al., 2008). When individuals disclose sensitive information such as their full name, birthday, or address, they inadvertently give potential thieves the tools to impersonate them. Identity theft can lead to financial loss, damaged credit scores, and a long, arduous process to reclaim one's identity.

Oversharing (Agger, 2012) can also result in long-term damage to one's reputation. Once information is posted online, it can be challenging to remove it altogether. Employers, colleagues, and friends may see posts that reflect poorly on an individual, affecting job prospects and personal relationships. Maintaining a positive online presence is essential in an age where digital footprints are scrutinized. Sharing personal details can make individuals vulnerable to cyberbullying and harassment. Malicious actors can exploit personal information to target individuals, leading to distressing experiences. This risk is particularly pronounced for younger users who may not fully understand the implications of their online activity.

Corporations often collect and analyze user data for targeted advertising and other purposes. Oversharing can lead to losing control over personal data, allowing companies to build detailed profiles that may be used in ways that individuals disapprove of. This exploitation raises ethical concerns about consent and privacy rights. Privacy settings are essential tools that provide users with control over their personal information. By adjusting these settings, individuals can determine who can see their posts, what information is shared, and how it is used. This control is vital in minimizing the risks associated with oversharing.

Utilizing privacy settings can help individuals manage their digital footprint. A digital footprint consists of the data trail left by online activity, which can reveal much about a person. By limiting what is shared publicly, individuals can reduce the amount of information available to potential thieves, marketers, and other entities. Understanding privacy settings empowers users to make informed decisions about their online presence. Many platforms provide tutorials and FAQs that explain how to adjust privacy settings. Users can better protect themselves and their information by taking the time to learn about these options.

With the rise of privacy regulations such as the General Data Protection Regulation (GDPR) and the California Consumer Privacy Act (CCPA), understanding privacy settings is more important than ever (Van der Sloot, 2020). These laws require organizations to obtain explicit consent from users for data processing. By utilizing privacy settings, individuals can assert their rights and ensure their preferences are respected. Privacy settings are not a one-time task; they require ongoing attention. Social media platforms and online services frequently update their privacy policies and settings. Users should regularly review and adjust their settings to align with their preferences and comfort levels.

Choosing the highest and most restrictive privacy settings available is advisable. Users should be cautious about sharing personal information, even with friends or family. Birth dates, addresses, and phone numbers should only be shared when necessary. Two-factor authentication adds an extra layer of security to online accounts. Users can protect their accounts from unauthorized access by requiring a second verification form and a password. This is particularly important for accounts that store sensitive information. Many online platforms offer additional privacy tools, such as dashboards, that allow users to manage their settings in one place. Familiarizing oneself with these tools can enhance privacy and security. Additionally, users should seek resources explaining how to use privacy settings effectively, including tutorials and guides from reputable sources.

Users should be vigilant about the permissions they grant when using third-party applications that connect to social media accounts. Many applications request access to personal information that is unnecessary for their functionality. Users should only provide the minimum amount of information required and regularly review the permissions granted to these applications. In conclusion, the risks associated with oversharing online are significant and multifaceted, ranging from identity theft to reputation damage. Embracing privacy settings is a critical step in mitigating these risks.

3. TEACHING DIGITAL CITIZENSHIP

3.1. Shaping Responsible Digital Citizens

Digital citizenship has gained paramount importance in the contemporary digital landscape as society increasingly relies on technology for communication, education, and social interaction. Digital citizenship encompasses the responsible and ethical use of technology, emphasizing the need for individuals—particularly children and teenagers—to navigate the online world safely and respectfully. This responsibility falls not only on educators but also on parents, who play a crucial role in shaping the values and behaviors of the next generation. Digital citizenship refers to the norms of appropriate, responsible behavior concerning technology use. It includes understanding how to protect personal information, recognizing the implications of one's digital footprint, and respectfully engaging with others online. As children and teenagers are often the most active users of digital platforms, instilling values of digital responsibility is essential for fostering a safe and positive online environment.

Educators are pivotal in promoting digital citizenship through structured programs and curricula. Schools can integrate digital literacy into their educational offerings, teaching students about the ethical use of technology, online safety, and the importance of maintaining a positive digital footprint. For instance, lessons can cover topics such as copyright laws, the consequences of cyberbullying, and the significance of critical thinking when evaluating online information. Moreover, educators can facilitate discussions that help students understand the impact of their online actions on themselves and others. By creating a classroom culture that values respect, empathy, and integrity, teachers can empower students to become responsible digital citizens. This approach prepares students for personal online interactions and equips them with the skills necessary for future professional environments where digital collaboration is often crucial.

While educators lay the groundwork for digital citizenship, parents must also be actively involved in reinforcing these values at home. Workshops and informational sessions can equip parents with the knowledge and tools necessary to support their children's safe use of technology. These initiatives can cover various topics, including establishing technology agreements within the home, which outline expectations for device usage and online behavior. Parents can also play a significant role by modeling responsible digital behavior themselves. Children often emulate the actions of adults, so when parents demonstrate respectful and ethical online interactions, they set a powerful example. Encouraging open communication about online experiences can help children feel comfortable discussing any challenges they face, such as cyberbullying or inappropriate content.

To effectively instill values of digital responsibility, a foundation of digital literacy is essential. This includes teaching children how to evaluate the credibility of sources, understand the implications of sharing personal information, and recognize the signs of online threats. As misinformation becomes increasingly prevalent, fostering critical thinking skills is vital for helping young people navigate the complexities of the digital world. Digital literacy education can also extend to understanding one's digital footprint. Children should learn that online actions can have long-lasting consequences, impacting their reputation and future opportunities. By teaching them to think critically about the content they share and the interactions they engage in, educators and parents can help cultivate a sense of accountability.

Digital citizenship is not solely about avoiding negative behaviors but also promoting positive engagement in online communities. Encouraging students to participate in civic activities, such as digital advocacy campaigns or community service projects, can foster a sense of social responsibility. By utilizing digital platforms for positive change, young people can learn the value of contributing to society and engaging in constructive discourse. Furthermore, educators and parents can guide children in understanding the importance of respecting diverse perspectives and engaging in respectful discussions. This is particularly crucial in an era where online interactions can often devolve into hostility or harassment. Teaching children to engage thoughtfully with others can help create a more respectful and inclusive online environment.

Cyberbullying (Harasymiw, 2012) remains a significant concern in the digital age, making it imperative for educators and parents to address this issue proactively. Schools should implement anti-bullying policies that encompass online behavior and provide resources for students who may be affected. Encouraging students to report cyberbullying incidents and providing them with strategies to handle such situations can empower them to take action. Parents can also play a critical role by maintaining an open dialogue about online experiences and encouraging their children to share any negative interactions. By fostering an environment of trust and support, parents can help their children navigate the challenges of online interactions more effectively. As technology evolves, so too must the strategies for promoting digital citizenship. Schools should regularly assess the effectiveness of their digital citizenship programs through student feedback and data analysis.

In conclusion, shaping responsible digital citizens requires a collaborative effort between educators and parents. By instilling values of digital responsibility, promoting digital literacy, and encouraging positive online engagement, both parties can empower the next generation to navigate the digital world safely and ethically. As society continues to embrace technology, fostering responsible digital citizens becomes not just an educational imperative but a societal necessity. Through these efforts, we can cultivate a generation that values integrity, respect, and responsibility in digital interactions, ultimately contributing to a healthier online community.

3.2. Strategies for Teaching Digital Citizenship

Strategies for teaching digital citizenship are essential in today's technology-driven society. Educators face the challenge of integrating digital citizenship education into curricula and community initiatives effectively. This involves not only teaching students about the responsible use of technology but also empowering them to navigate the complexities of the digital world critically. Here are several innovative approaches and resources for achieving this goal. Digital citizenship encompasses the norms of appropriate and responsible behavior when using technology. It includes understanding the impact of digital technologies on individuals and society and developing the skills necessary for effective participation in a digital society. As young people are increasingly immersed in digital environments, educational authorities must prioritize digital citizenship education to prepare them for active and responsible engagement in these spaces.

To effectively integrate digital citizenship into existing curricula, educators should consider a cross-disciplinary approach (Richardson & Milovidov, 2019). This means that digital citizenship should not be confined to a single subject; instead, it can be woven into various subjects such as social studies, language arts, and science. For example, discussions about the ethical implications of technology can be part of a science curriculum, while language arts can explore digital communication and its effects on interpersonal relationships. Another effective strategy is project-based learning, which engages students in exploring real-world issues related to digital citizenship. Through projects, students can create awareness campaigns about cyberbullying, develop apps to promote online safety or research social media's impact on mental health. This hands-on approach fosters critical thinking and problem-solving skills while making learning relevant and engaging.

Employing active pedagogies can significantly enhance the teaching of digital citizenship. One such method is the flipped classroom model, where students engage with digital content at home and participate in interactive, hands-on activities in class. This approach allows educators to focus on discussions and collaborative projects reinforcing digital citizenship concepts. Blended learning is another effective strategy, combining online and face-to-face learning to provide students with flexible learning environments. Online platforms can host discussions on digital ethics, while in-person sessions can facilitate workshops on practical skills like coding or digital content creation.

Schools should seek partnerships with local organizations, libraries, and community centers to extend digital citizenship education beyond the classroom. Collaborations can lead to workshops and training sessions for students and parents on online privacy, digital footprints, and responsible social media use. These sessions can foster a culture of digital responsibility and awareness within the community. Engaging students in community service projects that utilize digital tools can help them apply their digital citizenship skills in real-world contexts. For instance, students can assist local non-profits in creating their online presence or developing digital literacy programs for underserved populations.

Technology can be a powerful ally in teaching digital citizenship. Utilizing digital platforms for collaboration, such as Google Classroom or Edmodo, can facilitate discussions about digital citizenship issues. These platforms allow students to share resources, debate, and collaborate on projects, fostering a sense of community and shared responsibility. Incorporating gamification into the learning process can make digital citizenship concepts more engaging. Educational games that simulate online interactions can help students practice decision-making in a safe environment, allowing them to learn from their mistakes without real-world consequences.

With the aim of ensuring the effectiveness of digital citizenship education, it is crucial to implement robust assessment strategies. Continuous assessment methods, such as peer reviews and self-assessments, can help students reflect on their learning and understand their progress in digital citizenship competencies (Coklar & Tatli, 2020). Encouraging students to create digital portfolios showcasing their projects and reflections on digital citizenship can provide insights into their learning journey and growth. To effectively teach digital citizenship, educators must be well-versed in digital skills and the implications of technology. Schools should provide regular professional development opportunities focused on digital citizenship. This training can cover emerging technologies, digital ethics, and effective teaching strategies, ensuring that educators remain informed and capable of guiding their students. Creating networks for educators to share resources, experiences, and best practices in teaching digital citizenship can foster a supportive environment for professional growth.

4. EMERGING TRENDS AND FUTURE PERSPECTIVES

4.1. The Impact of Digital Citizenship on Society

In the digital age, the concept of digital citizenship has become increasingly important in shaping societal norms, values, and interactions. Digital citizenship encompasses the responsible use of technology and the ability to engage positively and productively in the digital world. As more aspects of our lives move online, the impact of digital citizenship on society is profound and far-reaching. Digital citizenship plays a crucial role in establishing societal norms in the online realm (Pangrazio & Sefton-Green, 2021). By promoting responsible and ethical behavior, digital citizenship helps create a more civil and respectful digital environment. When individuals practice digital citizenship, they set an example for others to follow, leading to the gradual adoption of positive norms. These norms include respecting intellectual property rights and avoiding plagiarism, engaging in constructive online discussions while steering clear of cyberbullying, protecting personal privacy, and respecting the privacy of others. Additionally, verifying information before sharing it is essential to combat the spread of misinformation. As these norms become more widely accepted, they contribute to a healthier, more inclusive digital society.

Digital citizenship also shapes the values that guide our interactions in the digital world. By emphasizing concepts such as digital literacy, digital etiquette, and digital rights and responsibilities, digital citizenship promotes values essential for thriving in the digital age. These values include a commitment to lifelong learning to keep up with rapidly evolving technologies, critical thinking to evaluate online information and sources, empathy and respect for diverse perspectives and cultures, and civic engagement that encourages active participation in digital communities. As individuals internalize these values through digital citizenship education and practice, they become more likely to make decisions and take actions that align with these values, further reinforcing their importance in society.

Digital citizenship also transforms the way we interact with each other in the digital realm. By fostering a sense of digital community and shared responsibility, digital citizenship encourages individuals to engage with others more meaningfully and productively. This shift can lead to increased collaboration and knowledge sharing across geographical boundaries, creating more inclusive and accessible digital spaces for people with diverse abilities and backgrounds. It also promotes greater transparency and accountability in online interactions and transactions and enhanced opportunities for civic engagement and political participation. As these transformations occur, the digital world becomes a more vibrant and connected space that better reflects the values and aspirations of a digitally literate and responsible society.

While digital citizenship has the potential to impact society positively, it also faces challenges and limitations (Kenner & Lange, 2019). The rapid pace of technological change can make it difficult to keep up with evolving norms and best practices. Additionally, the global and decentralized nature of the internet can complicate the enforcement of digital rights and responsibilities consistently across different jurisdictions and cultures. Furthermore, the digital divide—the gap between those with access to digital technologies and those without—can limit the reach and impact of digital citizenship initiatives. Without universal access to digital technologies and education, the benefits of digital citizenship may not be equally distributed throughout society.

4.2. Evolving Perspectives on Digital Citizenship

The concept of digital citizenship is undergoing significant transformation as technology becomes increasingly integrated into daily life. This evolution reflects not only advancements in digital tools but also shifts in societal norms and expectations regarding responsible technology use. Digital citizenship encompasses the ethical and responsible use of technology, which is crucial as individuals navigate an ever-expanding digital environment. This includes understanding online safety, privacy issues, and the importance of respectful communication. As technology evolves, so too do the skills and knowledge required to engage with it effectively. For instance, the COVID-19 pandemic accelerated the shift to digital platforms for education, highlighting the necessity for students to develop competencies in digital citizenship to thrive academically and socially in a digital-first world.

The implications of evolving digital citizenship extend to both individuals and communities. For individuals, possessing solid digital citizenship skills fosters critical thinking, enabling them to assess information sources, engage in meaningful online discussions, and navigate potential risks such as cyberbullying and misinformation. This empowerment enhances their ability to participate actively and responsibly in digital spaces, ultimately contributing to their academic success and personal growth. On a community level, promoting digital citizenship can bridge the digital divide by ensuring equitable access to technology and digital literacy resources. Communities prioritizing digital citizenship education can cultivate a culture of responsible technology use, leading to safer online environments and more informed citizens. This collective effort is essential in addressing the challenges posed by the digital landscape, such as privacy concerns and the ethical implications of technology use.

5. DIGITAL LITERACY AND CRITICAL THINKING

5.1. Developing Critical Thinking Skills

Developing critical thinking skills for evaluating online information is essential in today's digital age, where misinformation can quickly spread. By applying structured evaluation criteria and utilizing educational resources, individuals can enhance their critical thinking skills and become more discerning information consumers. This preparation is vital for navigating the complexities of the digital landscape and ensuring that the information consumed is accurate, reliable, and relevant.

When assessing online information, it is crucial to consider various factors. Accuracy is paramount, and verifying whether the information is factual and supported by evidence is crucial. Look for citations and references that substantiate the claims made. It is essential to cross-reference information from multiple reliable sources to ensure its validity. Authority is another key aspect, as reliable sources often have authors who are recognized experts in their respective fields. Checking the author's credentials, such as their educational background, professional experience, and any published works, can help determine the credibility of the information.

Objectivity is also essential, as it is crucial to examine the purpose of the information and determine if it is designed to inform or persuade. Potential biases that may influence the information presented should also be considered. Identifying political, ideological, or commercial interests that may shape the content can help assess its objectivity. Currency is another critical factor, as it is necessary to check the publication date of the information to ensure that it is up-to-date and relevant to the research topic. Outdated information can lead to inaccuracies and may not reflect current knowledge on a particular subject. Finally, coverage is an important consideration, as it is necessary to assess the comprehensiveness of the information and determine whether it addresses the topic thoroughly or leaves out significant details. A well-rounded source should provide a complete view of the subject, covering various aspects and perspectives.

To effectively evaluate online sources, it is recommended that structured frameworks such as the CRAAP Test and the WWWDOT framework be implemented (Wierzbicki, 2018). The CRAAP Test focuses on five key aspects: Currency, Relevance, Authority, Accuracy, and Purpose. The WWWDOT framework encourages users to ask questions such as "Who wrote this and what are their credentials?", "Why was it written?", "When was it written and last updated?", "Does it meet my needs?", and "Organization of the website: Is it user-friendly and well-structured?". These frameworks provide a systematic approach to evaluating online sources, helping individuals identify reliable information and avoid falling victim to misinformation.

Educators can utilize various resources to teach these critical evaluation skills, such as lesson plans and guides from websites like ReadWriteThink.org and Edutopia, as well as interactive activities like the "Bad News" game, which can help students understand the dynamics of misinformation and develop critical thinking skills. By incorporating these resources into their teaching practices, educators can empower students to become critical information consumers, fostering their ability to think critically and make informed decisions.

5.2. Strategies for Teaching Digital Literacy

In today's technology-driven world, digital literacy has become a crucial skill for students to navigate the vast amount of information available online. As educators, it is our responsibility to equip students with the necessary tools to find, evaluate, and utilize digital resources effectively. Before diving into advanced concepts, it's important to establish a solid foundation in digital literacy. Begin by teaching students the core components, such as online safety, search skills, and digital etiquette. Educate them on how to protect their personal information, recognize potential online threats, and communicate respectfully in digital spaces. Provide hands-on activities that simulate everyday digital tasks, such as creating strong passwords or identifying phishing emails.

One effective way to engage students and make digital literacy lessons more relevant is by incorporating real-world scenarios into your teaching. Assign research projects that require students to find and cite credible online sources, helping them practice their search skills in a structured way. Set up role-play activities where students create mock social media profiles, posts, and comments to teach digital etiquette and the impact of online actions. Introduce basic cybersecurity concepts through challenges that require students to identify phishing emails or create strong passwords.

Digital literacy should not be confined to a single lesson or unit; instead, it should be integrated into various subjects and everyday learning. Utilize digital tools across different disciplines to give students hands-on experience and reinforce their digital literacy skills. Encourage students to use gamified learning platforms that make literacy fun and engaging. Promote collaborative projects using cloud-based tools like Google Docs or Microsoft Teams, fostering collaboration and communication in a digital environment. Assign creative assignments that require students to create digital content, such as videos, blogs, or podcasts, helping them develop digital creation skills while expressing their creativity.

In the age of information overload, critical thinking is a key component of digital literacy. Teach students to question the information they encounter online and evaluate the credibility of online sources. Guide them through the process of checking the author's credentials, looking for bias, and verifying facts. Discuss the prevalence of fake news and misinformation, and create activities where students analyze news articles, social media posts, or websites to determine their reliability. Challenge students to research and propose solutions to real-world problems using credible online resources, fostering critical thinking and problem-solving skills.

Digital citizenship encompasses the responsible and ethical use of digital technologies. Encourage students to engage in positive online behavior, respect intellectual property rights, and understand the consequences of their digital actions. Teach them to think critically about the content they create and share online, considering the potential impact on their personal brand and future opportunities. Encourage open dialogues with parents, students, and other stakeholders about the use of technology in education, addressing any concerns or questions related to digital literacy.

Engage students as active participants in the digital literacy learning process by giving them opportunities to lead. Encourage them to conduct research on digital literacy topics, contribute to curriculum development, and plan and deliver lessons to their peers and the community. This not only reinforces their understanding but also fosters a sense of ownership and leadership in the digital literacy journey.

6. CYBERSECURITY AND PRIVACY

6.1. Best Practices for Protecting Personal Data Online

In an increasingly digital world, safeguarding personal data has become paramount. With the rise of cyber threats, individuals must adopt essential cybersecurity practices and privacy measures to protect their sensitive information. The digital landscape is fraught with risks, including data breaches, identity theft, and unauthorized access to personal information. Statistics reveal that billions of records are exposed each year due to cyber incidents, leading to significant financial and emotional repercussions for individuals. As such, implementing robust data protection strategies is not just advisable but necessary for anyone engaging with online platforms.

The foundation of online security begins with strong passwords. Users should create complex passwords that combine uppercase and lowercase letters, numbers, and symbols. Passwords should be unique for each account to prevent a single breach from compromising multiple accounts. Utilizing password managers can simplify this process by generating and securely storing these passwords, ensuring that users do not have to remember each one. Multi-factor authentication adds a layer of security by requiring users to provide two or more verification factors to gain access to their accounts. This could include a password and a code sent to a mobile device. MFA significantly reduces the risk of unauthorized access, even if a password is compromised.

Public Wi-Fi networks are often unsecured, making them prime targets for cybercriminals. When using public Wi-Fi, individuals should avoid accessing sensitive information, such as banking details. If necessary, using a VPN can encrypt data transmitted over these networks, providing an extra layer of security. Keeping software, operating systems, and applications updated is crucial in protecting against vulnerabilities that hackers can exploit. Regular updates often include security patches that address known threats. Additionally, installing reputable antivirus and anti-malware software can help detect and eliminate potential threats before they cause harm.

Social media platforms can inadvertently expose personal information. Users should regularly review and adjust their privacy settings to limit who can view their posts and personal details. Avoid oversharing sensitive information, such as location, birthdays, or family details, which cybercriminals can use for identity theft. Phishing attacks often come in the form of emails or messages that appear legitimate but are designed to steal personal information. Users should be cautious about clicking on links or downloading attachments from unknown sources. Always verify the sender's email address and look for signs of phishing, such as poor grammar or unusual requests. When registering for new accounts or making online purchases, consider using masked or temporary email addresses and phone numbers. This practice can help protect your real contact information from being exposed or sold to third parties, thus reducing the risk of spam and fraud.

For those managing sensitive data, establishing clear data usage policies is vital. These policies should outline who can access certain information, how it should be stored, and the protocols for sharing data. Adhering to the principle of least privilege ensures that individuals only have access to the data necessary for their roles, minimizing the risk of internal breaches. Regularly monitoring your online presence can help you identify any unauthorized use of your personal information. Search for your name online to see what information is publicly available and take steps to remove any sensitive data from public view. Being proactive can help mitigate the risks associated with data exposure. Understanding your privacy rights is crucial in the digital age. Familiarize yourself with the privacy policies of the websites and services you use, and know how to exercise your rights regarding personal data. This includes the right to access, correct, and delete your information and to withdraw consent for its use.

6.2. Understanding the Risks of Data Breaches

Data breaches have become a pervasive threat in today's digital landscape, affecting individuals and organizations alike. These incidents occur when unauthorized parties gain access to sensitive information, leading to severe implications such as identity theft, financial loss, and reputational damage. Understanding the nature of data breaches, their causes, and the potential consequences is essential for both individuals and organizations to safeguard against these risks. A data breach is defined as an incident where confidential, sensitive, or protected information is accessed, viewed, or shared without authorization. This can include personal data like Social Security numbers, banking information, and health records, as well as corporate data such as intellectual property and internal communications. The increasing interconnectedness of devices and the rise of the Internet of Things (IoT) (Greengard, 2021) have created more vulnerabilities, making it easier for hackers to exploit weaknesses in systems and networks.

Data breaches can occur through various methods, often exploiting human error or technological vulnerabilities. Phishing attacks, where cybercriminals use deceptive emails or messages that appear to be from trusted sources to trick individuals into providing sensitive information, are one of the most common causes of data breaches. Many breaches also result from weak or stolen credentials, with statistics indicating that a significant percentage of breaches are linked to compromised passwords. Malicious software can infiltrate systems through infected downloads or email attachments, allowing attackers to steal data or gain unauthorized access to networks.

Employees or contractors with access to sensitive information can intentionally or unintentionally cause data breaches, either through malicious actions or mistakes such as misconfiguring security settings or sharing data inappropriately. The loss or theft of devices containing sensitive data can lead to breaches if the information is not adequately protected, a risk that is exacerbated by remote work and the use of portable devices. Organizations often rely on vendors and partners for various services, and a breach at a third-party provider can compromise the data of all associated organizations, highlighting the need for comprehensive security measures across the supply chain.

The consequences of data breaches can be profound and far-reaching, affecting individuals and organizations in multiple ways. For individuals, one of the most immediate risks is identity theft, where stolen personal information can be used to open fraudulent accounts, make unauthorized transactions, or engage in other criminal activities under the victim's name. This can lead to significant financial loss and long-term damage to an individual's credit. The psychological impact of being a victim of identity theft can be severe, with individuals experiencing anxiety, stress, and a sense of violation as they navigate the complexities of recovering their identity and securing their financial accounts. Victims may also face direct financial losses due to unauthorized transactions or the costs associated with restoring their identity, and the time spent resolving these issues can lead to lost wages and other economic repercussions.

For organizations, data breaches often result in reputational damage, leading to decreased sales, customer attrition, and long-term harm to the brand's reputation (Zhu et al., 2024). Rebuilding trust after a breach can be a lengthy and challenging process. Organizations may also face lawsuits from affected individuals or regulatory penalties for failing to protect sensitive data adequately, and compliance with data protection regulations is essential to avoid legal repercussions. Data breaches can lead to significant operational challenges, including system outages and the need for extensive recovery efforts, making it difficult for organizations to maintain core functions while addressing the breach. Following a breach, organizations often need to invest heavily in cybersecurity measures to prevent future incidents, including upgrading technology, enhancing employee training, and implementing more robust security protocols.

Individuals and organizations can adopt several best practices to mitigate the risks associated with data breaches. Organizations should enforce strong password policies, utilize multi-factor authentication, and regularly update software to protect against vulnerabilities. Training employees on recognizing phishing attempts and understanding the importance of data security can significantly reduce the risk of insider threats and human error. Regular security audits should be performed to identify and address potential vulnerabilities before attackers can exploit them. Using encryption can protect sensitive information, ensuring that even if data is accessed without authorization, it remains unreadable without the appropriate decryption keys. Individuals should regularly check their financial accounts and credit reports for signs of unauthorized activity, while organizations should invest in breach detection tools to identify potential incidents quickly.

7. ONLINE ETIQUETTE AND BEHAVIOR

7.1. Netiquette

Respectful communication in online environments, often referred to as netiquette, is crucial for fostering positive interactions in forums, emails, and social media (Soler-Costa et al., 2021). Respectful communication helps establish and nurture relationships. When individuals engage with others online, they should remember that they interact with real people, not just screens. This awareness encourages empathy and consideration, which are essential for constructive dialogue. By treating others as they would like to be treated, individuals can create a more welcoming and supportive online community. Online communication lacks the non-verbal cues present in face-to-face interactions, making it easier for messages to be misinterpreted. Clear and respectful communication can mitigate misunderstandings. For instance, using proper grammar and avoiding excessive slang or abbreviations helps convey messages more effectively. Additionally, being mindful of tone—such as avoiding all caps, which can be perceived as shouting—can enhance clarity and prevent conflicts.

Respectful communication promotes constructive criticism rather than hostility. Participants in online discussions should aim to express differing opinions respectfully, focusing on the issue rather than personal attacks. This approach enhances the quality of discussions and encourages a culture of openness, where individuals feel safe to share their thoughts without fear of ridicule or aggression. In educational contexts, netiquette is particularly vital. Online classes and forums thrive on respectful exchanges, which help maintain a productive learning atmosphere. Students and instructors can engage in meaningful dialogues that enhance understanding and collaboration by adhering to netiquette. This is especially important in diverse environments where individuals may have varying backgrounds and perspectives.

Lastly, maintaining professionalism in online communications is essential, especially in academic and workplace settings. Using appropriate channels for communication, addressing recipients correctly, and keeping messages concise and relevant are all aspects of professional netiquette. This professionalism reflects well on the individual and upholds the standards of the community or organization they represent. In summary, respectful communication is a foundational element of effective online interactions. By adhering to netiquette guidelines, individuals can foster positive relationships, prevent misunderstandings, encourage constructive dialogue, enhance learning environments, and maintain professionalism.

7.2. Addressing Cyberbullying

Cyberbullying is a growing concern that requires a multifaceted approach to address effectively. Educating students, teachers, and parents about cyberbullying is crucial. Schools should organize workshops, seminars, and training sessions to create awareness and guide students and staff on recognizing, preventing, and responding to cyberbullying incidents. Schools should have clear policies outlining acceptable online behavior and procedures for reporting and handling cyberbullying incidents. They should work to create a culture of respect, kindness, and inclusion where all students feel valued and supported. Parents play a vital role in preventing cyberbullying. Schools should provide parents with information on recognizing and preventing cyberbullying, encourage them to monitor their children's online activities, and communicate with them about the dangers of cyberbullying.

Students should feel comfortable reporting incidents of cyberbullying to teachers and staff. Schools can create anonymous reporting systems where students can report cyberbullying without fear of retribution. Schools can use filtering software to block inappropriate content and monitor students' online activities. They can also use social media platforms and online forums to spread awareness about cyberbullying and provide resources for students experiencing it. Schools can organize anti-bullying campaigns and encourage students to speak out against cyberbullying. They can provide students with resources and tools to help them deal with cyberbullying incidents and create student-led organizations that prevent cyberbullying and promote a positive school culture.

Schools should have counseling services available for students who have been victims of cyberbullying. These services should provide support, compassion, and empathy to help the victims cope with the emotional impact of the bullying. Schools can use restorative practices, such as mediation and conferencing, to address cyberbullying incidents. These practices focus on repairing the harm caused and restoring relationships between the victim and the perpetrator. By implementing these strategies and promoting kindness and empathy in online interactions, schools and communities can work together to combat cyberbullying and create a safer and more inclusive environment for all students.

8. THE ROLE OF EDUCATION IN CYBERSECURITY AWARENESS

8.1. Understanding Cybersecurity Fundamentals

In today's interconnected and technology-driven world, understanding cybersecurity is no longer a specialized skill reserved for IT professionals—it is a fundamental necessity for everyone. With digital platforms increasingly integrated into every aspect of life, the risks associated with cyber threats, such as data breaches, phishing scams, and identity theft, are escalating. Cybersecurity education equips individuals with the knowledge and skills to navigate the digital landscape safely and responsibly. This education should begin at an early age, with schools integrating cybersecurity principles into their curricula to build a strong foundation. By doing so, students can develop the awareness and habits necessary to protect themselves in an environment where cyber risks are omnipresent (Rahman et al., 2020)

One of the first steps in cybersecurity education is familiarizing students with basic terminology. Understanding key terms such as “malware,” “phishing,” “firewalls,” and “encryption” helps demystify what can often seem like a complex field. For instance, malware refers to malicious software designed to harm or exploit computer systems, while phishing describes fraudulent attempts to obtain sensitive information by disguising it as a trustworthy entity in electronic communications. Firewalls are security systems that monitor and control network traffic based on predetermined rules, and encryption involves converting information into a secure code to prevent unauthorized access. When students comprehend these terms, they are better equipped to understand the mechanisms of cyber threats and how to counteract them (Branch, 2021). This foundational knowledge enables them to articulate concerns and recognize potential vulnerabilities in their digital interactions.

In addition to understanding terminology, students must learn how to identify common cyber threats (Barbu et al., 2023). Recognizing phishing emails is particularly important, as these are among the most prevalent forms of cyberattacks. Students should be taught to look for signs of phishing attempts, such as poor grammar, unsolicited requests for personal information, or suspicious links that do not match legitimate URLs. Similarly, understanding social engineering tactics—where attackers manipulate individuals into divulging confidential information—is crucial for avoiding scams that exploit human psychology rather than technical vulnerabilities. Furthermore, students should be trained to scrutinize links before clicking on them to avoid inadvertently downloading malware or visiting malicious websites. By empowering students with the ability to recognize these threats, educators can help them develop a proactive approach to online safety.

While threat recognition is essential, it must be complemented by instruction on safe online practices. One critical area is password security. Students should learn how to create strong passwords that combine uppercase and lowercase letters, numbers, and special characters while avoiding easily guessable information like birthdays or pet names. Another vital skill is recognizing secure websites by checking for HTTPS in the URL or verifying the presence of an SSL certificate icon (a padlock symbol). Additionally, understanding privacy settings on social media platforms is increasingly important in an era where personal information is often shared online without consideration of potential risks. By teaching students how to configure privacy settings appropriately, educators can help them safeguard their data from unauthorized access or misuse. These practices not only enhance individual security but also foster responsible digital habits that will benefit students throughout their lives.

8.2. Practical Workshops and Simulations

While theoretical knowledge provides a solid foundation in cybersecurity principles, hands-on experience is invaluable for reinforcing what students learn in the classroom (Cruz & Simões, 2021). Practical workshops and simulations allow students to apply their knowledge in controlled environments where they can safely experiment with identifying threats and responding effectively. These activities bridge the gap between theory and practice, enabling students to internalize cybersecurity concepts more deeply while building confidence in their ability to handle real-world scenarios.

Phishing simulations are particularly effective for teaching students how to identify fraudulent emails and other deceptive tactics used by cybercriminals. During these exercises, students might analyze mock emails designed to mimic phishing attempts. They would look for common indicators of fraud, such as generic greetings ("Dear Customer"), urgent language intended to provoke fear or action ("Your account will be suspended unless you respond immediately"), or suspicious links that redirect users to fake websites. After completing these exercises, group discussions can be held where students share their observations and insights about the tactics used in phishing scams. Such activities not only improve recognition skills but also foster critical thinking about how attackers exploit human behavior.

Another engaging way to teach cybersecurity is through password creation challenges (Younis & Alghamdi, 2021). These activities encourage students to devise strong passwords while emphasizing the importance of password security in protecting personal accounts and sensitive data. For example, students could participate in competitions where they create passwords that are evaluated based on criteria such as length, complexity, and uniqueness. Educators might also introduce mnemonic techniques for creating memorable yet secure passwords—for instance, using phrases like "I0veMyC@t!" (a variation of "I love my cat"). By combining creativity with practical instruction on what makes a password strong, these challenges help students internalize best practices for securing their online accounts.

Simulating data breaches or other cyber incidents prepares students for real-world scenarios by teaching them how to respond effectively under pressure. In these drills, participants might take on different roles—such as incident responders tasked with containing the breach or affected users who need guidance on the next steps—to simulate the dynamics of an actual cyber crisis. Afterward, debriefing sessions allow participants to reflect on what went well and identify areas for improvement in their response strategies. These exercises not only build technical skills but also emphasize the importance of teamwork and communication during high-stress situations.

SUMMARY

The chapter addressed the complexities of digital citizenship, emphasizing the importance of responsible online behavior, privacy, and security in an interconnected world. It explored the multifaceted nature of digital citizenship and highlighted individuals' need to cultivate ethical online practices while safeguarding personal information amid evolving threats. Digital literacy formed the core of navigating the online landscape, encompassing the skills required for effective and critical engagement with digital technologies. The chapter underscored that developing digital literacy was essential for personal empowerment and ethical conduct in online interactions. As individuals increasingly relied on digital platforms for communication, education, and commerce, understanding the ethical implications of their actions became imperative.

The chapter also highlighted the digital age's growing privacy and security concerns. It discussed the risks associated with oversharing personal information and emphasized the importance of utilizing privacy settings effectively. As cyber threats become sophisticated, individuals must be aware of potential dangers such as phishing and identity theft. The chapter aimed to empower readers by emphasizing best online security practices, including strong passwords and multi-factor authentication. Education plays a pivotal role in shaping responsible digital citizens. The chapter explored innovative strategies for educators and parents to instill values of digital responsibility in the next generation, fostering a culture of digital citizenship. It included practical approaches for integrating digital citizenship education into curricula and community initiatives, ensuring students possessed the knowledge and skills necessary for ethical engagement with digital technologies.

The chapter concluded by examining the current state of digital citizenship and emerging trends that would shape future online interactions. It emphasized the impact of digital citizenship on societal norms and values, encouraging individuals and communities to prepare for the challenges and opportunities in the digital age. Overall, the chapter provided a comprehensive overview of the essential components of digital citizenship, focusing on digital literacy, privacy, security, and ethical conduct while advocating for proactive and vigilant approaches to navigating the digital realm.

REFERENCES

Agger, B. (2012). *Oversharing: Presentations of self in the internet age.* Routledge. DOI: 10.4324/9780203125090

Barbu, M., Zamfiroiu, A., Marinescu, I. A., Iordache, D., & Bumbac, R. (2023). Empowering digital education: Understanding students' perceptions about risks and threats in the shifting educational paradigm. *Studies in Informatics and Control, 32*(4), 105–114. DOI: 10.24846/v32i4y202310

Branch, J. (2021). What's in a Name? Metaphors and cybersecurity. *International Organization, 75*(1), 39–70. DOI: 10.1017/S002081832000051X

Coklar, A. N., & Tatli, A. (2020). Evaluation of digital citizenship levels of teachers in the context of information literacy and internet and computer use self-efficacy. *Asian Journal of Contemporary Education, 4*(2), 80–90. DOI: 10.18488/journal.137.2020.42.80.90

Cruz, T., & Simões, P. (2021). Down the rabbit hole: Fostering active learning through guided exploration of a scada cyber range. *Applied Sciences (Basel, Switzerland), 11*(20), 9509. DOI: 10.3390/app11209509

Fabbri, M. (2019). Digital citizenship skills as an essential factor in Lifelong-Lifewide-Lifedeep Education. *Pedagogia Oggi, 17*(2), 285–299.

Gonzalez-Mohino, M., Rodriguez-Domenech, M., Callejas-Albiñana, A. I., & Castillo-Canalejo, A. (2023). Empowering critical thinking: The role of digital tools in citizen participation. *Journal of New Approaches in Educational Research, 12*(2), 258–275. DOI: 10.7821/naer.2023.7.1385

Greengard, S. (2021). *The internet of things.* MIT Press. DOI: 10.7551/mitpress/13937.001.0001

Guedes, I., Martins, M., & Cardoso, C. S. (2022). Exploring the determinants of victimization and fear of online identity theft: An empirical study. *Security Journal, 21*(1), 1–26. DOI: 10.1057/s41284-022-00350-5

Harasymiw, T. (2012). *Cyberbullying and the law.* The Rosen Publishing Group.

James, L. (2005). *Phishing exposed.* Elsevier.

Jones, L. M., & Mitchell, K. J. (2016). Defining and measuring youth digital citizenship. *New Media & Society, 18*(9), 2063–2079. DOI: 10.1177/1461444815577797

Kenner, S., & Lange, D. (2019). Digital citizenship education. Challenge and opportunity. *Scuola Democratica, 10*(4), 47–55.

Kim, J. J., & Hong, S. P. (2011). A method of risk assessment for multi-factor authentication. *Journal of Information Processing Systems, 7*(1), 187–198. DOI: 10.3745/JIPS.2011.7.1.187

Mossberger, K., Tolbert, C. J., & McNeal, R. S. (2007). *Digital citizenship: The Internet, society, and participation.* MIT Press. DOI: 10.7551/mitpress/7428.001.0001

Pangrazio, L., & Sefton-Green, J. (2021). Digital rights, digital citizenship and digital literacy: What's the difference? *Journal of New Approaches in Educational Research, 10*(1), 15–27. DOI: 10.7821/naer.2021.1.616

Pengl, J., & Li, C. W. (2022). Security breaches and modifications on cybersecurity disclosures. *Accounting and Management Information Systems*, *21*(3), 452–470.

Rahman, N. A. A., Sairi, I. H., Zizi, N. A. M., & Khalid, F. (2020). The importance of cybersecurity education in school. *International Journal of Information and Education Technology (IJIET)*, *10*(5), 378–382. DOI: 10.18178/ijiet.2020.10.5.1393

Richardson, J., & Milovidov, E. (2019). *Digital citizenship education handbook: Being online, well-being online, and rights online*. Council of Europe.

Shabahang, R., Shim, H., Aruguete, M. S., & Zsila, A. (2024). Oversharing on social media: Anxiety, attention-seeking, and social media addiction predict the breadth and depth of sharing. *Psychological Reports*, *127*(2), 513–530. DOI: 10.1177/00332941221122861 PMID: 35993372

Soler-Costa, R., Lafarga-Ostáriz, P., Mauri-Medrano, M., & Moreno-Guerrero, A. J. (2021). Netiquette: Ethic, education, and behavior on internet—a systematic literature review. *International Journal of Environmental Research and Public Health*, *18*(3), 1212–1236. DOI: 10.3390/ijerph18031212 PMID: 33572925

Spears, B. A. (2013). A review of initiatives using technology to promote cyber-safety and digital citizenship. *The Impact of Technology on Relationships in Educational Settings*, *15*(4), 188–203.

Tokovska, M., Ferreira, V. N., Vallušova, A., & Seberíni, A. (2023). E-Government—The Inclusive Way for the Future of Digital Citizenship. *Societies (Basel, Switzerland)*, *13*(6), 141. DOI: 10.3390/soc13060141

Trier, M., Kundisch, D., Beverungen, D., Müller, O., Schryen, G., Mirbabaie, M., & Trang, S. (2023). Digital Responsibility: A Multilevel Framework for Responsible Digitalization. *Business & Information Systems Engineering*, *65*(4), 463–474. DOI: 10.1007/s12599-023-00822-x

Van der Sloot, B. (2020). *The general data protection regulation in plain language*. Amsterdam University Press. DOI: 10.5117/9789463726511

Wierzbicki, A. (2018). *Web content credibility*. Springer International Publishing. DOI: 10.1007/978-3-319-77794-8

Wilson, D. C. (2021). *Cybersecurity*. MIT Press. DOI: 10.7551/mitpress/11656.001.0001

Younis, Y. A., & Alghamdi, M. Y. (2021). The use of computer games for teaching and learning cybersecurity in higher education institutions. *Journal of Engineering Research*, *9*(3), 16–29. DOI: 10.36909/jer.v9i3A.10943

Zhu, J. J., Tuo, L., You, Y., Fei, Q., & Thomson, M. (2024). A preemptive and curative solution to mitigate data breaches: Corporate social responsibility as a double layer of protection. *JMR, Journal of Marketing Research*, *61*(4), 778–801. DOI: 10.1177/00222437231218969

ADDITIONAL READING

Lu, C., & Gu, M. M. (2024). A systematic review and meta-analysis of factors and outcomes of digital citizenship among adolescents. *Asia Pacific Journal of Education, 20*(3), 1–16. DOI: 10.1080/02188791.2023.2296352

Mattson, K. (2024). *Digital citizenship in action: empowering students to engage in online communities.* International Society for Technology in Education.

Slavković, M., Pavlović, K., Mamula Nikolić, T., Vučenović, T., & Bugarčić, M. (2023). Impact of digital capabilities on digital transformation: The mediating role of digital citizenship. *Systems, 11*(4), 172–189. DOI: 10.3390/systems11040172

Tinmaz, H., Fanea-Ivanovici, M., & Baber, H. (2023). A snapshot of digital literacy. *Library Hi Tech News, 40*(1), 20–23. DOI: 10.1108/LHTN-12-2021-0095

Tomasello, F. (2023). From industrial to digital citizenship: Rethinking social rights in cyberspace. *Theory and Society, 52*(3), 463–486. DOI: 10.1007/s11186-022-09480-6 PMID: 35676929

KEY TERMS AND DEFINITIONS

California Consumer Privacy Act (CCPA): The CCPA is a law that gives California residents more control over the personal information that businesses collect about them. It requires businesses to disclose what data they collect, allow consumers to access and delete their data, and opt out of the sale of their personal information to third parties.

CRAAP Test: The CRAAP test is a framework for evaluating the reliability of information sources. CRAAP stands for Currency, Relevance, Authority, Accuracy, and Purpose. It provides a list of questions to help determine if a source is trustworthy enough to use in academic research.

General Data Protection Regulation (GDPR): GDPR is a set of rules that govern the collection and processing of personal data of individuals within the European Union. It gives EU residents more control over their personal data and places restrictions on how businesses can use it.

Internet of Things (IoT): IoT refers to the network of physical devices, vehicles, home appliances, and other items embedded with electronics, software, sensors, and network connectivity that enables these objects to connect and exchange data.

Multi-Factor Authentication (MFA): MFA is a security system that requires more than one method of authentication to verify a user's identity for a login or other transaction. It adds an extra layer of security to protect against unauthorized access.

Netiquette: Netiquette is the etiquette guidelines for online communication. It covers proper behavior and conduct when interacting with others on the internet, such as using appropriate language, respecting privacy, and avoiding flaming or trolling.

Virtual Private Network (VPN): VPN is a technology that creates a secure, encrypted connection over a less secure network, such as the public internet. It allows users to send and receive data across shared or public networks as if their devices were directly connected to the private network.

WWWDOT framework: The WWWDOT framework is a tool designed to help individuals, particularly students, evaluate the credibility of websites as information sources. The acronym WWWDOT stands for:- Who wrote it, and what credentials do they have? - Why was it written? - When was it written or updated? - Does it help meet my needs? - Organization of the site - To-do list for the future

Chapter 5
The Ethical Dimensions of Digital Interactions

Hakikur Rahman
https://orcid.org/0000-0002-2132-1298
Presidency University, Bangladesh

ABSTRACT

This chapter explores the ethical implications of digital interactions and challenges readers to critically examine their own online conduct as well as the wider effects of their decisions. In a time when social and personal values are shaped by digital contacts, this chapter discusses important ethical issues that come up in online settings, including privacy, respect, accountability, and openness. It provides a combination of theoretical understandings and useful recommendations to assist readers in responsibly and mindfully navigating the intricacies of the digital world, drawing on well-established ethical theories and real-life situations. The chapter promotes digital citizenship that values empathy, inclusion, and integrity and highlights the significance of comprehending how routine digital behaviors contribute to broader ethical frameworks and societal norms.

INTRODUCTION

Ethics are now a vital component of our online interactions in a society going digital. Beyond individual behavior, ethics have a significant impact on social norms that shape communication, information sharing, and community building in the digital sphere. This chapter will examine the moral dilemmas that occur in virtual spaces and examine the ways in which our online actions affect other people and society at large. Our online presence can be more responsible and courteous if we recognize the significance of ethical issues in digital interactions. Hence, to start with, significance of ethics in the digital world is presented next.

Significance of Ethics in the Digital World

According to Dewey (2008), ethics is the set of rules, conventions, and guidelines that control behavior in organizations, societies, and cultures as well as day-to-day interactions. A list of forbidden things that children are not allowed to do online is not a suitable way to handle digital ethics, which are

DOI: 10.4018/979-8-3693-6675-2.ch005

Copyright ©2025, IGI Global Scientific Publishing. Copying or distributing in print or electronic forms without written permission of IGI Global Scientific Publishing is prohibited.

the normative norms for behavior and engagement in digital settings. Knowledge and learning, civic engagement, employment, leisure, and regular social interaction with peers and others happen online for young people whose families and communities have affordable everyday access to the internet—many rural and remote, Indigenous, and economically marginalized communities do not have such access. Digital activities are by definition social acts; as such, they are utilized for goal-seeking with actual pragmatic outcomes and consequences. Examples of these actions include clicking, tweeting, posting, sharing, and liking (Wilden, 1972; Luke, 2018).

For information consumers and producers in the digital age, Elliott and Spence (2017) have created a succinct, instructive, and topical introduction on media ethics. An open-access, hyperlinked architecture that encourages self-reflective practice on the part of its users, the book explores and reflects the implications of the convergence of the Fourth and Fifth Estates. It is pitched at just the right depth of detail to provide a big picture contextualization of changing media practices grounded in concerns for democracy and the public good. Philip Gordon from Utah Valley University The quick and continuous development of digital technologies has changed how people communicate and process information globally.

Examining the social and ethical ramifications of media consumption and the technologies that support it has become more crucial than ever, as the 24-hour news cycle and post-truth politics have increased their influence over public opinion. A comprehensive examination of the ethical concerns that have surfaced as the digital revolution gains momentum is given in Ethics for a Digital Era. This includes an examination of journalistic practices that have an impact on the veracity, dependability, and credibility of information dissemination. In addition to outlining instructions for web-based news producers and users to conceive ethical dilemmas and examine ethically dubious acts, the collection investigates new approaches and models for ethical inquiry in a digital society.

Elliott and Spence (2017) consider changes in media ethics as modern mass communication blends old analog methods with new formats like blogs, vlogs, podcasts, and social media posts, and develops into an interactive medium with users who both create and consume the news in each of the three thematic sections. Subsequent chapters invite users to bridge their own thinking between the virtual and physical worlds of information and its conveyance by applying a normative decision-making process to some of the most significant challenges that arise in these encounters.

Say, for example, Media ethics in journalism refers to the conscientious application of the right to publish. Media ethics is the normative interpretation of practice as we reflect on our work. Normative interpretation begins with a social profession, like journalism or law, about which there is typically some consensus regarding the identity of practitioners and on specific instances of the practice. Interpretation presents the technique "in its best light," which explains its goal. It outlines the requirements that the practice must adhere to for this reason.

For instance, Ward (2014) uses two levels of criteria to assess journalism. The "base conditions" for creating a strong press are determined in the first tier. Newsroom independence, freedom of expression, and adequate economic stability are among the fundamental requirements. It is only once these prerequisites are partially met that we may begin to envision new possibilities. The "democratic" requirements make up the second tier. It states that media outlets should support egalitarian, participatory democracy once the right to publish is guaranteed.

Ward (2014) encourages public participation in media in all its forms, and also supports specific forms of participation. He searches for venues for dialogic journalism, or reasoned discourse across differences. Global thinking is something that he also wants in news media. The ethical impulse of journalism should be to enhance social justice, lessen conflict, foster human flourishing and democratic processes around the world, and establish bridges across cultural divides.

When anything goes wrong, research ethics as a matter of discussion in the public and academic arenas usually comes up again. Our collective endeavors to learn from our failures can serve as a window into the history of research ethics. The cruel experiments of World War II, the Tuskegee syphilis study, and Stanley Milgram's ground-breaking but unsettling studies of human behavior are ostensibly where we begin that history.

The UN Declaration of Human Rights (1948), the Nuremberg Code (1949), the Declaration of Helsinki (1964), and the Belmont Report (1979) were all influenced by (the revelation of) these errors, and they were intended to protect human subjects in biomedical and behavioral research. They are still relevant today in the ethical management of the majority of human subjects research (Tiidenberg, 2018).

As a result of a particular socio-cultural shift, ethics and its expression in moral behavior are subjects of "reformulation" or even "redefinition," rather than existing in a vacuum, sterile, or fixed human reality. The emergence of a global information-digital culture has had an impact on how ethics are perceived, understood, and applied as a moral code. One of the primary characteristics of this culture is its "artificiality," which results in a shift from face-to-face engagement to technologically mediated human communication. As a result, there is a "cultural distanciation" wherein the body and the message are separated, as well as perception and experience.

Another effect is the "transparency" at an ethical level, where a number of ethical distinctions are broken down, such as right from wrong and proper from unsuitable. One of the ethical issues of our day is community ethics, where a "commonality" is now created on artificial links such as loneliness, rejection, and powerlessness rather than traditional social ties. Another "strange" evolution of moral behavior is friendship in the digital age, where having a large number of friends just serves to validate one's loneliness. As a result, connection—the primary tenet of the architecture of our modern existence—has led us to encounter a paradox within culture, mediating but also dissociating our cultural experience (Piliang, 2015).

The first section of this chapter will highlight on ethical challenges in the digital world. Hence an overview of ethical challenges in online environments is given in Section-1.

METHODOLOGY

This chapter uses a qualitative approach to explore ethical dimensions in digital interactions, drawing from recent literature and thematic analysis. The research design involved a narrative literature review to analyze contemporary ethical frameworks and their application within digital citizenship, focusing on empathy, privacy, and ethical content creation.

Sources were selected based on their relevance to ethical theories and digital citizenship, specifically recent studies on deontological, consequentialist, and virtue ethics in online settings (Floridi & Taddeo, 2020; Reynolds & Cerutti, 2021). The review also considered emerging themes in privacy and security within digital interactions, analyzing how ethical challenges manifest across various online environments (Crawford, 2021). Through thematic analysis, key ethical dilemmas such as empathy and responsibility in virtual spaces were identified, with a focus on enhancing responsible digital behavior (Johnson et al., 2021).

To provide a global perspective, the methodology included a comparative review of cross-cultural studies on digital ethics (Kende et al., 2022). Case studies were analyzed to illustrate practical examples of ethical challenges in digital spaces, especially within social media platforms and online communities. This analysis emphasized empathy, cultural sensitivity, and community standards for responsible digital content.

This methodology seeks to connect ethical theories with practical guidelines, emphasizing how individuals and communities can foster a respectful and empathetic digital environment.

SECTION 1: ETHICAL THEORIES AND DIGITAL CITIZENSHIP

Traditional ethical theories, such Deontology, Virtue Ethics, and Utilitarianism, offer a fundamental structure for comprehending moral conduct in any situation, including the digital realm. These ideas provide insightful guidance for our online decisions and activities, helping us to effectively negotiate the difficulties of digital citizenship. Through the application of these ethical concepts to our digital interactions, we may foster the development of a pleasant online community, responsible behavior, and respect for others. In this context, let us explore a few classical ethical theories.

Exploration of Classical Ethical Theories (Utilitarianism, Deontology, Virtue Ethics)

In exploring the nature and place of virtue ethics within the modern moral landscape Wood (2019) in his book, Deontology, Consequentialism, and Virtue Ethics in the Contemporary Moral Landscape creates a more intricate framework for ethical theorizing.

Wood's (2019) book's main issue is how virtue ethics relates to other well-known normative ethical systems and how we can best comprehend these divergent viewpoints. Only philosphers are truly aware of the disagreement surrounding this subject, but the endeavor to resolve it is not just meant for a professional philosophical audience. Ethics is not mainly restricted to academic settings, and the study is being carried out in such a way that anyone looking for clarification on the current state of normative theories can learn something from the arguments and concepts presented.

According to the modern ethical standpoint that this study aims to address, utilitarianism, Kantian deontology, and virtue ethics are the three primary conflicting ethical systems. While virtue ethics is regularly questioned regarding whether it actually belongs in this category, the first two are commonly assigned this place with no disagreement. In a sense, the critics are right. It is incorrect to believe that virtue ethics is merely an addition to utilitarianism and Kantian deontology, even though virtue ethics does not precisely contradict these two unique normative theories.

Because these comparisons involve category errors on two levels, they are problematic. First, compared to more rigidly defined normative theories like utilitarianism and Kantian deontology, virtue ethics shares greater structural similarities with consequentialism and deontology. Second, it is a ploy to assume that utilitarianism and Kantian deontology can be effectively contrasted without identifying particular utilitarian theories and Kantian deontological interpretations. There is rarely any consensus among philosophers, and the consensus gets even more limited within a field of expertise.

One way to think of economics is as applied ethics. Once we get past the rhetoric, ethics is just economics when it comes to normative judgments about what is right and wrong. Since Robbins (1932), the majority of texts have defined the field of economics as the decisions that individuals make on the allocation of limited resources to achieve conflicting goals.

Furthermore, when evaluating people's decisions, economists frequently use normative standards like efficiency and marginal costs and benefits. Under this framework, economic choice theory can be understood as a practical subset of the consequentialist utilitarian school of moral philosophy. However, we don't often consider how this characterizes economics.

Traditionally, normative theories of ethics are divided into three primary groups: virtue ethics, consequentialist ethics, and deontological ethics.

Deontological ethics, or deontology, is essentially concerned with the intrinsic rightness or wrongness of our decisions and deeds. It emphasizes the moral principles of obligation and duty that direct righteous and correct behavior. On the other hand, consequentialist ethics, or consequentialism, places more emphasis on whether the outcomes of our activities are right or bad than on the inherent characteristics of the actions themselves. We can regard utilitarianism as the prototype of consequentialist ethics, despite the fact that consequentialist perspectives and arguments come in many other forms.

Lastly, virtue ethics emphasizes moral qualities. The premise is that our decisions shape who we are and who we want to be in addition to simply satisfying our desires. In this context, ethicists of virtue hold that eudaimonia, or true happiness, results from a sequence of decisions that uphold our self-worth, legitimize us, and enable us to coexist peacefully with others. Thus, virtue ethics emphasizes virtues like honor, decency, and a life well lived (Jonsson, 2011).

In the field of international relations (IR) and international politics, virtue ethics (VE) is a way of thinking about moral behavior that has not gotten much attention. The two main schools of thought in international ethics are consequentialist and deontological, which emphasize more or less the results of actions and moral standards or responsibilities, respectively. Contrarily, virtue ethics emphasizes the significance of moral character, particularly the possession of a combination of "virtues," such as practical wisdom, in determining what is right. It defines proper behavior in relation to specific notions of the good, or the flourishing of humans.

Since VE emphasizes individual character traits over the kinds of ethical concepts—such as rights, laws, and norms—that we are more familiar with in IR and political theory, the applicability of this to IR is not immediately apparent. But VE's rejection of rule-based ethics as well as virtue-based methods' emphasis on virtues, flourishing, and practical knowledge force us to consider international politics from fresh, frequently unconventional perspectives (Ainley, 2017).

The concept of ethics has so many definitions that it appears to have destroyed any universal definition. For instance in a classic study, Steiner and Steiner (1999) describe ethics as the study of good and bad, right and wrong, and just and unjust. Ethics is the study of fundamental human relationships; it is a set of roles that specify what action is considered appropriate and inappropriate. Similar to this at the same time, De George (1999) defines ethics as a productive effort to comprehend our distinct moral and social experiences in order to identify the values that should guide human action, the aspirations that are worthwhile to pursue, and the personal traits that are worthy of success in life (Benlahcene, et. al. 2018).

Understanding the moral aspects of life, such as social norms for proper behavior, is the goal of studying ethics, claim Beauchamp and Childress (1983). Additionally, according to Schwartz and Bilsky (1987), ethical values are individual concepts or beliefs that guide action selection or assessment.

Application of Utilitarianism, Deontology, Virtue Ethics to Digital Citizenship and Online Behavior

Online dangers like trolling, cyberbullying, revenge porn, and digital harassment are not sporadic events for many young people; rather, they are ongoing struggles that influence their digital lives (Livingstone et al., 2020). Despite being universal, these problems show themselves differently in diverse cultural contexts, hence a digital citizenship strategy that takes ethical heterogeneity into consideration is required. Cyberbullying frequently manifests as public shame (haji-katsu) in collectivist countries like South Korea and Japan, where victims are socially shunned (Huang & Chou, 2021). Conversely, cyberbullying may be more aggressive and anonymous in Western individualist cultures, emphasizing personal attacks over group exclusion (Kowalski, Giumetti, Schroeder, & Lattanner, 2019).

It is becoming more and more expected of teachers around the world to deal with these moral conundrums, yet many do not have the institutional support or culturally appropriate materials to do so (Selwyn, 2021). A utilitarian strategy for optimizing online well-being is in accordance with digital ethics education in nations such as Sweden, which places an emphasis on individual responsibility and autonomy (Johansson & Felmlee, 2022). In contrast, digital misconduct is perceived through the prism of morality and collective honor in Middle Eastern environments, where online conduct is frequently governed by religious and communal values (Al-Jenaibi, 2020).

This study makes the case for a novel method of teaching cyber-flourishing that incorporates ethical frameworks such as utilitarianism, deontology, and virtue ethics in addition to traditional digital literacy. For example, virtue ethics promotes youth to develop virtues like honesty and empathy online (Harrison, 2023). Deontology reinforces organized rules for cyber behavior by emphasizing digital rights and duties (Floridi, 2019). A utilitarian approach, on the other hand, advocates for proactive intervention in digital spaces in an effort to strike a balance between limiting harm and promoting online freedom (Buchanan, 2020). Teachers can develop more complex and successful digital ethics courses suited to a range of young people by fusing these ethical frameworks with regional cultural knowledge.

The model could be built by using a pragmatic approach to moral theory, taking the core ideas of utilitarianism, virtue ethics, and deontology and demonstrating how educators might combine and use these ideas to address the evils associated with the internet.

Harrison (2022) segmented three main parts of the model, as—cyber-rules, cyber-wisdom, and cyber-flourishing. He emphasizes that schools need to establish cyber-rules to lay the groundwork for the development of cyber-wisdom and character. Cyber-wisdom is the capacity to act morally and strategically when no one is watching, especially while oneself. Genuine enjoyment that results from leading a virtuous life online is known as "cyber-flourishing," and it's the utility value that we should all aim for.

A few moral psychology-informed research have investigated the degree to which users, depending on their emotions or perceptions of the ethical aspects of various online environments, make morally engaged or disengaged decisions when using the internet (D'Errico & Paciello, 2018; Ge, 2020). Research examining moral decision-making on the internet from a variety of moral theoretical perspectives is, nevertheless, scarce. In fact, we are aware of the degree to which teenagers make decisions on the internet that run counter to utilitarianism, deontology, and virtue ethics (Harrison, & Polizzi, 2021).

A few studies have looked into interactions that occur online via one of these channels. Using a deontological framework, Selwyn & Aagaard (2021) have examined how nations set regulations on the use of smartphones in classrooms, while other earlier studies (Lievens 2011; Lyu 2012) have focused on the rights and obligations of Internet users. It is hardly unexpected that deontological approaches predominate because in many fields of practical ethics, Kantianism—the source of deontology—is the preferred moral theory.

Cruz & Plaisance (2021) in their study offer a virtue theory framework to support particular normative norms while examining online activism from an ethical standpoint. They mention that digital media have developed into effective instruments for social movements in democracies throughout time, giving people places to gather, share ideas, and organize. However, the very digital infrastructure that fosters activism also, paradoxically, operates under unethical precepts that jeopardize the work of activists. Their study aims to develop a set of moral guidelines that help reorient online activism toward the pursuit of a mode of social change that is conscious of the ethical challenges ingrained in the digital sphere while simultaneously being concerned with the cultivation of the self through interactions with peers, drawing on scholarship on technomoral virtues and media participation.

To conclude this section, we will be highlighting the evolving nature of digital interactions and the ethical implications of new technologies, as such the impact of artificial intelligence, data privacy concerns, and the role of social media in shaping public discourse.

Digital interactions are continuously evolving, shaped by rapid advancements in technology, including artificial intelligence (AI), data privacy concerns, and the role of social media in public discourse. As digital citizenship extends beyond simple online behavior to complex ethical challenges, individuals and organizations must adapt to emerging technologies while upholding ethical responsibility.

i. The Impact of Artificial Intelligence on Digital Ethics

Artificial intelligence is transforming digital interactions, but its ethical implications remain a pressing concern. AI-driven algorithms influence decision-making in areas such as hiring, law enforcement, healthcare, and social media content moderation, raising issues of bias, accountability, and transparency (Floridi & Cowls, 2022). Some key ethical concerns include:

- **Algorithmic Bias**: AI systems can reinforce societal biases, leading to unfair discrimination (Binns, 2018).
- **Lack of Explainability**: Many AI models operate as "black boxes," making it difficult to understand how decisions are made (Mittelstadt et al., 2016).
- **Autonomous Decision-Making**: AI-powered surveillance and predictive policing may infringe upon civil liberties and privacy rights (Zuboff, 2019).

To mitigate these risks, ethical AI frameworks, such as the European Commission's *Ethics Guidelines for Trustworthy AI*, advocate for human-centered design, transparency, and fairness in AI systems (Floridi & Cowls, 2022).

ii. Data Privacy Concerns in a Hyperconnected World

With the exponential growth of digital data, privacy has become a fundamental ethical issue. Organizations collect vast amounts of personal information, often without explicit user consent, leading to concerns about surveillance, data breaches, and corporate misuse (Nissenbaum, 2019). Major privacy challenges include:

- **Mass Data Collection & Surveillance**: Governments and corporations increasingly rely on data tracking for security and marketing, raising ethical concerns about mass surveillance (Richards, 2013).
- **Data Ownership & User Control**: Many users lack control over their personal data, prompting legal responses such as the **General Data Protection Regulation (GDPR)** and the **California Consumer Privacy Act (CCPA)** (Voigt & Von dem Bussche, 2017).
- **Cybersecurity Risks**: Weak data protection measures lead to breaches, exposing sensitive personal and financial information (Solove, 2020).

To address these issues, privacy-enhancing technologies (PETs), such as encryption and decentralized identity management, are being developed to empower users and ensure ethical data governance (Mayer-Schönberger & Cukier, 2013).

iii. The Role of Social Media in Shaping Public Discourse

Social media platforms have revolutionized communication, allowing for free expression and global connectivity. However, they also introduce ethical dilemmas regarding misinformation, online harassment, and digital polarization (Sunstein, 2017). Critical concerns include:

- **Misinformation & Disinformation**: False information spreads rapidly on social media, influencing public opinion and even election outcomes (Wardle & Derakhshan, 2017).
- **Echo Chambers & Polarization**: Algorithm-driven content personalization reinforces ideological biases, leading to fragmented societies (Pariser, 2011).
- **Online Harassment & Hate Speech**: The anonymity of digital platforms enables cyberbullying, trolling, and other harmful behaviors, disproportionately affecting marginalized communities (Daniels, 2018).

To promote ethical digital citizenship, policymakers and social media companies must implement *responsible platform governance*, including stronger content moderation, transparency in algorithmic decision-making, and digital literacy programs for users (Gillespie, 2018).

As digital interactions continue to evolve, ethical considerations must remain at the forefront of technological advancements. The challenges posed by AI, data privacy, and social media require a *multi-stakeholder approach*, integrating ethical principles into policy, technology design, and public awareness. By fostering responsible digital citizenship, societies can harness the benefits of digital transformation while mitigating its ethical risks.

Henceforth, the next section will highlight the empathy and understanding in virtual spaces.

SECTION 2: EMPATHY AND UNDERSTANDING IN VIRTUAL SPACES

In order to promote meaningful and polite digital connections, empathy is essential. The capacity to empathize with others and comprehend their sentiments is especially important in virtual environments when in-person interactions are not possible. Through case studies, this section will examine the effects of empathic behavior online and stress the significance of cultivating emotional intelligence in digital contacts. We can make online spaces that are more inclusive and welcoming by practicing empathy.

The Importance of Empathy in Digital Interactions

Computer-mediated communication (CMC) exhibits empathy, but little is known about the situational factors that influence empathy in digital interactions. We investigated three types of empathy using a diary methodology: (1) the extent to which students experience cognitive, affective, and compassionate empathy in everyday (text- and image-based) dyadic digital interactions; (2) situational factors important for (different types of) empathy in CMC; and (3) the relationship between participants' perceptions of their empathy in CMC and face-to-face (FtF) contexts and their reported empathy in everyday CMC.

In a study conducted by Powell & Roberts (2017), a "digital interaction diary" was filled out by 100 student volunteers (50 of whom were women; Mage = 22.57 years) over the course of three days, producing 1939 observations. Individuals indicated considerably higher levels of affective empathy compared to cognitive empathy and affective empathy compared to compassionate empathy. While some situational factors—like the number of communications or the recipient—were linked to empathy as a whole, other factors—like the subject or mood—helped to create distinct contextual profiles for the different empathy subtypes. Particularly for those with lower baseline trait empathy, empathy recorded in the diaries predicted a more favorable ratio of perceived CMC to FtF empathy. These results contribute to the understanding of the complex feeling of empathy in CMC encounters.

Technologies for information and communication are employed in workplaces, leisure areas, and educational settings. People in those settings communicate with each other every day, often for several hours, through electronic devices. Through the internet, people and organizations may collaborate, learn, and communicate. When they communicate online, they feel, think, act, and decide. Recent advances in study have focused on the role that social and emotional competence, morality, and empathy—all of which are heavily utilized in informal communication—play in face-to-face interpersonal encounters. Although research on this area is still in its early stages, these competences may also be crucial in distinguishing between prosocial and antisocial interactions that occur online (Marín-López, et. al. (2019). In this perspectives, this chapter has tried to emphasize research on morality, empathy, and social and emotional skills in relation to cyberspace.

Verbal and nonverbal socioemotional cues can inspire empathy, which is essential for promoting healthy social interactions. While it can quickly connect people across distances, computer-mediated communication (CMC) has the potential to partially filter out nonverbal clues. We develop a coherent conceptual framework of online empathy, a tool the sector currently lacks, by drawing on the literature on emotion communication and telehealth that is currently available. The article describes the unique characteristics of online communication and how they affect empathetic interactions. We also go over tactics that users can take to help others feel, communicate, and be seen as empathic in CMC (Grondin, Lomanowska, & Jackson, 2019).

Friesem (2016) conducted another study and mentions that as part of a digital and media literacy lesson, a collaborative video was created highlighting social, emotional, and cognitive skills is the focus of a new concept called "digital empathy." Six unique empathy phenomena are enhanced by the structured five stages of video production, which include script writing, preproduction, production, postproduction, and screening (empathic concern, cognitive empathy, projective empathy, affective empathy, psychological empathy, and aesthetic empathy). As the procedure draws to a close, the majority of students exhibit increased empathy. His study explains the theoretical framework and shows how each production stage and the empathy phenomenon are related. Since digital empathy is a relatively new idea, more research is needed to examine these relationships and gain a better understanding of how to use digital devices in the classroom to develop students' social, emotional, and cognitive skills.

To continue the chapter, a few case studies are being discussed next.

Case Studies on the Impact of Empathetic Behavior Online

As part of their total patient support services, many healthcare organizations (HCOs), such as Johns Hopkins, MD Anderson Cancer Center, Kaiser Permanente, and Cleveland Medical Center, give their patients access to online health communities. The primary goal of creating and managing these virtual health communities is to provide patients with compassionate assistance. Empathy from patients is thought to play a crucial role in their rehabilitation, particularly in terms of improving their adherence to treatment plans and rate of healing.

The two primary purposes of the majority of online health groups are social and informational assistance. This study looks at how these two different roles—that is, as a place to get information and as a place to get social support—relate to how empathic patients feel in online health forums (Nambisan, 2011).

Salminen, et. al. (2021) in their study investigate the usefulness of personalities in assisting qualified advertisers in creating social media advertisements. They investigate whether giving online ad creators a personification user group (PUG) yields better ad performance than giving them a non-personified user group (NUG) with no name or face picture. Thirty people participated in our experiment and made Facebook ads with both PUG and NUG. It was discovered that the use of PUG did improve the advertising click performance of ads made by those with more ad and persona experience. Additionally, a study of the ad texts revealed that the use of PUG boosted the commercials' empathy, corroborating the fundamental empathy advantage mentioned in HCI research. PUG usage did not, however, considerably boost purchasing intent. The findings suggest that incorporating PUG into online ad design increases empathy and enhances click-through rates. Since more sympathetic advertisements seem to boost relevance, they may benefit social media users.

Empathy and social support are well established to improve mental health. Due to social media's extensive adoption, user behaviors during the formation of online communities may also be influenced by online social support and empathy. Few research, meanwhile, have looked at these impacts from the viewpoint of online mental health communities. Though it's unclear how online empathy spreads in these forums, they seem to be an important source of support for mental health issues. In order to match and compare users (1) who received social support with those who did not, and users (2) who received more empathetic social support with those who received less empathic social support, this study focused on 22 subreddits relating to mental health.

The findings of Chen & Xu (2021) demonstrated the "contagiousness" of empathy and social support. In other words, people who received social support for their first post were more inclined to do so again and to help others; people who received more empathic support would also be more likely to demonstrate empathy toward others in the future. Our results point to the possible domino effect of empathy and social support in virtual mental health networks. Our research also sheds light on how individuals might more effectively offer social assistance to those dealing with mental health issues by participating in online mental health groups.

Online donations to charity are common, therefore it's important for them to figure out how to get people to notice their offerings. The relationship between donor empathy and recipient vulnerability and donate intention is not well understood. People who were empathic were less likely to donate when they met highly vulnerable receivers; on the other hand, they were more likely to donate when they confronted lowly vulnerable receivers. This interaction effect was mediated by self-consciousness. In a study conducted by Wei, Yu, & Li (2021) clarifies the underlying mechanism of self-consciousness and finds a novel moderator of the influence of empathy on charitable giving. Based on these results, the authors show how empathy and vulnerability combine to influence contribute intention, which has practical consequences for charity.

Here are a few additional case studies that align with the impact of empathetic behavior online:

- **The #IAmAWitness Campaign (United States)**

The #IAmAWitness campaign, launched by the Ad Council in the United States, encouraged young internet users to counter cyberbullying through positive intervention. By using a special emoji to signal support for victims, the campaign promoted an empathetic digital culture. Studies showed that participation in such digital activism led to increased feelings of social responsibility among teens (Dredge et al., 2020).

- **The Kind Speech Movement (Japan)**

In response to the rising incidents of online harassment in Japan, particularly targeting public figures and young people, the Japanese government and NGOs initiated the Kind Speech Movement. This campaign encouraged social media users to leave positive comments and publicly challenge hate speech. Research found that positive reinforcement strategies significantly reduced the spread of online negativity (Nakamura, 2021).

- **"Reword" Anti-Cyberbullying Initiative (Australia)**

The Australian "Reword" initiative used artificial intelligence to detect and discourage harmful language before messages were sent. The tool prompted users to reconsider their words, fostering more empathetic digital interactions. A study revealed that students using the tool showed a 70% decrease in cyberbullying behaviors (Cross et al., 2019).

- **Online Communities Supporting Refugees (Germany & Europe)**

During the European refugee crisis, social media played a significant role in fostering digital empathy. Platforms like Facebook groups and Twitter hashtags (#RefugeesWelcome) helped refugees integrate by connecting them with housing, legal advice, and employment. Case studies demonstrated that online empathetic behavior facilitated offline social cohesion and reduced xenophobic rhetoric (Bhatia, 2020).

- **#YouAreNotAlone Mental Health Awareness (United Kingdom)**

This campaign encouraged social media users to share personal experiences with mental health challenges, providing emotional support and reducing stigma. Research found that participation in such online communities led to improved mental health outcomes and stronger digital solidarity (Bailey et al., 2022).

Now we like to discuss on developing emotional intelligence in digital communications.

Developing Emotional Intelligence in Digital Communications

This exploratory conceptual paper seeks to initiate a conversation on performance constraints that might be linked to leadership's failure to exhibit emotional intelligence in a way that is appropriate when relying solely on cutting-edge, Web-based communication technology. An extensive use of contemporary technology may make it more difficult for a leader to show their emotional intelligence (EI) traits when speaking with stakeholders, according to this theory and anecdotal data.

Performance may decline as a result of recipients of subpar communications from leadership accepting this as the status quo. There isn't much empirical data to support our embrace of novel communication technology, despite the fact that we have done so without hesitation and frequently with great excitement. These newer ways could improve or worsen the quality of leader-to-worker interactions, which could eventually impact worker performance and growth (Graham, 2009).

It is essential to expand our knowledge of emotional intelligence in a digital environment since emotional intelligence is a pervasive concept that permeates all part of our life, whether at work, school, or home. Additionally, the frequency of digital contacts is increasing. But because digital settings necessitate digital competency in interactions, the digital world is more than merely a contextual component to take into account. This research aims to define "digital emotional intelligence" as the combination of digital competence and emotional intelligence.

According to Audrin & Audrin (2023) states that in their proposed model, attitudes toward digital competence are predicted by trait emotional intelligence, whereas digital ability emotional intelligence is predicted by digital competence knowledge and abilities. A structural equation model demonstrates a favorable correlation between trait emotional intelligence and attitudes regarding digital competence using a self-reported questionnaire with 503 respondents.

The literature on business administration hardly ever discusses the notion of digital intelligence, despite its frequent mentions, particularly in the educational sciences. The purpose of this study was to examine the literature on digital intelligence in business administration and to debate whether or not businesses need both digital and emotional intelligence to compete, and whether or not these two phenomena should be taken into account simultaneously. In this instance, data were gathered through document analysis data collection methods from the films "The Internship" and the first three seasons of the television series "Silicon Valley."

Content Analysis was used to analyze the data that were obtained. The adoption of the grounded theory approach resulted from the creation of data-driven categories. Results indicated that if emotional intelligence and digital intelligence are assessed concurrently and given equal weight, businesses may benefit (Yeke, 2023).

Employers need candidates who can contribute to the firm not just with technical abilities but also with positive emotional intelligence and excellent communication. This is because today's workplace is increasingly digital and technological. Bhuvaneswari & Vijayakumar (2021), in their study focus on the relationship between emotional intelligence and communication adaptability because there is a deficiency of research on the soft skills employed by IT workers. The emotional intelligence and communicative adaptability of 111 Information Technology workers working in the United States are analyzed in this quantitative non-experimental correlational study. The study discovered a strong correlation between an information technology professional's emotional intelligence and communication flexibility. The favorable conclusion of this study implies that careers in information technology with a strong correlation between communication skills and emotional intelligence may benefit organizations in terms of relationship- and team-building within the organization.

The next section emphasizes not only on privacy, security, and ethical dilemmas, but also discusses on how to develop a more concrete guidelines or frameworks for individuals and organizations to follow.

SECTION 3: PRIVACY, SECURITY, AND ETHICAL DILEMMAS

Significant privacy and security concerns have arisen in the digital age, posing moral conundrums that call into question our conception of online rights and obligations. This section will look at how privacy standards are changing in a connected society, the moral ramifications of security procedures, such as ethical hacking, and the difficult choices we have to make when exchanging information. Striking a delicate balance between individual rights and group duty is necessary to navigate these conundrums. Now we discuss about privacy rights and expectations in the digital age considering digital citizenship.

Privacy Rights and Expectations in the Digital Age

The desire of individuals to control or have some influence over personal data is known as information privacy. Information systems researchers are investigating information privacy challenges, including technical solutions to address these concerns, as a result of growing concerns about information privacy and its effects brought about by information technology advancements.

In this context, Bélanger & Crossler (2011) suggest that multiple effects of information privacy and various degrees of analysis be taken into account in future study. Their study uses a tiered paradigm for information privacy concerns to demonstrate this. They demand that more design and action information privacy research be published in journal articles that can lead to IT artifacts for information privacy management or protection, as well as the utilization of a wider variety of sampling populations in information privacy studies.

People and society at large struggle in the digital age to weigh the benefits of digital connectedness against the fact that it eventually allows others access to private information, whether legally or illegally. Although there are hazards to privacy that are undoubtedly not exclusive to young people, it is possible to argue that people who have grown up with this technology have different conceptions of privacy.

It could be too soon to assess how technology will affect society's overall expectations for privacy. A manageable exploration of some of the concerns is achieved by concentrating on a subgroup of society: youth. This essay aims to investigate a Fourth Amendment issue that has never been explored previously by focusing on students and the possibility that young people as a group may have a different idea of privacy online (Leary, 2010).

People produce enormous volumes of data, almost all of it valuable, in an increasingly instrumented and networked digital environment. But there's still a great deal of disagreement over who can collect data, how best to protect it, and how much they can do with it. This position paper presents a socio-technical approach based on informed consent and collective awareness, in which users set data collection and protection standards through a cooperative, participatory argumentation process. This speaks to users' capacity to interpret privacy-related documents and their implications through visual clues, participatory processes, and wisdom-of-the-crowds methods (Patkos, et. al., 2015).

According to Ostanina & Titova (2023) people's right to privacy should be safeguarded in the current digital era just as much as it was in the past before digital technologies became widely used. It is crucial that privacy protection methods and practices change in light of the increased information interchange that has resulted from digitization. First and foremost, there should be greater focus on the application of techniques for the self-defense of privacy rights. Examples of these techniques include limiting access to information and modifying website settings to allow reviews and comments to be left by registered users only, rather than anonymously. Second, in order to both ensure that rights that have been infringed upon are duly restored and to stop future violations of private rights, the legal mechanisms of protection should be strengthened. The authors advise using class actions where there is a breach of personal data confidentiality. A claim for compensation for moral damage should be accessible when a commercial entity or the owner of a website violates the confidentiality of communication, medical information, or phone conversations.

Before proceeding further we discuss on security concerns in the digital age and ethical hacking.

Security Concerns in the Digital Age and Ethical Hacking

Maintaining cyber-security and safeguarding sensitive data is more crucial than ever in today's digital environment. But this needs to be weighed against another fundamental human right: the right to privacy. An overview of the moral and legal issues related to striking a balance between privacy and cyber-security in the digital age is given in this article. It addresses the current legal framework for cyber-security and privacy in various jurisdictions and looks at the difficulties of putting into practice efficient cyber-security measures while upholding privacy rights.

In addition, Allahrakha (2023) explores the moral ramifications of striking a balance between these two crucial principles and offers solutions for resolving privacy and cyber-security issues generally. This essay seeks to increase awareness of the necessity of ethical and legal concerns in the development of digital technologies and their regulation by highlighting the significance of striking a careful balance between cyber-security and privacy.

What is hacking with an ethical bent? In contrast to the computer science terminology, which only addresses vulnerability discovery and penetration/intrusion testing, my definition takes into account online civil disobedience, hacktivism, penetration/intrusion testing, counterattack/hackback, and security activism. The non-violent use of technology for a cause—political or otherwise—that is frequently morally and legally dubious is known as ethical hacking (Maurushat, 2019).

Security is a wide-ranging topic that encompasses many different sins. In its most basic form, it is about ensuring that sly individuals cannot read or, worse, alter messages meant for other receivers. People attempting to use remote services for which they are not permitted are the cause for concern. The majority of security issues are purposefully brought about by malevolent individuals attempting to hurt or obtain an advantage. Students enjoy prying into people's emails; hackers test out security systems or steal information; businessmen create security breaches to uncover competitors' strategic marketing plans; fired employees leak information in retaliation; and terrorists steal secrets related to germ warfare. Severe security issues arise from all of these unauthorized entry points to information systems.

The protection of a computer's or network's integrity falls under the large umbrella of data security, which encompasses all related tasks. Data security, in its broadest sense, refers to safeguarding a computer from risks that come from both within and external sources, including those who are not affiliated with the business and those who pose a risk to hardware or software. According to this view, disaster recovery is a component of data security since information managers' work to safeguard data against both man-made and natural disasters.

By following basic best practices, such as limiting access to computers and their files, and only using licensed copies of software that are unlikely to have viruses installed, organizations can increase the security of their systems. Data files should only be viewed by those who have a legitimate business need to, much like physical files have restricted access points. At this level, basic protection is provided by passwords and access codes, which will keep the merely curious out. The degree to which a company depends on information technology determines how critical information security is. However, the deployment of information security technology only addresses a tiny portion of the issue of information risk when an organization's information is exposed to danger. Actually, it appears more and more that information security technology is ineffective at lowering information risk (Tiwary, & Pradesh, 2011).

Researchers claim that "ethical hacking" has drawn interest from all over the world. In an effort to ascertain whether a corporate network has been compromised, identify potential vulnerabilities, and stop future hacking, many corporations support teaching their staff members how hackers think and operate. Furthermore, information technology personnel might receive training on ethical hacking techniques from consulting businesses.

The idea of training aspiring information technology professionals at the university level how to hack has also been proposed by proponents of ethical hacking, along with the moral and legal ramifications of such actions. Information security and assurance are becoming more and more significant in computer and information system students' educations, as many of them will be required to handle novel security and control difficulties (Subhashini, 2014). Now we like to navigate on ethical dilemmas related to data and information sharing.

Navigating Ethical Dilemmas Related to Data and Information Sharing

Emerging technologies in the contemporary technological period have a lot of promise. These include cybersecurity, big data and machine learning, artificial intelligence, autonomous vehicles, cloud computing, and autonomous vehicles. Before enterprises can use these technical breakthroughs in a production setting, ethical issues pertaining to data security and privacy must be addressed. Theories centered on the concepts of autonomy, fairness, beneficence, nonmaleficence, and faithfulness serve as the foundation for ethical thinking and concerns (Formosa, 2023).

Though it cannot be reduced to other disciplines, ethical reasoning is not wholly different from thinking in them. Specifically, ethical implications are not as easily proven as mathematical truths. This does not imply, however, that every ethical decision is equally sound. After all, the majority of philosophers of science would contend that all scientific results are provisional truths that cannot be conclusively proven. Certain findings are more likely to be true than others, regardless of the discipline—science, ethics, or any other. The claim that ethics has no right or wrong is a typical mistake made in ethics classes (Reiss, 2010).

Research on technology ethics has made significant use of emerging technologies, with an increasing focus on early-stage intervention in technological creation. Techno Ethics (TE) is an interdisciplinary field of study that combines methods and theories from several disciplines, including as sociology, communications systems, innovation, and ethical theories and principles (Choraś, et. al., 2019). For instance, the internet was not meant to be a secure or private space; rather, it was meant to be a global, free service available to all (Pawlicki, et. al., 2020). All illegal behaviors made possible by access to an IT infrastructure are collectively referred to as cybercrime, which includes system disruption, digital identity fraud, illegal data comparison filtration, unauthorized access, and more (Brey, 2017; Pawlicka, et. al., 2021).

The ethical considerations surrounding confidentiality and anonymity in the particular context of data sharing and secondary use of longitudinal research data in the United Kingdom (UK) are the focus of this chapter. Regarding data sharing for secondary use, all research data that is sponsored by the public is considered a public good that is created for the benefit of the public and should be made as freely accessible as feasible. In this context, funding bodies' and disciplinary associations' ethics rules typically discuss anonymity and confidentiality together.

However, in order to prevent the information that participants have disclosed from being connected to them, they are concerned with hiding the identities of those who have participated in research from secondary users of data sets (aside from the original researcher or research team), whether it be quantitative or qualitative (Edwards & Weller (2016).

In this context, Corti & Bishop (2020) mention that gathering current data for repurposing is frequently seen as being hampered by concerns about the state of informed permission or the unintentional usage of personal information of data acquired during the study process. It could appear that sharing research "data" runs the risk of undermining the participant-researcher relationship's foundation of trust. There may be misunderstandings about the functions that moral, legal, and ethical obligations play and represent as well as how open scientific principles and the idea of open data can coexist in light of the European General Data Protection Regulation (GDPR), which has received a lot of attention recently in the research community.

As mentioned earlier, before going to the next section, we like to provide a few concrete guidelines and frameworks that individuals and organizations can follow to address *Privacy, Security, and Ethical Dilemmas* in digital interactions. These frameworks enhance research acceptability by grounding ethical considerations in established principles.

i. Privacy Protection Frameworks

a. General Data Protection Regulation (GDPR) (European Union)

The GDPR (EU Regulation 2016/679) sets a global benchmark for data privacy by requiring organizations to implement principles such as data minimization, purpose limitation, and user consent (Voigt & Von dem Bussche, 2017). It emphasizes:

- **Right to be informed**: Transparency on how data is collected and used.
- **Right to access & rectification**: Users can request and correct their data.
- **Data Protection by Design**: Security and privacy must be integrated into system architectures.

b. California Consumer Privacy Act (CCPA) (United States)

The CCPA (2018) grants consumers more control over their personal data, allowing them to opt out of data sales and request disclosure of information collected about them (Smith, 2020). It mandates that businesses:

- Provide clear **privacy notices**.
- Implement **reasonable security** measures.
- Offer **opt-out mechanisms** for data tracking.

c. OECD Privacy Guidelines (Global Standard)

The OECD Guidelines on the Protection of Privacy and Transborder Flows of Personal Data (OECD, 2013) emphasize:

- **Collection Limitation**: Data should be gathered fairly and lawfully.
- **Accountability Principle**: Organizations must be responsible for compliance.
- **Security Safeguards**: Strong protection against risks such as unauthorized access or disclosure.

ii. Security Best Practices for Individuals and Organizations

a. NIST Cybersecurity Framework (United States)

The National Institute of Standards and Technology (NIST) Cybersecurity Framework (2018) provides a risk-based approach to security, consisting of five core functions:

- **Identify**: Understanding cybersecurity risks.
- **Protect**: Implementing safeguards like encryption and multi-factor authentication (MFA).
- **Detect**: Continuous monitoring for security threats.
- **Respond**: Developing response plans for cyber incidents.
- **Recover**: Ensuring resilience and data backup strategies (Keller, 2021).

b. ISO/IEC 27001 (Global Information Security Standard)

ISO/IEC 27001 sets international standards for information security management systems (ISMS). Key aspects include:

- **Risk assessment and mitigation**.
- **Access control measures** to restrict unauthorized data access.
- **Incident response planning** (Von Solms & Van Niekerk, 2019).

c. Zero Trust Architecture (ZTA) for Digital Security

Zero Trust (ZT) is a security framework that operates under the assumption that threats exist both inside and outside an organization's network. It recommends:

- **Least privilege access** (granting only necessary permissions).
- **Micro-segmentation** (dividing networks to limit attacks).
- **Continuous verification** (ensuring ongoing authentication) (Kindervag, 2020).

iii. Ethical Guidelines for Digital Conduct

a. The ACM Code of Ethics and Professional Conduct

The Association for Computing Machinery (ACM) outlines ethical responsibilities for technology professionals:

- **Honesty & Transparency**: Avoid deceptive practices in digital interactions.
- **Privacy Respect**: Protect individuals' personal data and avoid surveillance misuse.
- **Fairness & Inclusivity**: Promote accessibility and prevent algorithmic bias (Gotterbarn et al., 2018).

b. UNESCO's ROAM-X Model for Digital Ethics

UNESCO's ROAM-X framework ensures digital governance aligns with four core principles:

- **Rights-based**: Respecting human rights in online interactions.
- **Openness**: Encouraging transparency in algorithms and AI decision-making.
- **Accessibility**: Promoting equitable access to technology.
- **Multi-stakeholder engagement**: Collaborating with policymakers, businesses, and users to foster digital responsibility (UNESCO, 2019).

c. AI Ethics Guidelines by the European Commission

The European Commission's Ethics Guidelines for Trustworthy AI emphasize:

- **Human Agency & Oversight**: Ensuring AI does not undermine human rights.
- **Transparency & Explainability**: Users must understand AI decisions.
- **Accountability**: Organizations must document AI risks and take responsibility (Floridi & Cowls, 2022).

Hence it can be mentioned that by implementing privacy regulations (e.g., GDPR, CCPA), security frameworks (e.g., NIST, ISO 27001), and ethical guidelines (e.g., ACM, UNESCO) can enhance research acceptability by ensuring compliance, security, and moral responsibility in digital interactions. A multidisciplinary approach integrating these frameworks helps individuals and organizations navigate complex ethical dilemmas in an increasingly interconnected world.

In the next section we are going to discuss about creation of online ethical content.

SECTION 4: CREATING ETHICAL CONTENT ONLINE

We have a big say in the discourse and culture of the internet as content creators. The rules for producing and disseminating moral content will be covered in this part, with a focus on the authors' obligation to improve the digital environment. In order to cultivate an informed and moral online community, it is imperative to address problems like disinformation and advance digital literacy. We can create an online environment that is more reliable and trustworthy by upholding ethical norms.

Guidelines for Ethical Content Creation and Sharing

Social media are effective tools that let consumers interact with online communities directly. Plastic surgeons were among the first to use social media because they understood its possibilities for marketing and education. Social media use raises moral and legal questions. There have been requests for the American Society of Plastic Surgeons to create guidelines on the use of social media in medical practice due to the lack of professional society standards on the subject (Schoenbrunner, 2019).

Western civilizations today are characterized by massive and intensive data generation and storage. As platforms for information and communication technology (ICT) dedicated to the gathering of genomic and personal health data expand, so does the exchange of health data. Even when data is exchanged and publicized for scientific research, there are ethical issues because health information is private and sensitive.

In light of this, the COST Action CHIP ME 'Citizen's Health through Public-Private Initiatives: Public health, Market, and Ethical perspectives' (IS 1303) Science and Values Working Group identified six fundamental values that they believed were necessary for the moral exchange of health information via ICT platforms. In our opinion, applying this ethical framework will encourage deferential scientific conduct, which will preserve people's faith in research.

Along these perspectives, Riso, etl. Al. (2017) examine how new data sharing platforms are redefining the data sharing experience from a variety of angles by using these values to analyze five ICT platforms. The study go over the various principles, rights, and obligations that they encompass and incorporate into their perspective on the sharing of personal health information. They, further, examine new kinds of empowerment mediated by technology, as well as the design and development process of patient-oriented infrastructures and personal health data.

Young people can now engage in a variety of activities, such as creating and collaborating on interesting material for social media platforms, thanks to digital media and its tools. In light of the recent events in Egypt, digital media, as embodied by social media platforms like Facebook, has been essential in forming and influencing public opinion as well as user responses, particularly among youth, to a range of societal issues and circumstances.

However, social media is a free arena, and the creation and contribution of content is managed on an unspoken basis rather than by formal regulations or ethical considerations.

The significant ethical topic of "what is considered appropriate to post or share and what is not" was brought up by those engaging social media posts. to guarantee every citizen's involvement is equitable and safe. In order to reach findings and provide answers to the research objectives, the paper presents a qualitative study based on systematic reviews that addressed the ethical considerations of social media participation. The study's key conclusion is that, because users are still adjusting to the rapid information exchange, anonymity offered by social media, and the way that digital and social media work, there are still questions about the morality of producing, engaging in, and sharing content on these platforms (Bicher & Fathy, 2020).

However, practical dos and don'ts for sharing research data in ways that are ethical, efficient, and comply with the federal Common Rule are provided in this tutorial. I start by going over best practices for prospectively integrating data-sharing plans into research, talking about what to say and not say in consent forms and applications to the institutional review board, tools for de-identifying data and how to weigh the risks of re-identification, and things to take into account when choosing a data repository. Moving on to data that have already been gathered, I go over the moral and legal concerns that are brought up by sharing data when participants were either expressly assured on the consent form that the data would not be shared or the form was silent on the matter (Meyer, 2018).

Now it is important to discuss on the role of content creators in shaping online culture.

The Role of Content Creators in Shaping Online Culture

There is still a lack of knowledge about how content creators' brand subjectivities are impacted by the dynamic platform environment, despite the wealth of literature on their identities, tactics, and activities. In light of this, Arriagada & Ibáñez (2020) investigate how platform evolutions—such as ongoing adjustments to their affordances—influence the actions and cognitive processes of content producers. Based on a study of 35 fashion and lifestyle content creators in Chile and an examination of their Instagram photos (N = 165) and stories (N = 150), they demonstrate the way in which these creators are driven to embody various representations of their brand subjectivities.

Based on three interconnected levels—communicative styles, temporal acceleration, and ongoing negotiations with other actors in the social network that shape commercial activities— Arriagada & Ibáñez (2020) findings illustrate how producers perceive platform changes. Because of this, content creators have to modify their brand subjectivities and practices across platforms and affordances whenever platforms make modifications to increase their commercial viability. They also show how, given the constantly shifting technological, social, and commercial ecologies, the ideal of creation has been replaced with increased competitiveness, in addition to illuminating the new routines and increased economic demands required of today's digital workers.

In a recent study, Villegas-Simón, et. al. (2023) focus on "digital cultural communicators," or social media content creators who specialize in politics, culture, or the arts. The primary goal is to gain a deeper understanding of the perspectives held by digital cultural communicators regarding celebrity and the relationships and exchanges they have with their audiences. We use in-depth interviews with eighteen Spanish digital cultural communicators who upload content on many platforms (TikTok, YouTube, Instagram, and Twitch) to achieve this goal. Thematic analysis and ATLAS.ti software were utilized for the qualitative analysis of the transcribed interviews. The findings demonstrate that digital cultural communicators make an effort to set themselves apart from social media influencers and portray themselves as respectable cultural workers rather than "just" content producers. Their opinions about celebrities also influence the discussions they have with their educated audiences and how they respond to both positive and negative feedback.

We know, content producers create content centered on their passions, pastimes, and life events, which they then share and profit via digital channels. This study suggests an inverse U-shaped relationship between the frequency with which creators share their content and the engagement behavior of users on digital platforms, based on the two-factor hypothesis. The study also takes into account creator attributes including the quantity of followers, followership, and moderator competitors. Utilizing a panel dataset of content creators gathered through web scraping, the study calculates a fixed-effects regression model in order to evaluate its hypotheses.

The results support the inverted U-shaped relationship by showing that user engagement behavior rises initially but then declines after peaking as the frequency of authors' content sharing increases. Furthermore, the inverted U-shaped relationship is negatively moderated by the number of rivals, followers, and followers of creators. Tafesse & Dayan (2023). In their study adopts a creator-first strategy and offers insight into creators' frequency of sharing content and users' reactions to it on digital platforms, in contrast to earlier research that concentrated on creators' strategies as an extension of marketers' digital marketing efforts.

Before going to the next section on the role of ethics in online communities, we like to discuss a little on how to address misinformation and digital literacy in the digital arena.

Addressing Misinformation and Digital Literacy

We understand, policy and academic communities have voiced worries about potential hazards connected with digital channels for public communication and information throughout the 2010s and early 2020s. Given its possible effects on vaccine skepticism and adherence to public health measures, misinformation—that is, incorrect or misleading content—has come to light as a result of the Covid-19 outbreak. While it has occasionally been demonstrated that the amount of misinformation in larger online communications is overblown in comparison to other news sources (Allen et al., 2020), it is becoming more and more obvious that disinformation has the ability to magnify broader effects (Fletcher et al., 2018; Gagliardone et al., 2021).

Also, there is increasing evidence that the dissemination of false and misleading information online is linked to attitudes and behaviors offline. Examples of these include the US Capitol storming by Trump supporters, violence against Rohingyas in Myanmar, and the Brexit referendum campaign in the UK. Young people have emerged as a focal point for policy and study because, while they are viewed as "digital natives" capable of playing important roles as inventors and influencers online, they are also perceived as being vulnerable to negative effects associated with the internet (O'Reilly, 2020). A subset of misinformation has been discovered to specifically target women in politics and other fields in an effort to discourage their public engagement, raising serious concerns about gender-related harm (Jankowicz et al., 2021; Diepeveen, & Pinet, 2022).

The use, absorption, articulation, and dissemination of information have all undergone significant changes as society quickly moves into the digital era. People all over the world have access to unprecedented volumes of information thanks to the internet, yet because of its complexity, consumers must accept greater degrees of uncertainty because they prefer dependable facts to well-known ones (Gallagher, 2016). Controlling and confirming the information generated and shared in the digital world is a continuous challenge due to its limitless nature.

This quickly sparks a mass propagation of false information with disastrous consequences for the social and political facets of society. Additionally, a lot of people have turned to social media as a source of fun during the lockdown. Social media technologies, including blogs, social networking sites, and microblogging tools, are steadily evolving into dependable channels for fast data transmission with target audiences. Social media technologies, including blogs, social networking sites, and microblogging tools, are proving to be dependable channels for quickly disseminating information to target audiences.

Social media platforms facilitate the easy spread of information, even false information. Messages can be easily forwarded to several recipients with just a mouse click, in most cases. As a result, false information spreads swiftly through social media, potentially confusing and unnecessarily frightening large numbers of people (Guess et al., 2020; Jones-Jang et al., 2021).

Hence, disseminated false information on social media should be taken seriously. Misinformation can be troublesome and even dangerous, particularly when it is intended to correct information (Anthonysamy, & Sivakumar, 2024).

It is a fact that false information is not newly created. However, misinformation has evolved in the digital era and found new ways to propagate quickly. It has been challenging for social media platforms and internet companies to stop this spread, which usually only becomes noticeable after harm has already been done, for both pragmatic and cultural reasons. For example, anti-vaccination messages that were spread on Facebook (and possibly to a lesser extent on WhatsApp groups for parents) were not widely publicized until measles epidemics escalated.

As a result, take-down policies are severely constrained and frequently face opposition from the public and the legal system. Therefore, the task is to equip individuals with the critical literacy competencies that will enable them to recognize misinformation when it arises and stop it from spreading in the first place. Digital literacy is a subset of media literacy, which is the capacity to access, analyze, evaluate, and generate communications in a variety of forms, as it relates to digital media and the internet (Polizzi, & Taylor, 2019).

As we have discussed on creating ethical content online, including guidelines for ethical content creation and sharing, the role of content creators in shaping online culture and addressed misinformation and digital literacy, the next section will discuss on the role of ethics in online communities.

SECTION 5: THE ROLE OF ETHICS IN ONLINE COMMUNITIES

Any robust and active online community is built on ethics. The significance of moral behavior in creating and sustaining online communities, along with the functions of platform administration and moderation, will be discussed in this section. The tone, inclusivity, and sustainability of online interactions are all impacted by ethical leadership in virtual spaces. By placing a high value on moral behavior, we can build communities where everyone feels safe, empowered, and enriched.

Building and Maintaining Ethical Online Communities

As a result of social media's birth and explosive expansion, some government agencies across numerous nations have adopted it as a preferred means of communication with their constituents. We are investigating the potential to use social networks to support particular citizen groups in partnership with the Australian Department of Human Services. For this reason, we have created an online community called Next Step to assist those who are currently receiving welfare benefits in finding employment and achieving financial independence.

Paris, et. al. (2013) in their study examine that some of the ethical problems that come up when governments interact directly with the public, especially with communities who are facing hardships, and when researchers are involved. They list a few of the difficulties they encountered along with their solutions. The research demonstrates the difficulty of the situation when a government agency is involved in an online community along with a group of welfare recipients who are dependent on the agency. Maintaining community members' privacy while also carrying out the government's legal obligations becomes a balancing act.

The ethical concerns unique to studies on virtual communities are examined in this paper. We identify the following key issues, based on an empirical case with online forums of education experts: the definition of human subjects' research, participant recruitment, informed consent, publicity versus community privacy, and ethical questions related to observing virtual communities and reporting and disseminating research findings. We argue that analyzing global forums can be complicated by the disparate research cultures seen in many nations. This study suggests that ethical concerns should be more case-based rather than depending on a single paradigm that can solve all problems, acknowledging the transient nature of Internet environments.

Along this route, Kantanen & Manninen (2016) propose that, with their in-depth understanding of ethics, regional ethics committees or institutional review boards could offer invaluable assistance to researchers working in the dynamic and complicated world of Internet research, as well as in domains and research environments where an ethical review is not a customary step in the scientific method.

From an academic study perspective, social media platforms are attractive due to their growing popularity. Researchers can now access populations that could be difficult to reach otherwise and analyze and understand behaviors and beliefs in naturalistic circumstances at a low cost. In addition, new ethical dilemmas about what is private and what is public are brought up by social media's ease of access to search data. Furthermore, not every study environment will allow for the application of the current ethical norms.

Thus, protecting the autonomy of study subjects is one of the fundamental tenets of ethical research conduct. In the context of relatively new electronic environments, more discussion and understanding are required regarding the implications of subjects like avoiding harm, protecting privacy and data, and protecting data.

In an earlier study, Vanacker & Heider (2012) examine the circumstances under which ethically significant avatar injury takes place in virtual environments. The authors contend that players who view their avatars as integral parts of their identities and who perceive certain rules of appropriate behavior in virtual worlds are the most likely scenarios for this to happen. Apart from virtual spaces that have a "caveat emptor" mentality, Second Life is regulated by certain standards of appropriate conduct. Though residents of Second Life do not believe that their community needs a separate code of ethics, they do have ideas about what is appropriate and inappropriate behavior. However, there is often debate over the precise definition of norm-violating behavior and ethically meaningful avatar injury, as seen by the responses people have posted online to a Second Life avatar upskirt gallery.

When a norm is broken, players who regard their avatars as extensions of themselves are more likely to suffer ethical consequences than those who view their avatars as separate entities from themselves.

Keeping these in mind we argue for the moderation and governance in digital platforms, next.

Moderation and Governance in Digital Platforms

Wright (2022), in his research, raises a question, as what effects do algorithms have on content moderators and users, and how should academics interpret these effects? A large portion of the body of research on this topic regards moderation as a sequence of decision-making tasks and assesses moderation algorithms according to their accuracy. The author mentioned that he contend that content moderation is a complicated system of rules, mechanisms, and procedures rather than a set of discrete decisions based on literature from the subject of platform governance. Therefore, research needs to explain how automated moderation modifies the platform's overall governance structure. The authors uses interviews and trace ethnography to present the results of a qualitative study on the Reddit bot AutoModerator in order to illustrate this. The size of the bot, in his opinion, enables moderators to meticulously control content visibility and moderation, thereby changing the fundamental guidelines for platform governance.

In the past few years, TikTok, an app for short videos with content typically lasting between 15 and 60 seconds, has gained enormous popularity worldwide. Nonetheless, TikTok's global renown necessitates that the platform continually haggle with the laws, customs, and legal frameworks of the areas in which it conducts business. Its failure to do so has had serious repercussions. For instance, the platform has been (temporarily and permanently) prohibited in a number of nations, including India, Indonesia, and Pakistan, due to content-related issues.

In addition, the site has been under increased scrutiny and criticism because of its Chinese ownership and appeal to minors. Zeng & Kaye (2022) in their work present the concept of visibility moderation, which is the method by which digital platforms apply algorithmic or regulatory techniques to control the reach of user-generated material. They talk about specific steps TikTok takes to control visibility and the problems that result from it. The information gathered from user-centric interviews with content creators is presented in this study, which aims to comprehend how they interpret and deal with TikTok's visibility moderation.

The study's conclusions also point to issues that give these stakeholders a sense of confusion, frustration, or helplessness and provide crucial avenues for future investigation.

Content moderation is becoming a crucial feature of platform governance due to the rise in social media and online community participation. So far, the most important workforce for content control has been volunteer (human) moderators. In order to achieve scalable, desirable, and sustainable moderation, several online platforms have recently begun to deploy algorithm-based content moderation tools (bots). This is because volunteer-based content moderation confronts certain obstacles. It's unknown how volunteer moderators will respond in terms of community policing and nurturing when bots are added to platform governance.

In a recent study, He, et. al. (2024) use information gathered from 156 Reddit communities (subreddits) to perform an empirical study in order to comprehend the effects of these increasingly common bot moderators. They discover that bots encourage volunteer moderators to monitor more messages, which enhances their role as moderators. These effects are more noticeable in larger communities and are based on a number of econometric models. In particular, volunteer moderators enforce community standards 20.9% more strictly, especially when it comes to arbitrary ones. Furthermore, after their community implemented bots, volunteers in larger communities also made a greater effort to provide additional explanations and ideas.

Notably, the requirement for nurturing efforts to keep up with the growth in subjective policing is the primary driver of activity increases. Additionally, the use of bots in content moderation increases volunteer moderator retention. Overall, we demonstrate that maintaining digital communities benefits from the inclusion of algorithm-based content filtering in platform governance.

The Impact of Ethical Leadership in Virtual Environments

Organizations have always recognized the importance of ethical leadership in project teams, but as companies shift to more non-traditional and global work environments, the requirement for ethical leadership in virtual project teams has become even more crucial. Ethical leadership in the virtual workplace is referred to as e-ethics. The distinction between traditional project team management and virtual team management in terms of ethical considerations is still being researched. The number of virtual project teams in company is growing these days, and a key component of the new e-business ethics is knowing how to manage virtual teams ethically.

Lee (2009) reviews contemporary researches on ethical theories in relation to virtual project leadership and ethical leadership in the context of e-ethics, or virtual business environments. The application of political tactics, stakeholder theory, communitarian ethics, egalitarianism, Kantian ethics, motivation, and trust, participatory management, Theory Y, and its relationship to utilitarianism, and other ethical theories are reviewed in relation to virtual project management. The study presents obstacles to e-ethical leadership and discusses possible solutions. In the conclusion, four research directions are proposed. The research seeks to locate e-ethics secondary literature and explores the potential implications of this emerging field of corporate ethics for virtual project team leaders.

Teleworking's benefits and danger issues have been extensively studied. Among them, emotional weariness has received the greatest research attention. But the effect of moral leadership in online settings has gone entirely unrecognized. The question of whether an ethical leader slows down or speeds up processes connected to stress in teleworking remains unclear.

Furthermore, because the current results are unclear and conflicting, further empirical data is needed to support the significance of perceived job autonomy as a theoretical mechanism that guards against particular negative psychological effects. Accordingly, this study investigates the relationship between ethical leadership and emotional weariness through work autonomy and teleworking intensity (TI) at different times.

Santiago-Torner, et. al. (2024) in their research include 448 professionals from six Colombian electricity-related organizations made up the sample. Using statistical tools like PROCESS and AMOS macros for SPSS, the moderated mediation model was examined and the mediation impact was assessed using a four-step process. Their findings, surprisingly, showed a positive correlation between emotional tiredness and ethical leadership, with work autonomy possibly acting as a mediating factor in this association. It was discovered that teleworking intensity (TI) mediated this mediation process. In particular, when TI was high, the detrimental effects of moral leadership on emotional tiredness were neutralized.

However, a high TI level reduced the favorable association between emotional weariness and work autonomy while strengthening the one between ethical leadership and work autonomy. Finally, our research points to a possible correlation between emotional weariness experienced by followers and moral leadership. The quantity of teleworking days, however, has the power to affect and even reverse this tendency. Put differently, when teleworking intensity (TI) rises, it becomes a resource that can reshape the relationship between emotional weariness and ethical leadership as well as change the trajectory of work autonomy.

The traditional relationship between leadership, economic, and caring practices may no longer be adequate to address the needs of systems, doctors, patients, or providers in the current climate and environment. It is argued that a change toward humane caring principles and an ethic of genuine healing relationships is necessary in the current complex and challenging healthcare environment, particularly in light of the COVID-19 pandemic. For followers, the consequences of acting unethically may be considerably more severe. Assuming leadership responsibilities comes with ethical obligations. Keselman & Saxe-Braithwaite (2021) believe that there is a close relationship between the triangle of ethics and ethical behavior, followers, and patient results, based on their own experience.

Ineffective leadership can have a detrimental effect on patient outcomes and organizational effectiveness by inspiring followers to become disillusioned and mistrustful of the leader, which in turn can lead to a loss of enthusiasm and commitment.

SECTION 6: GLOBAL PERSPECTIVES ON DIGITAL ETHICS

Since digital contacts transcend national boundaries, it is critical to take the global ethical context into account. Cultural diversity is important because it shapes digital ethics because different societies have different values and viewpoints about what constitutes appropriate behavior online. This section will look at international laws and rules that regulate behavior online, emphasizing the necessity for coordinated efforts to create international moral guidelines. Accepting different points of view can result in digital practices that are more widely accepted and egalitarian.

Along the global perspectives on digital ethics, cultural diversity influences. Hence, this study likes to provide some input in terms research dialogue.

Cultural Diversity and Its Influence on Digital Ethics

The problems that digital young encounter on a daily basis are moral ones. In their daily lives on the internet, how do today's youth and young adults navigate the concepts of good and bad, truth and lie, and representation and misrepresentation? How do they prepare for, deal with, and navigate the actual repercussions of their online behavior and social interactions? How do they handle the complexity of their public exchanges and their private lives, and how do they engage with parental, corporate, and government surveillance? (Luke, 2018).

Lastly, how can they interact and take part in the public, political, cultural, and economic domains of the internet as citizens, customers, employees, friends, coworkers, and family? Current empirical research (such as, Livingstone & Sefton-Green, 2016; Quan-Haase, 2016) of young people's virtual and actual everyday lives in educational institutions and families address these topics.

Furthermore, designing digital infrastructure from the ground up with the ability to better support multicultural and diverse viewpoints is crucial. Without taking into account how technology may be modified to anticipate and promote pluralistic multicultural perspectives in its initial design, it is insufficient to focus just on how culture responds to and influences technology. This goes beyond just providing support for many languages and user interfaces; instead, digital and data infrastructure should be able to recognize and take into account pluralistic viewpoints, allowing for a variety of viewpoints to be represented and processed inside the data itself (Israel & Amer, 2021).

It is evident that the younger generation's trends in social media use are growing, but they are not being followed by the authority on digital ethics. The nation's diversity and unity are ultimately threatened by a large number of social media cases.

Basyah & Hayati (2020) conducted a study on high school students to learn about their online experience. The study's participants were high school students, ages 16 to 18, who were frequent users of social media. There were a total of 12 informants in all—10 students and 2 teachers. The case study approach was combined with a qualitative research design. According to research findings, a large number of students were accustomed to using social media as they pleased and were unaware of digital ethics. The character of pupils who use social media was just a place of presence and updating, psychologically unstable and weak control, and relates to the limitations of building integration and diversity, as well as the low awareness of digital ethics among peers, families, and teachers.

In a similar study, Colley (2014) discusses ethical concerns generated by the use of digital technology in archaeology and cultural heritage practice using a study of the literature and the author's Australian research experiences. The researcher claims that technology use gives new meaning to values previously embedded in professional codes of ethics, such as professional standards and the need to strike a balance between the public's right to know and other rights, such as intellectual property and other rights (e.g. through open-access data regulations). Technology has also brought up other ethical concerns, such as digital preservation and sustainability; the role of business and corporate interests in creating, marketing, and designing specific products; professional and community involvement in the digital public sphere; equitable access to technology and content; digital literacy; and philosophical and sociopolitical queries concerning the reality and representation of digital heritage.

We find that not only technology and behavior of the user are important, but also the international laws and regulations govern online behavior. Hence we discuss it next.

International Laws and Regulations Governing Online Behavior

The Internet poses a challenge to international law. The U.N. General Assembly and the International Telecommunication Union are two examples of international organizations that are getting more and more active in Internet regulation. This activity suffers from a deeper flaw than just the question of whether international regulation is desirable, which may now be inevitable. The current state of international law on freedom of expression and the implications of these regulatory choices for human rights are being ignored as international regulation moves forward.

In a classical study, Land (2013) mentions that the government and corporate interests will dominate regulatory decisions if international human rights legislation and principles are not taken into account. As a result, norms regarding things like censorship and data privacy are probably going to be at odds with each other and even with international human rights.

People are spending more and more time online these days. Researchers point out that new types of computer-mediated communication have been inspired by ever-improving virtual environments and intellectual technologies. Numerous facets of human existence are impacted by these new modes of connection. Web-based communication behavior is very different from traditional modes of communication. The establishment of norms for behavior on the internet is something we are currently seeing. The majority of online communication guidelines are unwritten.

However, Internet forums have been around for a while—the first web-based forums were created in 1994—and have amassed a wealth of experience in setting the standards of conduct that new users must abide by in order to participate actively in the community. The disciplinary actions that support these norms of behavior range from warnings from moderators to account bans that are permanent (Bylieva, et. al., 2019).

Additionally, the issue of Internet regulation is a microcosm of the discussion and ideas surrounding the issue of unilateralism in international affairs. Two particular insights that the issue of Internet regulation has to give the broader analysis of the issue of unilateralism were discussed by Benkler (2000). The first is whether the problem should be applied in the proper legal and normative arena. What means of transmitting a nation's values as behavioral limitations in another fall under the purview of international law is the specific question at hand.

Hence there is a need for collaborative efforts to establish global ethical standards and it deserves a little discussion.

Collaborative Efforts to Establish Global Ethical Standards

Although there has been human testing for several centuries, systematic measures to safeguard experiment subjects have only been in place for the past 50 years. The Nuremberg Code, which outlawed forced experimentation on humans in 1947, served as the impetus for the Helsinki Declaration, which was adopted a few years later. The notion of independent, individual informed consent was universally adopted as a result of the widespread support shown for these historic pronouncements by the world community. Effective human subject protection still varies greatly throughout nations, though, and the increasingly international character of research endeavors has brought some of these disparities to light, especially when developed and developing nations collaborate (Caballero, 2002).

In line with the Sustainable Development Goals of the UN, this commentary seeks to ascertain the extent to which the provisions of international health research guidelines foster fair collaborations and capacity building in global health research. Our evaluation indicates that collaborative research partnerships in resource-constrained settings lack governance; however, since The Council for International Organizations of Medical Sciences (CIOMS) in 2016 adopted the International Ethical Guidelines for Health-related Research Involving Humans, this has improved. However, in order to address the structural injustices and power inequalities present in international health research partnerships, more clarity is required in national legislation, industry and ethics norms, and regulatory provisions. Most notably, the International Conference on Harmonization Technical Requirements for Registration of Pharmaceutical for Human Use (ICH) Good Clinical Practice (ICH-GCP), the main industry ethics rules, do not allow ethical partnership governance (Ward, et. al. 2018; Meza, et. al. 2018).

Before concluding, we like to oversee this research by looking at interdisciplinary perspectives, incorporating insights from field such as, sociology, psychology, and law to provide a more holistic view of the ethical dimensions of digital interactions.

To provide a holistic understanding of digital ethics, it is essential to incorporate interdisciplinary perspectives from sociology, psychology, and law. These fields offer valuable insights into how digital interactions shape human behavior, social structures, and legal frameworks, influencing ethical considerations in the digital age.

i. Sociological Perspectives: Digital Society and Ethical Challenges

Sociology examines the impact of digital technologies on social structures, power dynamics, and cultural norms. The rise of digital platforms has led to the *restructuring of social interactions*, where individuals engage in online communities that shape identity, communication, and power relations (Castells, 2012).

Key ethical issues from a sociological standpoint include:

- **Digital Inequality & Accessibility**: The "digital divide" disproportionately affects marginalized populations, limiting access to technology, education, and economic opportunities (van Dijk, 2020). Ethical frameworks must ensure **digital inclusion** to prevent technological disparities.
- **Surveillance Capitalism & Power Structures**: Tech companies leverage data-driven business models that reinforce power imbalances between corporations and users (Zuboff, 2019). Ethical concerns arise regarding **consent, autonomy, and digital rights** in a surveillance-driven society.
- **Online Communities & Social Norms**: Digital spaces shape moral and ethical behavior, influencing social norms related to privacy, freedom of speech, and digital expression (Papacharissi, 2015). Ethical digital citizenship requires critical engagement with these evolving norms.

Sociological Frameworks for Ethical Digital Interactions:

- The concept of **network society** (Castells, 2012) highlights the transformation of human relationships in digital spaces.
- **Critical digital studies** (Noble, 2018) examine power dynamics in algorithmic decision-making and digital oppression.

ii. Psychological Perspectives: Human Behavior and Digital Ethics

Psychology provides insights into how digital environments influence cognitive and emotional processes, shaping online behavior and ethical decision-making.

Some major ethical concerns from a psychological perspective include:

- **Cyberbullying & Online Aggression**: Digital anonymity enables harmful behaviors, such as cyberbullying and trolling. Studies show that **deindividuation** (loss of self-awareness in groups) fosters unethical conduct online (Suler, 2004).
- **Digital Addiction & Mental Health**: Excessive engagement with digital technologies, especially social media, has been linked to **dopamine-driven reward mechanisms**, causing addiction-like behaviors (Alter, 2017). Ethical concerns arise regarding **user well-being and platform responsibility**.
- **Privacy & Psychological Safety**: Personal data exposure can lead to anxiety, stress, and digital paranoia. Individuals often experience the **privacy paradox**—valuing privacy but engaging in behaviors that compromise it (Acquisti et al., 2015).

Psychological Strategies for Ethical Digital Engagement:

- **Digital literacy programs** help users develop critical thinking skills to navigate misinformation and online manipulation (Lewandowsky et al., 2020).

- **Ethical design frameworks** (e.g., humane technology initiatives) promote digital well-being by reducing addictive design patterns (Harris, 2019).

iii. Legal Perspectives: Digital Rights and Governance

Law plays a crucial role in defining the ethical boundaries of digital interactions through policies, regulations, and enforcement mechanisms.

Ethical challenges from a legal standpoint include:

- **Data Protection & Privacy Laws**: Regulations such as the *General Data Protection Regulation (GDPR)* and the *California Consumer Privacy Act (CCPA)* establish legal frameworks for user data rights and corporate responsibilities (Voigt & Von dem Bussche, 2017).
- **Freedom of Expression vs. Hate Speech**: Legal systems struggle to balance **free speech rights** with the need to combat **online harassment, misinformation, and extremist content** (Citron, 2014). Ethical governance must ensure **content moderation aligns with human rights principles**.
- **Cybercrime & Digital Security**: Digital fraud, hacking, and identity theft necessitate robust cybersecurity policies to protect individuals and organizations (Brenner, 2010). Ethical cybersecurity practices must align with principles of **fairness, transparency, and proportionality**.

Legal Frameworks for Ethical Digital Governance:

- **International Human Rights Law** (United Nations, 2019) applies principles of privacy, digital rights, and freedom of expression to online spaces.
- **AI Ethics Regulations** (EU AI Act, 2023) address the ethical deployment of artificial intelligence in governance, employment, and security.

Therefore, the ethical dimensions of digital interactions require a *multidisciplinary approach*, integrating sociological, psychological, and legal perspectives to foster responsible digital citizenship. Moving forward, collaborative efforts among policymakers, tech developers, educators, and psychologists are essential to navigate emerging ethical challenges in the digital landscape.

CONCLUSION

The ethical aspects of our relationships will keep changing as we work through the challenges presented by the digital age. The main ethical principles that ought to direct our conduct on the internet have been emphasized in this chapter, with a focus on the significance of empathy, privacy, security, and a global viewpoint. As we look to the future, we must not waver in our commitment to supporting moral behavior in our digital lives and inspiring others to follow suit. By doing this, we can help create a digital society that is more equitable, civil, and responsible.

As a recap of the ethical considerations discussed, we can mention a few words here with a little hint on future of ethics and calls to action for readers to end the chapter.

The ethical dimensions of digital interactions are essential for fostering a responsible and empathetic online presence. This chapter has illustrated how ethical frameworks, including deontological, consequentialist, and virtue ethics, offer critical perspectives on digital citizenship and responsible behavior in virtual spaces (Floridi & Taddeo, 2020; Reynolds & Cerutti, 2021). By integrating empathy and cultural awareness into online interactions, individuals and communities can create more inclusive environments that respect privacy, uphold security, and promote respectful digital engagement (Crawford, 2021; Kende et al., 2022). Recognizing the ethical implications of digital actions not only benefits individual users but also contributes to the overall health and resilience of digital ecosystems.

As digital platforms continue to evolve and online communities grow, the role of ethics becomes increasingly significant. Future research and practice should address the emerging ethical challenges posed by rapid technological advancements and global interconnectedness, advocating for policies and educational initiatives that instill digital responsibility and ethical awareness (Johnson et al., 2021). By embracing a global perspective on digital ethics, users can help shape a more respectful, secure, and empathetic digital world that supports the rights and dignity of all individuals.

The Future of Ethics in an Increasingly Digital Society

A large portion of people's daily activities are mediated in one way or another by computers and algorithms as society becomes more digitally connected and digitalized. Because most individuals do not have personal experience of what digitalization feels like, it is difficult to appreciate the extent of digitalization and its effects. The abstract and challenging nature of digitalization causes people to feel disengaged from their digital environment.

In this work, we contend that an embodied understanding of digitalization is necessary to comprehend the nature and destiny of a digitalized society. Such an understanding ought to challenge established narratives, make use of non-rational knowledge sources, and shift from anticipating the future to investigating innovation.

We highlight the value of a more comprehensive understanding of digitalization in the context of education, learning including other aspects and talk about how a more varied perspective is necessary to enable individuals to participate in a digitalized society (Dufva & Dufva, 2019).

Calls to Action for Readers to Engage in Ethical Online Practices

Work enabled by digital platforms could eventually turn unfair or detrimental if a thorough and profound ethical analysis isn't done. In this context, "right things" refers to stakeholders in the "loop of work" acting in accordance with criteria that are logically justified. Making sure that no item in the loop is used in a way that could endanger people, the environment, or society is crucial. Each artifact would therefore need to adhere to specific legal, ethical, and regulatory frameworks.

Hence, in order to be ready for both a bright and a dark future, it is critical to consider questions about the relationship between humans and computational entities in terms of what would happen if humans were to be supplemented or replaced by machines. It is also important to attempt to imagine both utopian and dystopian future-of-work scenarios.

Given the current context, as compared to the current state of affairs, the function of the work platform in the relationship between the employers and employees needs to be more clearly defined and regulated.

In fact, platforms could serve a number of purposes, such as connecting workers and requesters, facilitating cash transfers between them, and providing a virtual workspace. These responsibilities must be clearly stated and adhere to all applicable legal requirements. In this setting, the concept of a work contract becomes complicated and frequently ill-defined because it involves three parties (e.g., job providers, workers, and platforms) and various contexts (e.g., online and offline). Additionally, requesters, workers, and platforms may reside in various nations, and it can be difficult to determine whose employment laws should be in effect (Bourhis, et. al., 2019).

REFERENCES

Acquisti, A., Brandimarte, L., & Loewenstein, G. (2015). Privacy and human behavior in the age of information. *Science*, *347*(6221), 509–514. DOI: 10.1126/science.aaa1465 PMID: 25635091

Ainley, K. (2017). *Virtue ethics*. Oxford University Press and the International Studies Association.

Al-Jenaibi, B. (2020). Digital ethics and social norms: A Middle Eastern perspective. *Journal of Media Studies*, *12*(3), 45–63.

Allahrakha, N. (2023). Balancing cyber-security and privacy: Legal and ethical considerations in the digital age. *Legal Issues in the digital. The Age (Melbourne, Vic.)*, (2), 78–121.

Allen, J., Howland, B., Mobius, M., Rothschild, D., & Watts, D. J. (2020). Evaluating the fake news problem at the scale of the information ecosystem. *Science Advances*, *6*(14), eaay3539. Advance online publication. DOI: 10.1126/sciadv.aay3539 PMID: 32284969

Alter, A. (2017). *Irresistible: The rise of addictive technology and the business of keeping us hooked*. Penguin.

Anthonysamy, L., & Sivakumar, P. (2024). A new digital literacy framework to mitigate misinformation in social media infodemic. *Global Knowledge, Memory and Communication*, *73*(6/7), 809–827. DOI: 10.1108/GKMC-06-2022-0142

Arriagada, A., & Ibáñez, F. (2020). "You need at least one picture daily, if not, you're dead": Content creators and platform evolution in the social media ecology. *Social Media + Society*, *6*(3), 2056305120944624. DOI: 10.1177/2056305120944624

Audrin, C., & Audrin, B. (2023). More than just emotional intelligence online: Introducing "digital emotional intelligence". *Frontiers in Psychology*, *14*, 1154355. DOI: 10.3389/fpsyg.2023.1154355 PMID: 37205063

Bailey, L., Chan, H., & Williams, M. (2022). Social media and mental health advocacy: The role of empathy in online communities. *Journal of Digital Psychology*, *8*(3), 112–129.

Basyah, N., & Hayati, E. (2020, March). Strengthening Integration and Diversity Through Digital Ethics. In *2nd Annual Civic Education Conference (ACEC 2019)* (pp. 558-562). Atlantis Press.

Beauchamp, T. L., & Childress, J. F. (1983). *Principles of biomedical ethics* (2nd ed.). Oxford University Press.

Bélanger, F., & Crossler, R. E. (2011). Privacy in the digital age: A review of information privacy research in information systems. *Management Information Systems Quarterly*, *35*(4), 1017–1041. DOI: 10.2307/41409971

Benkler, Y. (2000). Internet regulation: A case study in the problem of unilateralism. *European Journal of International Law*, *11*(1), 171–185. DOI: 10.1093/ejil/11.1.171

Benlahcene, A., Zainuddin, R. B., Syakiran, N., & Ismail, A. B. (2018). A narrative review of ethics theories: Teleological & deontological Ethics. *Journal of Humanities and Social Science (IOSR-JHSS), 23*(1), 31-32.

Bhatia, R. (2020). Digital empathy in crisis response: Social media and refugee support networks. *Global Media Journal, 18*(2), 55–73.

Bhuvaneswari, G., & Vijayakumar, S. (2021). Emotional intelligence and values in digital world through emoticons among indian students and faculty. *International Journal of Asian Education, 2*(2), 267–276. DOI: 10.46966/ijae.v2i2.142

Bicher, L. A., & Fathy, S. A. (2020). Ethical Considerations for Using Social Media:"Is it Okay to Share? [AJMCR]. *The Arab Journal of Media and Communication Research, 2020*(31), 172–187. DOI: 10.21608/jkom.2020.156744

Binns, R. (2018). Fairness in machine learning: Lessons from political philosophy. *Proceedings of the 2018 Conference on Fairness, Accountability, and Transparency*, 149-159.

Bourhis, P., Demartini, G., Elbassuoni, S., Hoareau, E., & Rao, H. R. (2019). Ethical challenges in the future of work. *A Quarterly Bulletin of the Computer Society of the IEEE Technical Committee on Data Engineering*.

Brenner, S. W. (2010). *Cybercrime: Criminal threats from cyberspace*. ABC-CLIO. DOI: 10.5040/9798400636554

Brey, P. (2017). Ethics of Emerging Technology. In *The Ethics of Technology Methods and Approaches*. Rowan & Littlefield International.

Buchanan, E. A. (2020). Digital ethics: A framework for online conduct. *Ethics and Information Technology, 22*(1), 23–37.

Bylieva, D., Lobatyuk, V., & Safonova, A. (2019). Online forums: Communication model, categories of online communication regulation and norms of behavior. *Humanities & Social Sciences Reviews, 7*(1), 332–340. DOI: 10.18510/hssr.2019.7138

Caballero, B. (2002). Ethical issues for collaborative research in developing countries12. *The American Journal of Clinical Nutrition, 76*(4), 717–720. DOI: 10.1093/ajcn/76.4.717 PMID: 12324282

Castells, M. (2012). Networks of outrage and hope: Social movements in the Internet age. *Polity*.

Chen, Y., & Xu, Y. (2021). Exploring the effect of social support and empathy on user engagement in online mental health communities. *International Journal of Environmental Research and Public Health, 18*(13), 6855. DOI: 10.3390/ijerph18136855 PMID: 34206719

Choraś, M., Pawlicki, M., & Kozik, R. (2019). The feasibility of deep learning use for adversarial model extraction in the cybersecurity domain. In Yin, H., Camacho, D., Tino, P., Tallón-Ballesteros, A., Menezes, R., & Allmendinger, R. (Eds.), Lecture Notes in Computer Science: Vol. 11872. *Intelligent Data Engineering and Automated Learning–IDEAL 2019*. Springer. DOI: 10.1007/978-3-030-33617-2_36

Citron, D. K. (2014). *Hate crimes in cyberspace.* Harvard University Press. DOI: 10.4159/harvard.9780674735613

Colley, S. (2014). Ethics and digital heritage. In *The ethics of cultural heritage* (pp. 13–32). Springer New York.

Corti, L., & Bishop, L. (2020). Ethical issues in data sharing and archiving. *Handbook of research ethics and scientific integrity*, 403-426.

Crawford, K. (2021). *Atlas of AI: Power, politics, and the planetary costs of artificial intelligence.* Yale University Press.

Cross, D., Lester, L., & Barnes, A. (2019). Digital interventions in cyberbullying prevention: Evaluating the "Reword" initiative. *Australian Journal of Educational Technology*, *35*(1), 65–80.

Cruz, J., & Plaisance, P. L. (2021). Virtue ethics and a technomoral framework for online activism. *International Journal of Communication*, *15*, 19.

D'Errico, F., & Paciello, M. (2018). Online moral disengagement and hostile emotions in discussions on hosting immigrants. *Internet Research*, *28*(5), 1313–1335. DOI: 10.1108/IntR-03-2017-0119

Daniels, J. (2018). The algorithmic rise of the "alt-right.". *Contexts*, *17*(1), 60–65. DOI: 10.1177/1536504218766547

De George, R. T. (1999). *Business ethics.* Prentice Hall.

Dewey, J. (2008). *The later works of John Dewey, volume 7, 1925-1953: Ethics.* J. Boydston (Ed.). Carbondale, IL: Southern Illinois University Press.

Diepeveen, S., & Pinet, M. (2022). User perspectives on digital literacy as a response to misinformation. *Development Policy Review*, *40*(S2), e12671. DOI: 10.1111/dpr.12671

Dredge, R., Gleeson, J., & de la Piedad Garcia, X. (2020). Social responsibility in online interactions: The impact of the #IAmAWitness campaign. *Cyberpsychology & Behavior*, *23*(2), 78–95.

Dufva, T., & Dufva, M. (2019). Grasping the future of the digital society. *Futures*, *107*, 17–28. DOI: 10.1016/j.futures.2018.11.001

Edwards, R., & Weller, S. (2016). Ethical dilemmas around anonymity and confidentiality in longitudinal research data sharing: the death of Dan. In *Qualitative ethics in practice* (pp. 97–108). Routledge.

Elliott, D., & Spence, E. H. (2017). *Ethics for a digital era.* John Wiley & Sons. DOI: 10.1002/9781118968888

Fletcher, R., Cornia, A., Graves, L., & Kleis Nielsen, R. (2018, February). *Measuring the reach of "fake news" and online disinformation in Europe.* Reuters Institute Factsheet. In *Reuters Institute for the Study of Journalism.* University of Oxford. https://reutersinstitute.politics.ox.ac.uk/our-research/measuring-reach-fake-news-and-online-disinformation-europe

Floridi, L. (2019). *The ethics of information.* Oxford University Press.

Floridi, L., & Cowls, J. (2022). Ethical guidelines for trustworthy AI: Principles, practices, and challenges. *AI & Society*, *37*(1), 1–18.

Floridi, L., & Taddeo, M. (2020). *The ethics of artificial intelligence: Principles, challenges, and opportunities*. Oxford University Press.

Formosa, P. (2023). A principlist-based study of the ethical design and acceptability of artificial social agents. *International Journal of Human-Computer Studies*, *2023*(172), 102980.

Friesem, Y. (2016). Empathy for the digital age: Using video production to enhance social, emotional, and cognitive skills. In *Emotions, technology, and behaviors* (pp. 21–45). Academic Press. DOI: 10.1016/B978-0-12-801873-6.00002-9

Gagliardone, I., Diepeveen, S., Findlay, K., Olaniran, S., Pohjonen, M., & Tallam, E. (2021). Demystifying the COVID-19 infodemic: Conspiracies, context, and the agency of users. *Social Media + Society*, *7*(3), 20563051211044233. DOI: 10.1177/20563051211044233

Gallagher, B. (2016), "What impact does fake news have on the real world?", Mary Jane. December 5th, available at: https://merryjane.com/culture/fake-news-internetimpact

Ge, X. (2020). Social media reduce users' moral sensitivity: Online shaming as a possible consequence. *Aggressive Behavior, 46*(5), 359–369. . 21904.DOI: 10. 1002/ ab

Gillespie, T. (2018). *Custodians of the internet: Platforms, content moderation, and the hidden decisions that shape social media*. Yale University Press.

Gotterbarn, D., Bruckman, A., Flick, C., Miller, K. W., & Wolf, M. J. (2018). ACM Code of Ethics and Professional Conduct. *Communications of the ACM*, *61*(1), 19–26.

Graham, C. M. (2009). Communications technology, emotional intelligence, and impact on performance: A conceptual exploration of connections. *Advances in Developing Human Resources*, *11*(6), 773–783. DOI: 10.1177/1523422309360695

Grondin, F., Lomanowska, A. M., & Jackson, P. L. (2019). Empathy in computer-mediated interactions: A conceptual framework for research and clinical practice. *Clinical Psychology : a Publication of the Division of Clinical Psychology of the American Psychological Association*, *26*(4), e12298. DOI: 10.1111/cpsp.12298

Guess, A. M., Lerner, M., Lyons, B., Montgomery, J. M., Nyhan, B., Reifler, J., & Sircar, N. (2020). A digital media literacy intervention increases discernment between mainstream and false news in the United States and India. *Proceedings of the National Academy of Sciences of the United States of America*, *2020*(27), 15536–15545. DOI: 10.1073/pnas.1920498117 PMID: 32571950

Harris, T. (2019). *The Center for Humane Technology: Ethical tech design principles.*

Harrison, J. (2023). The role of virtue ethics in digital citizenship education. *Journal of Digital Ethics*, *15*(2), 112–127.

Harrison, T. (2022). A new educational model for online flourishing: A pragmatic approach to integrating moral theory for cyber-flourishing. *Pastoral Care in Education*, *40*(2), 128–151. DOI: 10.1080/02643944.2021.1898665

Harrison, T., & Polizzi, G. (2021). drawing on moral theory to advance digital citizenship education.

He, Q., Hong, Y., & Raghu, T. S. (2024). Platform Governance with Algorithm-Based Content Moderation: An Empirical Study on Reddit. *Information Systems Research*, isre.2021.0036. DOI: 10.1287/isre.2021.0036

Huang, C., & Chou, C. (2021). Cyberbullying and cultural values: A cross-national comparison. *International Journal of Cyber Behavior, Psychology and Learning*, *11*(4), 1–15.

Israel, M. J., & Amer, A. (2021). Ethical implications of digital infrastructures for pluralistic perspectives. *Ethics and Information Technology*, *23*(3), 399–417. DOI: 10.1007/s10676-021-09582-2

Jankowicz, N., Hunchak, J., Pavluic, A., Davies, C., Pierson, S., & Kaufmann, Z. (2021). *Malign creativity: How gender, sex and lies are weaponized against women online*. Wilson Centre. https://www.wilsoncenter.org/publication/malign-creativity-how-gender-sex-and-lies-are-weaponized-against-women-online

Johansson, S., & Felmlee, D. (2022). Digital responsibility in Swedish education: A virtue ethics approach. *Scandinavian Journal of Educational Research*, *66*(2), 189–205.

Johnson, D. G., Powers, T., & Turner, J. (2021). Ethical considerations in online community management: Fostering inclusivity and respect in digital spaces. *Journal of Digital Ethics*, *13*(3), 145–158.

Jones-Jang, S. M., Mortensen, T., & Liu, J. (2021). Does media literacy help identification of fake news? Information literacy helps, but other literacies don't. *The American Behavioral Scientist*, *65*(2), 371–388. DOI: 10.1177/0002764219869406

Jonsson, P. O. (2011). On utilitarianism vs virtue ethics as foundations of economic choice theory. *Humanomics*, *27*(1), 24–40. DOI: 10.1108/08288661111110150

Kantanen, H., & Manninen, J. (2016). Hazy boundaries: Virtual communities and research ethics. *Media and Communication*, *4*(4), 86–96. DOI: 10.17645/mac.v4i4.576

Keller, J. (2021). A practical guide to the NIST Cybersecurity Framework. *Journal of Information Security*, *10*(2), 45–61.

Kende, A., Becker, J. C., & Hegedűs, P. (2022). Cross-cultural approaches to digital ethics: The impact of culture on online behavior. *Computers in Human Behavior*, *127*, 107060.

Keselman, D., & Saxe-Braithwaite, M. (2021, May). Authentic and ethical leadership during a crisis. [). Sage CA: Los Angeles, CA: SAGE Publications.]. *Healthcare Management Forum*, *34*(3), 154–157. DOI: 10.1177/0840470420973051 PMID: 33327784

Kindervag, J. (2020). Zero Trust architecture: Rethinking network security. *Cybersecurity Journal*, *15*(3), 34–49.

Kowalski, R. M., Giumetti, G. W., Schroeder, A. N., & Lattanner, M. R. (2019). Bullying in the digital age: A critical review. *Psychological Review*, *126*(2), 347–383.

Land, M. (2013). Toward an international law of the internet. *Harv. Int'l LJ*, *54*, 393.

Leary, M. G. (2010). Reasonable expectations of privacy for youth in a digital age. *Mississippi Law Journal*, *80*, 1035.

Lee, M. R. (2009). E-ethical leadership for virtual project teams. *International Journal of Project Management, 27*(5), 456–463. DOI: 10.1016/j.ijproman.2008.05.012

Lewandowsky, S., Ecker, U. K., & Cook, J. (2020). Beyond misinformation: Understanding and coping with the "post-truth" era. *Journal of Applied Research in Memory and Cognition*, *6*(4), 353–369. DOI: 10.1016/j.jarmac.2017.07.008

Lievens, E. (2011). Risk-reducing regulatory strategies for protecting minors in social networks. *Strategy for Telecommunications. Information and Media, 13*(6), 43–54. DOI: 10.1108/14636691111174252

Livingston, S., & Sefton-Green, J. (2016). *The class*. New York University Press.

Livingstone, S., Stoilova, M., & Nandagiri, R. (2020). Children's data and privacy online: Growing up in a digital age. *New Media & Society*, *22*(8), 1365–1382.

Luke, A. (2018). Digital ethics now. *Language and Literature*, *20*(3), 185–198.

Lyu, H. (2012). Internet policy in Korea: A preliminary framework for assigning moral and legal responsibility to agents in internet activities. *Government Information Quarterly, 29*(3), 394–402. . giq. 2011. 12. 008.DOI: 10. 1016/j

Marín-López, I., Zych, I., Monks, C. P., & Ortega-Ruiz, R. (2019). Empathy, morality and social and emotional competencies in interpersonal interactions online. *Thriving in Digital Workspaces: Emerging Issues for Research and Practice*, 217-233.

Maurushat, A. (2019). *Ethical hacking* (p. 370). University of Ottawa Press/Les Presses de l'Université d'Ottawa.

Mayer-Schönberger, V., & Cukier, K. (2013). *Big data: A revolution that will transform how we live, work, and think*. Houghton Mifflin Harcourt.

Meyer, M. N. (2018). Practical tips for ethical data sharing. *Advances in Methods and Practices in Psychological Science*, *1*(1), 131–144. DOI: 10.1177/2515245917747656

Meza, M. B., Luengo-Charath, X., Arancibia, M., & Madrid, E. (2018). Council for International Organizations of Medical Sciences (CIOMS) Ethical Guidelines: Advancements and unsolved topics in 2016 upgrade. *Medwave*, *18*(2), e7208. DOI: 10.5867/medwave.2018.02.7208 PMID: 29701729

Mittelstadt, B. D., Allo, P., Taddeo, M., Wachter, S., & Floridi, L. (2016). The ethics of algorithms: Mapping the debate. *Big Data & Society*, *3*(2), 1–21. DOI: 10.1177/2053951716679679

Nakamura, Y. (2021). Fostering digital empathy: The effectiveness of Japan's Kind Speech Movement. *Asian Journal of Communication*, *31*(4), 327–342.

Nambisan, P. (2011). Information seeking and social support in online health communities: Impact on patients' perceived empathy. *Journal of the American Medical Informatics Association : JAMIA*, *18*(3), 298–304. DOI: 10.1136/amiajnl-2010-000058 PMID: 21486888

Nissenbaum, H. (2019). *Privacy in context: Technology, policy, and the integrity of social life*. Stanford University Press.

Noble, S. U. (2018). *Algorithms of oppression: How search engines reinforce racism*. NYU Press. DOI: 10.18574/nyu/9781479833641.001.0001

O'Reilly, M. (2020). Social media and adolescent mental health: The good, the bad and the ugly. *Journal of Mental Health (Abingdon, England)*, *29*(2), 200–206. DOI: 10.1080/09638237.2020.1714007 PMID: 31989847

OECD. (2013). *Guidelines on the protection of privacy and transborder flows of personal data*. OECD Publishing.

Ostanina, E., & Titova, E. (2023). Legitimate Expectations of Privacy in the Era of Digitalization. *BRICS LJ*, *10*(1), 109–125. DOI: 10.21684/2412-2343-2023-10-1-109-125

Papacharissi, Z. (2015). A private sphere: Democracy in a digital age. *Polity*.

Paris, C., Colineau, N., Nepal, S., Bista, S. K., & Beschorner, G. (2013). Ethical considerations in an online community: The balancing act. *Ethics and Information Technology*, *15*(4), 301–316. DOI: 10.1007/s10676-013-9315-4

Pariser, E. (2011). *The filter bubble: What the Internet is hiding from you*. Penguin.

Patkos, T., Flouris, G., Papadakos, P., Bikakis, A., Casanovas, P., González-Conejero, J., & Stan, A. (2015, September). Privacy-by-norms privacy expectations in online interactions. In *2015 IEEE International Conference on Self-Adaptive and Self-Organizing Systems Workshops* (pp. 1-6). IEEE.

Pawlicka, A., Choraś, M., Kozik, R., & Pawlicki, M. (2021). First broad and systematic horizon scanning campaign and study to detect societal and ethical dilemmas and emerging issues spanning over cybersecurity solutions. *Personal and Ubiquitous Computing*, •••, 2021.

Pawlicki, M., Choraś, M., & Kozik, R. (2020). Defending network intrusion detection systems against adversarial evasion attacks. *Future Generation Computer Systems*, *2020*(110), 148–154. DOI: 10.1016/j.future.2020.04.013

Piliang, Y. A. (2015). Redefining Ethics and Culture in the Virtual World. *Melintas*, *31*(3), 236–251. DOI: 10.26593/mel.v31i3.1917.236-251

Polizzi, G., & Taylor, R. (2019). *Misinformation, digital literacy and the school curriculum. Media Policy Brief 22*. Media Policy Project, London School of Economics and Political Science.

Powell, P. A., & Roberts, J. (2017). Situational determinants of cognitive, affective, and compassionate empathy in naturalistic digital interactions. *Computers in Human Behavior*, *68*, 137–148. DOI: 10.1016/j.chb.2016.11.024

Quan-Haase, A. (2016). *Technology and society: Social networks, power, and inequality* (2nd ed.). Oxford University Press.

Reiss, M. (2010). Ethical Thinking. In *Ethics in the Science and Technology Classroom*. Brill. DOI: 10.1163/9789460910715_003

Reynolds, T., & Cerutti, E. (2021). *Digital ethics: A framework for responsible technology use*. Palgrave Macmillan.

Richards, N. M. (2013). The dangers of surveillance. *Harvard Law Review, 126*(7), 1934–1965.

Riso, B., Tupasela, A., Vears, D. F., Felzmann, H., Cockbain, J., Loi, M., Kongsholm, N. C. H., Zullo, S., & Rakic, V. (2017). Ethical sharing of health data in online platforms–which values should be considered? *Life Sciences, Society and Policy, 13*(1), 1–27. DOI: 10.1186/s40504-017-0060-z PMID: 28825221

Robbins, L. (1932). *An Essay on the Nature and Significance of Economic Science.* Macmillan.

Salminen, J., Kaate, I., Kamel, A. M. S., Jung, S. G., & Jansen, B. J. (2021). How does personification impact ad performance and empathy? An experiment with online advertising. *International Journal of Human-Computer Interaction, 37*(2), 141–155. DOI: 10.1080/10447318.2020.1809246

Santiago-Torner, C., Corral-Marfil, J. A., & Tarrats-Pons, E. (2024). Relationship Between E-Ethical Leadership and Emotional Exhaustion: A Moderated Mediation Model.

Schoenbrunner, A., Gosman, A., & Bajaj, A. K. (2019). Framework for the creation of ethical and professional social media content. *Plastic and Reconstructive Surgery, 144*(1), 118e–125e. DOI: 10.1097/PRS.0000000000005782 PMID: 31246836

Schwartz, S. H., & Bilsky, W. (1987). Toward a universal psychological structure of human values. *Journal of Personality and Social Psychology, 53*(3), 550–562. DOI: 10.1037/0022-3514.53.3.550

Selwyn, N. (2021). *Digital education and ethical challenges.* Routledge.

Selwyn, N., & Aagaard, J. (2021). Banning mobile phones from classrooms—An opportunity to advance understandings of technology addiction, distraction and cyberbullying. *British Journal of Educational Technology, 52*(1), 8–19.

Smith, A. (2020). The impact of the California Consumer Privacy Act on data protection. *Technology in Society, 42*(2), 113–128.

Solove, D. J. (2020). *Privacy law fundamentals.* IAPP.

Steiner, J. F., & Steiner, G. A. (1999). *Business, government, and society: A managerial perspective.* McGraw-Hill.

Subhashini, V. (2014). Ethical hacking and legal systems. [IJETCSE]. *Internal Journal of Emerging Technology in Computer & Electronics, 11*(4), 36–40.

Suler, J. (2004). The online disinhibition effect. *Cyberpsychology & Behavior, 7*(3), 321–326. DOI: 10.1089/1094931041291295 PMID: 15257832

Sunstein, C. R. (2017). *Republic: Divided democracy in the age of social media.* Princeton University Press. DOI: 10.1515/9781400884711

Tafesse, W., & Dayan, M. (2023). Content creators' participation in the creator economy: Examining the effect of creators' content sharing frequency on user engagement behavior on digital platforms. *Journal of Retailing and Consumer Services, 73*, 103357. DOI: 10.1016/j.jretconser.2023.103357

Tiidenberg, K. (2018). Ethics in digital research. *The SAGE handbook of qualitative data collection*, 466-479.

Tiwary, D. K., & Pradesh, U. (2011). Security and ethical issues in it: An organization's perspective. *International Journal of Enterprise Computing and Business Systems*, *1*(2), 2230–8849.

UNESCO. (2019). *Internet governance and ethical principles: The ROAM-X model*. UNESCO Publishing.

United Nations. (2019). *The right to privacy in the digital age*.

van Dijk, J. (2020). The digital divide. *Polity*.

Vanacker, B., & Heider, D. (2012). Ethical harm in virtual communities. *Convergence (London)*, *18*(1), 71–84. DOI: 10.1177/1354856511419916

Villegas-Simón, I., Anglada-Pujol, O., Lloveras, M. C., & Oliva, M. (2023). "I'm Not Just a Content Creator": Digital Cultural Communicators Dealing with Celebrity Capital and Online Communities. *International Journal of Communication*, *17*, 19.

Voigt, P., & Von dem Bussche, H. (2017). *The EU General Data Protection Regulation (GDPR): A practical guide*. Springer. DOI: 10.1007/978-3-319-57959-7

Von Solms, R., & Van Niekerk, J. (2019). Information security management: Standards and best practices. *Computers & Security*, *88*, 101–116.

Ward, C. L., Shaw, D., Sprumont, D., Sankoh, O., Tanner, M., & Elger, B. (2018). Good collaborative practice: Reforming capacity building governance of international health research partnerships. *Globalization and Health*, *14*(1), 1–6. DOI: 10.1186/s12992-017-0319-4 PMID: 29310698

Ward, S. J. (2014). Radical Media Ethics: Ethics for a global digital world. *Digital Journalism (Abingdon, England)*, *2*(4), 455–471. DOI: 10.1080/21670811.2014.952985

Wardle, C., & Derakhshan, H. (2017). *Information disorder: Toward an interdisciplinary framework for research and policy making*. Council of Europe.

Wei, C., Yu, Z., & Li, Y. (2021). Empathy impairs virtue: The influence of empathy and vulnerability on charitable giving. *Internet Research*, *31*(5), 1803–1822. DOI: 10.1108/INTR-07-2020-0407

Wilden, A. (1972). *System and structure: Essays on communications and exchange*. Tavistock.

Wood, N. (2019). *Virtue Rediscovered: Deontology*. Consequentialism, and Virtue Ethics in the Contemporary Moral Landscape.

Wright, L. (2022). Automated platform governance through visibility and scale: On the transformational power of AutoModerator. *Social Media + Society*, *8*(1), 20563051221077020. DOI: 10.1177/20563051221077020

Yeke, S. (2023). Digital intelligence as a partner of emotional intelligence in business administration. *Asia Pacific Management Review*, *28*(4), 390–400. DOI: 10.1016/j.apmrv.2023.01.001

Zeng, J., & Kaye, D. B. V. (2022). From content moderation to visibility moderation: A case study of platform governance on TikTok. *Policy and Internet*, *14*(1), 79–95. DOI: 10.1002/poi3.287

Zuboff, S. (2019). *The age of surveillance capitalism: The fight for a human future at the new frontier of power*.

Chapter 6
How Can Individuals Develop Critical Thinking Skills to Evaluate Online Information

Yuanlu Xiong
https://orcid.org/0009-0007-8134-2713
Chongqing Normal University, China

Zhanni Luo
https://orcid.org/0000-0002-5162-875X
Chongqing Normal University, China

ABSTRACT

The rise of the internet and digital platforms has revolutionized the production and sharing of information, allowing individuals to express their opinions quickly and effortlessly. However, the ease of creating and disseminating content has also resulted in the spread of misleading and irrelevant information. To address this challenge, this chapter outlines five criteria for evaluating credibility, three steps for assessing relevance, and six strategies to enhance the evaluation process. These tools aim to foster critical thinking and improve digital literacy, enabling individuals to manage the complexities of online information effectively.

SECTION 1: BACKGROUND

1. Why Do Individuals Need to Evaluate Online Information?

In the digital age, the advent of the internet and digital platforms has dramatically transformed how information is produced and shared, greatly simplifying the process for individuals to share their opinions and insights with minimal effort (Van Zyl et al., 2020). However, the ease of information creation and dissemination has also led to a significant increase in misleading and inaccurate content (Gong et al., 2023). Consequently, recognizing these factors of the rising amounts of disinformation is essential for highlighting the importance of rigorous online information evaluation, ensuring that users can effectively discern the reliability and accuracy of the content they consume.

DOI: 10.4018/979-8-3693-6675-2.ch006

Copyright ©2025, IGI Global Scientific Publishing. Copying or distributing in print or electronic forms without written permission of IGI Global Scientific Publishing is prohibited.

1) User proliferation and freedom of speech

User proliferation refers to an increase in ordinary users accessing and participating in information dissemination on digital platforms. In the digital age, the barriers to accessing devices and online platforms have significantly decreased, enabling a broader range of individuals to express their opinions freely (Kwak et al., 2021). Today, an increasing number of individuals have access to electronic devices, with statistics showing that 90% of the population can access digital platforms. Unlike in the past, when only educated or reputable professionals could publish their views, the current democratization of information production has presented a significant challenge in distinguishing reliable information from misinformation or false content.

2) Manipulative content creation

By utilizing the characteristics of user proliferation and freedom of speech, manipulators can more easily publish and spread misleading information. These manipulators, who may be individuals, organizations, or even states, exploit the openness of digital platforms and users' trust to influence public opinion, create confusion, or achieve specific goals with carefully designed content, such as creating polarization and advancing political motives. For instance, during the 2016 U.S. presidential election, fake news circulated more extensively on Facebook than authentic reports. Moreover, many commentators pointed out that online disinformation played a decisive role in that election. Consequently, political and ideological motivations frequently drive the production and dissemination of manipulative content for specific purposes, thus intensifying social polarization and political divisions.

3) Anonymity and identity masking

On digital platforms, anonymity allows individuals to freely express opinions and share information without revealing their identities in real life. However, creators of malicious information also exploit anonymity. By concealing their true identities, information creators can pose as experts or authorities to enhance the credibility of their information. Moreover, due to anonymity, creators can operate multiple accounts on various digital platforms, creating a false sense of “consensus” or “public opinion”. As a result, anonymity and identity masking complicate the efforts of digital platforms and law enforcement agencies to determine the real identities of information creators, thereby hindering the tracing of information origins and the prevention of the spread of misleading information

4) Algorithmic bias

An algorithm is a set of rules designed to perform a specific task or solve a problem. On social media platforms, the algorithms determine what information users see. Initially, platform algorithms recommend content based on the user’s past behavior and interests, creating “filter bubbles”. In these bubbles, users primarily encounter content that aligns with their existing preferences, which restricts users’ exposure to a broader range of information. At the same time, people tend to select information that reinforces their beliefs, leading to “information cocoons” that solidify their pre-existing views. Within social groups, the same views get repeated and amplified, forming “echo chambers” where only similar opinions are circulated and intensified, pushing users towards more extreme viewpoints. Driven by algorithms, a vicious cycle emerges that hinders users from encountering diverse and conflicting viewpoints, ultimately leading to the homogenization of information and an increase in misinformation.

5) Lack of review and supervision mechanisms

The lack of effective review and supervision mechanisms on digital platforms enables the rapid spread of misinformation and disinformation. In terms of feasibility, thousands of pieces of content are published daily, making it challenging for platforms to verify the credibility and relevance of each one. Additionally, the speed at which information is published is so fast that traditional review processes cannot keep up, resulting in false information being widely spread before it can be reviewed. In terms of efficiency, comprehensive review and supervision of content require substantial human, material, and financial resources. To reduce operating costs, many digital platforms rely on automated review tools, which often struggle to identify complex misleading information effectively. However, these tools frequently fall short of effectively detecting and mitigating complex forms of misleading information, leaving gaps in content accuracy and integrity.

Given the causes mentioned above, enhancing the credibility of online information has become a significant trend (Van Zyl et al., 2020). Two main approaches have been proposed to tackle the rising amounts of misleading or false content. The first approach is the "top-down" model, which involves restricting access from the platform owners. For example, on platforms like Facebook and Twitter, algorithms are used to analyze vast amounts of information to delete or flag suspicious content (Zannettou et al., 2019). However, the effectiveness of the "top-down" model is limited (Colliander, 2019), as the users are not motivated to collaborate in supporting this approach, making it probable that they will overlook the platform's warnings.

Therefore, another approach has been proposed: the "bottom-up" model, which aims to enhance the efficiency of information evaluation from the individual side (Marin & Copeland, 2022). This model emphasizes empowering individuals with the skills necessary for effective online navigation and critical assessment of content. It underscores the development of digital literacy and media literacy, as supported by literature highlighting the importance of equipping users with the strategies to discern the reliability and credibility of information (Hobbs & Frost, 2023). The shift towards individual empowerment reflects a broader educational goal to encourage users to develop and apply critical thinking skills, enable them to make informed decisions and engage constructively in digital spaces.

2. What is Critical Thinking?

Critical thinking is generally viewed as self-directed thinking (Paul & Elder, 2014). It involves a set of cognitive skills, such as curiosity, open-mindedness, or carefulness. In the context of information evaluation, critical thinking includes more specific cognitive modes such as reflective thinking and analytic thinking, as well as information literacy and media literacy. When individuals possess the skill of critical thinking, they are less likely to become participants in echo chambers (Marin & Copeland, 2022). Critical thinking enables them to question the validity of the information they encounter by recognizing biases, identifying underlying motives, and differentiating between well-supported arguments and mere opinions. Thus, exploring how to develop critical thinking to evaluate the vast amount of online information is indispensable for making informed personal decisions in the digital age.

3. How Can Individuals Develop Critical Thinking Skills to Evaluate Online Information?

Evaluating online information critically requires considering two factors: information credibility and information relevance. The evaluation of credibility focuses on five main criteria, while relevance evaluation involves three key steps. Due to the questioning essence of critical thinking skills, this chapter presents a series of questions to facilitate self-checks, helping us learn how to distinguish between factual information and disinformation in a more authentic environment. Additionally, given limitations such as time or energy, there are six cognitive shortcuts for information evaluation to help us quickly find the information we need through various irrelevant and false information. Notably, given the risks associated with each shortcut, we need to understand the considerations when individuals utilize the six cognitive heuristics to enhance the effectiveness of online information evaluation (see Figure 1).

Figure 1. The evaluation framework

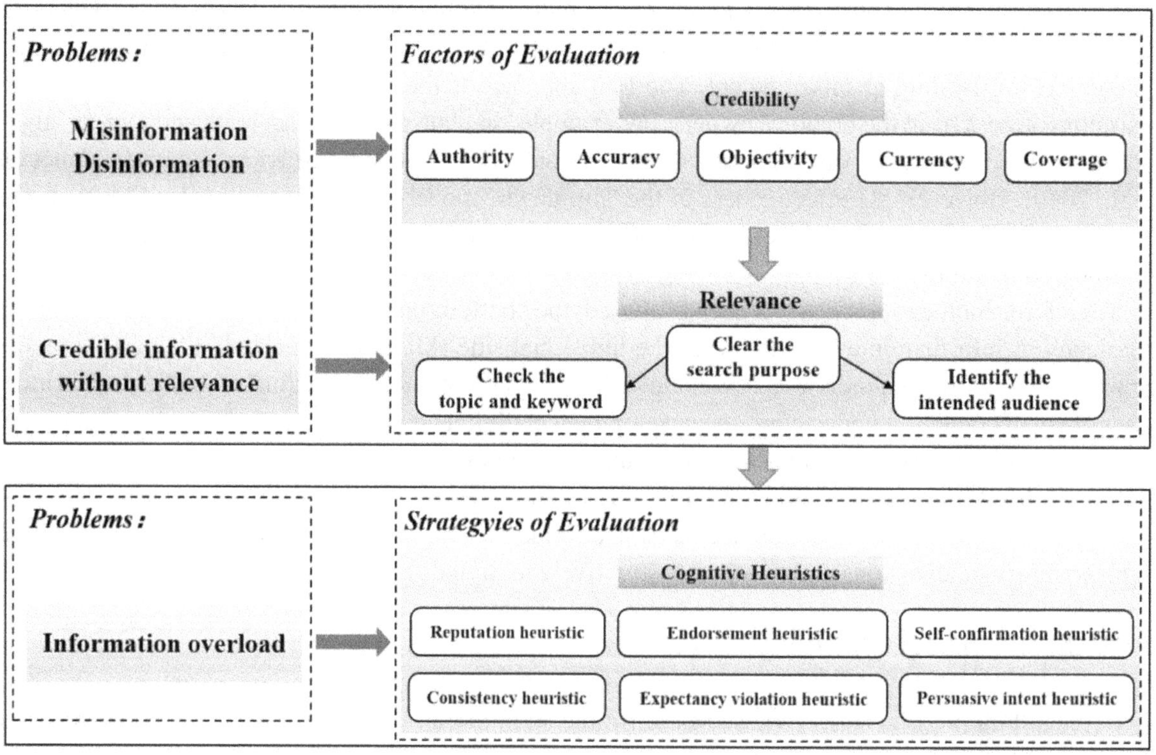

SECTION 2: FIVE CRITERIA FOR CREDIBILITY EVALUATION

The credibility of online information refers to the believability of a source, which is significantly influenced by how readers perceive the source's trustworthiness and expertise. Due to the prevalence of unverified information on the internet and the associated personal and social influence, several models have been developed to assess online credibility. Among them, the AAOCC model proposed by Kapoun (1998) is widely applied. The model includes five criteria: Authority, Accuracy, Objectivity, Currency, and Coverage. When evaluating online information, we need to consider whether the source is authoritative, the content is accurate, the perspective is objective, the information is up-to-date, and the coverage comprehensively addresses the necessary content (see Table 1).

1. Authority

When we evaluate information credibility, the first criterion to consider is the authority of information. Authority refers to the extent to which the content is persuasive based on the credibility of its authors and publishers. It is also concerned with whether the website providing the content is recommended or endorsed by a trusted source (Van Zyl et al., 2020). Although individuals lacking recognized authority can produce credible information, verifying authority is essential as it enables us to filter through vast amounts of data efficiently.

Specifically, authority evaluation mainly involves reviewing the authors of the information, the institution publishing the information, and the website where the information is found. Firstly, author evaluation needs to examine the author's social standing and professional background, helping us check whether the author is qualified to write on the topic. Secondly, institution evaluation involves reviewing the reputation and operational status of the institution that published the information. An institution with a good reputation and stable operation is more likely to provide reliable information. Lastly, website evaluation refers to checking whether the site providing the information is an official or trustworthy source, as official or professional websites usually have higher authority and reliability.

When assessing information authority, we can try asking ourselves several questions. For example, we can ask, *"Is there a note or paragraph describing the author's credentials?"* A note or paragraph describing the author's credentials can provide direct evidence of his or her educational background, work experience, and professional certifications, ensuring he or she can address the topic. If there is no relevant qualification description, the authority of the information may be questioned. Additionally, we can ask, *"Has the author been cited in other bibliographies?"* The author's citation in other bibliographies can reflect his or her influence. Generally, we believe that an author frequently cited in scholarly or professional contexts likely possesses greater authority in their field, thereby increasing the information's credibility. Furthermore, we can ask, *"Has the information been peer-reviewed or gone through the editing process?"* For instance, publishing a journal article requires a peer-review process, during which experts rigorously scrutinize the content. As a result, journal content generally possesses greater reliability and scientific validity, minimizing biases and enhancing the quality of information.

Addressing these questions can help us assess the authority of online information, enhancing our chances of using reliable content and reducing the risk of misguidance and errors from unreliable sources.

2. Accuracy

The second criterion to consider is the accuracy of information. Accuracy refers to the extent to which the content is devoid of mistakes and whether the information can be verified by other sources (Van Zyl et al., 2020). It is an important indicator of the reliability of online information.

Accuracy evaluation of the information involves assessing the correctness, consistency, completeness, and verifiability. Correctness refers to the number of errors or mistakes within the information. Consistency refers to the degree to which information aligns across multiple sources. Completeness indicates accurate information should cover all necessary aspects of a topic, such as relevant facts and data, rather than partial or selective details. Verifiability requires that the information be supported by tangible evidence, such as citations, references, and links to sources or data.

When we evaluate the accuracy of online information, we consider asking ourselves several targeted questions. To ensure correctness, we can ask, *"Does the information contain any spelling and grammar errors?"* For instance, if a professional article has misspellings and grammatical mistakes, such as confusing "their" with "there" or incorrect use of tenses, it might undermine the accuracy of the content. To assess consistency, we can ask, *"Can the information be searched elsewhere?"* If multiple credible sources report the same information, it indicates higher credibility. For instance, if several news agencies independently report the same details of an event, the consistency across different sources supports the accuracy of information. To check for completeness, we can ask, *"Does the information provided have missing details?"* Accurate information should be comprehensive and cover all necessary aspects of the topic. For example, if an article only discusses the benefits of physical fitness without mentioning its potential risks or precautions, it will lack completeness and raise questions about its accuracy. Lastly, to verify the information, we can ask, *"Are conclusions based on research or actual figures that can be verified offline?"* Accurate information should ideally be substantiated by data or research that can be validated. For example, if a report that provides original experimental data and detailed methodology allows other researchers to replicate the experiment to verify the results, it will enhance the information's verifiability.

In summary, evaluating accuracy is important because it involves multiple aspects of the information. It requires users to compare the consistency of information across multiple sources, assess whether the content is comprehensive and detailed, and verify the information through independent sources. The thorough approach is effective for differentiating reliable information from misleading or incorrect content, thereby enhancing the accuracy of the information used.

3. Objectivity

Objectivity is the third criterion for evaluating information, referring to assessing the impartiality and neutrality of the information provided (Metzger & Flanagin, 2013). Evaluating objectivity encompasses several key aspects. *The first aspect* is recognizing the purpose for sharing the information, as identifying the author's motivations can reveal potential biases. For example, information created to sell a product or service might be less objective. *The second aspect* is differentiating whether the content is based on fact or opinion. Facts can be checked and confirmed, whereas opinions are subjective and may reflect personal views and emotions. *The third aspect* is exploring potential conflicts of interest, such as whether the author or publisher gains financially from the content, which could influence objectivity.

When evaluating the objectivity of online information, we should consider, *"Does the writing use inflammatory or biased language?"* Typically, neutral and factual language will enhance the objectivity of information. For instance, a news article using terms like "radical" or "extremist" without context may be attempting to provoke an emotional response rather than providing objective information. Then, we can ask, *"Is there any advertising or solicitation for donations associated with the source?"* The advertisements or donation requests indicate a commercial motive, which might influence the objectivity of the information. For example, a blog post on a health supplement website may promote the benefits of their products while ignoring any potential side effects, driven by the intent to increase sales. Additionally, it's crucial to determine, *"Whether the content is based on fact or opinion?"* If information is mainly fact-based rather than opinion-based, its objectivity may be enhanced. For instance, an article that cites scientific research includes statistical data, and provides references generally maintains a higher degree of objectivity. Notably, we can also ask, *"Can the information be accessed fully without restrictions such as fees, browser technology, or software requirements?"* For example, educational websites that require subscriptions or specialized software to access key scientific articles can limit the accessibility and perceived objectivity of the content, as they may prevent users from verifying the information independently or comparing it with other sources.

By scrutinizing the author's intent, distinguishing between factual content and opinion, and identifying commercial motivations or conflicts of interest, the information we rely on will be more unbiased and objective, thereby protecting us from the influence of skewed or deceptive sources.

4. Currency

Currency refers to the timeliness of the information, including the frequency of updates and whether the date of the content is visible. In the digital age, information is updated at an unexpected pace, making it necessary to regularly verify the currency of the information we encounter, especially in rapidly changing fields such as technology, science, and current events.

Evaluating the currency of online information involves three key points. *The first* is the creation date, which helps assess the currency of the information. *The second* is the revision date. If the latest revision date is available, it indicates whether the content has been updated to reflect new changes or developments. *The last* point is link validity. Ensuring that all provided links are functional is key to avoiding unreliable or outdated information.

When evaluating the currency of online information, it is crucial to determine, *"Is the revision date appropriate for the subject (not out of date)?"* For instance, financial data from five years ago might be invalid for current market analysis, whereas historical research may remain valid much longer. Additionally, we should ask, *"Are all links provided current?"* Active or current links mean the content is regularly maintained and updated whereas broken links may indicate neglect and outdatedness. For instance, if a health information website links to resources on recent medical studies and all links lead to active, current research articles, this indicates that the website is up-to-date and reliable. Furthermore, it is beneficial to query, *"Is the website providing current information?"* For example, a news website that has not updated its featured articles for several months may not be the best source for breaking news or recent developments.

In the digital age, the rapid updating of online information presents unique challenges to its evaluation. In summary, the currency focuses on the timeliness of the content. By checking the creation and revision dates and ensuring links are functional, individuals are more likely to access current and credible information

5. Coverage

Coverage refers to the depth and breadth of the information provided. The rapid increase in online users indicates that a single event can generate a wide array of viewpoints. Therefore, coverage evaluation examines how thoroughly the topic is addressed and whether the information encompasses all necessary aspects for a complete explanation.

Typically, evaluating coverage involves considering two elements, the scope and depth of information. Information scope refers to the range of topics covered within the subject matter, while information depth refers to the level of detail provided about each topic. Notably, both coverage and accuracy are aimed to ensure thoroughness of information, but they approach it from different angles. Coverage focuses on breadth and depth, ensuring all relevant topics and perspectives are included, whereas accuracy requires the precision and correctness of the details, emphasizing evidence and fact.

When evaluating the coverage of online information, we should determine, *"Does the information source leave questions unanswered (ask the "five W and H" to check: who, what, when, where, why, and how)?"* For example, a news article about a natural disaster should detail who was affected, what occurred, when and where it took place, why it happened, and how it impacted the community. Additionally, when evaluating coverage of a political event, it is important to question, *"Are multiple perspectives included?"* A thorough analysis should incorporate views from various stakeholders and political affiliations to provide a balanced and equitable perspective. Thus, evaluating coverage involves not only ensuring comprehensiveness but also fairness, reflecting a diverse range of viewpoints and minimizing bias.

By rigorously assessing the coverage of online information, we enhance our ability to obtain content that fully addresses all essential aspects of the topic, providing us with the necessary detail for a more comprehensive understanding.

The AAOCC model provides a comprehensive framework for evaluating online information through five essential criteria (Authority, Accuracy, Objectivity, Currency, and Coverage). Each of these criteria targets a vital aspect of information quality. Authority focuses on the credentials and reputation of the content creator and publisher. Accuracy ensures that information is free from errors and aligns with verifiable facts. Objectivity examines the impartiality of content, checking for biases or underlying motives. Currency assesses the timeliness of the information, ensuring it remains up-to-date. Lastly, Coverage looks at the depth and scope of the topic to ensure a thorough exploration of all essential facets. By adhering to these standards, individuals can distinguish reliable sources from misleading ones, thereby enhancing the ability to make informed decisions.

Table 1. Checklist for credibility evaluation of information

Criteria to consider	Questions to ask
Authority	✓ Is the author's first and last name clearly identifiable? ✓ Is there a note or paragraph describing the author's credentials? ✓ Is there any indication of the author's education, other publications, professional affiliations, or experience? ✓ Is the author qualified to address the topic? ✓ Has the author been cited in other bibliographies? ✓ Is the publisher/institution identifiable? ✓ Has the information been peer-reviewed or gone through the editing process?
Accuracy	✓ Can the information be searched elsewhere? ✓ Does the information contain any spelling and grammar errors? ✓ Does the website contain links to reputable outside sources for additional information? ✓ Does the site have citations for borrowed materials? ✓ Does the information provided have missing details? ✓ Are conclusions based on research or actual figures that can be verified offline? ✓ Are sources of information listed in foot/end notes, bibliographies, or lists of references?
Objectivity	✓ Whether the content is based on fact or opinion ✓ Does the writing use inflammatory or biased language? ✓ Does the author provide more than one point of view? ✓ Is the site unbiased or are biases easy to identify? ✓ Is there any advertising or solicitation for donations associated with the source? ✓ Is the site free of advertisements? ✓ Can the information be accessed fully without restrictions such as fees, browser technology, or software requirements?
Currency	✓ Is the creation date provided? ✓ Is the latest revision date provided? ✓ Are all links provided current? ✓ Is the revision date appropriate for the subject (not out of date)? ✓ Is the web of information up-to-date?
Coverage	✓ Does the information source leave questions unanswered (ask the "five W and H" to check: who, what, when, where, why and how)? ✓ Is the information complete? ✓ Are multiple perspectives included? ✓ Does the information source provide sufficient coverage of the topic?

Putting it all together

✓ **Authority.** If the online information lists the author's credentials and the domain is preferred (.edu, .gov, .org, or .net), and...

✓ **Accuracy.** If the online information includes details about the author and the institution that published the content, providing a way to contact them, and...

✓ **Objectivity.** If the online information presents accurate data with minimal advertising and maintains an objective stance, and...

✓ **Currency.** If the online information is current, updated regularly, and all associated links (if any) are also up-to-date, and...

✓ **Coverage.** If the online information adequately covers the topic and multiple perspectives, then…

SECTION 3: THREE STEPS FOR RELEVANCE EVALUATION

However, does credible information always meet the users' needs? In reality, the internet is flooded with credible information that may not be relevant to our specific needs. Therefore, in online information evaluation, relevance is as important as credibility. Drawing from the evaluation model developed by librarians at California State University in 2004, we have proposed a three-step process for evaluating relevance. The first step is to clear the search purpose, the second step is to check the topic and keyword of the information, and the third step is to identify the intended audience of the content (see Table 2).

1. Clear the Search Purpose

Clearing the search purpose is the first step in evaluating the relevance of online information. It enhances the efficiency of our information search by helping us avoid wasting time on irrelevant or unsuitable content, thereby ensuring the chosen information effectively supports our decision-making, research, or other needs.

Basically, we need to ask, *"What specific topic do I want to seek?"* Limiting the topic helps narrow the search scope and increase the relevance of information. For example, if I am exploring the impact of artificial intelligence (AI) on language learning, specifying the topic helps me focus on relevant information without being distracted by broader discussions on AI. Moreover, we should ask, *"What format of information will best suit my needs (e.g., articles, videos, photos)?"* For example, if I prepare a visual presentation on AI-assisted language learning, incorporating media like figures and videos will be more effective. Conversely, writing an academic paper may prefer detailed articles and research reports for in-depth analysis. Similarly, we can also ask, "*What type of information do I need (e.g., statistical data, expert opinions, case studies)?"* For instance, I might prefer searching for specific data when conducting quantitative research, while qualitative research often requires more descriptive content to explore themes and patterns.

In conclusion, different search purposes require specific content to support. By limiting the information search to specific topics, formats and types, we can filter out irrelevant content at an early stage, thereby laying a solid foundation for relevance evaluation

2. Check the Topic and Keyword

Once the search purpose is clarified, the next step is to check the topic and keyword of the online information. By matching the topic of the information with our defined purpose, we can filter out unrelated content and ensure that the content we choose is more related to our specific needs.

During the process, we should assess, *"Are the keywords used in the information relevant to my search?"* For example, if I am researching the impact of AI on language learning, keywords such as "AI", "language learning", "education technology" and "learning outcomes" indicate that the information is likely relevant to my research. Additionally, we should consider, *"Is the content related to the particular variables or demographics I am studying?"* For instance, if I am focused on the impact of social media on teenagers in rural settings, I would seek out keywords like "rural teens' social media usage" instead of "urban adolescent social media trends" or "comparative studies on social media impacts". Moreover, to delve deeper into specific aspects, we can ask, *"Does the information address the specific issues or questions I have?"* For instance, if I want to explore how AI enhances students' writing skills, I need to look for more targeted keywords like "AI writing tools", "writing improvement" and "technology in writing education" to answer my question rather than just discussing the general impact of AI on skills enhancement.

By identifying and using relevant keywords to filter information, and analyzing the topic and content, we can effectively narrow down our search to the most relevant sources. This approach not only assures the credibility of the information but also guarantees its direct applicability to our specific research goals.

3. Identify the Intended Audience

Identifying the intended audience is the third step in evaluating information relevance. Knowing who the content is created for helps us determine whether it is relevant to our needs because different audiences require varying levels of detail, complexity, and presentation style. Of course, this step can only be effectively carried out once we have clearly defined our own search purposes, ensuring that we have a precise understanding of our own identity.

To effectively determine the suitability of content, we can ask, *"Is the language level appropriate for someone with my background and experience?"* For example, if a high school student is learning about AI, an article written with massive technical jargon and complex explanations will not be suitable. Additionally, it's helpful to consider the depth of the content by asking, *"Is the information presented suited to my preferences?"* Under the same topic, academic researchers prefer scholarly content to conduct in-depth research, while the general public may favor text that is easy to understand and includes illustrations. More directly, we can verify, *"Does the site have a section about its audience or mission statement?"* For example, many websites include an "About" page that outlines their intended audience and purpose, providing clear insights into who the content is aimed at.

In summary, assessing the relevance of information is needed, as even highly credible information may be of little use if it does not meet the user's specific needs or context. After completing these three steps—clarifying the search purpose, checking the topic and keywords of the information, and identifying the intended audience of the content—the relevance of the information we retrieve is significantly enhanced.

Table 2. Checklist for relevance evaluation of information

Steps to follow	Questions to ask
Clear the search purpose	✓ What specific topic do I want to seek? ✓ What format of information will best suit my needs (e.g., articles, videos, infographics)? ✓ What type of information do I need (e.g., statistical data, expert opinions, case studies)? ✓ Are there geographical or cultural considerations in the information I need?
Check the topic and keyword	✓ Does the topic align with what I am looking for? ✓ Are the keywords used in the information relevant to my search? ✓ Does the information address the specific issues or questions I have? ✓ Do the keywords and topics match the depth of detail I require? ✓ Is the content related to the particular variables or demographics I am studying?
Identify the intended audience	✓ Is the language and style appropriate for someone with my background and experience? ✓ Is the information written in a way that is understandable for a particular group? ✓ Does the site have a section about its audience or mission statement? ✓ Is the information presented suited to my preferences?

SECTION 4: SIX COGNITIVE HEURISTICS FOR EVALUATING ONLINE INFORMATION

However, can we thoroughly evaluate all the factors mentioned above when assessing online information? The answer is no. In an environment of information overload, we prefer to use more convenient and quick ways because of a lack of cognitive capacity or time to assess systematically. Consequently, Metzger and Flanagin (2013) utilized individuals' cognitive tendencies and proposed a series of shortcuts known as cognitive heuristics. What exactly is a heuristic? Heuristic is a general cognitive framework that enables people to find solutions quickly. For example, when students choose a subject to study at university, their intuitions will probably guide them toward the path they perceive as most satisfying, practical, and interesting. As such, cognitive heuristics play a significant role in influencing and shaping our daily and professional decisions.

Specifically, Metzger and Flanagin's (2013) cognitive heuristics included four positive heuristics (the reputation heuristic, the endorsement heuristic, the consistency heuristic, and the self-confirmation heuristic) and two negative heuristics (the expectancy violation heuristic and the persuasive intent heuristic). However, some researchers argue that cognitive heuristics can lead to biased or faulty information processing. Thus, individuals must be cautious and adhere to specific guidelines when employing these heuristics during information evaluation (see Table 3).

1. The Reputation Heuristic

1) What is the "reputation heuristic"?

The reputation heuristic signals a basic psychological tendency where individuals prefer options they recognize or are familiar with over those unfamiliar, aiming to minimize the cognitive effort required for decision-making. Thus, it is also known as a "recognition heuristic".

When applied to online information evaluation, the reputation heuristic manifests as a preference that individuals are more likely to trust information sources whose names they recognize. This trust is based on the assumption that a well-known reputation is built through long-standing positive interactions and consistent good performance, particularly visible on social networks. As such, a source's good reputation or name recognition is often taken as a primary indicator of its credibility, sometimes overshadowing the need for a thorough review of the actual content. Importantly, the reputation heuristic may also be "a subset of the 'authority' evaluation in credibility assessment" (Metzger et al., 2010, p. 426), because both concepts emphasize that the key aspect in evaluating information credibility is whether the source is recognized as official or authoritative.

2) How to use the "reputation heuristic" to evaluate information?

Employing the reputation heuristic to assess online information can be approached in several ways. First, it is essential to review the history and established reputation of the information source. For example, when assessing the reliability of a news website, users might consider the length of its operational history and its past performance. Established and widely recognized news organizations, such as The New York Times, are generally viewed as more reliable. Second, consulting public feedback and reviews is another way of assessing the trustworthiness of information sources. When choosing online shopping platforms, users frequently review comments and ratings from other consumers. Platforms like Amazon, with a wealth of positive feedback, are typically perceived as trustworthy. Third, the presence of third-party authoritative certifications can greatly enhance trust in information sources. For example, financial services or payment platforms that hold security certifications or high ratings from entities like the Better Business Bureau (BBB) are usually preferred by users, as these certifications contribute to the legitimacy and safety of the sites. Additionally, utilizing the influence of social media can provide insights into the credibility of information. For instance, information shared on Instagram or Twitter is usually seen as credible because these platforms have numerous followers and a high level of active interactions. By integrating these approaches, users can more effectively apply the reputation heuristic to evaluate the reliability of online information based on established reputations and public perception.

3) What are the considerations?

While the reputation heuristic is a commonly used shortcut that can efficiently filter information sources, it is necessary to exercise critical thinking to avoid misjudgments and the pitfalls of oversimplified decision-making. Here are several important considerations:

- **Beware of over-reliance on popular sources.** Although the reputation heuristic is fast and effective, over-dependence on well-known sources might lead to the neglect of emerging or less well-known but reliable sources. For example, an emerging tech blog with fresh insights might be ignored simply because it hasn't yet gained popularity, resulting in a loss of diverse perspectives.
- **Reassess information regularly.** Reputations are not static and can change rapidly in light of new events or scandals. For instance, if a company becomes involved in a scandal, its reputation might decline overnight. Consequently, it is essential for users to continually reassess the credibility of the information they rely on to maintain information reliability.
- **Be cautious of potential manipulation.** In some cases, reputation might be artificially manipulated by aggressive marketing strategies to enhance brand visibility and create a so-called public consensus we discussed before. For example, some vendors might pay for inflated high-star ratings, misleading consumers about the quality of a product or service. Thus, users should be cautious of such manipulative practices and make decisions based on more than just reputation.

2. The Endorsement Heuristic

1) What is the "endorsement heuristic"?

The endorsement heuristic suggests that people tend to evaluate the credibility of an event based on the recommendation of others, without deeply investigating the event itself. In the context of information evaluation, this heuristic implies that if a source is considered credible by others, an individual is also likely to view it as trustworthy, bypassing a detailed examination. This heuristic is sometimes called the "liking/agreement heuristic" (Chaiken, 1987), reflecting the principle that if respected or admired individuals believe something, it is perceived as more credible.

Importantly, the endorsement heuristic is based on two other concepts. Chaiken (1987) introduced the "consensus heuristic", which believes that people generally conform to the dominant opinion within a group, suggesting that a collective agreement can significantly sway individual judgments. For example, people often trust the information on Wikipedia because it is edited and reviewed by a broad community of users. This extensive group participation creates a consensus, leading people to believe in the reliability of the content. Additionally, Sundar (2008) pointed out the "bandwagon heuristic", highlighting that people are inclined to adopt behaviors or beliefs that seem popular or are quickly gaining traction, driven by the principle that "if everyone is doing it, it must be trustworthy or correct". For example, Apple's products, such as the iPhone or MacBook, often generate widespread attention and a buying frenzy upon release. Many consumers purchase these products partly because they see others buying them. In sum, Metzger et al. (2010) believed the essence of the endorsement heuristic is the group effect that widespread acceptance is correctness, emphasizing the power of collective influence and social validation in shaping individual perceptions and decision-making.

2) How to use the "endorsement heuristic" to evaluate online information?

The endorsement heuristic plays a significant role in the evaluation of online information, manifesting in several distinct ways. First, consider the transfer of trust. For example, in e-commerce, Amazon is a widely trusted platform. When consumers see the option "Pay through Amazon" on smaller and lesser-known websites, Amazon's established trustworthiness is implicitly transferred to these sites. The association makes consumers more confident in their information delivery and transactions on these platforms. Next is the impact of personal recommendations. For instance, on professional networking sites, personal endorsements are crucial for evaluating the credibility of potential partners or employees. If a professional receives recommendations from peers in their industry on their profile page, these endorsements significantly enhance their professional image and credibility, thereby attracting more cooperation or employment opportunities. Lastly, group behavior indicators on platforms such as YouTube provide insights into content quality and popularity. Videos that achieve high view counts and like ratios are often perceived as containing worthwhile or engaging content, influencing other users' decisions to watch and potentially shaping their decision-making process. These examples illustrate how the endorsement heuristic is effectively utilized across various online platforms to evaluate and choose information sources, highlighting its role in modern information evaluation strategies

3) What are the considerations?

While the endorsement heuristic aligns closely with individual cognitive tendencies, it can easily become a trap if people pay more attention to this heuristic than their first-hand information or feelings about the information they encounter online. Therefore, when using the endorsement heuristic, it is important to consider the following points.

- ***Scrutinize the Sources of Endorsements.*** A source might appear trustworthy if it is widely endorsed, but the authenticity of these endorsements must be verified. We need to ensure that the endorsements originate from reliable and reputable individuals or organizations, rather than being artificially inflated or manipulated.

- ***Be Wary of Going with the Crowd.*** Widespread agreement does not necessarily equate to correctness or quality. Popularity can sometimes lead to erroneous groupthink or decision-making biases.
- ***Embrace Contrasting Viewpoints.*** Encountering a widely accepted viewpoint should prompt a search for counterarguments or differing opinions, which can provide a more comprehensive understanding and prevent overlooking critical insights.
- ***Recognize Emotional Biases and Commercial Interests.*** Personal preferences, such as favoritism towards certain brands or influencers, can skew our judgments. Moreover, in the context of business, endorsements may often be motivated by financial incentives. Thus, identifying if there's a commercial incentive behind a recommendation can help us avoid being swayed by commercial manipulation.

Notably, to more comprehensively evaluate the reliability of the information, it is advisable to combine the endorsement heuristic with other heuristics, such as the reputation heuristic, to minimize the potential risks caused by oversimplified decision processes and ensure the breadth and depth of the information obtained.

3. The Consistency Heuristic

1) What is the "consistency heuristic"?

The consistency heuristic is the third shortcut that people use to simplify decision-making. In the context of online information evaluation, this heuristic is divided into two main types: interpersonal consistency and informational consistency. Interpersonal consistency refers to the consensus among people. If others agree with certain information sources or messages, people tend to trust those sources or messages. Essentially, if there is a widespread agreement that something is correct, it is generally considered credible. This form of consistency closely relates to the endorsement heuristic, as both rely on group validation to enhance the perceived reliability of information. Informational consistency, on the other hand, is the key focus of the consistency heuristic. It involves cross-verifying information across different websites to determine if the content is uniform across sources. This method is based on the principle that information supported by several reliable sources is more likely to be accurate and trustworthy, reducing the likelihood of encountering misinformation.

2) How to use the "consistency heuristic" to evaluate online information?

Using the consistency heuristic to evaluate online information typically involves a process known as cross-validation. This approach requires more cognitive effort than other heuristics but remains a relatively quick way to make credibility judgments. Here are two primary ways to apply cross-validation.

One way is comparing the new online information with other online information. This method validates the consistency of information by comparing new information on the same topic across different websites. For instance, to understand a new technological innovation like the application of VR technology in the medical field, we can examine articles across several tech news portals and professional blogs. If reputable sites such as "Science Daily", "TechCrunch" and "Wired" consistently report similar use cases and technical details, this consistency boosts the credibility of the information we are investigating. Another way is comparing the new online information with offline information. It involves comparing the online information found with books, academic journals, or the opinions of other authoritative experts. Taking medical research as an example, suppose we see a report online about a new drug that has significant effects on treating autism. To verify this, we can refer to recent medical journals to see if there are clinical trial results that support the claim and consult medical experts. Comparing the data in these authoritative resources with the online reports, we can more accurately assess the effectiveness of this new drug. By employing these two cross-validation methods, we can examine the consistency of information from multiple perspectives, thereby more effectively using the consistency heuristic to assess the reliability of online information.

3) What are the considerations?

While the consistency heuristic is a valuable tool for swiftly evaluating online information, it's important to consider several factors to avoid potential risk.

- ***Assess Each Source Individually.*** Consistency across multiple unreliable sources does not automatically indicate trustworthiness unless each source is independently verified for accuracy and reliability.
- ***Monitor Information for Changes.*** We need to regularly check if the information remains consistent over time. Information that frequently changes or contradicts previous statements may require additional verification to determine if new findings have emerged or if the information was initially incorrect. This consideration helps identify evolving insights that may affect the information credibility and relevance.
- ***Step Outside the Echo Chamber.*** We should be cautious of information that seems consistent merely because it is circulated within a closed group, called an echo chamber. It creates a false sense of consensus and reliability. By considering these points, we can more effectively utilize the consistency heuristic to assess the credibility and quality of information, enhancing our decision-making process based on informed judgments.

4. The Self-Confirmation Heuristic

1) What is the "self-confirmation heuristic"?

The self-confirmation heuristic describes a tendency where individuals are inclined to believe an event or accept a behavior if it aligns with their preexisting beliefs and dismiss it if it contradicts them. The vast amount of information available and the rapid pace of digital media mean that users often search online under time constraints and with limited motivation to thoroughly check all information. As a result, individuals place greater emphasis on information that supports their viewpoints, while often overlooking or undervaluing evidence that opposes them regardless of whether the information is accurate.

2) How to use the "self-confirmation heuristic" to evaluate information?

Let's explore a typical example to understand how the "self-confirmation heuristic" can be applied in online information evaluation. Imagine two friends debating whether coffee reduces the risk of heart disease. One friend, believing in coffee's benefits, searches online using terms like 'benefits of coffee for heart health,' which biases the results toward positive articles. When the friend finds a health blog that supports the claims, he quickly cites this as evidence in their debate. This example illustrates that selective searching plays an important role in online information evaluation. However, such selectivity might cause users to overlook a broader array of scientific studies, some of which might present mixed or even negative effects of coffee on heart health. In reality, the self-confirmation heuristic is usually applied regardless of the information's actual quality, depth of research, or credibility of sources. Thus, we should realize the importance of considering opposing viewpoints to achieve a more balanced perspective.

3) What are the considerations?

The self-confirmation heuristic is a highly subjective tool for evaluating information, making it a double-edged sword that, if misused, can lead to errors. To successfully navigate through a lot of information designed to reinforce our current perspectives, we need open-mindedness and critical thinking in digital environments.

- ***Evaluate Opposing Viewpoints Critically.*** We should make a concerted effort to validate information that contradicts our existing beliefs as rigorously as we can. This practice counteracts the natural tendency to dismiss conflicting data without sufficient scrutiny. By engaging deeply with opposing data, we can ensure a more balanced and comprehensive evaluation of information, which is vital for maintaining objectivity
- ***Engage in Dialogues to Challenge Biases.*** Understanding how our preferences and beliefs might skew our perception of credibility is crucial for unbiased information processing. This self-awareness helps in recognizing when we are dismissing valid information simply because it does not align with our views or overly crediting information that does. By considering these points, we can use the self-confirmation heuristic more effectively, ensuring that the information we choose is not only confirmed but also accurate and reliable.

5. The Expectancy Violation Heuristic

1) What is the "expectancy violation heuristic"?

The expectancy violation heuristic indicates that people evaluate the trustworthiness of a website based on whether it meets their anticipated standards related to appearance, layout, features, functionality, and content scope. If a website fails to meet these expectations, or if the information presented significantly deviates from what users anticipate, it is often perceived as unreliable.

2) How to use the "expectancy violation heuristic" to evaluate information?

The expectancy violation heuristic typically appears in three distinct forms, each significantly affecting how users judge information credibility. The first type occurs when websites request more information than necessary or provide more than what was asked for by users. This can be perceived as an invasion of privacy or an attempt at manipulation, which often leads to mistrust of the website. Imagine a website for a health app that, upon signing up, requests not only basic information like age and weight but also detailed medical history and family health data. This excessive data collection can be seen as unnecessary and even invasive. The second type arises when the information presented conflicts with users' personal beliefs or opinions. This encounter typically causes discomfort among users when their worldview is challenged, prompting them to consider such information unreliable. Type two is similar to the self-confirming heuristic, emphasizing how individuals favor information that conforms to their pre-existing beliefs and discount contradicting information. For example, when a health information website provides nutritional advice that matches well-established medical research, users will believe this information is reliable. Conversely, if the advice significantly deviates from conventional medical theories, such as promoting unproven extreme diets, users may question the accuracy of this information. The third type is related to the website's appearance and functionality. This type involves how aesthetic and functional aspects of a site can profoundly influence users' perceptions of its credibility. An appealing and user-friendly website design can instantly establish a baseline of trust, even before the content is thoroughly assessed. For instance, a site that is professionally designed features an intuitive interface and loads quickly is likely to be perceived as more credible compared to one that is poorly designed, difficult to navigate, or slow. As explained in the psychological "halo effect", the attractive external features of a website positively influence the perception of its content's credibility. Moreover, a financial services website with a professional design, clear layout, and quick loading features will likely be perceived as credible because these characteristics align with the expected standards of a legitimate financial institution.

3) What are the considerations?

Like the self-confirmation heuristic, the expectancy violation heuristic is highly subjective. To apply it effectively, several key points must be considered.

- ***Avoid Over-trust on Surface Characteristics.*** For instance, a well-designed website may attract users, but its content could be incorrect or biased. Therefore, while being drawn in by visual design, it's also essential to carefully check the source and accuracy of the information.
- ***Try to Evaluate the Depth of Content.*** When encountering information that contradicts our views, it's essential to delve deeper into the content to assess the strength of the evidence presented. For instance, a well-argued article that challenges our preconceptions that "anxiety is harmful" with solid data and credible references deserves careful consideration, even if it initially seems counterintuitive.
- ***Avoid Personal bias or Prejudice.*** For example, if a user holds a particular political viewpoint, they may tend to trust websites that align with these views while ignoring credible information that presents opposing viewpoints.

6. The Persuasive Intent Heuristic

1) What is the "persuasive intent heuristic"?

The persuasive intent heuristic is the last shortcut that leads people to doubt the credibility of information if they sense that the source has a motive to persuade. Thus, when individuals become aware of a source's persuasive intent, they often resist or discount the information. This heuristic often activates in response to commercial content online for two main reasons. On the one hand, the presence of advertisements makes people feel like they are being manipulated. This typically triggers an instant defensive reaction, causing them to distrust the information without further investigation. For instance, pop-up ads or sponsored content can immediately make users suspicious of the motives behind the information presented. On the other hand, when users browse information on websites, they sometimes encounter intrusive advertising. The annoyance caused by these intrusive ads can be transferred to the website or source itself. For example, if a user is trying to search for information online but finds themselves continually interrupted by ads for products, they may begin to doubt the primary purpose of the website- is it to inform or simply to sell?

2) How to use the "persuasive intent heuristic" to evaluate online information?

Using the persuasive intent heuristic effectively during online information evaluation involves recognizing and critically assessing the intent behind the information presented. Here are some methods and examples of how to apply this heuristic. Firstly, we need to distinguish between genuine content and advertisements or sponsored content, which is often designed to sell rather than inform. For example, on a news website, sections labeled as "sponsored" or "advertisement" should immediately signal that the content is paid for and may primarily serve the advertiser's interests, not necessarily providing unbiased information. Then, notice how many advertisements are present and how intrusive they are. A high volume of intrusive ads can be a red flag indicating that the site's primary goal might be revenue generation through clicks rather than providing credible information. For example, if a health advice website is cluttered with a number of ads for pharmaceutical products, it might suggest that the site's advice could be biased towards promoting certain products.

3) What are the considerations?

To sum up, advertising is a significant hurdle when searching for information online. However, it's not always easy to spot at first glance, necessitating a critical approach to discern potential commercial motives.

- ***Scrutinize the Subtlety of Advertisements***. We need to realize that not all advertisements are overt. For example, an article on a lifestyle blog may provide seemingly neutral advice on skincare but frequently mention specific brands, with links to purchase. This subtle placement of brand names within editorial content could indicate a commercial motive masked by helpful guidance.
- ***Balance Skepticism with Open-mindedness.*** To be honest, advertisements don't automatically discredit the information provided. Overly suspecting every piece of information can lead to cynicism and the rejection of useful content. For instance, a financial news website might display ads for investment tools or services; while these ads support the site's operations, the articles themselves can still offer valuable, well-researched insights into economic trends and investment strategies. In such cases, it's crucial to separate the site's commercial interests from the informational value of its content.

By being mindful of these points, users can more effectively apply the persuasive intent heuristic to assess the credibility of online information. This awareness helps in distinguishing between genuinely informative content and content primarily aimed at persuasion.

Cognitive heuristics, as simplified decision-making strategies or shortcuts, allow individuals to quickly navigate through vast quantities of data by focusing on familiar cues like reputation, consensus, or potential motives, thus expediting decision-making in environments overloaded with information. Compared to a thorough evaluation of credibility and relevance, these heuristics make the evaluation process more time-efficient and energy-saving. However, it's crucial to remember that heuristics are not foolproof methods, as they simplify complex decisions but do not guarantee accurate outcomes every time. Thus, individuals should be mindful of each heuristic's limitations and actively pay attention to the associated considerations to mitigate potential errors.

Importantly, these strategies are not mutually exclusive or isolated in their application, rather, they can be effectively combined to facilitate decision making (Gigerenzer & Todd, 1999). Some of them can build upon one another to increase both information efficiency and objectivity. For instance, the reputation heuristic and the consistency heuristic can be combined for a more thorough evaluation of information sources. By using the reputation heuristic to identify trusted sources and the consistency heuristic to verify that these sources consistently present similar information, the credibility of the information is strengthened. Similarly, the synergistic use of the persuasive intent heuristic with the expectancy violation heuristic offers deeper insights into the intentions behind the information presented. This combination allows users to discern whether the content aims to inform or persuade, based on how well it aligns with or deviates from expected norms and standards. Overall, while cognitive heuristics provide essential shortcuts for navigating vast amounts of information daily, their true value is unlocked through critical and synergistic application. By strategically combining these heuristics, users can achieve a balance of efficiency and thoroughness in their information evaluation practices.

Table 3. Definition and consideration of six heuristics

Cognitive heuristic	Definition	Consideration
The reputation heuristic	The reputation heuristic is the first shortcut that people prefer options they recognize or are familiar with over unfamiliar ones when making decisions.	✓ Beware of over-reliance on popular sources. ✓ Reassess information regularly. ✓ Be cautious of manipulation and misinformation.
The endorsement heuristic	The endorsement heuristic is the second shortcut that people tend to evaluate the credibility of an event based on the endorsement or recommendation of others, without deeply investigating the event itself.	✓ Scrutinize the source of endorsements. ✓ Be wary of going with the crowd. ✓ Seek different opinions. ✓ Recognize the emotional biases. and commercial interests.
The consistency heuristic	The consistency heuristic is the third shortcut that people use to simplify decision-making by assuming that if something is consistent with their previous beliefs or behaviors, it is likely to be true.	✓ Assess each source individually. ✓ Step outside the echo chamber. ✓ Monitor information for changes.
The self-confirmation heuristic	The self-confirmation heuristic is the fourth shortcut that individuals are more likely to accept information as credible if it aligns with their preexisting beliefs and dismiss it if it contradicts them.	✓ Validate opposing data rigorously. ✓ Engage in dialogues to challenge biases.

continued on following page

Table 3. Continued

Cognitive heuristic	Definition	Consideration
The expectancy violation heuristic	The expectancy violation heuristic is the fourth shortcut that people determine the trustworthiness of a website according to how well it aligns with their anticipated standards	✓ Avoid over-trusting surface characteristics. ✓ Try to evaluate the depth of content. ✓ Avoid personal bias.
The persuasive intent heuristic	The persuasive intent heuristic is the fifth shortcut that people tend to feel that information that may be biased is not credible.	✓ Scrutinize the subtlety of advertisements. ✓ Balance skepticism with open-mindedness.

CONCLUSION

This chapter has systematically explored the multi-faceted approach required to develop critical thinking skills for evaluating online information, encompassing credibility evaluation, relevance evaluation, and the application of cognitive heuristics. Each section contributes uniquely to the final goal of enhancing our ability to discern trustworthy and pertinent information in a digital age.

Initially, we delved into the principles of credibility evaluation, underscoring the necessity of assessing the source and content's trustworthiness. Techniques such as verifying author credentials and cross-referencing information sources are foundational to preventing the acceptance of misinformation. Following this, the relevance evaluation section highlighted the importance of aligning information with specific needs and contexts. By clarifying personal objectives and ensuring the information directly addresses these objectives, individuals can significantly enhance the efficiency and applicability of the data they gather.

The discussion then transitioned to cognitive heuristics, which are mental shortcuts used to simplify decision-making processes. While heuristics such as reputation, endorsement, and consistency heuristics facilitate quicker information evaluation, they also require a balanced approach to guard against potential biases and errors inherent in quick judgments.

Crucially, the development of critical thinking skills involves an iterative process of questioning, analysis, and reflection. The ability to integrate and apply the techniques from these three key areas—credibility, relevance, and cognitive evaluations—not only prepares individuals to better understand and critique the information they encounter but also empowers them to manage the complexities of the digital information landscapc with a morc informcd and skcptical lcns.

In conclusion, fostering critical thinking skills in online information evaluation is not merely about adopting specific strategies but about cultivating a comprehensive, adaptive mindset that remains ever-curious and rigorously analytical. This mindset is indispensable for anyone looking to thrive in today's dynamic information environments, ensuring that interactions with digital content are both discerning and insightful.

REFERENCES

Chaiken, S. (1987). The heuristic model of persuasion. In *Social influence: The Ontario symposium* (Vol. 5, pp. 3–39). Lawrence Erlbaum Associates, Inc.

Colliander, J. (2019). "This is fake news": Investigating the role of conformity to other users' views when commenting on and spreading disinformation in social media. *Computers in Human Behavior, 97*, 202–215. DOI: 10.1016/j.chb.2019.03.032

Gong, Y.-C., Wang, M., Liang, W., Hu, F., & Zhang, Z.-K. (2023). UHIR: An effective information dissemination model of online social hypernetworks based on user and information attributes. *Information Sciences, 644*, 119284. DOI: 10.1016/j.ins.2023.119284

Kapoun, J. (2020). Teaching undergrads WEB evaluation: A guide for library instruction I Kapoun I College & Research Libraries News. *College & Research Libraries News, 59*(7), 522–523. DOI: 10.5860/crln.59.7.522

Kwak, K. T., Lee, S. Y., Ham, M., & Lee, S. W. (2021). The effects of internet proliferation on search engine and over-the-top service markets. *Telecommunications Policy, 45*(8), 102146. DOI: 10.1016/j.telpol.2021.102146

Marin, L., & Copeland, S. M. (2022). Self-Trust and Critical Thinking Online: A Relational Account. *Social Epistemology*, 1–13. DOI: 10.1080/02691728.2022.2151330

Metzger, M. J., & Flanagin, A. J. (2013). Credibility and trust of information in online environments: The use of cognitive heuristics. *Journal of Pragmatics, 59*, 210–220. DOI: 10.1016/j.pragma.2013.07.012

Metzger, M. J., Flanagin, A. J., & Medders, R. B. (2010). Social and Heuristic Approaches to Credibility Evaluation Online. *Journal of Communication, 60*(3), 413–439. DOI: 10.1111/j.1460-2466.2010.01488.x

Paul, R., & Elder, L. (2014). *The Miniature Guide to Critical Thinking Concepts & Tools* (Elder, L., Ed.). The Foundation for Critical Thinking.

Van Zyl, A., Turpin, M., & Matthee, M. (2020). How Can Critical Thinking Be Used to Assess the Credibility of Online Information? In Hattingh, M., Matthee, M., Smuts, H., Pappas, I., Dwivedi, Y. K., & Mäntymäki, M. (Eds.), *Responsible Design, Implementation and Use of Information and Communication Technology* (Vol. 12067, pp. 199–210). Springer International Publishing., DOI: 10.1007/978-3-030-45002-1_17

Zannettou, S., Sirivianos, M., Blackburn, J., & Kourtellis, N. (2019). The Web of False Information: Rumors, Fake News, Hoaxes, Clickbait, and Various Other Shenanigans. *ACM Journal of Data and Information Quality, 11*(3), 1–37. DOI: 10.1145/3309699

Chapter 7
Implementing Digital Literacy in the Context of People-Centered Governance in Kazakhstan:
The Case of Astana IT University

Mergen Dyussenov
https://orcid.org/0000-0002-5808-2067
Astana IT University, Kazakhstan

Nurkhat Ibadildin
https://orcid.org/0000-0002-6352-8713
Astana IT University, Kazakhstan

Ha Jin Hwang
Astana IT University, Kazakhstan

ABSTRACT

The focus of this book chapter is to explore digital literacy (DL) skills in the context of people-centered governance in Kazakhstan. DL has become important, particularly in developing countries, as it is seen as a key factor in enhancing individual participation. The aim is to examine DL in people-centered governance, noting the importance of putting individuals at the center of decision-making processes. To provide a deeper understanding of these concepts, the case of Astana IT University will be utilized to analyze the intersection of DL and people-centered governance in Kazakhstan. By employing a case study methodology, this research seeks to generate insights that can apply to other IT-focused universities in Kazakhstan and Central Asia. Through a detailed analysis, this research aims to contribute to existing research on DL and people-centered governance in the context of higher education institutions. By highlighting these concepts, the research seeks to inform both policy theory and practice to enhance DL skills and promote a more inclusive governance approach in the region.

DOI: 10.4018/979-8-3693-6675-2.ch007

Copyright ©2025, IGI Global Scientific Publishing. Copying or distributing in print or electronic forms without written permission of IGI Global Scientific Publishing is prohibited.

I. INTRODUCTION

Much of the developing world has increasingly witnessed digital development in various areas, and many countries, including those in Asia, have embraced policies that favor information technology (IT) applications (Ameen & Gorman 2009). Digital literacy, in particular, has been noted as a crucial aspect of national development in various areas, such as sustainable growth (Radovanović et al. 2020; Sharma et al. 2016), education (e.g. Martin & Grudziecki 2006; Shopova 2014; Gruszczynska et al. 2013), labor market development (e.g. Bejaković & Mrnjavac 2020) and professional growth (Andriushchenko et al. 2020), in the context of knowledge economy (e.g. Erlanitasari et al. 2020), E-governance (Rajput & Nair 2013), among other areas.

The concept of 'digital literacy' (DL) should be viewed through the prism of its central competencies, including an ability to read and comprehend digital texts, to develop and communicate digital data and information, an ability to evaluate digital information, and having information literacy skills (Lankshear & Knobel 2008). Scholars increasingly view DL as a factor conducive to boosting individual participation in education (e.g. Martin & Grudziecki 2006; Mohammadyari & Singh 2015). Specifically, as Martin & Grudziecki (2006) note, the DL concept is important as it helps individuals achieve their personal goals in their particular life circumstances, and "focus on the individual's engagement with and articulation of the symbols and meanings of daily life" (p. 250). Furthermore, the fast-paced advancement of digital technologies necessitates individuals to possess knowledge and skills to effectively address specific issues in digital settings (Fu, 2013, Sarkar, 2012 as cited in Reddy et al. 2020). These skills are now referred to as digital literacy skills.

The above-listed definitions of DL stress the importance of an individual mastering DL skills in order to navigate the digital landscape effectively, and to build a personal growth path in education, employment, and other areas in the modern society. This particular context necessitates analyzing the DL notion under the broader notion of people-centered governance as it highlights the need for individuals to have the required skills to engage with digital technologies in meaningful ways. In the context of governance, DL thus becomes a tool for personal empowerment, enabling individuals to access information, communicate effectively, and make informed decisions.

Based on the above, the review of existing scholarly research should focus on two distinct subparts – an overview of digital literacy publications in developing nations and, specifically, across the Central Asian region (CA), and an overview of major sources related to people-centered governance. Such an approach forms the essence of an emerging framework focusing on DL in CA in the context of people-centered governance (Fig. 1). Finally, the case of Astana IT University, discussed in the following sections, seeks to advance DL skills within higher education in CA.

Figure 1. DL in Central Asia in the context of people-centered governance

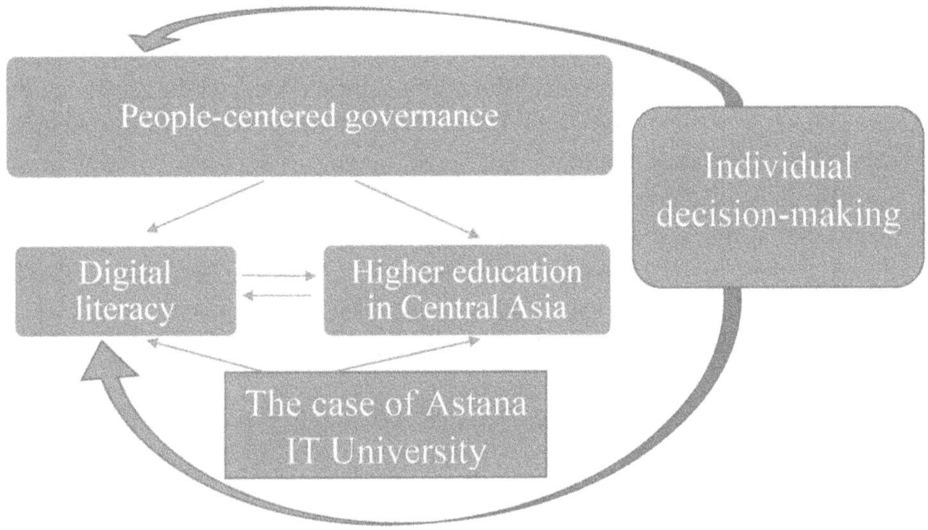

Source: The authors' own analysis

The overarching framework of digital literacy in CA within the context of moving towards people-centered governance is crucial in fostering individual empowerment and participation in decision-making processes. By equipping individuals with the necessary skills to navigate digital technologies effectively, DL promotes individual-driven education, employment, and personal growth in the modern society. As digital technologies continue to advance rapidly, the need for individuals to possess DL skills becomes increasingly important. By examining the intersection of DL and people-centered governance in Kazakhstan, particularly through the case study of Astana IT University, this research aims to generate insights and recommendations that can be applied to other IT-focused universities in CA. Through a detailed analysis, this research seeks to contribute to the existing body of knowledge on DL and people-centered governance, ultimately promoting a more inclusive and participatory governance approach in the region.

II. LITERATURE REVIEW

The key focus of the literature review section is to analyze existing scholarship on DL in developing nations with a narrower focus on CA, and to develop an overview of people-centered governance. The review seeks to shed light on key concepts, frameworks, and empirical studies on DL and people-centered governance, stressing their significance in boosting individual participation in education. This section aims to contextualize the research within the theoretical framework related to the intersection of Digital Literacy and People-centered Governance, and lay the foundation for the case study analysis of Astana IT University in Kazakhstan.

2.1. A Review of Digital Literacy Research in Central Asia

Over the past two decades, DL has increasingly become a scholarly focus across much of the developing world. These include parts of Latin America (e.g. Gómez-Galán et al. 2021, Huerta & Sandoval-Almazán 2007), African nations (e.g. Andema 2014, Lemphane & Prinsloo 2014), ASEAN states (e.g. Kusumastuti & Nuryani 2020), nations of the South Pacific (Reddy et al. 2020) etc. Furthermore, scholars note the context specific nature of the DL notion both in terms of policy area specificity, e.g. school education (Gruszczynska & Pountney 2013) and languages spoken across nations (Pangrazio et al. 2020). Specifically, as Pangrazio et al. (2020) note, the English-speaking context faces a debate between the terms 'digital literacy' and 'media literacy' in regards to the skills young people need to be engaged citizens in the digital era. Some scholars (e.g. Erstad 2010a, Erstad et al. 2007, as cited in Pangrazio et al. 2020) advocate for the use of media literacy, as it encompasses a wider range of skills related to production, representation, and audience interaction in a media-saturated society, rather than just technical skills related to analyzing ICT cases.

In this context, the review of existing research on DL in Central Asia is essential to provide insights into the current state of digital literacy initiatives in the region, as well as to identify gaps and areas for further research. By focusing on CA, this research paper aims to develop a regional context specific framework focusing on the intersection of DL and people-centered governance. First, the DL in Central Asian context research is analyzed. The basic methodological approach to analyzing existing studies aligns with the approach utilized by Pangrazio et al. (2020), for two reasons. First, their study focuses on analyzing definitions of digital literacy. Second, more importantly, in doing so it recognizes the importance of context specificity. In their analytic search for definitions of 'digital literacy' across language contexts of Australia, Sweden and Argentina, Pangrazio et al. (2020) employ the Google Scholar engine to select those research publications with a major focus on DL conceptualization. Similarly, this study uses Google Scholar to select some of the relevant publications related to DL research in the context of DL scholarly discourse in CA. As a result, the total of 12 major publications were selected for analysis (see Appendix 1 for the list of selected publications).

To begin with, when referring to the notion of digital literacy as applied to the regional context, some scholars use terms such as 'digital literacy' (Balgabayeva et al. 2024, Khamdamov 2024, Kairat et al. 2024, Marav 2016, Bykov et al. 2019), 'media literacy' (e.g. Muratshina & Valeeva 2020, Akhmetova & Shorokhov 2018, Abdikarimov 2024, Chelysheva 2019), and 'information literacy' (Kairat et al. 2024, Muratshina & Valeeva 2020, Akhmetova & Shorokhov 2018), while Dadakhonov in a series of his publications (2023, 2024a, 2024b), as well as Akhmetova & Shorokhov 2018, and Muratshina & Valeeva 2020 refer to digital literacy through the prism of both media and information literacy (Fig. 2).

Figure 2. The use of the DL notion in Central Asian scholarly discourse

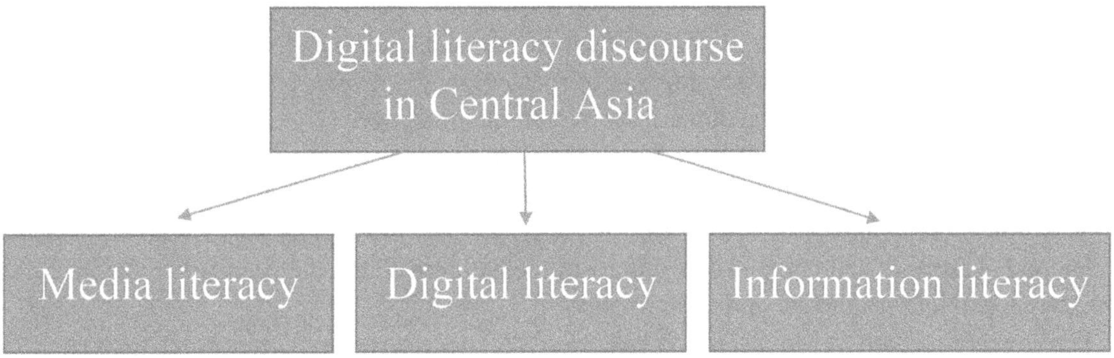

Source: The authors' own analysis

First, major sources of the broad media and information literacy (MIL) research is analyzed. To begin with, Muratshina & Valeeva (2020) look into processes related to the recent humanitarian cooperation between Russia and CA states, in the area of information and media space. Specifically, they use a case of the Digital dictation, an annual contest initiated by Russia aimed to assess the Russian proficiency among various age participants from a wide range of countries, and to boost digital literacy among local populations (Muratshina & Valeeva 2020), but also to promote Russian culture and values (IKBFU 2024). This short case study shows how initiatives such as this can improve digital literacy while fostering cultural ties between Russia and CA nations. It also raises certain ethical implications related to the possible use of similar educational initiatives as soft power tools across post-Soviet space. On the other hand, Abdikarimov (2024) views media literacy as a means against disinformation in Uzbek media space. In doing so, the recommended measures include the need to boost media literacy awareness among the population, and to increase the number and quality of fact-checking platforms.

The research of Muratshina & Valeeva (2020) and Abdikarimov (2024) should be compared and contrasted in detail. Both works examine the significance of media literacy in contemporary contexts but approach the issue from different, if not opposite, angles, focusing on distinct aims, methodologies, and implications.

1. *Focus on Digital Literacy and Cultural Ties*: Muratshina & Valeeva (2020) emphasize the role of the Digital Dictation initiative to boost DL and to promote Russian culture and values in other countries, including the CA nations. They view this as a way to foster intercultural links and strengthen ties through shared language and values. Abdikarimov (2024) views media literacy as a defensive tool against disinformation in Uzbekistan, highlighting the growing need for critical awareness and fact-checking amid challenges due to false information. His focus is more on safeguarding the citizenry rather than promoting cultural affiliation.
2. *Methods and Approaches*: Muratshina & Valeeva's (2020) case study method focuses on the initiative within a broader education context, with certain benefits and potential ethical concerns on using soft power. Abdikarimov (2024), on the other hand, uses a prescriptive approach, recommending measures to boost media literacy awareness and fact-checking in Uzbekistan.

3. *Ethical Considerations*: The case study by Muratshina & Valeeva (2020) raises important ethical questions regarding the utilization of educational initiatives as tools of soft power, suggesting a complex interplay between cultural promotion and potential political implications. Abdikarimov's (2024) work is more cautionary in nature; by focusing on disinformation, it implicitly critiques the current media landscape's shortcomings and advocates for proactive measures to combat these issues without delving into the ethical dimensions of cultural promotion.
4. *Geopolitical Context*: Muratshina & Valeeva (2020) frame their research within the context of Russian-Central Asian relations, suggesting that initiatives aimed at boosting DL can serve broader geopolitical aims. In contrast, Abdikarimov (2024) focuses specifically on Uzbekistan, analyzing internal challenges and complexities related to media literacy and the need for resilience against external and internal threats of misinformation.
5. *Implications for Policy*: Muratshina & Valeeva (2020) develop policy implications supporting cultural initiatives or soft power strategies focusing on CA. Abdikarimov (2024) offers a more local perspective, advocating for policies aimed at empowering individuals through education and resource availability to counter misinformation effectively.

Thus, while both studies underscore the importance of MIL, they diverge in focus, methods, and implications. Muratshina & Valeeva (2020) examine literacy initiatives in terms of cultural diplomacy and soft power across the region, whereas Abdikarimov (2024) views media literacy as a critical tool for combating disinformation specifically in Uzbekistan. Together, these perspectives provide a more nuanced understanding of the multifaceted role of DL and MIL in both empowering individuals and navigating complex geopolitical landscapes.

Other major MIL sources worth analyzing include a series of research publications by Dadakhonov (2023, 2024a, 2024b). First, while analyzing MIL theoretical frameworks, Dadakhonov (2023) notes that COVID-19 spurred the need to boost digital and information literacy among population in CA, which then led to the further spread of social news networks and online media websites. More interestingly, in his second publication (2024a), analyzing the sustainability landscape of MIL projects across CA, he stresses the importance of developing local partnerships, contextualization of foreign aid interventions, efficient resource allocation, technology integration, and policy advocacy in promoting sustainable MIL practices. These observations are interesting: first, Dadakhonov (2024a) emphasizes the context specificity of the region, in line with this research, and second, while supporting the argument of Abdikarimov (2024) in terms of the need to promote MIL sustainability, he also stresses the need to build partnerships across the region, in line with Muratshina & Valeeva (2020). In other words, building MIL sustainability and resilience, and promoting regional partnerships, is not necessarily a zero-sum game. In his final publication selected for this analysis, Dadakhonov (2024b) stresses the need to build MIL into school curricula across CA, in order to better equip local students with critical skills to navigate "the complex media landscape of the 21st century" (p. 66).

Second, with regard to research focusing directly on DL in CA, four major studies are analyzed. To begin with, Khamdamov (2024) recommends integrating DL into university curricula. This includes measures related to embedding DL learning outcomes and activities into curricula, from general education to discipline-specific courses, so that students develop these skills in a contextual and meaningful manner (Ng 2012, as cited in Khamdamov 2024). DL should be embedded across all subjects, not just taught as a standalone course. For example, students can learn to use digital tools for research in science, digital storytelling in language arts, and data analysis in mathematics. These observations are vital as they point to the importance of context specificity, similarly to Dadakhonov (2024a), but also provide the rationale for using a specific university for case study analysis. Next, Balgabayeva et al. 2024 analyze DL as a cognitive part of young students' creativity in mastering a foreign language, and find that DL has a positive effect on building speech-thinking skills among younger school-age children in Northern Kazakhstan, thus increasing the effectiveness of teaching a foreign language in a classroom. The third study (Bykov et al. 2019) compares PR education programs at two universities, one in Kazakhstan and one in Russia, based on a combined survey of 60 students (83% female). As its findings suggest, the majority of students remain rather unsatisfied with the degree of IT technologies currently implemented in higher education, while advocating for the need to move some traditional courses into the massive open online course (MOOC) format. However, the limitations of this study (Bykov et al. 2019) include a small and unrepresentative number of students and possible gender bias. Finally, Marav (2016), based on a survey of 98 public university students in Ulaanbaatar, explores the ways Mongolian students majoring English use digital technologies, especially the Internet. This study finds that DL use among Mongol students was mostly motivated by boosting English skills, securing better employment or study abroad opportunities upon graduation, and upgrading their social status.

To summarize, while Khamdamov (2024) advocates for an integrated DL approach within university curricula, Balgabayeva et al. (2024) focus on DL applications in language education among younger school students. Bykov et al. (2019) highlights dissatisfaction with current DL implementation, suggesting a gap between educational aspirations and reality, though with certain limitations. Finally, Marav's (2016) study provides insights into Mongol students' motivations for using DL, indicating personal and socio-economic factors. These motivation-related observations might help CA universities to better improve their digital technology implementation, again taking into account context specificity of the higher educational landscape in the region. This comprehensive approach may lead to more successful DL outcomes and a more substantial impact on educational practices in the digital age.

2.2. International Comparisons of Digital Literacy Practices

In the realm of digital literacy, different regions of the world have exhibited unique approaches shaped by their local contexts, educational systems, and technological infrastructure. While Central Asia (CA) has made strides in promoting DL, it is crucial to compare these practices with those in other regions, such as the European Union (EU) and the Association of Southeast Asian Nations (ASEAN), to identify common trends and regional variations.

To begin with, the EU has placed a strong emphasis on digital literacy as a cornerstone of its digital transformation agenda. The European Commission's Digital Education Action Plan (2021) stresses the need for citizens, especially in EU member states and partner countries, to acquire digital skills, not only for personal development but also for economic growth. This initiative has stressed the need to continue investing in digital infrastructure, skills and DL across various sectors (European Commission, 2020). Moreover, AI integration in DL practices within the EU has been robust, with educational tools incorporating AI-driven personalized learning platforms that adjust to individual student needs (Laak & Aru, 2024). AI's role in DL is seen as part of the broader effort to enhance both the accessibility and efficacy of digital education, enabling students to engage in more adaptive and responsive learning environments. The EU also promotes open-access educational resources and platforms like Erasmus+, which facilitate cross-border collaboration and the exchange of digital education best practices (Fumasoli & Rossi 2021).

Next, In the ASEAN region, the promotion of digital literacy is also closely tied to regional development goals, particularly in terms of economic modernization and social cohesion. ASEAN nations, such as Singapore and Malaysia, have invested heavily in DL policies that integrate technology into education, focusing on creating a skilled workforce to drive digital economies (Yuen & Hew, 2018). In Singapore, for instance, the Smart Nation initiative has transformed the educational system to ensure that students not only acquire traditional academic skills but also develop strong digital literacy capabilities, with an increasing focus on the integration of AI-based educational tools. AI is actively used in ASEAN nations for personalized learning environments, assessment systems, and skill-building platforms designed to close the digital divide. However, countries like Cambodia, Myanmar, and Laos face challenges in terms of digital infrastructure and access (Chen 2020), thus hindering widespread adoption of AI in educational settings. Despite these challenges, efforts are being made to develop AI-driven educational programs, such as AI literacy workshops and initiatives, which aim to equip students and educators with the skills to navigate and utilize AI technologies in their learning environments (Isono & Prilliadi 2023).

2.3. An Overview of Transition Toward People-Centered Governance

Scholars have emphasized the need to move toward people-centered governance mainly due to the criticism of good governance theoretical models. First, such criticism stems from the abuse of the 'good governance' notion by international organizations such as special conditions imposed by the World Bank, International Monetary Fund (IMF) and other agencies on developing countries (Parthasarathy 2005), the World Bank's use of structural adjustment (Okechukwu 2011) as a pre-condition for a developing country to be eligible for a loan etc. It can be argued that emphasizing political, administrative, and economic reforms overlooks social structures and divisions, which have significant effects on the poverty and equity aspects of development programs, and thus significant implications for the economic and political growth of developing countries, further denying basic human rights to local citizens (Parthasarathy 2005).

Another point of criticism relates to the universal model of good governance, which implies that all effective governments must conform to a similar standard (Andrews 2010). As Andrews further suggests, governments deemed "good" or "more effective" (p. 12) display significant diversity in their structures and practices, particularly in the realm of public financial management (PFM). Thus, such good governance models fail to reflect on the reality of diverse governance structures across different contexts. Furthermore, the key characteristics of good governance practices are found to differ considerably between countries, even among those regarded as successful (ibid). Contrary to the assumptions of the universal model, each country chooses from "different menu selections" (p. 29) of governance structures tailored to its specific context.

Somewhat in a similar manner, Grindle (2007) notes that good governance frameworks frequently lack specific guidance for practical implementation, particularly in fragile states. The discourse around good governance includes various institutional preconditions deemed critical for economic and political development (reminiscent of the IMF and World Bank approach), yet the translation of these general principles into actionable strategies remains challenging. Grindle further contrasts findings of large-N studies, which tend to support the links between effective governance and development, and small-N case studies, focusing on the historical context and specific conditions of individual countries. She recommends adopting a more nuanced approach to governance interventions, proposing frameworks that evaluate both the current state capacities and the specific characteristics of proposed reforms.

Finally, Evans (2019) emphasizes the need for understanding the processes and actors involved in policy transfer, which resonates with Andrews' critique of the universal model of governance. The variability in governance structures, as highlighted by Andrews, suggests that sound policy transfer cannot rely on rigid normative frameworks; rather, it should adapt to the unique institutional and cultural contexts of each nation. Furthermore, Evans points to the lack of rigorous frameworks in policy transfer studies and suggests refining the methods to better evaluate lessons from past policies. Evans, similarly to Andrews (2010) and Grindle (2007), also emphasizes the significance of historical context in informing policy decisions. As the last point, the processes of learning Evans describes, i.e. copying, emulation, hybridization, and inspiration, reflect a need for flexibility in governance. In attempts to move toward people-centered governance, these learning mechanisms can better engage local actors and respond to the community's needs, ensuring that policy adaptations resonate more effectively with citizens' experiences and expectations.

Thus, the evolution toward people-centered governance reflects a significant shift in response to criticisms of traditional models of good governance. Among the key critiques are issues of universality, the inadequacy of standard frameworks, and neglect of local social structures and context. Parthasarathy (2005) highlights the misuse of the 'good governance' notion by international agencies, showing how terms and conditions are exploited to enforce structural adjustment measures that often disregard social divisions and contribute to human rights violations. Andrews (2010) emphasizes that the universal model's assumption does not hold true in practice, as governance varies in form and structure across countries. Grindle (2007) adds to this by noting that while good governance frameworks list prerequisites for development, they often provide vague guidelines that do not translate into concrete action, particularly in fragile states.

However, while all scholars seem to agree that governance approaches must be context-sensitive, their focal points differ. Parthasarathy (2005) underscores the social injustices due to a narrow focus on political and economic reforms, arguing for a broader lens that includes human rights. In contrast, Andrews (2010) critiques the theoretical uniformity of governance models, calling for recognition of diverse structures and practices. Next, Grindle (2007) emphasizes the difficulty in translating governance principles into action, calling for pragmatic, contextualized frameworks. Finally, Evans (2019) adds another dimension by focusing on the learning processes in policy transfer, suggesting that flexibility and local engagement are crucial for effective governance.

To summarize, the transition to people-centered governance emerges from a recognition that traditional good governance models are inadequate, often failing to incorporate the needs of local communities. Scholars argue for a nuanced understanding that accounts for local context and local engagement, emphasizing tailored governance solutions. By moving beyond rigid frameworks to embrace flexibility and local perspectives, governments can better respond to the needs of their citizens, enhancing the effectiveness of their governance practices.

Finally, as this section shows, there is an increasing scholarly interest in DL research across CA, where contextual specificity in definitions and applications is critical. Scholars employ terms such as digital, media, and information literacy to convey the multifaceted nature of DL (Fig. 2). Notable studies, e.g. Dadakhonov (2024b), Khamdamov 2024, point to the importance of embedding DL within educational curricula to meet local needs, while also addressing challenges such as disinformation (Abdikarimov 2024) and the implications of cultural initiatives for broader geopolitical strategies (Muratshina & Valeeva 2020). Furthermore, the review notes the shift towards people-centered governance (e.g. Parthasarathy 2005, Andrews 2010, Grindle 2007) critiquing traditional good governance models for their failure to account for local social structures, historical contexts, and the unique needs of individual nations. Scholars advocate for flexible, context-sensitive governance frameworks (e.g. Evans 2019) that engage local communities and address human rights concerns (Parthasarathy 2005), thus underlining the importance of localized approaches in enhancing governance effectiveness and responsiveness to citizens' needs. Based on the analysis, Fig. 3 below outlines a revised framework focusing on addressing digital literacy issues within a broader context of a shift toward people-centered governance in Central Asia.

Figure 3. A revised DL framework in the context of people-centered governance in CA

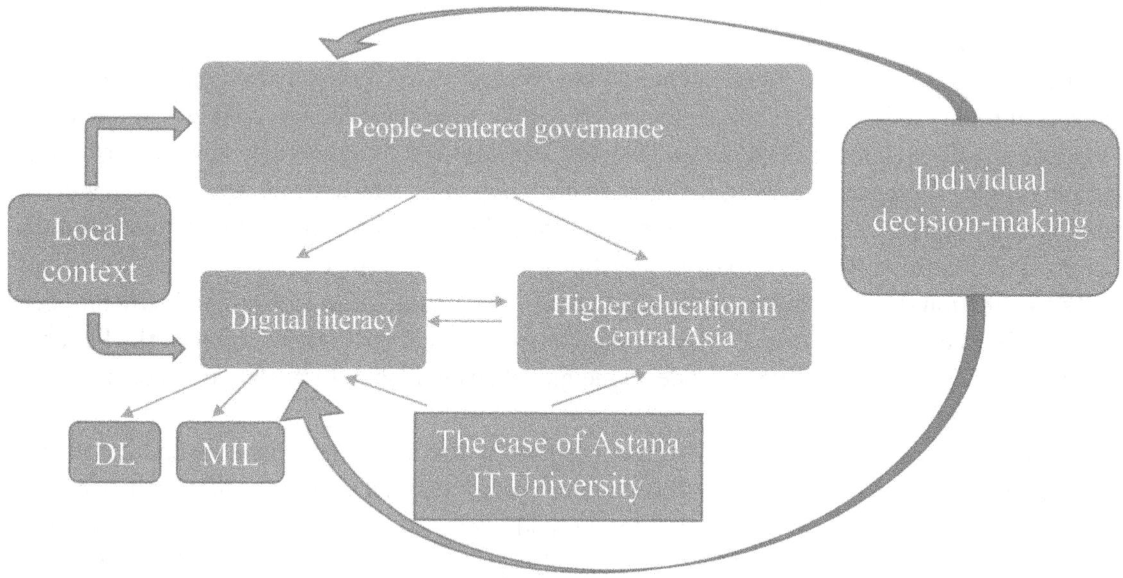

Source: The authors' own analysis

As Fig. 3 suggests, the case study of Astana IT University illustrates how embedding DL skill formation and implementation in the context of higher education development in CA can pave the way for further educational transformation in Kazakhstan and the region. The University serves as a model for integrating DL into higher education curricula, emphasizing practical applications of digital skills that resonate with the local context and societal needs. This integration not only cultivates a more digitally literate workforce but also encourages active civic participation in governance processes. Thus, what follows below is a description of case study methodology, followed by an analysis of Astana IT University as the case study of this research.

III. CASE STUDY AS A RESEARCH METHOD

The case study method can be defined in terms of its three features, as follows: 1) its ability to triangulate information from a variety of sources such as documents, interviews, observations etc. (Yin 2000); 2) its assumption of including rich data as this method analyzes "…a phenomenon in its real-life context" (p. 186); and 3) its focus on research encompassing either a single case study or multiple cases (ibid). Thus, the key characteristics of this method, as applied to the case study of Astana IT University, include triangulation of various research sources and documents (AITU 2024a, AITU 2024b, AITU 2024c), and the reflection on local context, i.e. higher education development in Kazakhstan and CA. It is worth noting that the contextual nature of case study method allows for flexibility in its design, as "...data gathering and analyses are largely determined by the subject matter" (Fidel 1984, Abstract). Finally, regarding the third feature as in Yin (2000), this research focuses on analyzing a single case study.

The case study method has found application in analyses of DL research across various contexts and jurisdictions. For instance, Zulkarnain et al. (2020) employed a single case study approach to assess student DL skills using an Indonesia-based university as the focal point. In contrast, Uršej (2019) adopted a multiple case study approach, examining three primary schools in Slovenia to explore the varying methods and outcomes of DL applications in school educational settings. These examples illustrate the adaptability of the case study method in addressing distinct research contexts, reinforcing its significance in the field of educational research.

IV. THE CASE OF ASTANA IT UNIVERSITY

Astana IT University opened its doors in 2019 (Shayakhmetova 2019), aimed to maintain high standards for educational services and establish AITU as a leader, both in Kazakhstan and Central Asia, in producing highly skilled professionals in the digital economy and the ICT sector, utilizing interdisciplinary educational programs (AITU 2024a). It offers a range of programs at Bachelor's, Master's and Doctoral levels. Some of the Bachelor's degree programs offered include Software engineering, Big data analysis, Big data in healthcare, Computer science, Cybersecurity etc. The Master's degree programs are Applied data analytics, Computer science and engineering, Project management, Digital public administration, and Media technologies. Finally, the two doctoral programs currently available are Computer science and Project management (ibid).

The AITU's website (AITU 2024a) contains relevant details on the programs offered, information for students including the learning process and grants, research projects related to big data, blockchain technologies, smart city research etc. Fig. 4 provides a snapshot of the website.

Figure 4. A snapshot from the website of Astana IT University

Source: AITU (2024a)

Among the plethora of education programs, this case study focuses on 6B03201 Digital Journalism education program offered for undergraduate students, as this specifically reflects on the development of DL skills (AITU 2024b), and 7M04102 Digital Public Administration and Services master's degree program (AITU 2024c) as this relates to the scholarly discourse on a shift toward people-centered governance in the context of higher education development in the region.

Educational program titled '6B03201 Digital Journalism' is analyzed on three levels of disciplines such as introductory, basic and advanced. Introductory disciplines focus on developing foundational digital and entrepreneurship skills (AITU 2024b). Basic level disciplines emphasize technical skills in programming, databases, web technologies, and data visualization. Advanced level disciplines provide specialized skills in graphics, marketing, and live digital media (ibid).

Introductory or general education disciplines provide foundational knowledge and skills essential for all students, regardless of their major. They include Information and Communication Technologies (ICT) and Technological Entrepreneurship (TE). The ICT course serves as an introduction to ICT and computer systems, fostering an understanding of cybersecurity, smart technologies, and human-computer interaction. This course concentrates on essential ICT skills, including DL, computing, and the use of common software applications. The TE course introduces tools of entrepreneurship, theoretical knowledge and practical skills needed to launch a startup, considering the basics of business (AITU 2024b).

Basic level or fundamental disciplines provide more specialized skills necessary for students' chosen fields of study such as Database Management Systems, Introduction to Programming, Web Technologies, Data Visualization for Journalism, Machine Learning, Information and Digital Literacy, and Web Design. For instance, in the Database Management Systems course, students learn about the design, implementation, and management of databases, gaining skills in mastering the Structured Query Language (SQL) and database optimization. More important, the Information and Digital Literacy course enhances students' ability to find, evaluate, and use information effectively in a digital context. This course plays a crucial role in equipping students with the essential skills needed to navigate the complexities of the modern information landscape, reminiscent of the complexities related to disinformation as noted by Abdikarimov (2024).

Finally, the advanced level, or major, disciplines provide in-depth knowledge and specialized skills relevant to the students' major areas of Computer Graphics and Visual Effects, Digital Marketing, Monetization & Search Engine Optimization (SEO), Live Broadcasting and Streaming in Digital Media.

To summarize, the Digital Journalism program underscores the vital role of DL skills at least across two of the three levels of disciplines: introductory and basic. This structured approach ensures that students progressively build a set of skills that prepares them for research and real-world applications. Particularly notable is the Information and Digital Literacy course, which equips students with the competencies needed to critically evaluate the vast amount of information in modern life. This skill is essential where disinformation potential is prevalent (Abdikarimov 2024), ensuring that students can discern credible sources and effectively utilize information. This also echoes Khamdamov's (2024) call for embedding DL skills into various disciplines, not just as a single course. Yet, further efforts should be put into expanding DL-related courses into advanced-level (major) disciplines within the Digital Journalism program, as well as across other programs of the Bachelor's level e.g. Computer science, Big data analysis, Mathematical science, the Master's level e.g. Computer science and engineering, Media technologies, and possibly Doctoral-level Computer science program (AITU 2024a).

Next, 7M04102 Digital Public Administration and Services master's degree program seeks to equip managers with the new skills in digital economy transformation and an evolving information society (AITU 2024c). The competencies should enable the graduates to engage in professional and research roles in government, local administration, state enterprises, international agencies, and private business. They will be prepared to undertake various tasks and to contribute to government projects focused on digital governance.

Basic level or fundamental disciplines include Digital (Local) Public Administration, Audit of Processes in Information Security etc. Specifically, Digital Local Public Administration seeks to foster an understanding of regional development strategies and develop effective ICT solutions that can enhance the operations of local executive authorities and community initiatives, while Digital Public Administration, with a focus on governance, seeks to develop sound IT solutions in nationwide government activities. The course syllabus is based on historical, contextual and theoretical perspectives focusing on a bridge between theory and practice of public management and administration. Finally, the Audit of Processes course aims at studying methods and procedures for auditing processes in information security, including an identification of vulnerabilities and risks, analyzing control measures and developing measures to eliminate deficiencies and threats.

Finally, the advanced level, or major, disciplines include Public Policy and E-government, Legal Aspects of Digitization, and Social Innovation and Digital Transformation of Society, among others including elective courses. First, the Public Policy and E-government course examines the interaction between the public and private sectors with an emphasis on E-government, ICT, public values and social activity, co-creation of public services, the role of citizens as co-producers and consumers, as well as social learning and feedback mechanisms in E-government. Next, the purpose of Legal Aspects of Digitization is to provide an overview of e-governance from the viewpoint of law and legal policy, and to analyze legislative acts of e-governance. Third, the Social Innovation course explores social innovations of collaborative creativity and live labs as platforms for their support. Joint co-creation of social innovations is studied, with a focus on creating value for society and engaging the local citizenry in social innovation and digital transformation processes.

To summarize, this master's degree program is meticulously designed to educate leaders and managers in public administration in a rapidly evolving digital landscape. This program emphasizes human-centeredness, focusing on creating public value and driving social activity, while engaging citizens as integral participants in the co-creation of public services and social innovations. Thus, students should be able to reflect on diverse voices and needs of the communities they serve. By prioritizing citizen engagement and collaboration, the program fosters an inclusive environment where public values drive digital transformation efforts. Furthermore, it stresses the importance of historical context (e.g. Grindle 2007, Evans 2019). Finally, this program underscores the need to promote participatory governance in digital public administration, as the basis for a further shift toward people-centered governance in Kazakhstan and across the region.

V. DISCUSSION OF FINDINGS

This research points to a number of vital findings. First and foremost, it seeks to map out an emerging theoretical framework focusing on DL implementation in the context of a shift toward people-centered governance in Central Asia, first as outlined in Fig. 1 following an introductory analysis, then further revised in Fig. 3 as a result of a two-level literature review. Its key defining features include the following:

1. An ongoing shift in scholarly debates from good governance to people-centered governance models (e.g. Parthasarathy 2005, Andrews 2010, Grindle 2007);
2. The multi-faceted nature of the 'digital literacy' notion (Pangrazio et al. 2020, Balgabayeva et al. 2024, Abdikarimov 2024, Khamdamov 2024, Dadakhonov 2023 etc.);
3. The context specific nature of both DL (Gruszczynska & Pountney 2013, Pangrazio et al. 2020) and the general framework (e.g. Ng 2012, as cited in Khamdamov 2024; Evans 2019);
4. The importance of individual engagement in developing DL skills and boosting individual decision-making (e.g. Martin & Grudziecki 2006; Mohammadyari & Singh 2015); and
5. The inclusion of a unique context-specific case study (e.g. AITU 2024a, AITU 2024b).

The second finding, which stems from the above description of the framework, is the multi-faceted notion of digital literacy (first noted by Pangrazio et al. 2020, then as applied to DL research in Central Asia). Specifically, as analyzed earlier, some scholars use terms such as 'digital literacy' in various country contexts (e.g. Balgabayeva et al. 2024 in Northern Kazakhstan, Khamdamov 2024 in Uzbekistan, Marav 2016 in Mongolia), 'media literacy' (e.g. Muratshina & Valeeva 2020 on cooperation between Russia and CA states, Akhmetova & Shorokhov 2018 in Kazakhstan, Abdikarimov 2024 on Uzbek media space), and 'information literacy' (e.g. Muratshina & Valeeva 2020, Akhmetova & Shorokhov 2018), etc. as outlined in Fig. 2. In summary, although these findings generally support the arguments developed by Pangrazio et al. (2020), specific nuances differ as related to the intricate context-specific DL processes unfolding in Central Asia.

Third, the analysis of Astana IT University as the case study fits well into the framework, as it informs and refines scholarly debates by embedding DL implementation in the context of higher education development in CA (Fig. 3). Specifically, the Digital Journalism program reflects the implementation of DL skills, particularly at the introductory and basic levels. The Information and Digital Literacy course is key in teaching critical thinking and evaluation skills useful in combating disinformation (Abdikarimov

2024). This approach aligns with calls for integrating DL across various disciplines (Khamdamov 2024). However, there is a need to expand DL-related courses into other Bachelor's, Master's, and Doctoral-level programs, to enhance overall digital competency. Furthermore, the Digital Public Administration master's program emphasizes human-centeredness and public value, engaging citizens as active participants in the co-creation of services. By prioritizing participation, the program fosters an inclusive environment that aligns public values with digital transformation efforts. It also focuses on local context and advocates for participatory governance to shift toward people-centered governance in Kazakhstan and the region.

The summary table of key findings and recommendations is presented below:

Table 1. Findings and recommendations

Findings	Recommendations
1. Shift from Good Governance to People-Centered Governance: The study stresses a transformation in scholarly debates from good governance models to people-centered governance, emphasizing the need for local context and citizen engagement in decision-making processes.	**1. Foster Local Governance Models:** Policymakers should encourage the growth of governance frameworks that prioritize community needs and individual engagement to ensure inclusivity in governance.
2. Multifaceted Nature of Digital Literacy (DL): The study identifies that DL encompasses terms such as digital, media, and information literacy, reflecting its diverse applications across contexts.	**2. Integrate Diverse DL Concepts:** Educational institutions should reflect various DL frameworks into curricula to address the specific needs of students and local context.
3. Context-Specific Nature of DL Implementation: The analysis stresses that DL initiatives must be tailored to fit the unique cultural, social, and technological contexts of Central Asia.	**3. Localize DL Strategies:** Develop localized DL initiatives that consider the cultural and technological landscape of Central Asia, ensuring relevance and effectiveness.
4. Importance of Individual Engagement: The findings emphasize the role of individual engagement in fostering DL skills, enhancing decision-making and participation in governance.	**4. Encourage Active Participation:** Educational and governance programs should involve citizens in decision-making processes to empower them and boost their digital skills.
5. Case Study of Astana IT University: The case study shows sound implementation of DL skills in higher education, through the Digital Journalism and Digital Public Administration programs.	**5. Expand DL Curriculum:** AITU and similar institutions should broaden the scope of DL courses across all levels of education to boost overall digital competency.
6. Critical Thinking and Evaluation Skills: The Information and Digital Literacy course at AITU is crucial for developing skills, and evaluating digital information and combating disinformation.	**6. Prioritize Critical Literacy Education:** Academic programs should foster DL to equip students with the skills to navigate and discern information in the digital age.
7. Emphasis on Human-Centeredness in Public Administration: The Digital Public Administration program focuses on creating public value and engaging citizens in governance processes.	**7. Promote Participatory Governance:** Encourage educational programs and policies that emphasize participatory governance and citizen engagement to advance towards people-centered governance in Kazakhstan.

This table synthesizes the key findings from the research and provides corresponding recommendations to more specifically guide policy and educational strategies.

VI. CONCLUDING REMARKS

As this research seeks to offer an emerging framework at the intersection of DL and evolution toward people-centered governance in the context of higher education development in Central Asia, other scholarly efforts should be directed toward refining this theoretical framework both as applied to Kazakhstan and other countries in CA. Furthermore, other studies might shed light onto the multi-faceted notions of DL, MIL and, possibly, other similar notions reflecting the context specificity of the region. Third, efforts should focus on adding new institutional case studies in comparative perspective in order to improve the generalization of this research findings, to refine the scope of the emerging DL framework within a shift toward people-centered governance, and to improve external validity. By doing so, this would also help to address the major limitation of this research, i.e. being limited to a single case study.

ACKNOWLEDGEMENT

This work was supported by the Kazakh Ministry of Science and Higher Education project grant, registration number AP23490441.

The authors' short bios:

Mergen Dyussenov (PhD), Assistant professor at Astana IT University, Kazakhstan (full-time), Research fellow at University of Illinois Urbana-Champaign, United States (2023-2024). A former advisor to the Kazakh Minister of Culture and Sports for anti-corruption policy (2020-2021), deputy director of the Institute of Management at the Academy of Public Administration under the President of Kazakhstan (2019-2020).

Nurkhat Ibadildin (PhD), Associate professor at Astana IT University, Kazakhstan (full-time). Former Director of R7 Group Design and Construction Company (2013-2018), senior lecturer at Satpayev Kazakh National Technical University (2009-2010).

Prof. Ha Jin Hwang is the Director of the School of creative industries at Astana IT University, Kazakhstan. Prior to that, he served as Professor and Head of Department of Business Analytics at Sunway University, Malaysia (2016-2023), Professor at KIMEP University, Kazakhstan (2010-2016), Professor at Catholic University of Daegu, Korea (1991-2010), Associate Professor of MIS, Minnesota State University, Mankato, Minnesota, USA (1989-1991).

REFERENCES

Abdikarimov, K. (2024). Media literacy as a means of protecting Uzbekistan's media environment from disinformation. *International Journal of Media and Communications in Central Asia*. Tashkent. *UzJOKU*, (Special issue), 47–55.

Akhmetova, L.S. & Shorokhov, D.P. (2018). Media and Information Literacy: The solution of questions of formation of a civic stand in Kazakhstan. *Публічне урядування, 3*(13), 14-27.

Ameen, K., & Gorman, G. E. (2009). Information and digital literacy: A stumbling block to development? A Pakistan perspective. *Library Management*, *30*(1/2), 99–112. DOI: 10.1108/01435120910927565

Andema, S. (2014). *Promoting digital literacy in African education: ICT innovations in a Ugandan primary teachers' college* (Doctoral dissertation, University of British Columbia).

Andrews, M. (2010). Good government means different things in different countries. *Governance: An International Journal of Policy, Administration and Institutions*, *23*(1), 7–35. DOI: 10.1111/j.1468-0491.2009.01465.x

Andriushchenko, K., Rozhko, O., Tepliuk, M., Semenyshyna, I., Kartashov, E., & Liezina, A. (2020). Digital literacy development trends in the professional environment. *International Journal of Learning. Teaching and Educational Research*, *19*(7), 55–79.

Astana, I. T. University (2024a). *About us*. Retrieved from: https://astanait.edu.kz/en/about-2/

Astana, I. T. University (2024b). *Digital Journalism*. Retrieved from: https://astanait.edu.kz/en/digital-journalism-2/

Astana, I. T. University (2024c). *Digital Public Administration and Services*. Retrieved from: https://astanait.edu.kz/en/digital-public-administration-and-services-2/

Balgabayeva, A.E., Karataeva, T.O, Karabulatova, I.S. et al. (2024). Digital Literacy as a Meta-Cognitive Component of Younger Students' Intellectual and Creative Potential in Foreign Language Lessons. *Rupkatha Journal, 16*(1).

Bejaković, P., & Mrnjavac, Ž. (2020). The importance of digital literacy on the labour market. *Employee Relations*, *42*(4), 921–932. DOI: 10.1108/ER-07-2019-0274

Bykov, I., Gladchenko, I. A., Ibrayeva, G. Zh., & Myssayeva, K. N. (2019). *Digital Literacy and Strategy of PR-education in Russia and Kazakhstan: Comparative analysis. In 2019 Communication Strategies in Digital Society Workshop (ComSDS)*. IEEE.

Chelysheva, I. (2019). Media literacy education in Uzbekistan, Kyrgyzstan and Turkmenistan. *International Journal of Media and Information Literacy*, *4*(1), 3–10.

Chen, L. (2020). Data connectivity in the Greater Mekong Subregion. *Subregional Development Strategy in ASEAN after COVID-19: Inclusiveness and Sustainability in the Mekong Subregion (Mekong 2030)*, 42.

Dadakhonov, A. O. (2023). Theoretical approaches to Media and Information Literacy in Foreign Aid Projects (Model for Central Asian countries). *Communication Papers.Media Literacy and Gender Studies*, *12*(25), 23–46.

Dadakhonov, A. O. (2024a). Sustaining foreign aid initiatives in the field of media and information literacy within the Central Asian context. *Medical Education*, *15*(1), 57–70. DOI: 10.36253/me-15703 PMID: 38212802

Dadakhonov, A. O. (2024b). Main Issues of Implementing Media and Information Literacy in School Education in Central Asian Countries (The Case of Kazakhstan, Kyrgyzstan, and Uzbekistan). *Media Literacy and Academic Research*, *7*(1), 51–71. DOI: 10.34135/mlar-24-01-03

Erlanitasari, Y., Rahmanto, A., & Wijaya, M. (2020). Digital economic literacy micro, small and medium enterprises (SMES) go online. *Informasi*, *49*(2), 145–156. DOI: 10.21831/informasi.v49i2.27827

European Commission. (2020, September 9). *Communication from the Commission to the European Parliament, the Council, the European Economic and Social Committee and the Committee of the Regions on the Digital Education Action Plan. COM(2020) 624 final*. https://eur-lex.europa.eu/legal-content/EN/TXT/?uri=celex:52020DC0624

Evans, M. (2019). International policy transfer: Between the global and sovereign and between the global and local. In Stone, D., & Moloney, K. (Eds.), *The Oxford Handbook of Global Policy and Transnational Administration* (pp. 94–110). Oxford University Press., DOI: 10.1093/oxfordhb/9780198758648.013.3

Fidel, R. (1984). The case study method: A case study. *Library & Information Science Research*, *6*(3), 273–288.

Fumasoli, T., & Rossi, F. (2021). The role of higher education institutions in transnational networks for teaching and learning innovation: The case of the Erasmus+ programme. *European Journal of Education*, *56*(2), 200–218. DOI: 10.1111/ejed.12454

Gómez-Galán, J., Martínez-López, J. Á., Lázaro-Pérez, C., & Fernández-Martínez, M. D. M. (2021). Usage of Internet by University Students of Hispanic Countries: Analysis Aimed at Digital Literacy Processes in Higher Education. *European Journal of Contemporary Education, 10*(1), 53–65.

Grindle, M. S. (2007). Good enough governance revisited. *Development Policy Review, 25*(5), 533–574. DOI: 10.1111/j.1467-7679.2007.00385.x

Gruszczynska, A., Merchant, G., & Pountney, R. (2013). "Digital futures in teacher education": Exploring open approaches towards digital literacy. *Electronic Journal of e-Learning, 11*(3), 193–206.

Gruszczynska, A., & Pountney, R. (2013). Developing the concept of digital literacy in the context of schools and teacher education. *Enhancing Learning in the Social Sciences, 5*(1), 25–36. DOI: 10.11120/elss.2013.05010025

Huerta, E., & Sandoval-Almazán, R. (2007). Digital literacy: Problems faced by telecenter users in Mexico. *Information Technology for Development, 13*(3), 217–232. DOI: 10.1002/itdj.20071

Immanuel Kant Baltic Federal University. (2024). *Russian Dictation*. Retrieved from: https://kantiana.ru/en/international/russian-dictation/

Isono, I., & Prilliadi, H. (2023). *Accelerating Artificial Intelligence Discussions in ASEAN: Addressing Disparities, Challenges, and Regional Policy Imperatives*. Economic Research Institute for ASEAN and East Asia.

Kairat, A., Sultanbayeva, G. S., & Kuash, R. (2024). Ways to solve the problems of information and digital literacy within the framework of the "One Belt, One Road" initiative. *Bulletin of LN Gumilyov Eurasian National University.JOURNALISM Series, 146*(1), 57–73.

Khamdamov, N. N. (2024). Strategies to Improve Students' Digital Literacy in Higher Education. *Educational Research in Universal Sciences, 3*(6), 243–248.

Kusumastuti, A., & Nuryani, A. (2020, March). Digital literacy levels in ASEAN (comparative study on ASEAN countries). In *IISS 2019: Proceedings of the 13th International Interdisciplinary Studies Seminar, IISS 2019,30-31 October 2019,Malang, Indonesia* (p. 269). European Alliance for Innovation. DOI: 10.4108/eai.23-10-2019.2293047

Laak, K. J., & Aru, J. (2024). AI and Personalized Learning: Bridging the gap with modern educational goals. *arXiv preprint arXiv:2404.02798.*

Lankshear, C. J., & Knobel, M. (2008). *Introduction: digital literacies: concepts, policies and practices.* Peter Lang Publishing.

Lemphane, P., & Prinsloo, M. (2014). Children's digital literacy practices in unequal South African settings. *Journal of Multilingual and Multicultural Development, 35*(7), 738–753. DOI: 10.1080/01434632.2014.908894

Marav, D. (2016). Mongolian students' digital literacy practices: The interface between English and the internet. *Trabalhos em Lingüística Aplicada, 55*(02), 293–318. DOI: 10.1590/010318134962176441

Martin, A., & Grudziecki, J. (2006). DigEuLit: Concepts and tools for digital literacy development. *Innovation in teaching and learning in information and computer sciences, 5*(4), 249-267.

Mohammadyari, S., & Singh, H. (2015). Understanding the effect of e-learning on individual performance: The role of digital literacy. *Computers & Education, 82*, 11–25. DOI: 10.1016/j.compedu.2014.10.025

Muratshina, K. G., & Valeeva, M. V. (2020). Russia—Post-Soviet Central Asia Cooperation in Information and Media. *Communication trends in the post-literacy era: polylingualism, multimodality and multiculturalism as prerequisites for new creativity. Ekaterinburg, 2020*, 476-495

Okechukwu, A. A. (2011). PEOPLE-CENTERED GOVERNANCE AS THE BED-ROCK OF SUSTAINABLE DEVELOPMENT. *ISIG JOURNAL*, 108.

Pangrazio, L., Godhe, A. L., & Ledesma, A. G. L. (2020). What is digital literacy? A comparative review of publications across three language contexts. *E-Learning and Digital Media, 17*(6), 442–459. DOI: 10.1177/2042753020946291

Parthasarathy, D. (2005). *Taking responsibilities and participation seriously: A Critique of 'Good Governance. Droit, gouvernance et développement durable, numéro spécial des Cahiers d'Anthropologie du Droit.* Éditions Karthala.

Radovanović, D., Holst, C., Belur, S. B., Srivastava, R., Houngbonon, G. V., Le Quentrec, E., Miliza, J., Winkler, A. S., & Noll, J. (2020). Digital literacy key performance indicators for sustainable development. *Social Inclusion (Lisboa), 8*(2), 151–167. DOI: 10.17645/si.v8i2.2587

Rajput, A., & Nair, K. (2013). Significance of digital literacy in e-governance. *The SIJ transactions on industrial financial & business management, 1*(4).

Reddy, P., Sharma, B., & Chaudhary, K. (2020). Digital literacy: A review of literature. [IJT]. *International Journal of Technoethics, 11*(2), 65–94. DOI: 10.4018/IJT.20200701.oa1

Sharma, R., Fantin, A. R., Prabhu, N., Guan, C., & Dattakumar, A. (2016). Digital literacy and knowledge societies: A grounded theory investigation of sustainable development. *Telecommunications Policy, 40*(7), 628–643. DOI: 10.1016/j.telpol.2016.05.003

Shayakhmetova, Z. (2019, Sep 14). *Astana IT University opens in Kazakh capital.* Retrieved from: https://astanatimes.com/2019/09/astana-it-university-opens-in-kazakh-capital/

Shopova, T. (2014). Digital literacy of students and its improvement at the university. *Journal on Efficiency and Responsibility in Education and Science, 7*(2), 26–32. DOI: 10.7160/eriesj.2014.070201

Uršej, K. (2019). Digital literacy in the first three years of primary school: Case study in Slovenia. *International Journal of Management. Knowledge and Learning, 8*(1), 61–77.

Yin, R. K. (2000). Case study evaluations: a decade of progress? In *Evaluation models: Viewpoints on educational and human services evaluation* (pp. 185–193). Springer Netherlands.

Yuen, A. H., & Hew, T. K. (2018). Information and communication technology in educational policies in the Asian region. *Handbook of information technology in primary and secondary education*, 1-20.

Zulkarnain, Z., Heleni, S., & Thahir, M. (2020, October). Digital literacy skills of math students through e-learning in COVID-19 era: A case study in Universitas Riau. [). IOP Publishing.]. *Journal of Physics: Conference Series, 1663*(1), 012015. DOI: 10.1088/1742-6596/1663/1/012015

APPENDIX 1. THE LIST OF DL MAJOR PUBLICATIONS RELATED TO CENTRAL ASIAN DEBATES

Table 2.

#	Author's name(s), year	Publication title	Published by
1	Abdikarimov, K. (2024)	Media literacy as a means of protecting Uzbekistan's media environment from disinformation	International Journal of Media and Communications in Central Asia
2	Akhmetova, L.S. & Shorokhov, D.P. (2018)	Media and Information Literacy: The solution of questions of formation of a civic stand in Kazakhstan.	Публічне урядування (journal)
3	Balgabayeva, A.E. et al. (2024)	Digital Literacy as a Meta-Cognitive Component of Younger Students' Intellectual and Creative Potential in Foreign Language Lessons.	Rupkatha Journal
4	Bykov I. et al. (2019)	Digital Literacy and Strategy of PR-education in Russia and Kazakhstan: Comparative analysis.	IEEE
5	Chelysheva, I. (2019).	Media literacy education in Uzbekistan, Kyrgyzstan and Turkmenistan.	International Journal of Media and Information Literacy
6	Dadakhonov, A. O. (2023).	Theoretical approaches to Media and Information Literacy in Foreign Aid Projects (Model for Central Asian countries).	Communication Papers. Media Literacy and Gender Studies.
7	Dadakhonov, A.O. (2024a).	Sustaining foreign aid initiatives in the field of media and information literacy within the Central Asian context.	Media Education
8	Dadakhonov, A.O. (2024b).	Main Issues of Implementing Media and Information Literacy in School Education in Central Asian Countries (The Case of Kazakhstan, Kyrgyzstan, and Uzbekistan).	Media Literacy and Academic Research
9	Kairat, A. et al. (2024).	Ways to solve the problems of information and digital literacy within the framework of the "One Belt, One Road" initiative.	Bulletin of LN Gumilyov Eurasian National University. JOURNALISM Series
10	Khamdamov, N.N. (2024).	Strategies to Improve Students' Digital Literacy in Higher Education.	Educational Research in Universal Sciences
11	Marav, D. (2016).	Mongolian students' digital literacy practices: the interface between English and the internet.	Trabalhos Em Linguística Aplicada
12	Muratshina, K.G. & Valeeva, M.V. (2020).	Russia—Post-Soviet Central Asia Cooperation in Information and Media.	Communication trends in the post-literacy era: polylingualism, multimodality and multiculturalism as prerequisites for new creativity

Source: The authors' own analysis

Chapter 8
Fair Use in the Digital Era: Balancing Creativity and Copyright Protection

Puneet Chadha
https://orcid.org/0009-0001-7374-9157
School of Liberal Studies, Pandit Deendayal Energy University, Gandhinagar, India

Pooja Shukla
https://orcid.org/0009-0006-7229-4783
Amity College of Commerce and Finance, Amity University, Ranchi, India

ABSTRACT

In the digital age, the idea of fair use has evolved as an important tool for balancing copyright holders' rights with the need for artistic expression. This chapter explores the changing interpretations of fair use in the context of digital content, where accessibility and distribution have reshaped traditional copyright protection. The study examines how fair use promotes innovation and public discourse while protecting intellectual property via the analysis of major cases and legal frameworks. The chapter also looks at the issues that technology breakthroughs like artificial intelligence and social media provide, as well as the ramifications for both producers and consumers. The objective is to present an equitable approach that protects original works while promoting innovation and knowledge exchange in the digital realm.

1.0 INTRODUCTION

The digital era has significantly changed how material is created, disseminated, and consumed throughout the world. The internet has democratized content production, providing individuals with unparalleled chances to express their ideas through blogs, podcasts, social media platforms, and streaming services (Goldman, 2020). Similarly, technology such as artificial intelligence (AI), machine learning, and automated content development have introduced new forms of art and communication. However,

DOI: 10.4018/979-8-3693-6675-2.ch008

Copyright ©2025, IGI Global Scientific Publishing. Copying or distributing in print or electronic forms without written permission of IGI Global Scientific Publishing is prohibited.

the ease of access and ubiquity of digital information have posed novel issues for copyright protection, intellectual property (IP), and the extent of fair use in this new context (Samuelson, 2019).

In traditional copyright regimes, the creator of an original work has exclusive rights to its use and distribution, which prevents unlawful reproduction, modification, or dissemination (Davison & Hugenholtz, 2017). These safeguards are critical to ensure that artists, authors, and other creators are properly rewarded for their contributions. However, there has always been a counterbalance in the shape of fair use, which permits the restricted use of copyrighted works for certain purposes without requiring permission or license. Fair use allows for scholarly inquiry, commentary, parody, and criticism, resulting in a more dynamic cultural and intellectual interchange (Litman, 2021).

However, the lines defining fair use have blurred more and more in the digital age, since massive amounts of material may be instantly shared across countries and platforms (Aufderheide & Jaszi, 2018). User-generated content (UGC), artificial intelligence (AI) materials, and imaginative remixes that defy the conventional understanding of fair usage give rise to problems. Social networking sites, for example, promote the sharing and resharing of photographs, music, and movies, frequently in violation of copyright regulations (Sunder, 2018). Similar to this, AI systems may create and analyse content based on previously published works, which raises concerns regarding copyright and originality (McCutcheon, 2021).

India, with its expanding digital economy, faces unique challenges in balancing the interests of content creators and consumers. The landmark case of Super Cassettes Industries Ltd. v. MySpace Inc. (2011) highlighted the complexities of intermediary liability and fair use principles in the Indian context. In this context, cross-border copyright enforcement has become a significant challenge as different jurisdictions have different interpretations of fair use, resulting in inconsistencies in legal enforcement and compliance. The European Union's Directive on Copyright in the Digital Single Market introduced stricter content filtering measures, while the United States continues to emphasize a flexible, case-by-case analysis under the fair use doctrine (European Commission, 2019; U.S. Copyright Office, 2021).

Furthermore, copyright holders find it difficult to monitor and regulate the use of their work in a virtual environment that is continually growing due to the sheer size of digital platforms (Hughes, 2019). This has hampered enforcement attempts. A great deal of controversy has resulted from this authors and copyright holders contend that in order to preserve their moral and financial rights, tougher enforcement measures and protections are necessary (Bently & Sherman, 2014). Conversely, digital activists, content producers, and the general public tout the value of innovation, cooperation, and unrestricted information exchange in support of a more accommodating interpretation of fair use (Lessig, 2015).

Indian courts have been grappling with the meaning of fair use in the digital environment. In the case of Chancellor Masters and Scholars of the University of Oxford v. Rameshwari Photocopy Services (2016), the Delhi High Court ruled in favor of educational institutions, highlighting the value of knowledge access. These ruling highlights the particular socioeconomic factors in India that impact fair use determinations.

This chapter seeks to explore the complicated dynamics of fair use in the digital era, namely how current legal frameworks interact with the new realities of digital information. It will look at major legal issues that have affected the meaning of fair use in the twenty-first century, emphasizing the continuous tension between protecting copyright holders and encouraging creation. It will also investigate how new technologies, notably AI and automated systems, are altering the debate over intellectual property rights (Katz 2020).

1.1 Overview

Fair use has long been a cornerstone of copyright law, striking a delicate balance between preserving authors' intellectual property and enabling others to expand on existing works for educational, commentary, criticism, research, and parody purposes. This balance is even more important in the digital era, due to the unparalleled ease of accessing, sharing, and remixing material. Individuals may produce and distribute content internationally within seconds utilizing platforms like as YouTube, Instagram, TikTok, and new artificial intelligence (AI) capabilities, frequently employing copyrighted material in ways that challenge traditional conceptions of fair use (Aufderheide & Jaszi, 2018).

This chapter explores how the fair use concept is evolving to meet the challenges of the digital era, in which intellectual property borders are more flexible and challenged than ever before. The digital revolution has democratized content production, allowing ordinary people to create and distribute media alongside companies and conventional media institutions. However, this democratization has raised significant worries about copyright infringement, as user-generated content usually involves the unlawful use of copyrighted information, often without the creator's intent to harm or steal intellectual property (Hughes, 2019).

The main purpose of this chapter is to explain how fair usage works in today's digital environment. It presents a detailed assessment of the legal frameworks that regulate fair use, with a focus on user-generated content (UGC) and AI-assisted creativity. Landmark court decisions, such as Lenz v. Universal Music Corp. (2015), which addressed a takedown request for a YouTube video under the Digital Millennium Copyright Act (DMCA), demonstrate the changing nature of fair use in the digital environment (McCutcheon, 2021). The intricacies of fair use in India are highlighted by cases like Civic Chandran v. Ammini Amma (1996), which acknowledged the transformative use of copyrighted work for the public interest, and Super Cassettes Industries Ltd. v. MySpace Inc. (2011), which dealt with intermediary liability for copyright infringement. These examples demonstrate the contradiction between the need to protect copyright holders and the need to promote creative expression, public conversation, and cultural growth.

Furthermore, the fast advancement of AI technology creates new and difficult issues for copyright law and fair use. AI-generated works, which are frequently developed by algorithms that draw on massive datasets of pre-existing content, blur the distinction between originality and ownership. While AI tools provide new opportunities for content production, they also pose problems regarding who owns the copyright to AI-generated works and how fair use applies in these cases (Katz, 2020). This chapter investigates how AI-driven innovations may disrupt current intellectual property frameworks, as well as what legal safeguards may be required to ensure that fair use principles can accommodate the rise of such technologies.

The ethical aspect of fair usage is equally important in the digital age. Ethical issues include ensuring that content providers are fairly recognized and compensated for their efforts, as well as creating an atmosphere that promotes cooperation and information sharing. For example, ethical quandaries occur when AI-generated material closely resembles the manner of human artists, resulting in questions over authenticity and artistic integrity (Boucher, 2022). Striking an ethical balance entails connecting legal rules with moral concerns in order to minimize exploitation while encouraging innovation.

The subject matter also covers the rising dispute between copyright holders and digital innovators. Copyright holders call for tougher enforcement procedures, particularly on big platforms where content may spread quickly and infringe on copyrights (Samuelson, 2019). At the same time, digital artists and activists call for a broader definition of fair use, particularly given the collaborative and transformative character of most of today's digital work (Lessig, 2015). This chapter underlines the need of copyright legislation striking a balance between preserving intellectual property rights and fostering the creativity and innovation that digital platforms enable.

User-generated content (UGC), AI-generated works, and moral quandaries have made fair use a crucial problem in the digital age, posing new difficulties for copyright law. In the context of social media, the intricacies of fair use are best shown by the seminal case of Lenz v. Universal Music Corp. (2015) in the United States. In one instance, Universal Music Corporation requested that a video showing her toddler dancing to Prince's song "Let's Go Crazy" be taken down from YouTube by the mother, in accordance with the Digital Millennium Copyright Act (DMCA). The court decided that before Universal could issue a takedown notice, it had to determine whether the song's use constituted fair use. This case demonstrates how, in the era of user-generated content, copyright holders must consider fair use before bringing infringement charges (Aufderheide & Jaszi, 2018). In a similar vein, Super Cassettes Industries Ltd. v. MySpace Inc. (2011) in India highlighted the function of platforms in overseeing content published by users. The Delhi High Court upheld platform responsibility while recognizing the right to fair use by ruling that MySpace, as an intermediary, was not responsible for copyright violations until it was notified of them (Sengupta, 2011). Even while fair use encourages artistic expression, the legal system must find a balance between copyright protection and content production, as seen by the millions of DMCA takedown requests that sites like YouTube continue to receive.

Particularly with regard to the ownership of works created by algorithms, the emergence of AI-generated works has presented significant difficulties for copyright law. The AI-generated image "Edmond de Belamy," for instance, which was produced by the art group Obvious, sold for $432,500 in 2018. This raised concerns over authorship and whether AI-generated works should be covered by conventional copyright regulations (Katz, 2020). The Sham Lal Jain v. Union of India (2020) lawsuit in India also looked at this matter, determining if copyright protection may be given to poetry produced by artificial intelligence. The Delhi High Court asked for revisions to copyright rules to reflect this new creative reality and acknowledged the need for a more precise legal framework for works produced by AI (Jain, 2020). A more sophisticated approach to fair use and ownership is needed to account for these technical breakthroughs as AI continues to influence the creative sectors.

The ethical aspect of fair usage is another urgent problem in the digital age, especially when it comes to deepfakes and AI-generated material. Deepfake technology raises serious ethical concerns since it uses pre-existing information to produce incredibly lifelike audio or video of humans saying or doing things they never did. In 2018, a deepfake video of Mark Zuckerberg confessing to managing user data was shared on Instagram by an artist by the name of Bill Posters. The film aroused concerns about misrepresentation and the possible harm caused by false information, even if it was intended as an artistic reflection on the dominance of digital giants (Boucher, 2022). This reflects the ethical discussions in international cases like Indian Performing Right Society Ltd. v. Sanjay Dalia & Anr. (2011), in which the Delhi High Court upheld the IPRS's right to perform copyrighted music in public without permission, emphasizing the value of protecting creator rights (Ranjan, 2011). The ethical issues surrounding fair use are becoming more complicated as a result of the usage of AI in deepfakes and other creative expressions, which push the limits of authenticity and artistic integrity. Similar issues were brought up in India in M/s. Ramesh D. Naik v. M/s. Rajhans (2017), where the Delhi High Court decided that, even when a painting is transformative, commercial exploitation of it without permission cannot be justified as fair use (Singh, 2017). For digital producers operating in a world where remix culture and collaborative production are ubiquitous, this case brought to light the necessity of defining the boundary between infringement and inspiration.

These instances show how fair use is changing in India and beyond to meet the demands of consumers and artists in the digital age. Fair use rules must be modified to meet the new problems presented by developing technologies as AI, user-generated content, and ethical concerns continue to reshape our understanding of intellectual property. A well-balanced approach to copyright law may safeguard the rights of artists while promoting the type of ground-breaking and revolutionary output that propels cultural advancement in the digital era.

1.2 Emergence

Fair use originated in classical copyright law, but its importance has grown exponentially in the digital era. Fair use was originally created to strike a compromise between copyright holders' rights and the public's right to free speech, criticism, and education. However, the internet, digital media, and rapid technical improvements have expanded the breadth and complexity of fair use (Aufderheide & Jaszi, 2018).

In the pre-digital age, fair use was largely used by historians, journalists, and artists to reference or critique previous works without infringing on the rights of the original creator. It was a well-defined legal principle that was used in a specific set of situations, such as citing a section of a book for literary criticism or utilizing a short film clip for academic reasons (Davison & Hugenholtz, 2017). However, with the growth of digital technologies, notably the internet and social media platforms, fair use has had to change to accommodate the enormous number and accessibility of information accessible today.

One of the most significant changes has been the identification of transformative works as an essential component of fair use. Transformative usage happens when a user adds fresh expression, meaning, or value to the original work, resulting in something new and different (Hughes, 2019). The historic decision Campbell v. Acuff-Rose Music, Inc. (1994) was essential in establishing this concept, especially in the context of satire and criticism. In the digital age, this notion has broadened to encompass remixes, mashups, memes, and other types of user-generated content that largely rely on pre-existing content.

At the same time, digital technologies have made copyright enforcement more important and complex. Copyright holders must increasingly monitor a vast volume of online material for infringements, while automated methods such as YouTube's material ID seek to find a balance between preserving artists' rights and permitting fair usage (Litman, 2021). However, these methods are frequently chastised for being unduly restrictive and occasionally misidentifying valid fair usage, restricting creativity in the process.

The growth of artificial intelligence (AI) has complicated the evolution of fair usage in the digital age. AI systems, such as content generators and creative algorithms, sometimes rely on large databases of previously published works to produce new material, raising concerns about originality and copyright. AI-generated art, literature, and music call into question traditional conceptions of authorship and fair use since these technologies might copy or change existing works without the original creator's involvement (Katz, 2020). This new frontier in content production has prompted requests for more specific criteria on how fair use applies to AI-generated material and who owns the copyright to such works.

As the digital world evolves, fair use becomes an increasingly important instrument for balancing innovation, creativity, and intellectual property rights. The continuous conflict between copyright holders and digital producers highlights the importance of flexible, adaptable interpretations of fair use that can accommodate new forms of expression and technical breakthroughs. By adapting to the difficulties of the digital era, fair use can maintain a dynamic and inclusive cultural environment in which innovation thrives.

2.0 REVIEW OF LITERATURE

The issue of fair use has received a lot of attention in the legal, intellectual, and creative communities, particularly as it relates to the fast proliferation of digital media. Scholars and practitioners have long disputed how to reconcile fair usage with copyright laws, given the ever-changing nature of technology and content creation. The literature review gives an overview of significant works in this field, with an emphasis on how fair use is understood, its limitations in the digital age, and its ability to stimulate innovation.

Fair Use and its Evolution

Several researchers have investigated the historical evolution of fair usage and its place in copyright law. Bently and Sherman (2014) trace the roots of fair use back to early copyright legislation, highlighting its importance as a legal instrument that prevents copyright from stifling creativity. They contend that the idea was first used narrowly, solely in circumstances of teaching, criticism, and parody. Similarly, Davison and Hugenholtz (2017) define fair use as a balancing act between copyright holders' interests and public access to works. In their research, the authors emphasize that the fundamental premise of fair use is to avoid idea monopolization and promote the free movement of information.

Fair Use in the Digital Era

With the growth of digital technology and the internet, the application of fair use has received more attention. Aufderheide and Jaszi (2018) analyze how the digital world has blurred the distinction between permitted and unauthorized uses of copyrighted content. Their findings highlight the need for more flexible interpretations of fair use, especially given the prevalence of user-generated content (UGC) and remixes. They contend that, while copyright rules remain restrictive, digital artists require greater legal clarity around what constitutes fair usage.

Litman (2021) delves deeper into digital copyright concerns, examining the contradiction between copyright holders' economic interests and the public's capacity to use and change information. In her book, Digital Copyright: Law and Policy, she contends that although the digital era has democratized content production, copyright regulations have fallen behind, resulting in widespread misapplications of fair use. She criticizes automated copyright enforcement methods, such as YouTube's Content ID, for incorrectly flagging transformative works as infringing content, restricting creativity.

Transformative and Non-Transformative Use in User-Generated Content

In fair use research, the focus on transformative usage has been one of the biggest developments. The U.S. Supreme Court defined transformative use in Campbell v. Acuff-Rose Music, Inc. (1994) as when a work is used in a way that gives the original a new meaning or expression that makes it unique enough to be considered fair use. In the digital age, this idea has proven essential in court cases involving fair use.

According to Hughes (2019), transformative usage is relevant when it comes to user-generated content (UGC). He talks on how remixing and changing preexisting works has allowed regular people to produce new, important material thanks to the digital revolution. His case study research demonstrates that this type of information frequently resides in a legal limbo where judges find it difficult to distinguish between derivative and genuinely transformative content.

The concept of transformative usage is expanded by McCutcheon (2021) to encompass works produced by artificial intelligence. He contends that because AI systems usually rely on enormous databases of copyrighted information to create new content, they go beyond the bounds of fair use. In a time when artificial intelligence is becoming more and more prevalent in creative output, his research poses important questions regarding originality and ownership. McCutcheon argues that in order to handle the unique difficulties presented by AI-generated information, fair use laws may need to change.

McCutcheon (2021) broadens the idea of transformative usage to include AI-generated works. He contends that AI technologies push the bounds of fair use since they frequently rely on vast databases of copyrighted material to generate new work. His research raises critical considerations regarding originality and ownership in a future when AI plays a greater role in content production. McCutcheon contends that fair use regulations may need to adapt to handle the specific issues given by AI-generated information. Not all situations, nevertheless, are eligible for transformative usage. When an original work is replicated with little alteration and for the same purpose, it is referred to as non-transformative usage. For example, the U.S. Supreme Court decided against The Nation in Harper & Row Publishers, Inc. v. Nation Enterprises (1985) for publishing unapproved passages from former President Gerald Ford's memoirs. By simply republishing significant parts of the book without providing further context or comments, the court determined that the snippets were utilized without alteration (Litman, 2020). Super Cassettes Industries Ltd. v. Chintamani Rao (2012), a case in India, similarly stated that unapproved commercial copying of songs without modification is not fair use. Because it only reproduces the original work for profit rather than generating new meaning, the Delhi High Court decided that playing copyrighted tunes in public places and restaurants did not amount to transformative usage (Singh, 2013).

Simple content reposting, in which people or organizations upload or disseminate copyrighted information unaltered, is another important illustration of non-transformative usage. This was made clear in the 1984 case of Sony Corp. of America v. Universal City Studios, Inc., in which the U.S. Supreme Court considered whether time-shifting—the practice of recording television shows for later viewing—constituted fair use. Even though Sony won the lawsuit in the end, it highlighted questions about how copyright owners' market rights may be affected by unaltered reproduction (Samuelson, 2018).

In the field of AI-generated material, non-transformative usage is especially controversial. Copyright rules may be violated by AI systems that simply duplicate old works without offering a fresh interpretation. For instance, the Sham Lal Jain v. Union of India (2020) case in India raised questions about whether AI-generated poetry could be protected by copyright, highlighting the difficulties with originality when AI merely mimics preexisting works rather than producing something essentially unique (Jain, 2020).

Non-transformative usage frequently tends toward copyright infringement, especially when it doesn't offer fresh context, meaning, or commentary, whereas transformative use fosters creative innovation. The line between fair use and infringement will continue to be a contentious legal issue as digital media, artificial intelligence, and user-generated content develop further, necessitating ongoing copyright policy adaption.

Similar concerns are highlighted by Indian legal precedent. In Super Cassettes Industries Ltd. v. Chintamani Rao (2012), the Delhi High Court ruled that businesses exploiting copyrighted music without modification or licensing were causing financial harm to music producers and artists. This ruling emphasized the economic value of licensing regimes, which ensure that innovators and copyright holders are adequately rewarded.

AI-generated content also poses economic issues when used for non-transformative purposes. AI models trained on existing copyrighted work may diminish employment possibilities for human creators, raising worries about job displacement. The Sham Lal Jain v. Union of India (2020) lawsuit called into question whether AI-generated poetry should be protected by copyright, echoing wider economic worries about how AI may replace conventional content creators. If AI can lawfully duplicate previous works without significant alteration, it may reduce economic incentives for original authors, resulting in a decrease in creative sector earnings.

The economic benefits of fair usage must be weighed against the need to preserve copyright holders' financial interests. Governments and regulators are progressively enacting copyright rules that address both creative rights and innovation demands. For example, the Indian Copyright Act (2012) establishes fair dealing exemptions that permit limited, transformative usage while guaranteeing royalty-based licensing structures to compensate original artists. Similarly, platforms such as YouTube's Content ID system enable copyright holders to pay rather than delete illegal content, resulting in a hybrid economic model that benefits both original creators and remix artists (Kumar, 2021).

Furthermore, the emergence of blockchain and non-fungible tokens (NFTs) opens up new economic prospects for copyright protection. Artists and content producers are now using NFT technology to secure unambiguous ownership rights and receive royalties from secondary sales, guaranteeing that revolutionary works reap long-term financial rewards (Chatterjee, 2022).

The contrast between transformative and nontransformative usage has important economic implications. While transformative use fosters creativity, innovation, and market development, non-transformative usage can cause financial losses for copyright holders and discourage investment in intellectual property production. The Indian legal environment, like international frameworks, strives to strike a balance between commercial incentives and fair use safeguards, ensuring that digital content ecosystems remain economically viable and culturally vibrant. As technology evolves, new issues in AI-generated works, content replication, and copyright enforcement will necessitate ongoing legal and economic changes to ensure equitable and sustainable growth in the creative industries.

The Role of Legal Frameworks and Case Law

Throughout history, various court cases have affected the legal frameworks governing fair use. Samuelson (2019) gives a thorough examination of significant legal cases that have affected fair use in the digital era. Her study underscores that courts have generally been hesitant in broadening the definition of fair use, frequently deferring to copyright holders' economic interests. However, Samuelson notes that recent instances have indicated a change, with judges growing more prepared to acknowledge the transformative nature of digital creations.

The case of Lenz v. Universal Music Corp. (2015) is noteworthy for defining how fair use applies to digital property. The case, which included the removal of a YouTube video including a small sample from a song, was a pivotal court struggle in establishing the boundaries of automatic copyright enforcement. Samuelson emphasizes how the judgment reflected the courts' growing appreciation of the need of balancing copyright enforcement with digital producers' rights to participate in creative usage.

Economic Implications of Transformative and Non-Transformative Use in User-Generated Content

The controversy around the transformative vs non-transformative use of copyrighted content has important economic ramifications for digital platforms, content producers, AI developers, and businesses. In order to decide whether producers can freely utilize preexisting works, the fair use doctrine is essential. This decision has a direct impact on market competitiveness and the economic worth of intellectual property.

By promoting digital innovation, broadening the market, and facilitating the production of new material, transformative usage stimulates economic growth. According to Hughes (2019), platforms that encourage producers to remix and reinterpret preexisting material, including YouTube, Instagram, and TikTok, earn billions of dollars in income. This propels the growth of creative economies, advertising income, and consumer engagement. The worldwide UGC business was estimated to be worth $10.5 billion in 2022 by PwC, mostly because to revolutionary fair use policies that let content producers expand on previously created works without facing legal constraints.

As McCutcheon (2021) points out, transformative usage also helps AI-generated material. Artificial intelligence (AI) systems that produce original works of art, literature, or music save production costs and increase productivity in sectors including marketing, gaming, and entertainment. AI is also booming in the Indian industry; by 2025, it is expected that AI-powered content automation technologies would boost the country's GDP by $500 billion (NASSCOM, 2022).

A work that significantly expands on existing meaning may be considered fair use in India, according to the Delhi High Court's 1996 ruling in Civic Chandran v. Ammini Amma. This decision has affected cultural industry-related economic policies by enabling Indian singers, filmmakers, and satirists to expand on their previous works without worrying about excessive litigation, which has boosted the creative industries' economies (Singh, 2017).

However, non-transformative usage can cause market distortions and frequently results in financial losses for copyright holders. Reproduction of content without substantial alteration can diminish the value of original works and discourage investments in the development of intellectual property. Global piracy costs the entertainment sector more than $29 billion a year, according to the World Intellectual Property Organization (WIPO, 2020), since illegal material replication reduces demand for official releases.

Similar concerns are highlighted by Indian legal precedent. The Delhi High Court banned the unapproved commercial use of copyrighted songs in Super Cassettes Industries Ltd. v. Chintamani Rao (2012), ruling that companies that exploited copyrighted music without alteration or licensing were hurting musicians and producers financially (Singh, 2013).

When used in non-transformative ways, AI-generated material also poses financial difficulties. Concerns about job displacement may arise if AI models trained on protected content decrease employment options for human producers. Reflecting larger economic worries about how AI may replace conventional content producers, the Sham Lal Jain v. Union of India (2020) case questioned whether AI-generated poetry should be afforded copyright protection (Jain, 2020). AI may lessen financial incentives for original producers and cause the creative industry's earnings to drop if it is able to lawfully replicate previous works with little to no alteration.

Challenges to Fair Use in the Digital Landscape

The digital realm has posed substantial difficulties to traditional concepts of fair usage. Sunder (2018) investigates these issues, especially in the context of global content sharing. She points out that the transnational character of the internet complicates the enforcement of fair usage because copyright laws differ greatly between states. Her research underlines the importance of international collaboration and the harmonization of copyright regulations to ensure that fair usage is not disproportionately limited in specific locations.

Katz (2020) explores another rising issue the role of artificial intelligence and machine learning in copyright law. Katz contends that AI's ability to assess and develop content based on previous works raises concerns about the scope of fair usage. He claims that present legal systems are unprepared to deal with the complicated concerns of ownership, originality, and creativity that AI-generated content raises.

Future Directions for Fair Use

The future of fair usage in the digital era is a topic of constant discussion. Lessig (2015), in his seminal work Free Culture, proposes for a broader definition of fair use that takes into account the collaborative and participatory character of digital media. He contends that stringent copyright rules stifle creativity and innovation, and that fair use should be widely defined to promote remix culture, online cooperation, and open interchange of ideas.

Similarly, Goldman (2020) considers the prospect of new legal frameworks that better reflect the reality of digital content generation. He believes that the advent of platforms, automation, and AI will continue to test the conventional limitations of copyright, necessitating a more dynamic and flexible approach to fair use to keep up with technological breakthroughs.

The literature on fair use illustrates the constant conflict between the necessity to preserve intellectual property and the need to promote creativity, especially in the digital age. As content production advances through digital platforms, user-generated material, and AI technology, researchers have increasingly urged for a more adaptable understanding of fair use that can accommodate these new forms of expression. The issue for legal regimes and legislators will be to achieve a balance that protects copyright holders while also allowing for the continuous innovation and creativity that distinguishes the digital world.

2.1. Research Gap

Despite substantial study on fair use, major gaps persist, particularly in its applicability to developing digital platforms and artificial intelligence (AI). Current literature, such as Aufderheide and Jaszi (2018) and Goldman (2020), focuses on established platforms like YouTube and Instagram, with less emphasis on younger, fast expanding platforms like TikTok and Twitch. These platforms have distinct content-sharing dynamics that challenge standard fair use interpretations. Furthermore, while researchers such as Katz (2020) and McCutcheon (2021) have begun to investigate AI's influence on copyright, thorough research is absent. The use of AI to produce and reuse material from large databases brings new complications around originality and infringement that existing fair use rules may not fully handle. This changing landscape emphasizes the need for current research to better understand and apply fair use rules to these new technology realities.

3.0 STATEMENT OF THE PROBLEM:

The fast growth of digital technology, as well as the proliferation of digital platforms, have fundamentally altered the landscape of content generation and delivery, calling into question traditional interpretations of copyright law and fair use. Despite fair use's fundamental role in balancing intellectual property rights and public interests, existing legal frameworks sometimes fail to handle the complications posed by new digital environments and future technologies. Specifically, present legal rules do not effectively address the applicability of fair use to modern digital platforms such as TikTok and Twitch, as well as the role of artificial intelligence (AI) in content creation. The issue is the necessity for a thorough examination and adaption of fair use principles to meet these changing technical realities, ensuring that they appropriately safeguard copyright holders' rights while encouraging innovation and creativity in the digital era. Addressing this gap is critical to creating a fair and effective legal system that reflects the complexity of modern content generation and dissemination.

4.0 RESEARCH QUESTIONS

1. How has the fair use doctrine evolved over time, and what is its current state in the context of the digital era?
2. What are the specific impacts of digital technologies on the application and interpretation of fair use principles?
3. How can a balance be struck between fostering creativity and ensuring robust copyright protection in the digital landscape?
4. What are the economic implications of fair use for content creators and rights holders in the digital age?
5. What policy recommendations can be made to update fair use guidelines in response to rapid technological advancements?

5.0 OBJECTIVES OF THE STUDY

i) To examine the evolution and current state of fair use doctrine, particularly in the context of the digital era.
ii) To analyse the impact of digital technologies on the application of fair use principles.
iii) To explore the balance between fostering creativity and ensuring copyright protection in the digital landscape.
iv) To assess the economic implications of fair use for content creators and rights holders.
v) To provide policy recommendations for updating fair use guidelines in the context of digital advancements.

6.0 RESEARCH METHODOLOGY

This section delineates the methodological framework employed to comprehensively investigate the fair use of copyright protection in digital era.

6.1. Research Design

The study will employ a qualitative research approach, including a thorough analysis of current literature and case law, to better understand the applicability and limitations of fair use in digital environments. The study will look at how fair use is applied to burgeoning digital platforms like TikTok and Twitch, as well as the complications that artificial intelligence brings to content production. The design will also involve an evaluation of automated copyright enforcement technologies to assess their impact on fair usage and copyright protection. The findings will help to shape proposals for legislative reforms and revised interpretations that better reflect current concerns.

6.2. Data Collection

Data will be gathered through an extensive review of relevant academic literature, legal case studies, and current fair use rules. This will entail examining legal texts, court judgments, and scholarly publications to better understand the contemporary use and issues of fair use.

6.3. Data Analysis

Data analysis will include a thematic review of the literature and case law to identify common patterns, trends, and concerns with the implementation of fair use. This will entail determining how fair use is applied to digital platforms and AI-generated material, as well as examining the efficacy of automated copyright enforcement systems. The report will also compare current fair use regimes to new difficulties, emphasizing areas where reform is required. The findings will be aggregated to make suggestions for upgrading fair use interpretations and legal frameworks to better reflect current concerns in digital media and technology.

6.4. Research Limitations

This study has various limitations that may affect its conclusions. First, the body of literature on the application of fair use, particularly in relation to future digital platforms and AI technology, may be restricted, thus limiting the depth of the examination. Furthermore, the quick rate of technical breakthroughs might render the research obsolete when new digital media and AI technologies arise. Geographic and jurisdictional variances in copyright rules may restrict the findings and suggestions' global applicability. Finally, limited access to private data from automated copyright enforcement systems may hinder the thoroughness of the examination of their influence on fair use.

7.0 DISCUSSION

7.1 Specific evolution and current state of fair use doctrine, particularly in the context of the digital era.

The fair use concept, created by the United States Copyright Act of 1976, was intended to strike a balance between copyright holders' rights and public interests by allowing limited use of copyrighted content for criticism, commentary, and education (U.S. Copyright Office, 2021). Historically, its breadth was limited, but the landmark Campbell v. Acuff-Rose Music, Inc. case (1994) broadened its application by acknowledging transformative usage as an important criterion in fair use, especially in commercial circumstances. The digital era has put this concept to the test, with platforms like YouTube and TikTok introducing new obstacles, such as automated copyright enforcement systems that occasionally misidentify transformative uses as infringements (Goldman, 2020). Current legal interpretations are responding to technological changes, with courts increasingly considering the impact of digital media on fair use, emphasizing the need for a flexible approach to modern content creation and distribution (Aufderheide & Jaszi, 2018; Hughes, 2019; Samuelson, 2019).

The economic repercussions of fair use are significant, affecting content producers, copyright holders, and digital platforms. The United States Copyright Act of 1976 sought to strike a compromise between copyright owners' financial interests and public access to information and creativity (US Copyright Office, 2021). Courts have gradually enlarged the fair use concept, particularly in commercial circumstances, as seen by Campbell v. Acuff-Rose Music, Inc. (1994), which established transformative use as a crucial criterion for deciding fair use.

Economic innovation and digital entrepreneurship rely heavily on fair usage principles. Platforms such as YouTube, Instagram, and TikTok depend on user-generated content (UGC), which supports advertising income, influencer marketing, and digital media growth (Goldman 2020). According to a PwC (2022) estimate, the worldwide UGC industry was worth $10.5 billion, proving its economic importance. Fair use promotes market competitiveness and decreases obstacles for new content providers, resulting in the emergence of new sectors such as remix culture, gaming streams, and digital satire (Hughes, 2019).

Fair use supports the Indian economy as well, particularly through educational institutions, research papers, and digital entertainment. Section 52 of the Indian Copyright Act of 1957 has fair dealing provisions that enable restricted usage for purposes such as criticism, education, and news reporting (Singh, 2017). The Delhi High Court's judgment in Civic Chandran v. Ammini Amma (1996) upheld this approach, stating that a derivative work can be considered fair use if it adds significant new meaning, hence encouraging cultural and economic development. This has been especially true in Indian film, literature, and online content production, where adaptation and reinterpretation are critical to economic viability.

Despite its advantages, fair usage can pose economic hurdles, especially when automated copyright enforcement systems perceive lawful transformative content as copyright infringement (Samuelson, 2019). Many content providers risk demonetization or legal battles, which result in financial losses. This is obvious in India's music and film industries, where copyright claims on YouTube frequently obstruct revenue sources for artists doing remixes, criticisms, or instructional versions.

In Super Cassettes Industries Ltd. v. Chintamani Rao (2012), the Delhi High Court ruled against unlawful commercial exploitation of copyrighted recordings, emphasizing the economic significance of licensing systems. Similarly, in Amarnath Sehgal v. Union of India (2005), the Delhi High Court ruled that moral rights of creative works must be protected, which affects how derivative works are monetized in India.

Furthermore, AI-generated material complicates the economics of fair usage. As McCutcheon (2021) puts out, AI systems rely on large databases of copyrighted information to generate new content, creating concerns about economic ownership and revenue-sharing arrangements in a fast-changing digital ecosystem.

A balanced approach to fair use policy is required for economic viability. Many digital platforms now utilize hybrid licensing arrangements, such as YouTube's material ID system, which allows copyright holders to sell their material rather than restrict it (Kumar, 2021). Furthermore, the emergence of blockchain and NFTs (Non-Fungible Tokens) is altering economic frameworks for copyright management, allowing artists to keep custody of their works while earning digital resale royalties (Chatterjee, 2022).

The Indian Copyright Act of 2012 has evolved to meet current economic problems, ensuring that fair dealing rules promote economic growth while preserving intellectual property rights. However, with an increased dependence on AI, automated copyright enforcement, and digital media, authorities must constantly review how fair usage helps to both economic development and cultural creativity in India.

Despite its advantages, fair usage can pose economic hurdles, especially when automated copyright enforcement systems perceive lawful transformative content as copyright infringement (Samuelson, 2019). Many content providers risk demonetization or legal battles, which result in financial losses. This is obvious in India's music and film industries, where copyright claims on YouTube frequently obstruct revenue sources for artists doing remixes, criticisms, or instructional versions.

In Super Cassettes Industries Ltd. v. Chintamani Rao (2012), the Delhi High Court ruled against unlawful commercial exploitation of copyrighted recordings, emphasizing the economic significance of licensing systems. Similarly, in Amarnath Sehgal v. Union of India (2005), the Delhi High Court ruled that moral rights of creative works must be protected, which affects how derivative works are monetized in India.

Furthermore, AI-generated material complicates the economics of fair usage. As McCutcheon (2021) puts out, AI systems rely on large databases of copyrighted information to generate new content, creating concerns about economic ownership and revenue-sharing arrangements in a fast-changing digital ecosystem.

Fair use has emerged as a critical economic driver, promoting content creation, digital entrepreneurship, and market competitiveness. However, misapplication of copyright rules, particularly in the digital era, can cause financial hardship for creators. Indian legal precedents have attempted to strike a balance between cultural expression and economic viability; nevertheless, new legal frameworks may be necessary to handle future concerns brought by AI and digital automation. Moving forward, fair use policies must be adaptable to support economic development, copyright protection, and ongoing creative progress.

7.2 Impact of digital technologies on the application of fair use principles.

The implementation of fair use principles has been significantly influenced by digital technologies, which have also changed the ways in which information is produced, disseminated, and used. User-generated content has become more widespread due to the emergence of digital platforms like YouTube, TikTok, and Instagram. These sites frequently creatively integrate copyrighted information into their work. Due to their inability to accommodate the widespread and quick distribution of digital media, traditional fair use frameworks have been put to the test by this change.

The growing use of automated copyright enforcement tools, like YouTube's Content ID, which tracks and analyzes user-uploaded infringing material, is one notable effect. Although these mechanisms are designed to safeguard copyright holders, they might mistakenly label transformative applications as violations, which can cause disagreements and disrupt the equilibrium that fair use is meant to achieve (Goldman, 2020). This emphasizes how fair use guidelines must change to accommodate new content-sharing strategies and technological advancements.

The ease with which digital information may be copied and distributed is another complexity brought about by digital technology that puts into question the traditional fair use requirements, such as the usage's intent and nature. For example, user-generated material is now often monetized by digital platforms via sponsorships and advertising, blurring the line between commercial and non-commercial use. Due to this, there is legal ambiguity over whether some uses are sufficiently transformative to be considered fair use, especially where there are financial interests at stake (Aufderheide & Jaszi, 2018).

The global application of fair usage is another issue brought up by the digital era. The enforcement of rights across borders is made more difficult by the disparate ways that different governments interpret copyright and fair use laws. International harmonization of fair use standards is becoming more and more necessary as digital platforms expand internationally in order to handle these cross-border issues (Samuelson, 2019).

The rules of fair use ought to be reevaluated in light of digital technologies to guarantee that they continue to foster innovation and creativity while suitably safeguarding the rights of copyright holders. Legal frameworks must constantly adapt to the changing digital world in order to handle the particular difficulties presented by new media and technology (Hughes, 2019).

For example, the United States maintains the fair use theory established by the Copyright Act of 1976, which allows for transformative usage under certain conditions. In contrast, the United Kingdom follows fair dealing, which places stronger limits on authorized usage. A major decision, Ashdown v. Telegraph Group Ltd. (2001), revealed that UK copyright law does not support the liberal transformative use criteria found in the United States.

Similarly, in the European Union (EU), the InfoSoc Directive (2001/29/EC) specifies fair use with severe fair dealing exclusions, restricting transformative applications that are generally authorized under US copyright law. Geiger et al. (2020) discovered that EU member states use varying conceptions of fair use, creating uncertainty in copyright compliance for digital content providers working in various countries.

Section 52 of India's Copyright Act of 1957 permits fair dealing exemptions for educational, research, and journalistic purposes. However, its interpretation remains narrow when compared to the US model. In India TV Independent News Service v. Yashraj Films Pvt. Ltd. (2013), the Delhi High Court found that using a copyrighted music in a news program was not transformative enough to qualify for fair dealing. This case demonstrates Indian courts' limited view to fair use when compared to US legal precedence.

Several comparative legal studies have examined cross-border applications of fair usage and its problems in the digital age:

Australian Fair Use Study (2018): The Australian Law Reform Commission (ALRC) investigated whether Australia should adopt a US-style fair use regime. The study discovered that Australia's fair dealing regime frequently prevents digital content makers from lawfully exploiting copyrighted assets in ways that would be allowed under US law (ALRC, 2018).

WIPO Report (2021): WIPO conducted global research on how countries with various copyright laws manage fair use claims in cross-border digital content disputes. The research noted difficulties in harmonizing fair use regulations due to national legal disparities, notably between common and civil law nations (WIPO, 2021).

YouTube's Global Copyright Enforcement Study (2022): This research, conducted by YouTube's Legal Compliance Division, looked at copyright takedown requests in several regions. It discovered that films marked for copyright infractions in one jurisdiction were frequently considered fair use in another, resulting in uneven enforcement outcomes for digital artists (YouTube, 2022).

To address the rising difficulties of cross-border copyright issues, academics have proposed international collaboration on fair use policies. According to Samuelson (2019), WIPO and trade agreements such as the Trans-Pacific Partnership (TPP) might provide a consistent approach to digital fair use, resulting in more predictable outcomes for global content creators. Similarly, Geiger et al. (2020) urge for the harmonization of fair use rules across the EU in order to decrease legal difficulties for enterprises and people operating in various member states.

Fair use regulations must be reevaluated in light of digital technology to ensure that they continue to promote innovation and creativity while safeguarding copyright holders' rights. The transnational nature of digital material production necessitates increased worldwide harmonization of fair use norms, since legal discrepancies generate uncertainty for global producers and enterprises. Moving forward, legal frameworks must continually adapt to the changing digital world in order to deal with the specific issues offered by new media and technology.

7.3 Balancing Creativity and Copyright Protection in the Digital Landscape

In the digital era, striking a balance between encouraging creation and enforcing copyright protection has become increasingly difficult. The growth of digital technology and platforms has democratized content production, enabling individuals and small artists to make and distribute their work worldwide. This accessibility has fueled innovation and artistic expression, but it has also presented substantial hurdles to copyright enforcement.

On the one hand, fair use regulations promote innovation by allowing the use of copyrighted content for purposes such as commentary, criticism, parody, and education. This freedom enables authors to expand on previous works, participate in transformational applications, and contribute to a thriving cultural conversation. For example, remix culture and transformative works frequently rely on the capacity to include copyrighted information in novel and original ways, therefore enriching the creative landscape and encouraging a wide range of artistic expressions (Aufderheide & Jaszi, 2018).

On the other hand, copyright protection is critical for protecting the rights of original artists and encouraging the development of new works. Digital technologies have made it simpler to replicate and distribute intellectual information without authorization, possibly jeopardizing copyright holders' economic interests. Automated solutions, such as YouTube's Content ID, are intended to handle these difficulties by detecting and correcting unlawful uses, but they may also lead to overreach, in which legitimate, transformative uses are incorrectly classified as infringements (Goldman, 2020). This emphasizes the necessity for a sophisticated approach to distinguishing between infringement and fair usage.

The problem is to keep copyright laws in existence and interpret them in light of these changing dynamics. Legal frameworks must strike a balance between protecting intellectual property and encouraging creativity and public access to new works. This entails tailoring fair use rules to the reality of digital material creation and dissemination, while ensuring that copyright enforcement measures do not hinder innovation. Policymakers and courts must work together to develop solutions that respect authors' rights while encouraging and protecting innovative and transformative uses of material (Samuelson, 2019; Hughes, 2019).

In the digital age, finding a balance between fostering creation and enforcing copyright protection has become increasingly difficult. The advancement of digital technology and platforms has democratized content production, allowing people and small artists to create and disseminate their work globally. This accessibility has spurred innovation and artistic expression, but it has also posed significant challenges to copyright enforcement.

On the one hand, fair use laws encourage innovation by permitting the use of copyrighted information for purposes such as commentary, criticism, parody, and education. This independence allows authors to build on prior works, take part in transformational applications, and contribute to a vibrant cultural debate. For example, remix culture and transformative works typically rely on the capacity to include copyrighted content in fresh and unique ways, expanding the creative landscape and fostering varied artistic expressions (Aufderheide & Jaszi, 2018).

On the other hand, copyright protection is critical for protecting original artists' rights and encouraging the development of new works. Digital technologies have made it simpler to reproduce and distribute intellectual property without permission, possibly harming copyright holders' economic interests. Automated enforcement methods, like as YouTube's Content ID, are intended to discover and rectify illicit uses, but they sometimes result in overreach, with acceptable, transformative applications incorrectly categorized as infringements (Goldman, 2020). This emphasizes the critical need for a more sophisticated strategy to differentiating between copyright infringement and fair usage.

Challenging Dynamics in the Digital Copyright Landscape.

Blurred Lines Between Transformative and Infringing Use:

Courts and legislators grapple with situations involving transformational works that have commercial components. For example, many YouTubers and independent artists monetize their material, making it difficult to discern between fair use and commercial exploitation (Samuelson, 2019).

Complexities of Cross-border Copyright Enforcement:

With digital platforms functioning internationally, copyright rules differ between regions, resulting in conflicting interpretations of fair use. While the United States adheres to a liberal fair use concept, nations such as India and the European Union have stricter fair dealing laws, causing legal uncertainty for international content providers (Geiger et al., 2020).

AI-generated Content and Copyright Ownership:

The proliferation of AI-generated content has prompted serious concerns regarding ownership and originality. AI algorithms trained on massive datasets of copyrighted information call into question existing fair use frameworks by remixing and generating derivative output with no obvious authorship or license (McCutcheon, 2021).

Overreliance on automated copyright enforcement:

While digital platforms use automated copyright filters like Content ID, these methods frequently fail to account for context. According to research, lawful educational, satirical, and journalistic uses are commonly detected as copyright infringement, resulting in unfair material removals (Goldman, 2020).

Actionable insights for a balanced approach.

To solve these difficulties, stakeholders should consider regulatory interventions and technology advancements that strike a compromise between copyright protection and creative freedom.

Develop more context-aware copyright enforcement systems:

- Platforms should utilize hybrid models that combine AI-powered copyright identification with human review panels to prevent improper takedowns of fair use material.
- Enhancing machine learning algorithms to spot transformative components like satire, criticism, and remix culture might help to avoid over-enforcement (Hughes, 2019).

Harmonize the International Fair Use Standards.

- Policymakers should work to develop global fair use principles that ensure cross-border uniformity in copyright enforcement.
- International frameworks such as WIPO and regional trade agreements should include flexible fair use norms that apply across all digital platforms (Geiger et al., 2020).

Reevaluate Copyright Protections for AI-Generated Works.

- Legal frameworks must clearly define copyright ownership for AI-generated content, differentiating between AI-assisted and wholly AI-created works.
- Introducing new licensing structures in which AI-generated works pay royalties to original copyright holders may provide a long-term solution (McCutcheon, 2021).

Increase transparency in content moderation decisions.

- Platforms should establish explicit appeal methods for producers whose material has been wrongly flagged.
- Establishing independent review boards to resolve transformative and fair use allegations would prevent arbitrary enforcement (Goldman, 2020).

Encourage licensing and revenue sharing models.

- Instead of enforcing rigid takedown measures, platforms might enable licensing agreements that allow artists to utilize protected material in exchange for income sharing.
- The music industry's YouTube monetization model, in which artists receive a portion of ad revenue from videos featuring their material, might serve as a model for larger copyright deals (Samuelson, 2019).

Balancing fair usage and copyright protection need ongoing coordination between creators, copyright holders, technology firms, and legislators. The complex dynamics of transformative works, cross-border enforcement, AI-generated material, and automated copyright moderation necessitate flexible legal frameworks that support both intellectual property rights and creative expression. As digital content production advances, flexible regulations, enhanced enforcement methods, and international collaboration will be required to guarantee that copyright laws foster innovation while not impeding creative creativity (Hughes, 2019).

7.4 Economic implications of fair use for content creators and rights holders in the digital age

The economic ramifications of fair use in the digital era varied by country, affecting content producers, rights holders, and digital platforms differentially. Fair use promotes innovation by allowing producers to build on existing copyrighted material without incurring high licensing fees, hence supporting the creation of remix culture, instructional materials, and commentary-driven works (Aufderheide & Jaszi 2018). However, inconsistency in legal interpretations between jurisdictions has created economic issues, notably in monetization and enforcement. Automated copyright enforcement systems, such as YouTube's Content ID, routinely misclassify transformative works as infringing, demonetizing digital artists and causing financial instability (Goldman, 2020).

In the United States, the landmark case Campbell v. Acuff-Rose Music, Inc. (1994) established transformative use as a key factor in fair use determination, influencing cases such as Google LLC v. Oracle America, Inc. (2021), in which the Supreme Court ruled that Google's use of Java APIs in Android was fair use, citing its transformative nature and market impact. This decision emphasized the necessity of striking a balance between innovation and copyright protection, with implications for software development and the digital economy. Similarly, in Authors Guild v. Google, Inc. (2015), the court determined that Google Books' digitization of copyrighted works for search functionality was fair use, benefiting both scholars and copyright owners by improving accessibility (U.S. Court of Appeals, 2015).

The EU Copyright Directive (Directive 2019/790) governs fair usage in the European Union with a more stringent "fair dealing" approach. The Deckmyn v. Vandersteen (2014) decision demonstrated that parody can be considered fair dealing, but the copyright owner's rights must be balanced with freedom of speech (Court of Justice of the European Union, 2014). The regulation also included Article 17, which requires digital platforms to proactively screen copyrighted information, posing hazards to producers who rely on transformative works. In Pelham GmbH v. Hütter (2019), the European Court of Justice decided that unlicensed music sampling might violate copyright if significantly changed, strengthening tighter interpretations of fair dealing (ECJ, 2019).

In the United Kingdom, Ashdown v. Telegraph Group Ltd. (2001) underlined that commercial usage outweighs fair dealing, which has an influence on journalistic and educational reuse of protected materials. Similarly, in Sport Five GmbH v. Litman (2011), the court dismissed fair dealing arguments for broadcasting football match clips, confirming stringent commercial restrictions (UK High Court, 2011).

In India, the fair dealing clause under Section 52 of the Copyright Act of 1957 permits limited exclusions but lacks defined criteria for digital works. In Super Cassettes Industries Ltd. v. MySpace Inc. (2011), the Delhi High Court ruled against MySpace, stating that posting copyrighted information without proactive monitoring violates copyright rules, underscoring the difficulties in aligning Indian law with worldwide digital norms. In India TV v. Yashraj Films (2013), the Delhi High Court found that news reporting is fair dealing, hence enhancing media rights under fair usage principles.

In Australia, the University of New South Wales v. Moorhouse (1975) decision established that organizations that facilitate copyright infringement might be held accountable, affecting the obligations of digital platforms. However, in Roadshow Films Pty Ltd. v. iiNet Ltd. (2012), the High Court of Australia found that ISPs are not accountable for user infringement unless actively participating, which has implications for platform responsibility.

From an economic standpoint, severe copyright enforcement can result in financial losses for artists owing to fraudulent infringement claims, whereas permissive fair use regulations risk compromising rights holders' revenues (Samuelson, 2019). Digital platforms suffer higher expenses when implementing compliance procedures, which affects their business models. Some nations, such as the United States, have implemented revenue-sharing models in which copyright holders receive a portion of advertising money from commercialized material, therefore balancing economic interests (Hughes, 2019). Meanwhile, India and the EU have implemented stronger licensing regulations, moving economic authority to rights holders (Singh, 2021).

Given these jurisdictional disparities, worldwide harmonization of fair use rules remains difficult. Policymakers must strike a balance between innovation and copyright enforcement, ensuring that fair use regulations promote economic development while preserving intellectual property rights. Standardized worldwide norms and equitable income sharing systems can help to build a more sustainable digital economy that benefits both content producers and rights holders.

7.5 Policy recommendations for updating fair use guidelines in the context of digital advancements.

As digital technologies continue to change the way information is generated, transmitted, and consumed, updated fair use principles are becoming increasingly critical. Several significant policy proposals may be made to handle the difficulties of the digital age while balancing the promotion of innovation and the protection of copyright. One critical step is to define the requirements for transformative use, which is an important part of fair use. Transformative works bring fresh meaning, expression, or insight to original copyrighted materials, distinguishing them from basic copies. However, the definition of transformative usage is frequently ambiguous, causing difficulty for both producers and judges. Creators would benefit from more defined definitions of transformative usage, which would provide better direction and reduce the uncertainty that frequently leads to legal issues, particularly in situations involving digital material remixing, parody, and commentary (Samuelson, 2019).

Another critical area for development is adapting fair use standards to the realities of digital platforms. In the digital world, sites such as YouTube, Instagram, and TikTok have transformed the way material is shared, often at lightning speed and across boundaries. Because these platforms are worldwide in scope, international harmonization of fair use norms is required. Inconsistent interpretations of fair use across jurisdictions cause legal ambiguity, especially for content providers that operate in numerous countries. A more uniform worldwide framework will assist to eliminate cross-border legal issues and offer artists with better clarity on how to legally utilize copyrighted content (Aufderheide and Jaszi, 2018).

Furthermore, the digital environment has resulted in the proliferation of automated copyright enforcement tools, such as YouTube's Content ID, which are intended to identify unlawful use of copyrighted material. While these algorithms have been effective in protecting content owners' rights, they frequently misclassify content that qualifies for fair use, resulting in unjustified takedowns or monetization claims. This presents enormous hurdles to authors who rely on fair usage to develop transformative creations. To properly balance innovation and copyright protection, these automated systems must be enhanced with more complex algorithms capable of distinguishing between infringement and acceptable fair usage (Goldman, 2020). Policymakers should support the development of such systems by increasing openness in their operations and establishing a clear method for authors to challenge wrong judgments.

In addition to improving automated enforcement methods, there is a need for cost-effective and accessible dispute resolution tools for fair use problems. Currently, legal conflicts concerning copyright and fair use may be costly and time-consuming, making it difficult for smaller producers to assert their fair use rights. Introducing a simplified, cost-effective resolution method, such as third-party mediation or arbitration, will assist to resolve these disagreements more effectively. This would allow creators to protect their rights without the need for prolonged litigation, resulting in a more balanced ecology in which artists and copyright holders can coexist (Hughes, 2019).

Furthermore, education and awareness programs are critical for keeping both content creators and the general public informed of fair use. Many authors accidentally infringe on copyright because they don't grasp what constitutes fair usage. Governments and internet platforms should work together to develop public awareness campaigns that educate users about their rights and obligations under fair use laws. This will not only lower the frequency of inadvertent infringements, but also give innovators the confidence to utilize copyrighted content in transformative ways (U.S. Copyright Office, 2021).

As emerging technologies such as artificial intelligence (AI) and machine learning gain traction, fair use standards must adapt to new ways of creating material. AI systems are rapidly using large volumes of copyrighted content to train models and produce new works, posing difficult problems concerning what constitutes fair use. Clear parameters must be developed for how AI may exploit copyrighted material, allowing innovation in AI-driven content creation to flourish while not infringing on the rights of original authors. Fair use standards may remain relevant in an ever-changing technological context by proactively considering the consequences of artificial intelligence in content development.

Finally, policymakers should aim for legislative flexibility in fair use rules. The digital world is fast evolving, and inflexible legal frameworks may struggle to keep up with emerging technology and innovative behaviors. A more flexible, principles-based approach to fair usage can assist ensure that the legislation is adaptive and capable of meeting future difficulties. This flexibility would enable fair use regulations to grow in tandem with technological improvements, ensuring that both artists and copyright holders are sufficiently protected in a changing environment (Hughes, 2019).

Finally, revising fair use standards to properly reflect the problems of the digital world is critical for striking a balance between promoting innovation and protecting copyright. Policymakers may build a stronger and more egalitarian legal environment by defining transformative use requirements, improving automated enforcement systems, offering accessible dispute resolution processes, and adapting to developing technology. Such measures would benefit both creators and rights holders, creating an atmosphere in which innovation and intellectual property protection may live together.

8.0 CASE STUDIES

Warhol Foundation v. Goldsmith (2023)

Overview: In this landmark Supreme Court case, photographer Lynn Goldsmith sued the Andy Warhol Foundation for using her shot of musician Prince in a series of silkscreen prints. Goldsmith alleged copyright infringement because Warhol did not get permission to use her photos as the foundation for his artwork.
Fair Use Argument: The Warhol Foundation said that Warhol's paintings were transformational and so eligible for fair use since he transformed the original image to create something new, thereby adding to the pop art trend.
Outcome: The United States Supreme Court found in favor of Goldsmith, ruling that Warhol's commercial use of Goldsmith's image was insufficiently transformative to qualify as fair use. This case demonstrated the limitations of the transformative use concept in commercial circumstances, emphasizing the importance of balancing creative expression with copyright protection for the original inventor.

Google LLC v. Oracle America, Inc. (2021)

Overview: This lawsuit concerned Google's use of Oracle's Java programming language to create the Android operating system. Oracle sued Google for copyright infringement, alleging that Google stole more than 11,000 lines of Java code without authorization.
Fair Use Argument: Google contended that their use of the code was fair use since it altered the code by applying it in a new environment (smartphone development) and was required for Android operation, benefitting developers and the general public.
Outcome: The United States Supreme Court found in favor of Google, concluding that their use of the Java API was transformative and constituted fair use. The court determined that Google's usage was highly transformative since it repurposed the code for a new and distinct platform, therefore encouraging innovation and benefitting the public. This case is noteworthy because it has wide-ranging ramifications for the technology sector and the future of software development under fair usage.

McGucken v. Newsweek LLC (2022)

Overview: Photographer Elliot McGucken filed a lawsuit against Newsweek for copying his shot of a lake without permission in a story on the same lake that he had uploaded on Instagram. McGucken filed a copyright infringement action after Newsweek used the Instagram post in the piece.
Fair Use Argument: Newsweek claimed that their use of McGucken's photo was transformative since it added context to the piece and was utilized for journalistic reporting, which is a recognized area of fair use.
Outcome: The court found against Newsweek, concluding that embedding the Instagram photo did not constitute fair use. The court determined that just embedding the photo did not fundamentally alter its purpose or meaning, and that utilizing the photograph for commercial reasons (as part of an article) without proper permission violated McGucken's copyrights. This case emphasized the

issues that digital media providers confront when combining the demand for material with individual authors' copyright rights.

Deckmyn vs. Vandersteen (2014)

Overview: This lawsuit examined whether a parody of a comic book artwork violated copyright.
Fair Use Argument: The defendant claimed that the parody was a lawful exercise of free expression.
Outcome: The Court of Justice of the European Union (CJEU) found that parody must strike a balance between copyright and free expression. The ruling influenced fair use guidelines for creative works in the EU (Deckmyn v. Vandersteen, 2014).

elham GmbH v. Hütter (2019).

Overview: This case included the unlawful use of a sound sample from the band Kraftwerk's song Metall auf Metall in a hip-hop track.
Fair Use Argument: The defendant claimed that the brief sample was transformative and important for musical creativity.
Outcome: The CJEU found that unlawful sampling was an infringement unless adequately adjusted, restricting the scope of fair use in music creation (Pelham GmbH v. Hütter, 2019).

Ashdown v. Telegraph Group Ltd. (2001).

Overview: The Telegraph released passages from classified political papers without permission.
Fair Use Argument: The newspaper argued that publishing the snippets was important for public interest reporting.
Outcome: The UK Court of Appeal decided against the Telegraph, concluding that commercial purpose undermines fair dealing allegations (Ashdown v. Telegraph Group Ltd., 2001).

Sport Five GmbH vs. Litman (2011)

Overview: In this instance, a third party used sports broadcasting portions without permission.
Fair Use Argument: The defendant claimed fair dealing for informative reasons.
Outcome: The UK High Court rejected the fair dealing argument, emphasizing the importance of proper authorization for commercial usage (Sport Five GmbH v. Litman, 2011).

Super Cassettes Industries Ltd. vs. MySpace Inc. (2011)

Overview: MySpace was sued for hosting illegally copyrighted music.
Fair Use Argument: MySpace claimed that it was only a middleman and not directly liable for user-generated material.
Outcome: The Delhi High Court decided against MySpace, stating that platforms must actively monitor and delete illegal content (Super Cassettes Industries Ltd. v. MySpace Inc., 2011).

India TV vs Yashraj Films (2013)

Overview: India TV utilized film snippets in their news coverage without authorization, resulting in a copyright conflict.
Fair Use Argument: The station claimed fair dealing in news reporting.
Outcome: The Delhi High Court decided in favor of India TV, holding that restricted use of copyrighted information for news reporting is permitted (India TV v. Yashraj Films, 2013).

University of New South Wales v. Moorhouse (1975).

Overview: This case considered whether libraries might be held accountable for copyright infringement by users.
Fair Use Argument: The institution claimed that giving access to copyrighted content did not constitute direct infringement.
Outcome: The High Court of Australia ruled that institutions that facilitate infringement might be held accountable, including implications for digital platforms (University of New South Wales v. Moorhouse, 1975).

Roadshow Films Pty Ltd vs. iiNet Ltd. (2012)

Overview: Film studios sued iiNet, an internet service provider, for failing to prevent piracy among its subscribers.
Fair Use Argument: iiNet claimed that it had no direct control over user activities and so could not be held accountable.
Outcome: The High Court of Australia decided in favor of iiNet, finding that ISPs are not always accountable for users' copyright infractions (Roadshow Films Pty Ltd v. iiNet Ltd., 2012).

9.0 FINDINGS

The chapter presents major discoveries about the use of fair usage in the digital era. First, while the notion of transformative use is central to fair use findings, recent judicial cases demonstrate that commercial reasons can restrict its breadth. The Warhol Foundation v. Goldsmith decision is especially instructive, demonstrating that, while alteration is required for fair use, it must also include a significant contribution of new meaning or value to the original work. The court's opinion underlines that for commercial purposes, the standard for proving change is greater, reflecting a cautious approach to awarding fair use protection in profit-driven circumstances.

Another significant result is the expanded application of fair use in the technology industry. The Google LLC v. Oracle America case demonstrates how fair use may spur technical innovation by allowing developers to improve old code in ways that benefit the public. This decision indicates that fair use is not limited to conventional media but also includes technical breakthroughs, which allow for new applications of old works that foster innovation and growth. This wide view emphasizes the importance of fair use in promoting the development of new technology and applications.

Conversely, the McGucken v. Newsweek LLC case highlights the complexity of digital material and fair use. The court's decision that embedding copyrighted social media information without authorization does not constitute as fair use emphasizes the importance for digital media companies to get authorized licensing. This conclusion emphasizes that fair use rights are not automatic when dealing with information from sites such as Instagram, and it emphasizes the significance of following copyright rules in digital publication. It demonstrates that credit is insufficient to prevent copyright infringement.

Beyond the United States, different countries have approached fair use (or its equivalent, such as fair dealing) in unique ways. In the United Kingdom, the decision of Ashdown v. Telegraph Group Ltd (2001) demonstrated the limited nature of fair dealing, concluding that unlawful publishing of sensitive political material did not fulfill the conditions of fair use under UK copyright law. In Canada, the landmark decision in SOCAN v. Bell Canada (2012) expanded the definition of fair dealing by acknowledging the value of user-generated material in research and teaching, notably in the digital music sector. Similarly, the Australian case of University of New South Wales v. Moorhouse (1975) established precedent for institutional liability by ruling that libraries that facilitate unauthorized copying could be held liable for copyright infringement, influencing modern copyright policies in educational institutions. Similarly, the Australian case of University of New South Wales v. Moorhouse (1975) established precedent for institutional liability by ruling that libraries that facilitate unauthorized copying could be held liable for copyright infringement, influencing modern copyright policies in educational institutions. Meanwhile, in Super Cassettes Industries Ltd v. MySpace Inc. (2011), India's Delhi High Court stressed platform accountability, declaring that MySpace's inability to prevent copyright infringement on its platform excluded it from fair use protection.

These cases demonstrate the diverse global approaches to fair use, with some governments adopting a more liberal interpretation to accommodate digital innovation and others prioritizing rigorous copyright enforcement. As technology reshapes content production and transmission, courts throughout the world must navigate the complexity of balancing intellectual property rights with the public interest in information access, creative reuse, and technical advancement. The changing nature of fair use necessitates continuing legislative changes to guarantee that copyright rules remain relevant and successful in the digital era.

Overall, these findings show that fair use rules are developing to meet the new difficulties posed by digital and technical breakthroughs. The legal environment is becoming more dynamic, with courts weighing the rights of original inventors against the need to encourage creativity and innovation. The shifting meaning of fair use represents a sophisticated approach to ensure that copyright protection, as well as the promotion of new forms of expression and technical advances, are properly addressed.

10.0 CONCLUSION AND POLICY RECOMMENDATION

The evolving framework of fair use in the digital era highlights the challenges of reconciling copyright protection with creativity and innovation. Recent case law, including Warhol Foundation v. Goldsmith (2023), Google LLC v. Oracle America, Inc. (2021), McGucken v. Newsweek LLC (2022), and cases from other jurisdictions such as SOCAN v. Bell Canada (2012, Canada), Super Cassettes Industries Ltd

v. MySpace Inc. (2011, India), and University of New South Wales v. Moorhouse (1975, Australia), demonstrates fundamental shifts in how fair use is applied across different contexts.

The Warhol case demonstrates that, while transformative usage remains an important consideration, commercial motivation can have a major influence on fair use rulings, necessitating a higher level of transformation for profit-driven objectives. Meanwhile, the Google case demonstrates how fair use may spur technical innovation by allowing developers to build on prior works, so benefiting the public and promoting progress. In contrast, the McGucken case highlights the issues associated with digital content, highlighting that simply embedding or acknowledging anything from social media does not fulfill fair use criteria. Similarly, SOCAN v. Bell Canada reiterated that fair dealing exclusions in Canada should not jeopardize the financial viability of original works, while Super Cassettes v. MySpace stressed that digital platforms cannot utilize fair use as a blanket defense for illicit material distribution.

As digital platforms and artificial intelligence continue to transform content creation and delivery, courts will be more called upon to understand the convergence of copyright law, technical breakthroughs, and creative expression. Fair use must remain adaptive, protecting the rights of original artists while also promoting public access and creativity. To do this, legislators and legal professionals should address the emerging difficulties with specific policy adjustments.

Policy Recommendations

Short-Term Policy Adjustments

In the immediate term, there is a pressing need to define fair use norms in digital and commercial settings. Establishing clearer differences between personal, educational, and commercial uses can assist to reduce legal uncertainty and minimize the abuse of fair use to circumvent copyright rights. A specific digital fair use framework should be created to address growing issues including AI-generated content, social media embedding, and algorithm-driven content selection. AI models that use copyrighted works for training must follow fair licensing policies to guarantee that the original artists are not unjustly exploited. Furthermore, digital platforms and news organizations should impose required license agreements to compensate photographers, artists, and content producers whose work is exploited commercially.

Enhanced automatic fair use identification technologies, such as updated versions of YouTube's Content ID, should be implemented to better discern between true infringement and legally authorized fair use. Finally, public awareness efforts should be conducted to educate consumers about responsible digital material sharing and the legal bounds of fair use, decreasing unintentional copyright breaches.

Long-Term Legislative Reforms

In the long run, complete fair use legislation adapted to digital content must be implemented. Current copyright rules must be modernized to clearly recognize fair use in scenarios like AI-driven content development, deep learning, data mining, and remix culture. A formalized revenue-sharing approach should be implemented to compensate authors when their works are used under fair use exceptions, especially if digital platforms earn significantly from user-generated material. Given the worldwide character of digital information, organizations such as WIPO should work to establish international fair use norms that provide uniformity while recognizing regional legal variances.

As AI-generated material becomes more common, legal frameworks must provide clear criteria for the extent to which AI-created works are eligible for fair use and if license agreements are necessary. In addition, judicial and policy review panels should be formed on a regular basis to analyze the impact of fair use rules on content creators, technology improvements, and public access. These committees should suggest timely revisions to keep up with the quickly changing digital world.

The digital era brings both opportunities and challenges for fair use, necessitating a flexible strategy that combines copyright protection with creative and technical innovation. By making short-term policy changes and pursuing long-term legislative improvements, policymakers may guarantee that fair use remains a viable tool for stimulating innovation while protecting content producers' economic interests. A forward-thinking regulatory framework, together with industry collaboration and public awareness, will aid in striking this balance in the rapidly expanding digital economy.

REFERENCES

Ashdown v. Telegraph Group Ltd, [2001] EWCA Civ 1142 (UK).

Aufderheide, P., & Jaszi, P. (2018). *Reclaiming Fair Use: How to Put Balance Back in Copyright*. University of Chicago Press., DOI: 10.7208/chicago/9780226374222.001.0001

Aufderheide, P., & Jaszi, P. (2018). *Reclaiming fair use: How to put balance back in copyright*. University of Chicago Press. DOI: 10.7208/chicago/9780226374222.001.0001

Australian Law Reform Commission (ALRC). (2018). *Fair Use and Copyright Reform in Australia: A Legal Perspective*. ALRC.

Bently, L., & Sherman, B. (2014). *Intellectual Property Law* (4th ed.). Oxford University Press., DOI: 10.1093/he/9780199645558.001.0001

Boucher, D. (2022). *The Ethics of Artificial Intelligence in Content Creation: Fair Use in the Age of Deepfakes*. MIT Press.

Campbell v. Acuff-Rose Music, Inc., 510 U.S. 569 (1994).

Campbell v. Acuff-Rose Music, Inc., 510 U.S. 569 (1994).

Chancellor Masters and Scholars of the University of Oxford v. Rameshwari Photocopy Services, (2016) 160 DRJ 679 (Del).

Chatterjee, A. (2022). The Role of Blockchain and NFTs in Copyright Protection. *Journal of Digital Economy*, *15*(3), 87–102.

Court, D. H. (2011). Super Cassettes Industries Ltd. v. [Del]. *MySpace Inc.*, *48*(PTC), 49.

Court of Justice of the European Union. (2014). Deckmyn v. Vandersteen. CASE (Philadelphia, Pa.), C-201(13).

Court of Justice of the European Union. (2014). Deckmyn v. Vandersteen. *CASE (Philadelphia, Pa.)*, *C-201*(13).

Court of Justice of the European Union. (2019). Pelham GmbH v. Hütter. *CASE (Philadelphia, Pa.)*, *C-476*(17).

Davison, M., & Hugenholtz, P. B. (2017). *Research Handbook on Copyright Law*. Edward Elgar Publishing.

Court, D. H. (2011). Super Cassettes Industries Ltd. v. [Del]. *MySpace Inc.*, *48*(PTC), 49.

Delhi High Court (2013). India TV v. Yashraj Films, (2013) SCC Online Del 500.

Delhi High Court (2013). India TV v. Yashraj Films, (2013) SCC Online Del 500.

ECJ. (2019). Pelham GmbH v. Hütter. *CASE (Philadelphia, Pa.)*, *C-476*(17).

European Commission. (2019). Copyright in the Digital Single Market. Retrieved from https://ec.europa.eu/digital-strategy

Geiger, C., Hilty, R., Griffiths, J., & Sganga, C. (2020). Fair Use in the European Union: Inconsistencies and Legal Uncertainties. *European Intellectual Property Review*, *42*(2), 97–112.

Goldman, E. (2020). Content ID and the Fair Use Dilemma in Digital Platforms. *Journal of Internet Law*, *23*(4), 12–18.

Goldman, E. (2020). Content ID and the Fair Use Dilemma in Digital Platforms. *Journal of Internet Law*, *23*(4), 12–18.

Goldman, E. (2020). Content moderation remedies. *Michigan Technology Law Review*, *26*(2), 215–267.

Goldman, E. (2020). Copyright in the Age of Automated Content Moderation. *Stanford Law Review*, *72*(4).

Goldman, E. (2020). *Internet Law: Cases and Problems* (9th ed.). Semaphore Press.

Google LLC v. Oracle America, Inc., 593 U.S. ___ (2021).

High Court of Australia (1975). University of New South Wales v. Moorhouse, (1975) 133 CLR 1.

High Court of Australia (1975). University of New South Wales v. Moorhouse, (1975) 133 CLR 1.

High Court of Australia (2012). Roadshow Films Pty Ltd. v. iiNet Ltd., (2012) HCA 16.

High Court of Australia (2012). Roadshow Films Pty Ltd. v. iiNet Ltd., (2012) HCA 16.

Hughes, J. (2019). *Copyright and Its Limits: Privacy, Publicity, and the Legal Uses of Fair Use*. Cambridge University Press.

Hughes, J. (2019). The role of transformative use in copyright law. *Harvard Journal of Law & Technology*, *33*(1), 45–79.

Hughes, J. (2019). *Transformative Use and Digital Copyright: Navigating User-Generated Content*. Oxford University Press.

Hughes, J. (2019). Transformative Use in the Digital Age: The Expansion of Fair Use and Its Implications. *Columbia Journal of Law & Arts*, *42*(3), 245–267.

Hughes, J. (2019). Transformative Use in the Digital Age: The Expansion of Fair Use and Its Implications. *Columbia Journal of Law & Arts*, *42*(3), 245–267.

Jain, S. L. (2020). Sham Lal Jain v. Union of India. Delhi High Court.

Katz, A. (2020). Artificial Intelligence and the Future of Copyright: Who Owns AI-Generated Works? *Journal of Intellectual Property Law*, *28*(3).

Katz, S. (2020). AI and Copyright: Challenges and Opportunities. *Journal of Intellectual Property*, *57*(2), 115–132.

Kumar, P. (2021). Content ID and Fair Use: A Case Study of YouTube's Monetization Model. *Indian Journal of Media Studies*, *27*(1), 54–71.

Lessig, L. (2015). *Free Culture: The Nature and Future of Creativity*. Penguin Books.

Litman, J. (2020). Harper & Row and the Limits of Fair Use: Analyzing Copyright Protection in Journalism. *Columbia Law Review*, *38*(2).

Litman, J. (2021). *Digital Copyright: Law and Policy*. Wolters Kluwer.

McCutcheon, D. (2021). *AI-Generated Works and the Future of Fair Use: Challenges and Opportunities*. MIT Press.

McCutcheon, J. (2021). *AI and Intellectual Property: New Horizons*. Oxford University Press.

McCutcheon, J. (2021). Artificial intelligence and the evolving boundaries of fair use. *Journal of Intellectual Property Law & Practice*, *16*(4), 239–258.

McGucken v. Newsweek LLC, 464 F. Supp. 3d 594 (2022).

McGucken v. Newsweek LLC, U.S. District Court for the Southern District of New York (2022) Aufderheide, P., & Jaszi, P. (2018). Reclaiming Fair Use: How to Put Balance Back in Copyright. University of Chicago Press.

Ranjan, A. (2011). Indian Performing Right Society Ltd. v. Sanjay Dalia & Anr. Delhi High Court.

Samuelson, P. (2018). The Evolution of Copyright in the Digital Age: Sony Corp. and Its Legacy. *Harvard Law Review*, *131*(5).

Samuelson, P. (2019). Copyright's Evolution in the Digital Age. *Harvard Law Review*, *132*(5), 1530–1570.

Samuelson, P. (2019). Rethinking fair use in the digital age. *Columbia Journal of Law & the Arts*, *42*(2), 145–176.

Samuelson, P. (2019). The Blurred Lines of Fair Use: Copyright, Creativity, and Digital Innovation. *Harvard Law Review*, *132*(6).

Samuelson, P. (2019). The Complexities of Fair Use: The Role of Market Impact and Transformative Use. *Berkeley Technology Law Journal*, *34*(2), 567–590.

Samuelson, P. (2019). The Evolution of Copyright in the Digital Age. *Harvard Law Review*, *131*(5).

Sengupta, R. (2011). *Super Cassettes Industries Ltd. v. MySpace Inc*. Delhi High Court.

Singh, M. (2017). M/s. Ramesh D. Naik v. M/s. Rajhans. Delhi High Court.

Singh, R. (2013). Super Cassettes Industries Ltd. v. Chintamani Rao: A Case Study on Public Performance Rights. Journal of Indian Intellectual Property Law, 22(4).

Singh, R. (2021). Fair Dealing and Digital Content in India: Challenges and Prospects. *Indian Journal of Intellectual Property Law*, *13*(1), 98–120.

Singh, T. (2017). Copyright and Fair Dealing in Indian Law: Analyzing Civic Chandran v. Ammini Amma (1996). *Economic and Political Weekly*, *52*(18), 34–40.

SOCAN v. Bell Canada, [2012] 2 S.C.R. 326 (Canada).

Sunder, M. (2018). From Goods to a Good Life: Intellectual Property and Global Justice. Yale University Press.Supreme Court of the United States, Google LLC v. Oracle America, Inc. (2021)

2011. Super Cassettes Industries Ltd v. [Delhi HC] [India]. MySpace Inc., 48(PTC), 49.

Super Cassettes Industries Ltd. v. MySpace Inc., (2011) 174 DLT 279 (Del).

Supreme Court of the United States, Warhol Foundation v. Goldsmith (2023)

UK Court of Appeal (2001). Ashdown v. Telegraph Group Ltd., (2001) EWCA Civ 1142.

UK Court of Appeal (2001). Ashdown v. Telegraph Group Ltd., (2001) EWCA Civ 1142.

UK High Court (2011). Sport Five GmbH v. Litman, (2011) EWHC 684 (Ch).

UK High Court (2011). Sport Five GmbH v. Litman, (2011) EWHC 684 (Ch).

University of New South Wales v. Moorhouse, (1975) 133 CLR 1 (Australia).

U.S. Copyright Office. (2021). Copyright and Fair Use: Historical and Legal Perspectives. Washington, D.C.

U.S. Copyright Office. (2021). Fair use index: A resource on fair use cases. Retrieved from https://www.copyright.gov/fair-use

U.S. Copyright Office. (2021). Fair use index. Retrieved from https://www.copyright.gov/fair-use

U.S. Copyright Office. (2021). Fair Use. Retrieved from https://www.copyright.gov/fair-use

U.S. Court of Appeals (2015). Authors Guild v. Google, Inc., 804 F.3d 202 (2d Cir. 2015).

U.S. District Court (2022). McGucken v. Newsweek LLC, No. 1:20-cv-05063 (S.D.N.Y. 2022).

U.S. Supreme Court (1994). Campbell v. Acuff-Rose Music, Inc., 510 U.S. 569 (1994).

U.S. Supreme Court (1994). Campbell v. Acuff-Rose Music, Inc., 510 U.S. 569 (1994).

U.S. Supreme Court (2021). Google LLC v. Oracle America, Inc., 141 S. Ct. 1183 (2021).

U.S. Supreme Court (2021). Google LLC v. Oracle America, Inc., 141 S. Ct. 1183 (2021).

U.S. Supreme Court (2023). Warhol Foundation v. Goldsmith, 598 U.S. 310 (2023).

Warhol Foundation v. Goldsmith, 598 U.S. ___ (2023).

World Intellectual Property Organization (WIPO). (2021). *Global Copyright and Fair Use Policies: A Comparative Analysis*. WIPO.

YouTube. (2022). *Global Copyright Enforcement and Fair Use: A Digital Study of Cross-Border Disputes*. YouTube Legal Compliance Division.

Chapter 9
Towards Technological Literacy:
Fostering Digital Citizenship and Sustainable Development Education in Next-Generation Teaching

Nicole Simon
https://orcid.org/0009-0004-0678-1300
SUNY Nassau Community College, USA

ABSTRACT

Empowering the next generation with digital citizenship is paramount in navigating the complexities of the online world. By instilling principles of responsibility, ethics, and safety, we equip young individuals with the necessary tools to thrive in the digital age. Education plays a pivotal role in this process, teaching learners to critically evaluate information, engage respectfully in online discourse, and protect their privacy. Through interactive workshops, seminars, and curriculum integration, educators can cultivate a generation of informed digital citizens who understand their rights and responsibilities in virtual spaces. Integrating the United Nations Sustainable Development Goals (SDGs) into higher education offers a powerful framework for teaching digital citizenship while fostering global awareness and engagement (United Nations. 2021). One strategy involves aligning digital citizenship education with specific SDGs, such as Quality Education (SDG 4) and Gender Equality (SDG 5), which are closely linked to digital access, inclusion, and empowerment.

DIGITAL CITIZENSHIP AND EDUCATIONAL FRAMEWORKS

Fostering digital citizenship promotes active participation and collaboration within online communities (Aldahshan, 2016). By embracing inclusive practices and valuing diverse perspectives, learners can leverage digital platforms to drive positive change. Through social activism, creative expression, or knowledge sharing, individuals equipped with digital citizenship skills can make a meaningful impact.

Through their research, Alqayed (2014) and later Alselehat, Alfalloh, and Alsarhan (2018) delve into the complex nature of digital citizenship among undergraduate learners. They focus on eight of the nine recognized components, including digital commerce, communication, literacy, etiquette, law, rights and responsibilities, health and wellness, and security. By evaluating students' comprehension

DOI: 10.4018/979-8-3693-6675-2.ch009

Copyright ©2025, IGI Global Scientific Publishing. Copying or distributing in print or electronic forms without written permission of IGI Global Scientific Publishing is prohibited.

and application of these concepts, their study emphasizes the urgent need for effective digital citizenship education within higher education institutions. Although many students possess a basic understanding of these components, there are still gaps in their ability to put them into practice.

This research underscores the significance of creating an environment where students can cultivate essential skills for responsibly navigating the digital realm. As technology continues to advance, the repercussions of inadequate digital citizenship knowledge can have far-reaching consequences, potentially resulting in ethical dilemmas and unsafe online behaviors. The authors advocate for implementing improved educational strategies that engage students in meaningful practices, promoting the safe and responsible use of technology. These findings contribute to the ongoing discourse surrounding digital citizenship education, highlighting the importance of equipping students with the necessary tools to thrive in a digital society.

Based on the findings of this study, it becomes evident that the level of Internet usage experience among undergraduate learners does not directly affect their understanding and implementation of digital citizenship. However, an interesting observation arises when considering the influence of educational technology courses on this matter. The inclusion of educational technology courses, specifically those that emphasize the use of technology, plays a significant role in enhancing students' comprehension and application of digital citizenship principles. These courses provide learners with the necessary tools and knowledge to navigate the digital landscape responsibly and securely.

Building on these insights, it is imperative for educational institutions to consider incorporating digital citizenship training within their curriculum. By doing so, they can cultivate a culture of good digital citizenship among their learner body, ensuring that the principles of digital commerce, communication, literacy, etiquette, law, rights and responsibility, health and wellness, and security are not only understood but also actively practiced. Further research should be conducted to explore the most effective methods of integrating digital citizenship education into various academic disciplines. This research can provide valuable guidance for policy development and practical application, enabling educational institutions to tailor their approaches to suit the unique needs of their learners. This study highlights the importance of training undergraduate learners in the secure and responsible utilization of information technology. It emphasizes the need to go beyond mere awareness and comprehension of digital citizenship components and focus on practical implementation. By doing so, we can equip future generations with the skills and mindset necessary to navigate the digital world ethically and responsibly.

The nine components of digital citizenship offer a thorough framework for responsibly navigating the digital landscape. Digital access underscores the necessity of ensuring equitable access to technology and the internet for all individuals, thereby bridging the digital divide that can impede societal participation. Digital commerce emphasizes the significance of understanding online transactions, promoting safe and informed purchasing practices. Other components include digital communication, which focuses on respectful and effective interactions online; digital literacy, ensuring individuals can critically evaluate information; and digital rights and responsibilities, highlighting the ethical implications of digital engagement. Together, these elements foster a responsible online presence, empowering individuals to contribute positively to the digital world.

The nine components of digital citizenship offer a thorough framework for responsibly navigating the digital landscape. Digital access underscores the necessity of ensuring equitable access to technology and the internet for all individuals, thereby bridging the digital divide that can impede societal participation. Digital commerce emphasizes the significance of understanding online transactions, promoting safe and informed purchasing practices. Other components include digital communication, which focuses on respectful and effective interactions online; digital literacy, ensuring individuals can critically evaluate information; and digital rights and responsibilities, highlighting the ethical implications of digital engagement. Together, these elements foster a responsible online presence, empowering individuals to contribute positively to the digital world.

Collaborating with parents and caregivers allows educators to extend the learning beyond the classroom and provide consistent messaging about digital citizenship at home. This can be achieved through regular communication, workshops, and resources that empower parents to navigate the digital world alongside their children. By involving the broader community, such as local organizations and experts, educators can access additional expertise and resources to enhance digital citizenship education.

Furthermore, assessment plays a crucial role in effective digital citizenship education. It is important to design assessments that not only measure knowledge and understanding but also evaluate the application of digital citizenship principles in authentic situations. This can be done through performance-based assessments, where learners demonstrate their ability to make ethical decisions, protect their privacy, and engage responsibly online. By providing constructive feedback and encouraging self-reflection, educators can guide learners towards continuous improvement in their digital citizenship skills.

It is essential to stay up to date with the evolving digital landscape and adapt teaching strategies accordingly. Technology is constantly changing, and educators need to be initiative-taking in their professional development to effectively address emerging digital citizenship issues. This can involve attending conferences, participating in online courses, and engaging in ongoing discussions with colleagues. By staying informed and adaptable, educators can ensure that their digital citizenship instruction remains relevant, impactful, and responsive to the needs of learners in the digital age.

In higher education settings, promoting active learning through experiential activities and real-world applications is particularly impactful. Utilizing case studies, group discussions, and collaborative projects that reflect the complexities of digital life, educators can challenge learners to analyze, evaluate, and respond to ethical dilemmas and issues related to digital citizenship. Encouraging learners to reflect on their own online practices, engage in ethical decision-making, and consider the broader societal implications of their digital interactions foster deeper learning and empowers them to navigate the complexities of the digital landscape with confidence and integrity.

Creating Partnerships for Digital Citizenship

Developing partnerships with industry professionals, alumni networks, and relevant organizations can enrich digital citizenship education in higher education settings. By providing opportunities for learners to connect with industry experts, participate in internships, or engage in service-learning projects focused on digital literacy and citizenship, educators can enhance learners' understanding of real-world applications and career readiness skills. Additionally, collaborating with campus resources such as counseling services, IT departments, and learner organizations can provide additional support and resources for addressing learners' digital well-being and promoting a culture of responsible digital citizenship across the campus community.

By contextualizing digital citizenship within the broader framework of sustainable development, educators can inspire learners to recognize the interconnectedness of digital technologies and social progress, motivating them to leverage their digital skills for positive social change. Another strategy is to incorporate experiential learning opportunities that enable learners to address real-world challenges related to digital citizenship within the context of the SDGs. This could involve designing service-learning projects, research initiatives, or social entrepreneurship ventures that address issues such as digital inequality, online safety, or digital rights. By providing learners with opportunities to apply digital citizenship principles in practical settings, educators can enhance their understanding of global issues, encourage critical thinking, and foster a sense of social responsibility. Moreover, collaboration with local communities, NGOs, and international partners can enrich learners' learning experiences and facilitate meaningful contributions to sustainable development efforts.

Leveraging digital technologies themselves can enhance the effectiveness of teaching digital citizenship within the context of the SDGs in higher education. Interactive online platforms, multimedia resources, and virtual collaboration tools can facilitate engaging and immersive learning experiences that transcend geographical boundaries and cultural differences. Educators can utilize digital storytelling, virtual simulations, and online discussions to explore complex topics related to digital citizenship and sustainable development, encouraging learners to develop empathy, cultural competence, and digital fluency. By harnessing the power of digital technologies to support active learning and global citizenship, higher education institutions can empower learners to become ethical and responsible digital citizens who contribute positively to the achievement of the SDGs.

In today's rapidly evolving digital landscape, fostering a generation of technologically literate individuals is paramount. The integration of technology into education has transformed how students interact with information, offering new avenues for learning and problem-solving. However, technological literacy extends beyond mere proficiency with digital tools; it encompasses the ability to navigate digital environments responsibly and effectively. As educators, there is an increasing need to cultivate these skills in students, ensuring they not only use technology effectively but also understand its impact on their lives and the world around them.

As educational paradigms shift to include digital and sustainability aspects, the role of educators is evolving. Teaching students to be proficient in digital technologies is no longer enough; they must also be guided to use these tools in ways that contribute to sustainable development and ethical practices. This includes instilling a sense of digital citizenship that emphasizes responsible behavior online and an understanding of how technology can be harnessed for positive environmental and societal outcomes. By embedding these principles into the curriculum, educators can prepare students to become informed and conscientious digital citizens.

The convergence of digital literacy and sustainable development is essential for preparing the next generation for a future where technology plays a central role in addressing global challenges. Educators are tasked with the challenge of integrating these concepts into their teaching practices, fostering a deeper understanding of how technology intersects with sustainability. This approach not only equips students with critical skills for the digital age but also encourages them to consider the broader implications of their technological choices. By embracing this dual focus, education can play a pivotal role in shaping a future that balances technological advancement with environmental stewardship and ethical responsibility.

Understanding Digital Citizenship in Contemporary Education

Defining digital citizenship encompasses a comprehensive understanding of how to engage with technology responsibly and effectively. It involves not only the technical skills to navigate digital environments but also the ethical considerations of online behavior. This includes understanding the impact of one's digital footprint, respecting privacy and intellectual property, and recognizing the importance of maintaining a positive online reputation. Digital citizenship is about equipping individuals with the ability to critically assess digital content, discern credible sources from unreliable ones, and use technology in ways that are both legally and morally sound.

Furthermore, digital citizenship is ultimately about fostering a balanced approach to technology use. It involves integrating knowledge of digital tools with an awareness of their broader implications for personal and social well-being. As digital landscapes continue to evolve, the principles of digital citizenship must adapt, emphasizing the importance of continuous learning and ethical reflection. By embedding these practices from an early age, educators and parents can help learners develop the skills to navigate the digital world responsibly and constructively, ensuring they are prepared to contribute positively to the digital community.

According to Common Sense Media (2009), on page 1, "the ability to use technology competently, interpret and understand digital content, assess its credibility and create, and research and communicate with appropriate tools" is the definition of digital literacy given in "Digital Literacy and Citizenship in the 21st Century". This definition highlights not only the technical skills required for effective technology use but also the critical thinking necessary to navigate the complex digital landscape. Despite the focus on middle school students and older individuals in many recent digital citizenship initiatives, evidence suggests that learners often develop their own informal guidelines for technology use long before reaching this age. This underscores the necessity of introducing digital literacy at an earlier stage in education.

Given that children are frequently exposed to technology at home even before they enter formal schooling, it is crucial for parents to begin teaching digital citizenship from an early age. Parents play a pivotal role in laying the groundwork for responsible technology use through proactive engagement and guidance. They need to foster awareness of digital safety, ethical behavior, and the potential consequences of online actions. This early preparation equips children to handle technology thoughtfully and securely as they progress through their education. Once children are in school, educators - including teachers, media specialists, and IT specialists - must build upon this foundation by integrating digital citizenship principles into the curriculum. Educators should continually enhance their own understanding of digital issues and be prepared to address emerging challenges, ensuring that students receive consistent and comprehensive instruction in digital literacy throughout their academic journey.

In higher education, digital citizenship becomes increasingly nuanced as students are expected to engage with technology in more sophisticated ways. At this level, digital literacy extends beyond basic skills to include advanced research techniques, critical evaluation of digital sources, and ethical considerations in digital creation and collaboration. Students must navigate complex academic databases, manage their online presence professionally, and understand issues such as plagiarism and intellectual property rights. Institutions of higher learning have a critical role in providing targeted training and resources to help students develop these advanced competencies. This preparation is essential for ensuring that students can contribute meaningfully to their fields of study and the broader digital society, while upholding academic integrity and professional standards.

The Significance of Digital Citizenship in Technological Literacy

The idea of "digital citizenship," which reflects the changing nature of civic engagement in the digital era, has gained popularity since the invention of the Internet (Choi, 2016). According to Hollandsworth et al. (2011), putting the idea of digital citizenship into practice in educational contexts can be difficult. According to their comparison to character education, it entails developing into a responsible member of the online community while taking into account student safety, educational advantages, moral and legal behavior, and productive engagement in online communities (Hollandsworth et al., 2011, p. 37). This viewpoint is consistent with the fundamental notions of citizenship found in American schools.

The concept of digital citizenship traces its roots back to the late 1990s and early 2000s. The following section will provide an overview of key digital citizenship frameworks developed over the past two decades. Among the earliest works is Pruitt-Mentle's C3 Framework from 2000, followed by Ribble and Bailey's Nine Elements of Digital Citizenship from the mid-2000s. Later, Project Zero and Common Sense Education introduced the Five Core Dispositions of Digital Citizenship and their associated curriculum. Most recently, Choi developed a Digital Citizenship Scale. Understanding these frameworks is essential for a comprehensive grasp of digital citizenship.

Digital citizenship encompasses the responsible use of technology and the internet, guiding individuals in creating a positive online presence. At the heart of this concept are the nine components of digital citizenship: digital access, commerce, communication, literacy, etiquette, law, rights and responsibilities, health and wellness, and security. Each component plays a vital role in shaping how we interact with others online and how we manage our digital footprints.

Equal access to technology and information is crucial because digital access guarantees that everyone has the chance to engage in the digital world. The ability to locate, assess, and use information efficiently is the main focus of digital literacy, which empowers people to make wise decisions when using the internet. A culture of kindness and understanding is fostered by courteous interactions that are encouraged by effective communication and decorum. People can navigate the digital world more responsibly and recognize their rights and duties as users by being aware of the legal aspects of online activity, such as copyright rules and privacy rights. People can develop a responsible online persona and help create a more secure and civil online community by actively interacting with these elements.

Crafting Curriculum to Embed Digital Citizenship Into Educational Frameworks

A deliberate approach that incorporates these ideas into the main curriculum and institutional policies is necessary to incorporate digital citizenship into higher education pedagogical frameworks. This entails creating an atmosphere where responsible technology use is a fundamental expectation in addition to integrating digital literacy and ethical standards into different courses. Institutions can accomplish this by creating specialized courses or modules on digital citizenship that cover subjects like data protection, online research ethics, and the effects of digital footprints. In order to ensure that students comprehend the significance of these concerns within their particular fields of study, faculty members should also be urged to integrate themes of digital citizenship into their subject areas.

Crafting a curriculum that includes digital citizenship involves designing assignments and projects that require students to practice and reflect on responsible technology use. This can be achieved by creating opportunities for students to engage in digital collaboration, evaluate online sources critically, and produce ethical digital content. Institutions should also offer resources and support services, such as digital literacy centers or online tutorials, to help students navigate these challenges effectively. By embedding digital citizenship into the curriculum and offering continuous professional development for educators, higher education institutions can prepare students to be conscientious digital citizens who are capable of navigating and contributing to an increasingly complex digital landscape.

Digital citizenship is essential in today's interconnected world, emphasizing the importance of responsible online behavior and engagement. As we navigate various digital landscapes, it's crucial to understand the principles outlined in the next generation standards, which advocate not only technical skills but also ethical usage of technology. This holistic approach prepares individuals to thrive in a digital society while fostering respect and integrity among users.

Since the nine elements of digital citizenship foster the abilities necessary to prosper in a digital world, they strongly correspond with the standards of the next generation. Fair opportunities for everyone to use technology are guaranteed by digital access, which is essential for promoting inclusivity in learning settings. Students are taught about safe and responsible online buying through digital commerce, which highlights the value of making educated decisions in online purchases. In digital teamwork and collaborative learning, polite interactions are created through communication and etiquette. Teachers may equip pupils to be responsible, conscientious digital citizens by incorporating these elements into their curricula.

Integrating digital citizenship into education prepares learners for the workforce of tomorrow (Almasry & Shaath, 2013; Alqayed, 2014; Alselehat et al., 2018). In an interconnected world, digital communication, collaboration, and problem-solving skills are essential. Early development of these skills ensures students are ready for the digital workplace and can adapt to technological advancements. By fostering digital literacy, we promote personal success and contribute to a society that values ethics, inclusivity, and innovation (Kara, 2018). Digital citizenship involves various behaviors, some with risks and downsides. Lack of knowledge and guidance can lead to harmful behaviors, potentially impacting the future digital landscape. Educators must help students become responsible digital citizens, drawing on expert insights.

We are at a critical moment for enhancing digital citizenship awareness. As Weigel, James, and Gardner (2009) note, young people's lack of skills for effective online navigation limits the educational potential of the Internet. With more students online, it is essential to guide them towards safe, productive digital engagement. The future of society depends on students' ability to make responsible technology decisions, including understanding ethical, legal, and safety issues. As Blackshaw (2009) reports, millions of young students engage with digital media more than previous generations, underscoring the need for guidance during critical developmental stages.

Cultivating Digital Citizenship in the Next Generation Classroom

In 2010, Common Sense Education partnered with Project Zero at Harvard to create the Digital Literacy and Citizenship Curriculum, now used in over 60,000 U.S. schools. The curriculum defines digital citizenship as the "responsible use of technology to learn, create, and participate" (James et al., 2019, p. 5). Updated in 2019, it emphasizes five core dispositions to help learners become "reflective, responsible, and ethical decision-makers in their connected lives" (James et al., 2019, p. 10), as outlined in Table 4 (James et al., 2019, p. 11).

Ribble's work on digital citizenship has deeply influenced the International Society for Technology in Education (ISTE), shaping its framework for integrating technology into education. His approach, covering areas like digital access, commerce, communication, and etiquette, aligns with ISTE's focus on promoting responsible technology use. ISTE's standards now emphasize digital citizenship education, fostering ethical behavior and critical thinking. By incorporating Ribble's framework, ISTE supports educators in teaching students how to engage positively with the digital world.

Ribble's influence also extends to Project Zero at Harvard, where his focus on the social and ethical dimensions of technology enhances the initiative's emphasis on critical thinking and ethical decision-making. This alignment is evident in Project Zero's incorporation of digital citizenship into its educational practices, helping students understand the implications of their digital actions and promoting critical thinking about technology's role in society.

Common Sense Education's Five Core Dispositions of Digital Citizenship—Respect, Responsibility, Empathy, Safety, and Digital Footprint—reflect a synthesis of Ribble's principles and current educational needs. These dispositions guide students to engage with technology in an ethical and thoughtful manner. The emphasis on respect and empathy fosters understanding in digital communication, while the focus on safety and digital footprint encourages responsible management of personal information. This framework helps students navigate the complexities of digital interactions, aligning with broader goals of digital citizenship education.

Connecting Digital Citizenship With Sustainable Development

The integration of digital technologies offers a unique opportunity to promote environmental and social sustainability. Digital citizenship, which emphasizes responsible technology use, must consider the environmental impact of digital activities, such as the energy consumption of data centers and the challenges of electronic waste management. Educators and policymakers can guide individuals toward eco-friendly choices, ensuring technology supports sustainable development goals and aligns with broader environmental objectives. To promote sustainability, digital citizenship education must incorporate environmental considerations. This includes teaching energy-efficient practices, like using energy-saving settings and reducing unnecessary data storage, as well as encouraging responsible disposal of electronic waste. Educators can raise awareness and foster sustainable habits, while policymakers can create guidelines and incentives to influence consumer behavior and industry practices.

Integrating sustainability into digital citizenship can also spur innovative solutions for environmental challenges. By encouraging technology use for environmental monitoring, resource management, and advocacy, educators and policymakers can drive positive change. For instance, digital tools can help track energy usage, monitor pollution, or support conservation efforts. These applications not only highlight technology's potential to contribute to sustainability but also empower individuals to align their digital practices with environmental and social goals.

Aligning Digital Skills With Environmental and Social Goals

Incorporating digital skills with a focus on sustainability offers a powerful way to address global challenges. By aligning digital competencies with environmental and social goals, education can empower individuals to use technology to tackle issues such as climate change, resource depletion, and social inequality. For example, data analytics skills can drive environmental monitoring, helping students analyze climate data or evaluate conservation efforts. These skills can also support social justice initiatives by analyzing inequality trends and advocating for policy changes.

Integrating sustainability into digital literacy education is key to fostering a responsible and effective use of technology. This involves teaching not just how to use digital tools, but also how they impact the environment and society. Students might learn about energy-efficient technologies or explore how digital platforms can promote sustainable practices and ethical consumption. By embedding these principles into the curriculum, educators can inspire students to use technology creatively in support of sustainability.

This integration also drives innovation by motivating students to tackle global issues. Understanding the link between digital skills and sustainability encourages students to pursue careers focused on reducing environmental impact, improving resource management, or enhancing social equity. By preparing students to apply their digital expertise toward long-term ecological health and social justice, educational institutions help develop a generation capable of addressing 21st-century challenges.

Utilizing Digital Tools to Promote Sustainable Practices

The application of digital tools to promote sustainable practices is a practical extension of digital citizenship principles. Technologies like energy-efficient devices, smart grids, and platforms for environmental advocacy contribute to sustainable development goals. By demonstrating how these tools reduce carbon footprints, optimize resource use, and foster global collaboration, educators and organizations can inspire users to adopt and advocate for sustainable practices. This reinforces the link between technology and environmental stewardship, motivating meaningful actions toward a sustainable future.

Educators and organizations play a crucial role in highlighting how digital tools help reduce carbon footprints and optimize resource use. By incorporating case studies and practical examples into educational programs, they can illustrate how technology supports sustainability. For example, students can learn how smart grids minimize energy waste or how digital platforms promote recycling. This direct approach not only enhances understanding but also motivates individuals to apply similar technologies in their lives and communities.

By highlighting the practical use of digital tools, educators reinforce the connection between technology and environmental responsibility. When individuals see how their digital choices impact sustainability, they are more likely to adopt eco-friendly habits. This application of digital citizenship encourages users to think critically about the environmental impact of their technology use and inspires them to act. By emphasizing these connections, educators and organizations foster a culture of environmental responsibility and drive efforts toward a more sustainable future.

Overcoming Challenges in Digital Citizenship Education by Setting Standards

Digital citizenship standards, established by organizations like ISTE (International Society for Technology in Education) and the American Association of School Librarians (AASL), are critical for guiding educators and learners. As of April 2021, all 50 U.S. states and territories, including Puerto Rico and Guam, had adopted ISTE's 2006 or 2017 Standards for Learners and/or Teachers, emphasizing digital citizenship nationwide (ISTE, 2021).

ISTE first introduced the National Educational Technology Standards in 1998 (for Learners) and 1999 (for Teachers) to integrate emerging educational technologies. Digital citizenship was included in revisions in 2007 and 2008 (International Society for Technology in Education, 2007; International Society for Technology in Education, 2008), reflecting growing concerns about technology use outside school. The standards were updated in 2016, renaming them the ISTE Standards for Learners and Educators.

The updated ISTE Standards for Learners (2016b) include Standard 2 on digital citizenship:

- Digital Citizens: Learners understand their rights, responsibilities, and opportunities in a connected world, acting safely, legally, and ethically.

- 2a: Cultivate digital identity and reputation, acknowledging the permanence of online actions.
- 2b: Engage in positive, safe, legal, and ethical online behavior.
- 2c: Respect intellectual property rights.
- 2d: Manage personal data to protect privacy and security and understand data collection practices.

Similarly, the ISTE Standards for Educators (2016a) focus on:

- Citizen: Educators inspire responsible participation in the digital world by:
  - 3a: Creating opportunities for learners to contribute positively and empathetically online.
  - 3b: Fostering a learning culture that promotes curiosity, critical evaluation of online resources, and digital literacy.
  - 3c: Guiding learners in safe, legal, and ethical digital practices, including protecting intellectual property.
  - 3d: Modeling effective management of personal data and ensuring student data privacy.

These standards challenge educators not only to teach digital citizenship but also to model these principles in their professional practices.

Components of Digital Citizenship

This study focuses on eight key components of digital citizenship: digital commerce, digital communication, digital literacy, digital etiquette, digital law, digital rights and responsibilities, digital health and wellness, and digital security. It aims to assess undergraduate students' understanding and application of these concepts. While many students demonstrate ethical practices in these areas, they express concerns related to security and safety, such as verifying the legitimacy of digital resources, ensuring information accuracy, understanding digital laws, reporting irresponsible behavior, and managing daily device usage. The findings suggest that students' understanding of digital citizenship is not significantly influenced by their internet usage experience, though educational technology, particularly in technology-focused courses, enhances this understanding. The study offers recommendations for further research, policy development, and practical applications.

The internet's widespread adoption has sparked the "digital revolution," transforming communication, information exchange, public services, commerce, entertainment, social interaction, health, and education. However, alongside its benefits, this revolution introduces significant social and ethical challenges, including issues like identity theft, fraud, hacking, phishing, plagiarism, unauthorized access, and health concerns like excessive screen time and internet addiction. Given these risks, educators must instruct students not only about responsible and effective internet use but also about the potential dangers, while emphasizing the advantages of digital technologies. With increasing evidence of technological misuse, there is an urgent need for policies that address these risks and promote responsible use to maximize the benefits.

In response, the concept of "digital citizenship" has gained global traction and is increasingly integrated into educational curricula in countries like the United States, Canada, and the United Kingdom. Notable initiatives include Australia's "Connecting with Confidence," which trains parents and educators in digital citizenship, and France's national initiative to prioritize digital citizenship as a key issue.

Global Digital Literacy and the United Nations Sustainable Development Goals

Integrating the United Nations Sustainable Development Goals (SDGs) into higher education enhances digital citizenship education while cultivating a sense of global responsibility among students. By aligning digital literacy with SDGs such as Quality Education (SDG 4) and Gender Equality (SDG 5), educators can emphasize the importance of digital access and inclusion. This approach encourages students to explore how digital tools can bridge educational gaps and empower marginalized communities. Highlighting technology's role in expanding learning opportunities fosters awareness, motivating students to advocate for inclusive policies and contribute to equitable educational practices.

Situating digital literacy within sustainable development also empowers students to critically engage with technology's societal impact. As they explore issues like digital divide and access to information, they gain insights into how these challenges intersect with broader social and environmental goals. This understanding inspires students to leverage their digital skills to address global issues, positioning digital citizenship as a tool for promoting sustainable development. In turn, students are encouraged to take active roles in their communities, advocating for digital equity and supporting initiatives that contribute to the SDGs.

The components of digital citizenship align closely with the SDGs, particularly in promoting inclusivity and equity. Digital access corresponds to SDG 10, which seeks to reduce inequalities and ensure universal participation in the digital world. Addressing the digital divide fosters social equity and empowers marginalized communities. Digital literacy supports SDG 4 by providing the skills needed for inclusive, quality education. Additionally, digital commerce and security resonate with SDG 8, which focuses on sustainable economic growth. Digital law and rights reinforce ethical online behavior, key to achieving SDG 16, which promotes peaceful and just societies. Finally, digital health and wellness contribute to SDG 3 by supporting mental well-being in the digital age.

Integrating the SDGs with the Next Generation Standards enhances digital literacy by emphasizing critical thinking, creativity, and collaboration—skills essential for solving global challenges like poverty, inequality, and climate change. SDG-focused projects enable students to develop digital skills while engaging in problem-solving, reinforcing the role of technology in driving sustainable solutions. This integrated approach deepens students' understanding of digital literacy and global citizenship, helping them see the connection between technology and societal progress.

Aligning digital literacy with the Next Generation Standards fosters an interdisciplinary approach that broadens students' perspectives. By exploring themes such as Sustainable Cities (SDG 11) or Responsible Consumption and Production (SDG 12), students can investigate how digital tools promote sustainability in their communities. Projects involving data analysis and digital communication empower students to assess local environmental issues or advocate for sustainable practices, enhancing their technical competencies and encouraging real-world application.

Embedding SDGs within the Next Generation Standards framework fosters a culture of innovation and social responsibility. Through collaboration on real-world challenges, students learn to collaborate with peers, stakeholders, and communities to create impactful solutions. This emphasis on teamwork and social engagement cultivates purpose and drives students to apply their digital skills meaningfully. This approach equips students with essential digital skills and the awareness to contribute to a sustainable, equitable future.

Fostering Digital Citizenship and Sustainable Development Education in Next-Generation Teaching

Integrating digital literacy into the Next Generation Standards is essential for preparing students to thrive in a rapidly evolving, technology-driven world. These standards prioritize competencies like critical thinking, collaboration, and effective communication, all of which are fundamental in the digital age. By embedding digital literacy within these standards, educators equip students with the skills to navigate and critically assess digital content, ensuring they are responsible consumers of information. For instance, lessons that guide students in evaluating sources for credibility not only enhance their research skills but also promote a discerning mindset, a vital aspect of responsible digital citizenship.

Additionally, aligning digital literacy with the Next Generation Standards fosters creativity and innovation in students' use of technology. Engaging students in projects that challenge them to create multimedia presentations, design digital campaigns, or develop apps to address real-world problems empowers them to leverage their digital skills in creative, purposeful ways. This experiential learning not only deepens their understanding of digital tools but also inspires out-of-the-box thinking, cultivating a mindset that embraces technology as a tool for exploration and innovation. Through collaborative projects, students also hone their ability to communicate effectively across digital platforms, preparing them for professional environments in an increasingly interconnected world.

In the context of next-generation teaching, integrating digital citizenship and sustainable development education is critical for equipping students to navigate and contribute to a world characterized by rapid technological advancements and complex global challenges. Digital citizenship highlights responsible and ethical behavior in online environments, emphasizing privacy, security, and respectful communication. Next-generation teaching methods, which utilize innovative technologies and pedagogical approaches, can make these concepts relevant and engaging for students. By embedding principles of digital citizenship into modern teaching practices, educators ensure that students are not only proficient in technology but also aware of their responsibilities as digital citizens, contributing to a well-rounded educational experience.

Promoting digital literacy within the Next Generation Standards also provides a platform to address the digital divide, which affects many communities around the world. By incorporating lessons about equitable access to technology and information, students gain insight into the social implications of digital disparities. Discussions around digital equity, such as the influence of socioeconomic status on technology access, motivate students to advocate for change within their communities. This approach nurtures their digital skills while fostering a sense of social responsibility, empowering them to drive initiatives that promote inclusiveness and access for all. This dual focus on digital literacy and social justice prepares students to be proactive leaders in the fight for greater equality in the digital realm.

Cultivating Digital Citizenship in the Next Generation Classroom

Ribble's work on digital citizenship continues to inform best practices for technology integration in education. In today's digital world, fostering responsible online behavior is more important than ever. Project Zero, an initiative of the Harvard Graduate School of Education, and Common Sense Education have identified five core dispositions of digital citizenship that guide both educators and learners in navigating the digital landscape. These dispositions, respect, responsibility, empathy, courage, and honesty—serve as foundational principles for ethical engagement with technology. By emphasizing these core values, educators empower students to become informed, respectful, and responsible digital citizens.

The five core dispositions of digital citizenship outlined by Project Zero and Common Sense Education include:

1. **Slow Down and Self-Reflect**
   - Encourage students to pay attention to their initial reactions and reflect on their responses.
   - Promote self-awareness in recognizing complex digital situations.
   - Foster habits of regularly assessing their online behaviors and identifying any "red flag feelings."
2. **Explore Perspectives**
   - Cultivate curiosity and an open mind in considering diverse viewpoints.
   - Encourage empathy and understanding of others' emotions and experiences.
   - Reflect on the broader ethical and civic responsibilities that guide online behavior, including respecting others' digital rights and privacy.
3. **Seek Facts and Evidence**
   - Develop students' skills in gathering information from credible sources and evaluating it from multiple perspectives.
   - Teach them to distinguish between facts, opinions, and misinformation, reinforcing the importance of media literacy.
4. **Envision Options and Possible Impacts**
   - Help students consider different courses of action in digital interactions, reflecting on how their choices align with their values.
   - Guide students in understanding the potential impacts of their decisions on themselves and others.
5. **Take Action**
   - Empower students to take constructive actions that promote positive change in their digital environments.
   - Encourage students to function as allies and advocates, standing up for others when they encounter digital harm.
   - Promote reflection on how digital habits can enhance overall well-being.

In addition to these core dispositions, the curriculum addresses six key topics identified through current research, ensuring developmentally appropriate coverage across K-12 education:

1. **Media Balance and Well-Being** – Examines students' online time and digital activities, fostering healthy habits.

2. **Privacy and Security** – Focuses on data privacy and responsible management of personal information.
3. **Digital Footprint and Identity** – Explores how students create and manage their online presence.
4. **Relationships and Communication** – Teaches students how to build and maintain positive relationships and communicate effectively online.
5. **Cyberbullying, Digital Drama, and Hate Speech** – Addresses harmful digital behaviors, providing strategies for addressing them responsibly.
6. **News and Media Literacy** – Encourages critical consumption of media, teaching students to understand copyright issues and create media responsibly.

By integrating these principles into the curriculum, educators can create a learning environment that not only promotes digital proficiency but also emphasizes the ethical and responsible use of technology. This approach ensures that students are prepared to navigate the digital world thoughtfully and contribute to a sustainable, equitable digital future.

Integrating the United Nations Sustainable Development Goals (SDGs) into the Next Generation Standards

Sustainable development education plays a vital role within the Next Generation Standards framework by encouraging students to examine the environmental and societal impacts of their actions. Next-generation teaching employs interactive and interdisciplinary methods to help students understand complex global issues such as climate change, resource management, and social equity. This approach empowers students to recognize how technology intersects sustainability goals and fosters a mindset of environmental stewardship and social responsibility. Through project-based learning, simulations, and real-world problem-solving, students can witness the tangible effects of sustainable practices and understand the importance of integrating these principles into their daily lives.

By combining digital citizenship and sustainable development education with next-generation teaching strategies, educators can cultivate a well-rounded skill set in students. This integrated approach not only prepares students to use technology effectively but also equips them with the knowledge and values necessary to address global challenges. As educational practices evolve, it is essential to focus on both the ethical use of technology and its role in promoting sustainability, ensuring that students are ready to contribute positively to a future where technological and environmental considerations are seamlessly intertwined.

Shifting focus to the United Nations Sustainable Development Goals (SDGs) within the Next Generation Standards framework enhances global awareness and ethical engagement among students. By addressing SDGs like Clean Water and Sanitation (SDG 6) and Climate Action (SDG 13), educators can create learning experiences that connect local actions to global outcomes. For instance, students can engage in projects that analyze water usage in their communities, explore solutions for promoting sustainable water practices, and connect their actions to larger global environmental efforts. This contextualization helps students recognize the real-world relevance of their learning, fostering a sense of purpose and responsibility in their educational journey.

Integrating the SDGs into the Next Generation Standards promotes systems thinking, enabling students to understand the interconnections between various global challenges and appreciate how their actions can have broader, far-reaching implications. For example, discussions around Sustainable Industry, Innovation, and Infrastructure (SDG 9) can lead to explorations of how technological advancements can drive sustainable economic growth. By emphasizing these connections, educators empower students to envision themselves as change-makers who can use their skills to create a more sustainable future. This systems-thinking approach reinforces the idea that education is a powerful tool for addressing global challenges and generating solutions that align with sustainable development.

CONCLUSION

Incorporating digital citizenship and sustainable development into education ensures that students are not only proficient in using technology but also equipped to make informed and ethical decisions in a digital world. By emphasizing responsible online behavior and the broader environmental and societal impacts of technological choices, educators can nurture a generation that values both innovation and sustainability. This comprehensive approach prepares students to navigate the complexities of the modern world, balancing digital proficiency with a commitment to making positive societal and environmental contributions.

As educational practices evolve to meet the demands of the digital age, integrating technological literacy with sustainability education becomes increasingly important. Educators can guide students in how to engage with technology in ways that support long-term global goals, ensuring their actions benefit both the planet and society. This shift in education not only enhances students' immediate capabilities but also empowers them to address future challenges with a well-rounded perspective, aligning technological progress with the overarching objectives of environmental preservation and social responsibility.

Ultimately, fostering a blend of digital proficiency and sustainability awareness in students prepares them for a more thoughtful and responsible approach to technology use. As we look ahead, the educational focus must remain on equipping students to be ethical and informed digital citizens who can leverage technology to drive innovation for the greater good. By embedding these principles into teaching practices, we can cultivate a future generation capable of advancing technology while upholding values that contribute to a sustainable and equitable world.

REFERENCES

Aldahshan. (2016). Digital citizenship as an introduction to Arab education in the digital age. *Journal of Criticism and Enlightenment*, 5/(2), 72-104.

Almasry, M. W., & Shaath, A. H. (2013). The level of digital citizenship of a sample of Palestine university learners from their point of view. *Palestine University Journal for Research and Studies*, 7/(2), 171-203.

Alqayed, M. (2014). The concept of digital citizenship. Retrieved from https://www.neweduc.com/definition-of-digital-citizenship

Alselehat, Y. R., Alfalloh, F. R., & Alsarhan, A. K. (2018). The degree of awareness of the concept of digital citizenship for undergraduate learners in the Faculty of Educational Sciences at the University of Jordan. *Dirasat: Educational Sciences*, 45/(3), 19-33.

Alsuhaim, A. A., & Alibrahim, A. A. (2019). The extent to which computer teachers activate the digital citizenship standards in the secondary school. *Specialized Educational Journal*, 8/(4), 1-12.

Blackshaw, P. (2009). A pocket guide to social media and kids. [Web log comment]. Retrieved from http://blog.nielsen.com/nielsenwire/consumer/a-pocket-guide-to-social-media-and-kids/

Bleazby, J. (2006). Autonomy, democratic community, and citizenship in philosophy for children: Dewey and philosophy for children's rejection of the individual/community dualism. Analytic Teaching, 26(1), 30-52. https://journal.viterbo.edu/index.php/at/article/view/832

Choi, M. (2016). A concept analysis of digital citizenship for democratic citizenship education in the internet age. *Theory and Research in Social Education*, *44*(4), 565–607. DOI: 10.1080/00933104.2016.1210549

Common Sense Media. (2009Pru). Digital citizenship. Retrieved from www.common sense.org/education/digital-citizenship

Hollandsworth, R., Dowdy, L., & Donovan, J. (2011). Digital Citizenship in K-12: It Takes a Village. *TechTrends*, *55*(4), 37–47. DOI: 10.1007/s11528-011-0510-z

International Society for Technology in Education. (2007). *National Educational Technology Standards for Students (NETSS*). https://iste.org/standards/students

International Society for Technology in Education. (2008). *National Educational Technology Standards for Teachers (NETSS*). https://iste.org/standards/educators

International Society for Technology in Education. (2016a). *ISTE Standards for Students (NETSS*). https://iste.org/standards/students

International Society for Technology in Education. (2016b). *ISTE Standards for Educators (NETSS*). https://iste.org/standards/educators

International Society for Technology in Education. (2021). *ISTE Standards adoption by state (NETSS*). https://iste.org/adopting-the-standards

James, C., Weinstein, E., & Mendoza, K. (2019). Teaching digital citizens in today's world: Research and insights behind the Common Sense K–12 digital citizenship curriculum. Project Zero. http://www.pz.harvard.edu/resources/teaching-digital-citizens-in-todays-world

Kara, N. (2018). Understanding university learners' thoughts and practices about digital citizenship: A mixed methods study. *Journal of Educational Technology & Society*, 21/(1), 172-185.

Lenkaitis, C. A. (2019). Valuing technology in the L2 classroom: Learner and teacher perceptions, preferences, and digital identity. In Montebello, M. (Ed.), *Embracing Digital Learners in an Age of Global Educational Change and Rapid Technological Advancements* (pp. 173–189). IGI Global., DOI: 10.4018/978-1-5225-9304-1.ch011

Mahdi, H. R. (2018). The awareness of the digital citizenship among the users of social networks and its relation to some variables. *International Journal of Learning Management Systems*, 6/(1), 11-25. https://doi.org/DOI: 10.18576/ijlms/060102

National Council for the Social Studies. (2013). *The college, career, and civic life (C3) framework for social studies state standards: Guidance for enhancing the rigor of K-12 civics, economics, geography, and history*. Author.

Pring, R. (2016). Preparing for citizenship: Bring back John Dewey. Citizenship, Social and Economics Education. 15(1), 6-14. https://journals.sagepub.com/doi/pdf/10.1177/2047173416646467

Ribble, M. (2010). *Digital citizenship in schools: Nine elements all students should know*. International Society for Technology in Education.

Ribble, M. S., Bailey, G. D., & Ross, T. W. (2004). Digital citizenship: Addressing appropriate technology behavior. *Learning & Leading with Technology*, 32/(1), 6-12.

Suson, R. L. (2019). Appropriating digital citizenship in the context of basic education. *International Journal of Education, Learning and Development*, 7/(4), 44-66.

United Nations. 2021, January 14. SDG knowledge: Sustainable Development Goals. https://sdgs.un.org/goals

Weigel, M., James, C., & Gardner, H. (2009). Learning: Peering backward and looking forward in the digital era. *International Journal of Learning and Media*, *1*(1), 1–18. DOI: 10.1162/ijlm.2009.0005

Chapter 10
Amplifying Unheard Voices:
Exploring Representation Gaps in Indian Social Media Spaces

Dheera C. Sasidharan
https://orcid.org/0000-0001-9639-9375
Bennett University, India

ABSTRACT

With the emergence of new digital technologies and the enhanced efficacy of social networking platforms in offering interactive content sharing, revenue generation, and fame, social media (SM) has grown to become an indispensable part of human life. In this discourse, equitable representation is essential for any democratic debate, where these platforms have become integral to communication, shaping perceptions and decision-making. In India, a conservative Asian society, there is an underrepresentation of marginalized voices across almost all social media platforms based on gender, ethnicity, culture, or political viewpoints. This chapter sheds light on the Indian perspective of digital citizenship, emphasising the challenges marginalised communities face. The study also aims to uncover the underlying societal effects, including the perpetuation of stereotypes and their impact on public opinion. It also emphasizes the necessity for media stakeholders and governments to prioritize diversity and inclusion in cyberspaces.

INTRODUCTION

Inequality and underrepresentation have been inherent aspects of human societies since the formation of organized groups, reflecting long-standing social hierarchies and power imbalances. Oppression, exploitation and discrimination by those in power have prevailed throughout history. Indian society is no different, underrepresentation and inequality are deeply rooted in its historical, social, and economic structures, manifesting in various forms such as caste, class, gender, religion, and regional disparities. These inequalities ingrained within are perpetuated by social exclusions, segregation, customs, and taboos (Mishra, 2017). India being an extremely diverse and conservative society has successfully transcended these social norms and inequalities to its digital spaces in the last decade. The digital era has brought about an enormous shift in the media landscape. With the emergence of new digital technologies and the enhanced efficacy of social networking platforms in offering interactive content sharing, revenue generation, and fame (Duffy, 2019). Social media (SM) has grown to become an indispensable part of

DOI: 10.4018/979-8-3693-6675-2.ch010

Copyright ©2025, IGI Global Scientific Publishing. Copying or distributing in print or electronic forms without written permission of IGI Global Scientific Publishing is prohibited.

human day-to-day life. This increased usage comes with several disadvantages on the other hand as marginalized sections of society tend to get more marginalised at some point. The phrase “missing voices” as used in the paper signifies the marginalized sections of society which go missing/ under-represented which includes women, people of colour, sexual minorities, and subaltern communities (Fraser, 1990) whereas, this research study specifically focuses on women, LGBTQ+ community, Dalits and Muslim minorities of India. These subaltern communities are oppressed groups that are excluded from dominant social, political, and economic power structures.

In an age where media content is not merely a reflective mirror of societal values but also a potent shaper of public perceptions, these representation gaps assume paramount significance. They are of import to those whose voices remain unheard, to those subjected to partial narratives, and to society at large, contending with the ramifications of an asymmetrical digital discourse. In this regard, social media is often regarded as an ‘Alternative media’ (Thakur, 2019) for the marginalised communities in India. Bypassing traditional media gatekeepers, which are often controlled by elite interests and may overlook or misrepresent the issues of marginalized groups, social media allows for more democratic and citizen-based participation. Individuals from various backgrounds can create and distribute content, express opinions, and engage in activism without the constraints of corporate or state-controlled media outlets.

For marginalised communities freely expressing their opinions may result in cyberbullying, criticism, and hate (Duffy, 2019). According to the notion of “counterpublic” introduced by Fraser (1990) the marginalised sections of society who are left behind consciously or unconsciously form their own spaces based on their shared experiences of rejection. In this sense; women, people of colour, sexual minorities, and subaltern communities form "counter-publics" that are continuously trying to fight hegemonic notions yet are deprived of voices (Thakur, 2019). Marginalized or excluded groups come together to express their voices, challenge dominant societal norms, and foster collective identity and resistance. In contrast to the mainstream "public sphere" dominated by powerful and privileged groups, counterpublics provide marginalized communities with spaces to articulate their experiences of exclusion and oppression, and to resist hegemonic narratives that often silence or misrepresent them.

Hence digital media offers a platform for self-expression and resistance, it becomes a powerful media with social impact in a country like India with a high internet usage and penetration rate. India is one of the most formidable powers in the global digital space with more than 600 million total internet users as of 2023 following China. India's internet penetration rate increased to more than 52% in 2024, up from around 14% in 2014. Although these figures appear to be low, they represent more than half of the 1.4 billion people who had access to the Internet that year (Statista, 2024). According to the statistics of Data Reportal and OOSGA published in 2023, India accounts for a total of 467 million social media users. There was an increase of 4.2 percent which is approximately equal to 19 million users between 2021 and 2023. As per the data released by Meta, Facebook is the leading SM platform in India with 314.6 million users with a male user population accounting for approximately 72.99% and female users accounting for 27.01%. In early 2023, Instagram had 229.6 million users in India, with 65.8% of them being male and 34.2% being female. According to the data given by Twitter's advertising resources, Twitter had 27.25 million users in India in early 2023, with 68.1% of Twitter users being male and 31.9% being female. As for marginalized gender minorities (LGBTQ+), religious or caste-based communities, there is no credible and accurate data available to show their underrepresentation as most of them chose to hide their identities on public platforms for fear of getting isolated.

CONCEPTS OF 'HATE SPEECH' AND 'VOICE' IN DIGITAL SPHERES

While social media gives freedom to individuals to express their views and opinions without any restrictions, it also makes it easy for some people to post anti-social (Chetty & Alathur, 2019) content hiding behind the mask of anonymity. Individuals who post anti-social, derogatory or defaming content about others on social media platforms often make unfair use of this anonymity feature of digital platforms. 'Hate speech' refers to communication, whether spoken, written, or behavioural, that attacks, demeans, or discriminates against individuals or groups based on characteristics such as race, religion, ethnicity, gender, sexual orientation, disability, or nationality. The interwoven nature of repressive ideologies, historical injustices, and power disparities contribute to the targeted victimisation of marginalised populations on and off the internet. The concept of the 'spiral of hate speech' highlights this vicious circle of discriminatory and dehumanising narratives spread throughout various digital platforms including social media and also addresses the systemic absence of voices from marginalized communities within these platforms. Concurrently, the pervasive repression of these marginalised voices in digital media contributes to the reinforcement of dominant discourses and the persisting social inequities. The article authored by Banaji (2023) provides an overview of the prevalence and dissemination of hateful content on digital platforms across the globe. They propose the 'hierarchy of hate', a hierarchical paradigm that accounts for the diverse character of discriminating and dehumanising information shared online. Certain communities and groups attract more hate and cyber harassment through online platforms compared to others which are predominantly based on the political ideologies and power structure of a country. Towards a more accepting and inclusive digital society, the focus must be on the ways to break down such 'hierarchy of hate' narratives. There are two hazardous effects to such targeted victimisation, 1) For young people who are initially exposed to digital platforms, media exposure is crucial in shaping public opinion among minority populations. Images, narratives, and storylines shown on media platforms have the capacity to influence their impressions and opinions about various minority groups. Representing specific religious or ethnic groups as potential risks to society on a regular basis may foster prejudice and lead to social strife. Furthermore, media images influence a country's overall social fabric by shaping intergroup relations and promoting or hindering social cohesiveness. 2) Hate speech, misinformation, or discriminatory content, against the marginalised communities has the potential to fuel radicalism. When individuals are repeatedly exposed to hateful ideologies in digital spaces, particularly in echo chambers where opposing views are minimized, these messages can become normalized and reinforced. Over time, such exposure may contribute to the radicalization of individuals, driving them to adopt extreme beliefs or engage in violent actions. The sense of anonymity and the rapid reach of the internet amplify the spread of hateful content, making it easier for extremist groups to mobilize, and marginalise vulnerable individuals and sections of society. Studies suggest that hate speech, misinformation and fabricated content spread faster than that of their factual counterparts. This process of hate-mongering is often accelerated by social media algorithms that promote divisive content for engagement, further deepening polarization and creating fertile ground for extremist ideologies to take root. Therefore, unchecked digital hate poses a serious threat, as it can transform online intolerance into real-world radicalism and violence.

Keeping both the implications of hate speech in check, and ensuring digital representation of marginalized communities is still significant. As per the concept of 'voice' theorised by Mitra (2004) in Indian digital spheres, 'voice' is a metaphor for the representation of one's presence and agency in public discourse, emphasizing the importance of empowering underrepresented communities to articulate their own narratives. The concept underlines the role of social media platforms to act as an alternative media and to give voice to the voiceless. By this, the marginalised should be able to reclaim their status and space in society both online and offline. 'Social inclusion' is the related concept that comes after a certain marginalised group gets equal recognition in the social fabric. It refers to allowing all groups and individuals to participate in society (Kumar et al., 2021) without discrimination. Whereas in the context of social media, inclusion should facilitate the idea of the creation of digital spaces where people from all backgrounds, identities, and perspectives can engage, openly share their experiences and participate in discourse and discussions without facing discrimination or exclusion.

BARRIERS TO INCLUSIVITY ON DIGITAL PLATFORMS

The reason for digital inequality and underrepresentation of certain communities is not just the overall technological divide, it has undermining reasons of social and economic conditions (Laskar, 2023). Social inequalities deep-rooted in Indian history play a significant role in the exclusion of minorities, but economic disparities make it impossible to bridge the gap. Wealth in hand enables the possession of material resources, as named by Newman et al. (2016) as the economic digital capital. This means all sections of the population have the economic stability to acquire digital inclusion given the basic requirement for it is the availability and accessibility of digital devices and internet connectivity. However, internet penetrations have been uneven around the world (Jamil, 2021). In a country like India which has high economic disparities with more than 90% variation in its wealth distribution factors (Roy et al., 2024), it is hard to attain 100 per cent digital inclusion in the near future. The purchase and maintenance costs of digital devices in the country are unaffordable (Botelho, 2021) to large sections of the population. The digital revolution caused by the stepping up of Jio into the telecommunication market of India is huge. Jio brought affordable data plans, and high-speed internet connectivity (Jyoti & Priyanka, 2022) along with low-cost mobile phones. This initiative of the Reliance group helped promote the 'Digital India' initiative of the Indian government (Mukherjee, 2018) by introducing infrastructure development and cost-effective services. According to the study conducted by Daga et al. (2018), more than 80 per cent of the Indian population believes that Jio has a significant impact on the digital change in India.

Another major challenge to inclusivity in the digital era is the lack of algorithmic awareness of social media algorithms and echo chambers created by these algorithms. Social media algorithms often reinforce existing biases by prioritizing content that reflects users' established preferences, which can create 'Echo chambers' that exclude diverse perspectives. Echo chambers are created when individuals are always exposed to the type of content that aligns with their pre-conceived notions, beliefs and ideologies which reinforces their existing views without engaging with differing perspectives. Individual views are always created and deeply rooted in their livid experiences, social upbringings and over-confidence in personal beliefs. This phenomenon is often amplified by social media algorithms that curate and show content based on user's previous interactions, creating a closed-off environment where similar viewpoints are echoed. As a result, echo chambers can limit critical thinking, reduce exposure to diverse ideas, and contribute to social and political polarization.

What can be done? The incorporation of digital media literacy education into the curriculum can be a game changer. This chapter intends to give practical insights on the potential solutions through policy regulations on algorithmic bias, digital media literacy education and inclusive media representation, by analyzing the current scenario using case studies of the underrepresentation of marginalized groups in India such as Dalits, Women, LGBTQ+ Individuals and Muslims. In a diverse country like India, accurate and inclusive media depictions are important to promote inter-community understanding, empathy, respect and peaceful co-existence.

Systemic Challenges in Digital Representation: Algorithmic Bias

Algorithmic bias is the systemic and repeated errors in computer systems that produce unfair outcomes (Jonker & Rogers, 2024). As a result of algorithmic bias, various social groups, generally based on race, gender, and class are further marginalised (Kordzadeh & Ghasemaghaei, 2021). Algorithmic bias occurs due to the usage of biased data in training algorithms, which reinforces the stereotypes and completely excludes under-represented communities in digital spaces. This socio-technical phenomenon (Favaretto et al., 2019) has the power to guide real-world decisions resulting in far-reaching consequences that negatively affect society. In a country like India, where marginalized minorities like Dalits, Muslims, women, and the LGBTQ+ population confront large-scale discrimination, this risk of algorithmic bias becomes even greater (Arora, 2019). While one of the primary causes of AI-based algorithmic biases is a lack of access to data, at the same time biased people developing algorithms pose a greater threat (Stinson, 2022). Due to their historical exclusion from studies and statistics, these populations are under-represented in the data used for training AI algorithms (Sambasivan et al., 2021). As a result, the algorithms may require human assistance to effectively comprehend and respond to these individuals' specific requirements and traits and give them equal visibility and importance.

Major tech companies such as Facebook, Google, and Twitter, which dominate India's social media scene are responsible for building and operating algorithms that shape online experiences. These platforms frequently use algorithms based on Western user data or assumptions, which may fail to account for the unique cultural and social dynamics of Indian society. This results in the structural exclusion experienced by marginalized groups in real life getting mirrored, and in some circumstances aggravating it in their digital environments (Kharitonova et al., 2021).

Furthermore, algorithms designed to increase engagement frequently prioritize sensational material available on digital platforms (Metzler & Garcia, 2023). As sensational content is often exaggerated in order to attract more readers to websites, blogs and online news portals, social media networks also want users to spend more time interacting and discussing issues on their platforms. Thus, the algorithms on these platforms trigger divisive and polarizing narratives over productive discourse (Lewis-Kraus, 2022). This disproportionately affects marginalized voices, as their appeals for justice or equality may be swamped by viral content that does not necessarily reflect their interests. Transparency in algorithmic design is vital for public scrutiny and accountability. Non-governmental organizations, governments, and activists must advocate for reforms that prioritize accountability and inclusivity in the digital spaces as well as spread awareness on algorithmic literacy.

Moreover, content generated by or about Dalits may not obtain the same visibility as information created by upper-caste groups as the engagement-based algorithms are trained to prioritize popular and mainstream viewpoints. For instance, Dalit activist hashtags such as #DalitLivesMatter or #JaiBhim are frequently deprived of visibility by more widely used and supported causes, limiting the reach of essential talks about caste prejudice. During politically charged moments, such as elections or communal upheavals, algorithms may promote divisive or inflammatory content that reinforces perceptions about the Muslim community. This creates unfavourable images while drowning out more nuanced or sympathetic narratives, further marginalizing the community in public discourse.

Such algorithms that follow the dominant discourse often elevate misogynistic or transphobic content while downplaying or restricting positive representation of women and LGBTQ+ perspectives. For example, Instagram and Facebook's content moderation algorithms have been rebuked for discarding posts about body positivity or LGBTQ+ rights under the pretension of "community standards," while leaving damaging content available. Apart from this, if a certain content is being mass reported by a large number of people, platforms like Facebook and Instagram automatically remove it without further analysis. Sometimes such content may be challenging the mainstream narratives but could be the voice of marginalized communities and not necessarily malicious.

DIGITAL REPRESENTATION GAPS ACROSS MARGINALIZED COMMUNITIES

This chapter discusses four highly underrepresented minority communities in India, the Dalits, Women, LGBTQ+ individuals and Muslims. These groups frequently face constraints that limit their visibility and influence in mainstream media. As a result, their experiences and viewpoints are frequently disregarded, resulting in a limited comprehension of their situations. Addressing these representation disparities is critical to building inclusivity. These are comprehensive case studies based on the digital campaigns conducted by marginalized groups in India. It emphasizes the significance of amplifying the perspectives and voices of under-represented communities, and the necessity to reflect on their contributions to modern society and to understand the problems faced by them.

Case I: Dalits in Digital Spaces

According to the census of 2011, Dalits constitute approximately 17 per cent of India's total population. Despite the fact that caste discrimination was abolished in India 75 years ago, Dalits continue to experience pervasive mistreatment, with many of their efforts at upward advancement meeting with violent repression offline and cancellation on social media platforms. To fully understand the oppression of Dalits in the digital sphere, understanding the caste-based social hierarchy in India is important. The caste system in India is depicted in Rigveda (credited to be the longest surviving Indian text) which consists of four categories: the Brahman (priest), the Kshatriya (warriors), the Vaishya (traders), and the Shudra (labourers/workers). The first three categories were considered 'upper castes' in the respective order of their power and importance whereas the last category evolved as the 'lower caste' in the modern world (Shubham, 2022). This category of "Shudhras" is the Dalits.

Barriers to Dalit Representation Online

Dalits have barriers to internet access due to variables such as shelter, economic livelihood, basic education, unemployment and lack of knowledge of English. Thus, the majority of these socially disadvantaged urban and rural communities are under-represented in the public realm (Ippili & Venkateswarlu, 2017). The underrepresentation of Indian Dalits in SM platforms is not only because of the online environment, the barriers of illiteracy and the digital divide contribute equally (Thakur, 2019). Rather than reducing the barriers of inequality, social-digital platforms are re-enforcing the status quo and privilege enjoyed by the majority.

Dalit Activism and Counter Narratives

When compared to the discrimination and ignorance faced in mainstream media regarding their issues SM platforms have provided Dalits with an arena for expressing themselves (Thakur, 2019). Representations of religious and caste-based communities in social media have often been viewed as political and socio-cultural moments. Explorations of digital Dalit activism typically focus on Dalit organisations' websites, which are termed as the Dalit cyberspace (De Kruijf, 2015). According to the studies conducted by A. Verma and Misra (2017) even though Dalit internet users accept that there is an inclination towards Dalit social issues on social media platforms, it still has an inherent tendency to portray negative images of Dalits.

A positive step towards equality is marked by the inhabiting of journalists and activists from marginalised castes in the digital arena, questioning the mainstream casteist narrative. For instance, the #DalitLivesMatter campaign, gained momentum in 2020 following the tragic death of a 19-year-old Dalit woman in Hathras, Uttar Pradesh, India, after a brutal assault. The incident sparked widespread outrage, and the hashtag was used to demand justice for the victim and raise awareness about the systemic caste-based violence and oppression that Dalits face in India. The #DalitLivesMatter campaign drew parallels with the global #BlackLivesMatter campaign, highlighting the shared struggles of marginalised communities against deeply ingrained societal discrimination.

As Dalit oppression offline has spurred an increase in casteist slurs and hate speech against the Dalit community online, the central and state governments in India along with the UN have organised Dalit-targeted programmes through social media mobilisations for the better representation of the community (Verma, 2019). Addressing Dalit under-representation in digital spaces requires a structural change in platform accountability, educational and economic reforms in providing digital access as well as caste-based advocacy and mobilisations.

Case II: Women's Digital Media Representation

Gender identity plays a crucial role in online activism for gender-related issues, such as feminism awareness campaigns and combating predominant misogyny in digital spaces (Majumdar et al., 2022). As a result of long-standing socioeconomic and gender disparities and the resulting limited access to Information and Communication Technologies (ICTs), women's voices on digital platforms remained underrepresented (Shockley et al., 2020).

Challenges to Women's Digital Participation

According to the study conducted by Jain (2020) women who reported online abuse or harassment have accounts of having reduced self-confidence or self-esteem. Hate speech, physical threats, and vulgar language have a significant impact on women's online voices, leading to self-censorship, adopting an anonymous identity or a pseudonym, or withdrawing from online domains entirely. According to Titus (2018), cyberspace has a notion of victim blaming in cases of violence and harassment against women to an extent when heinous crimes against women such as rape and murder are being discussed.

Prior studies by Karusala et al. (2019) have also found that the underrepresentation of women on digital platforms also aligns with their privacy and safety concerns contending with the difficulties of self-representation and worries about external depictions. The common trend of spreading hatred by picking on a specific incident or quote by famous women to spread hatred and tarnish the general image of women across social media is an emerging trend. The hashtag #women is being used widely in a sarcastic manner in the same context.

Women Digital Activism aka 'Cyberfeminism'

Fourth-wave feminism which focuses on sexual violence, rape culture, body shaming and related issues with the use of digital tools has been impactful in giving women voices on social media. With the advent of 'Cyberfeminism', a concept where individuals use technology and the internet to challenge the existing gender norms, in short, it is the intersection of feminism and digital technologies (Issac, 2023). In the cyberfeminism approach, women not only consume information but also actively participate in movements through engagement.

Several feminist advocacy organisations created campaigns using hashtags (#) to increase awareness and win community support. Tarana Burke's 2007 women's activism campaign, #MeToo, is one such movement which became a popular hashtag and later about revolutionary social change. The #MeToo campaign in India gained significant traction in late 2017 and early 2018, inspired by the global movement against sexual harassment and assault. It provided a platform for individuals, particularly women, to share their experiences of harassment and abuse in various sectors, including entertainment, journalism, academia, and corporate environments. The movement gained traction after numerous notable persons, including actresses and journalists, openly accused well-known celebrities of sexual assault. Social media played an important part in amplifying these voices, as people posted their accounts with the hashtag #MeToo, resulting in broad debates about gender-based violence and workplace harassment.

Another important cyberfeminism movement in modern digital history is the Arab Spring. Arab women were instrumental in altering people's perception of women's digital participation during the Arab Spring movement. Arab Spring was an anti-government protest and rebellion against corruption and economic recession that started in the early 2010s. Starting from Tunisia, the movement took over five other Arab countries: Libya, Egypt, Yemen, Syria and Bahrain. Women's involvement in the Arab Spring extended beyond physical protest participation to include cyber activism, emphasising the importance of voice and representation through digital platforms.

Similar to Arab women, Indian women empowerment activists also highly rely on social media (Jain, 2020) as the swiftness of social media and blogging has democratised feminist movements in the Indian digital landscape. In an interesting study, it has been observed that the majority of female social media users are in support of women leaders in the socio-political sphere, whereas male users tend to be less supportive and more oppressive towards them (Shockley et al., 2020).

Digital Misogyny is a real threat to equitable online presence. To build a responsible online community, teaching and learning should begin from the early stages of Individual development, along with including gender sensitization courses in schools and colleges, women and girls should be equipped with digital skills to navigate digital spaces carefully and safely.

Case III: LGBTQ+ Individuals and Online Visibility

The LGBTQ+ community refers to the Lesbian, Gay, Bisexual, Transgender, and Queer with the "+" symbol representing additional sexual orientations and gender identities, such as Intersex, Asexual, Pansexual, and others. 'Homophobia' and 'Transphobia' are referred to as the irrational fear, aversion, or hostility towards individuals who identify as homosexual and transgender respectively, often manifesting in discriminatory attitudes, behaviours, and practices. 'Heteronormativity' is the root cause of both homophobia and transphobia in digital and social spheres (Mkhize et al., 2020). It is a common belief that heterosexuality is the only normal sexual orientation; that gender is strictly binary and that all individuals should live by aligning with traditional male and female gender roles. This perspective marginalizes and stigmatizes other sexual orientations and gender identities, such as LGBTQ+ individuals, by positioning them as deviant, outside of societal norms or "abnormal". According to Freeman & Acena (2022), being queer entails taking an anti-normative attitude to sexuality and/or gender identities, which may include rejecting purely heterosexual identity and/or embracing non-binary or expanding gender identities.

Barriers to LGBTQ+ Representation Online

'Heteronormativity' influences cultural, social, and institutional practices, often leading to discrimination and exclusion of non-heteronormative identities in various aspects of life, including media, education, and law. This limits the voices of LGBTQ+ individuals on digital platforms and their eagerness to depict and explain their identities on digital media platforms. As a result, individuals prefer to express themselves in terms of fluid identities online such as 'queer' rather than assertive identities such as 'lesbian' (Patel, 2019) or 'gay'.

Often the ideologically driven or politically motivated opinions of the LGBTQ+ community are reduced to their gender identities and are considered valid on social media platforms (Sharma, 2020) which adds to their reduced social voice. The queer online visibility is a "double-edged sword" because some are represented while others are left behind, resulting in crucial silences - silences that expose assumptions, values, and omissions that need questioning (Das & Farber, 2020). Most LGBTQ+ individuals who were reluctant to voice their opinions on social networking platforms are found to have encountered homophobic backlash online (Fox & Warber, 2014).

Social Backlash and challenges to Digital Inclusion of LGBTQ+ community

In June 2024, when Thailand passed the revolutionary bill to legalise same-sex marriages, there was a vast digital hate campaign and public outrage against it on social media platforms. Even though Thailand legalised the Marriage Equality Act which came into effect later on 23rd January 2025 becoming the third Asian country after Taiwan and Nepal to do so, the hate on SM platforms is concerning. The same scenario circled back to the Indian context in July 2024 when the plea to legalise same-sex marriage was under consideration by the Supreme Court of India. There were widespread digital media mobilizations on X (formerly known as Twitter), where users retweeted identical tweets against the law claiming that it would harm the family ecology of Indian culture and that it is a Western notion. This indicates the lack of proper and accurate depiction of LGBTQ+ communities through mainstream media which contributed to the increase in stereotypes and prejudice in the society. Despite the growing visibility of LGBTQ+ issues, social media and other digital platforms often amplify hate speech, trolling, and exclusionary narratives, particularly from conservative and traditionalist sections of society.

The adverse effects of online homophobic and transphobic speech on LGBTQ+ individuals who are exposed to discriminatory and homophobic/transphobic remarks have a major emotional impact (Chakravarthi, 2023). This often leads to mental health problems such as depression, anxiety, self-harm, suicidal mentality etc. Depression rates among LGBTQ+ individuals are 1.5 to 2 times higher than those among their heterosexual counterparts as per the reports of Escobar-Viera et al. (2018). There is an urgent need for inclusive digital policy fostering acceptance for LGBTQ+ individuals.

Case IV: Muslims as religious minorities in digital media

The Lok Sabha (lower house of the parliament) election campaigns in India for the past decade have marked an enormous increase in the deliberate propagation of hate speech and cyberbullying which calls for violent extremism and abuse against the Muslim community on Indian Digital spaces. Ideologues speeches and hired digital trolls by Hindutva extremist organisations have perpetuated the myth that India's secular and intellectual traditions, as well as tolerance for people of different faiths, will lead to political weakness (Banaji, 2018). There are widespread incidences of Islamophobia and discriminatory discourses directed against Muslim persons on online platforms. This type of content against Muslim minorities can be termed as 'Islamophobic' content and is often seen as a part of Indian politics (Chetty & Alathur, 2019). Islamophobia is a subset of Orientalist ideology that is more of a 'common sense' version of racism with roots in imperialist practices. The 2014 elections were one of a kind in the history of Indian politics as it used social media as a tool to gain votes (K. Patel et al., 2020).

As per the reports released by the Islamic Council of Victoria (ICV), 86 per cent of the Islamophobic posts on social media platforms originate from the US, the UK and India. Among this India topped the list with 871,379 Islamophobic tweets between a period of two years from 28 August 2019 and 27 August 2021. Not just on social media platforms like X, Facebook and Instagram, hate speech was deliberately propagated through the comments sections of news portals and other digital platforms (Sinha, 2017). Anti-Muslim hate speech in India increased by 62% in the second half of 2023 compared to the first six months of the year, with the Israel-Gaza conflict playing a significant influence in the final three months. According to reports from the United Nations and other organizations, Islamophobia is more common online than in real life. It contributes to a vicious cycle in which online expressions of hatred drive offline attacks against Muslims, which in turn provokes more online hate.

Digital Hate and Societal Impact on Muslims

Additionally, forced internet shutdowns in the different parts of the country by the government also limit the voices of the marginalised. In December 2019 when the government passed the Citizen Amendment Act (CAA), a large number of people from all over the country protested at the Shaheen Bagh in the Delhi National Capital Region (NCR) of India, as a result, the internet was terminated for days. The law establishes India's first religion-based citizenship test, which discriminates against Muslims and certain other refugees. Internet connectivity was shut down once again for more than 365 days in Kashmir during the abolition of Article 370, which gave special rights to Kashmir. Due to the pandemic lockdown when physical protests were no longer feasible, this movement shifted to online platforms through virtual get-togethers, WhatsApp groups and signal meetings (Chopra, 2021). Social media played an important role in gathering support for this movement locally, nationally, and internationally, effectively countering mainstream media's biased narratives (Bhatia & Gajjala, 2020). Furthermore, pictures that featured Muslim gatherings before the lockdown were falsely spread through social media during the COVID-19 pandemic lockdown to induce fear and hatred (Poole & Williamson, 2021).

Recently, the Bollywood movie *The Kerala Story* (2023) left the Indian digital space polarised with conflicting opinions and views about Muslims. The film, which claims to depict the alleged radicalization and forced conversion to Islam of young women in Kerala, led to a widespread outcry, particularly over its portrayal of sensitive religious and political issues. The movie spurred a sudden wave of hate speech and islamophobia in the country with #TheKeralaStory trending over a political and religious debate on X. Though #MyKeralaStory and #RealKeralaStory campaigns was launched as a counter-narrative to this propaganda hate campaign, the damage it already did to the social harmony was irreversible.

The Role of Social Media Platforms in Hate Amplification

Despite facing scrutiny for their role in promoting hatred and digital divide instances of radicalisation online in 2017, social media giants haven't put efforts into more platform accountability. Twitter and Facebook have faced criticism for ignoring polarising fake news reports and allowing 'bots' to spread them on their platforms. The algorithms utilised by these platforms can distort reality and create echo chambers (Mirchandani, 2018). Adolescents who consume news online are more likely to have negative opinions towards Muslims than those who prefer traditional media (Ahmed and Matthes, 2016), for instance, this shows the toxicity propagated through digital news platforms.

Studies suggest that Muslims, majorly based in metropolitan areas, do not seek seclusion, nor are they keen to respond to abusive words and behaviours in shattered scenarios on a regular basis (Eckert et al., 2019). The daily internet use patterns of Muslims given the considerable increase in Islamophobia within the social system (Eckert et al., 2019) have made them hyper differential counterpublic. They are now communities which make quick decisions on whether or not to engage in opposing actions on the basis of individual circumstances.

Intersectionality of Marginalisation in Digital Spaces

Intersectionality is a term that emphasises how different facets of a person's social and political identities—such as race, gender, class, sexuality, and ability—interact to produce distinct experiences of discrimination or privilege. Intersectionality, a term coined by researcher Kimberlé Crenshaw in 1989, emphasises that many types of oppression, such as sexism, racism, and classism, do not work separately. Instead, they overlap, resulting in intricate, multilayered disparities. In the backdrop of marginalised communities, intersectionality is critical for comprehending the full spectrum of inequality. Rather than treating these identities in isolation, intersectionality recognizes that individuals may experience multiple, simultaneous forms of inequality or privilege, depending on the complex interplay of their social categories. It helps broaden perspectives on discrimination, making it possible to develop more inclusive policies and advocacy initiatives. Intersectionality, which recognises the linked nature of identities and oppressions, asks for answers to the numerous and overlapping issues that individuals experience in society.

In India, Female content creators (Mehta & Kaye, 2019) and influencers belonging to the LGBTQ+ community (Das & Farber, 2020) are at high risk of character assassination, harassment and bullying in online environments which diminishes their voice and participation in such platforms to a large extent. According to Gabel et al., (2020), Even within these marginalised communities, several subgroups of people suffer more compared to others. Specific subcultures within the queer community, characterised by overlapping identities (e.g., transgender Dalit individuals), frequently face higher marginalisation than others (e.g., brahmin gay men), as they face compounded challenges as a result of intersecting variables. Muslim women globally as well as in India have been subject to constant demeaning and criticism on social media (Jaiswal & Kapoor, 2024). The controversies following their dressing styles, for instance, the use of headscarves, and the matter of triple talaq, have always been a topic of debate and scrutiny in the digital spaces. The intersectionality of this particular group of individuals being in two of the most marginalised groups in India – Women and Muslims, makes it tougher for them to navigate digital spaces.

According to Rao & Lingam (2020), the voices of women and other marginalised groups such as LGBTQ+ people, Dalits and Muslims are notably missing from social media narratives, as we take into account how these coverage gaps reflect underlying power dynamics. Another research on women's movements on social media by Sujatha Subramanium found that "subaltern voices are very prominent on social media in Kerala, especially sexual minorities and Dalit groups" (Subramanian, 2015). A survey conducted by the Centre for the Study of Developing Societies (CSDS) found that Hindus and religious minorities have different sentiments and perceptions about Indian media and topics important to its independence. The results found that more Muslims, Sikhs, and Christians than Hindus believe the media is less free than it formerly was. They are also less likely than Hindus to support government surveillance, the regulation of social media platforms, and internet shutdowns.

Social media on the other hand also allows marginalised views to be heard in public conversation. In recent years, the role of social media platforms in facilitating networked hatred and abuse has been brought to light. The representation of marginalised communities online should also imply social policy regulations as their invisibility is frequently used to justify their exclusion from rights. The content generation policies of platforms as well as the power to maintain complete anonymity in social media platforms such as Facebook, Twitter and Instagram should be revised. Often casteist slurs, words used as derogatory remarks on LGBTQ+ communities and misogynistic words are coined in regional languages, making it easier to bypass the cyberbullying and harassment surveillance methods set up the social media platforms. This makes it hard for them to report and get action taken against the offenders.

UNDERSTANDING 'DIGITAL CITIZENSHIP', 'DIGITAL ACTIVISM' AND 'DIGITAL MEDA LITERACY'

Since the case studies evidently outline the struggles of marginalised communities while navigating digital media spaces, it is crucial to understand the basic concepts of 'Digital citizenship' to contextualize their experiences within the broader framework of rights, responsibilities, and participation in the digital world. 'Digital citizenship' is the responsible and ethical use of technology and the internet, encompassing behaviours, rights, and responsibilities in the digital world. As per Ribble (2008), a digital citizen should be an individual who understands human, cultural, and societal issues related to technology and practises legal and ethical behaviour. They should also advocate for safe, legal, and responsible use of information and technology, have a positive attitude towards using technology for collaboration, learning, and productivity, take personal responsibility for lifelong learning, and lead digital citizenship. As per the Commonsense Media White Paper (2011), the main objective of Digital Citizenship for the 21st century is to educate, empower and protect (Isman & Ozlem, 2014). 'Protect' refers to safeguarding individuals and communities from potential risks and harms in the digital environment. This involves ensuring privacy, security, and well-being while using digital platforms, tools, and services. The well-being of all the individuals who use digital platforms is kept as a primary goal to reduce disparities in the representation of marginalised communities on digital media platforms. At present, the relationship between citizenship and the digital world has become much more complex, centred on collective identities and social networks with numerous possibilities (Fernández-Prados et al., 2021). With the rise of social media platforms as a popular mode of mass communication, Facebook and X have become areas of civic participation (Pangrazio & Sefton-Green, 2021) and political discourse. As the medium gained more power in debates and discussions regarding socio-political issues, the concepts of inequality and marginalisation followed.

'Digital activism' is a term closely related to digital citizenship which refers to the use of Information and Communication Technologies (ICT) and other digital platforms and tools to spread awareness, organise mass mobilisation, fight against inequality or fundraise for a cause. In light of growing polarization within the digital communities, protesting against inequality and unjust treatment is a life skill. Although ICTs and social media have been popular instruments for activism globally, they have just recently emerged as an alternative platform in India. The nationwide protests in India following the Delhi Nirbhaya rape case on December 16, 2012, made heavy use of digital technologies, resulting in one of the country's largest women's movements ever (Dey, 2019). Even with the rise of Hashtag activism in India, gender prejudices spread without consequences in the social media domain, where men comprise 68.6% of all users and women only account for 31.4% (Kemp, 2024). Digital Dalits frequently use social networking sites to combine political commentary with non-political discussion, addressing the topic of marginalisation and social injustice in the instance of Dalits and other marginalised groups. Their activism takes the form of a unique combination of original content and recycled thoughts and images of fellow Dalits and worldwide political giants (De Kruijf, 2015).

'Digital media literacy' is an essential life skill in today's digital era. It fuels the idea of digital citizenship and digital activism. In order to be responsible a digital citizen and to raise your voice against inequality and oppression on digital platforms, individuals must be well versed in the uses, working and power of digital platforms. Digital media literacy in simple terms can be defined as a third-space literacy that combines digital literacy skills and media literacy skills. In the context of digital media literacy, this "third space" serves as an area where digital skills (such as usage of digital accessories and platforms, navigating digital platforms, understanding algorithms, being aware of privacy and security issues online, source evaluation and content creation) blend with media literacy skills (media analysis and critical thinking) to navigate the complex digital world. This skill set allows individuals, especially those from marginalized communities, to negotiate their identities, challenge dominant narratives, and create new ways of interacting within digital spaces.

RELEVANCE OF SOCIAL MEDIA REPRESENTATION

Social media refers to online platforms and applications that enable users to create, share, and interact with content, including text, images, videos, and links, in a dynamic and interactive environment. Even though these platforms are used for a variety of purposes such as acquiring information, education, leisure and entertainment, etc., the main motive remains networking and communication between individuals. With posting, sharing, tagging and commenting with like-minded individuals every social media user creates an identity for themselves which often reflects their thoughts, opinions and ideologies. Consequently, the more an individual posts certain content, the more will be the influence of those in his/her networked communities. Therefore, social media works as a network of combined identities, which means an individual's visibility is constructed based on their networked identity (Bitman, 2021).

The Social Representation Theory (SRT) proposed by Serge Moscovici in 1961 which analyses how knowledge and meanings are constructed and shared within societies is affirmative of how marginalisation of communities happens in each society. Shared reality through communication shapes collective beliefs, values and social norms in order to maintain a social order. These social norms, if kept rigid, may result in the marginalisation of people who don't align with the set rules. Even within digital and social media platforms, these pre-set social norms create dominance of mainstream narratives, stereotyping and marginalization of individuals. The influence of social media representation extends to policymaking as well. Politicians and public figures are increasingly using social media to assess public sentiment, and the material and storylines that dominate these platforms have the potential to influence policy decisions. When marginalised voices are excluded from these places, the breadth of policy conversations is limited, potentially leading to policies that do not address the interests or concerns of all segments of society.

Although social media provides a platform for the marginalized sections of society (Bitman, 2021) who have been neglected by the mainstream media (Das & Farber, 2020) on the other hand it has also re-enforced the old societal norms along with rendering oppression across the globe divisions (Matamoros-Fernández & Farkas, 2021). Previous studies suggest that people with 'problematic' identities according to the conservative society tend to conceal their true identities on SM platforms in order to get included within the societal structure. This inclusion not only provides them with a sense of belonging but also enhances their chances of a better career and life opportunities (Duffy, 2019). At the same time for the majority of individuals belonging to marginalized communities, these so-called social standards put up by the social media platforms have forced them to refrain from openly accepting their identity as well as expressing their opinions.

Social media has a major influence on individuals' perceptions of themselves, their surroundings (Mishra, 2017) as well as the way they presume others. It also has a significant impact on how individuals, especially young people, project themselves in a certain socially acceptable way. The practices that individuals carry out such as posting, sharing and commenting as well as the way they interact in their closed groups and chats equally contribute to creating this social identity. This in turn means that the more hate and stereotypical misinformation posted on social media platforms against marginalized communities the more will be the resulting backlash and harassment they face in real society. The vast and swift reach of social media platforms has aroused the threat of misinformation, disinformation, propaganda and polarization in the society (K. Patel et al., 2020). India has witnessed instances of mob lynching, violence, terror, and communal and religious tensions in recent years as a result of fake news being produced and shared online on an alarming basis (Kanozia et al., 2022).

Psychological and Emotional Impact of Digital Marginalization

In the digital era where social identities are formed in digital media spaces, constant bullying and sidelining of selected communities can cause identity crises and low esteem in individuals. A single social media post glorifying discrimination against marginalized sections can have a multifold impact on hundreds and thousands of people. Apart from inducing physical violence and abuse offline as discussed earlier in this chapter, cyberbullying and harassment can cater to increased mental health issues among individuals. Cyberbullying and harassment may lead to serious mental health issues such as depression, anxiety, psychological discomfort, and even post-traumatic stress symptoms. The emotional toll of such exposure often results in trauma and long-lasting psychological harm (Nixon, 2014). Individuals may experience symptoms of depression, and difficulty sleeping, along with harmful behaviour patterns that affect their performance and even day-to-day activities. Over time, this can lead to feelings of detachment and a desire to withdraw from their surrounding environments (V. L. Kumar & Goldstein, 2020).

Harassment based on social discrimination can be even more lethal. Digital exclusions caused due to social discrimination or economic factors (Pavez et al., 2023) fuel mental health issues more, as the sense of inclusion in the digital sphere is vital to the social life of individuals these days - especially youngsters (Metherell et al., 2021). Cyberbullying reduces participation, discouraging victims from joining in online discourse and silencing different views and perspectives. This lack of representation exacerbates existing power disparities and normalises hostility towards marginalised groups, sustaining detrimental cultural norms. Furthermore, the fear of repercussions might weaken advocacy activities, making activists reluctant to speak out.

TOWARDS A RESPONSIBLE DIGITAL PRESENCE

Building an inclusive digital space requires collective efforts and systematic working to ensure fair representation, accessibility, and safety for all users. The efforts need to be participatory, from imparting digital media literacy education to addressing and rectifying biases on platform algorithms and creating moral and inclusive content on social and digital platforms. A truly inclusive digital environment demands accountability from all sectors of society to create a space where diversity, equity, and respectful discourse can thrive.

Digital Literacy Education as a Tool for Countering Online Marginalization

Digital literacy education forms the basics of all inclusion efforts. 'Critical Digital literacies' that can combine both digital skills required for navigating digital spaces along with critical thinking skills incorporated with democratic and moral social values can help digital users safeguard themselves from harmful content online as well as teach others with empathy and respect (Agapoglou et al., 2021). Educational interventions such as workshops, digital literacy courses and awareness campaigns can make individuals capable of assessing digital media and evaluating the content suitable for consumption while recognizing biases and challenging traditional norms. Along with digital skills, basic knowledge of AI skills creating an 'AI-literate' society is required to create an informed citizenry in the digital era (Shams et al., 2023). Increasing digital literacy plays a vital role towards the United Nations Sustainable Development Goal (SDGs) number 10 of reducing inequality (Shafira et al., 2024).

Digital literacy education should incorporate *Technical Skills* including effective usage of devices, software, and online platforms (Lee, 2014), *Critical Thinking* including the ability to evaluate online content for credibility and bias (Yustika & Iswati, 2020), *Privacy and Security awareness* such as being capable of protecting personal data and recognizing cyber threats (Beck et al., 2021), *Ethical usage skills* by understanding the responsible use of digital tools and platforms (Krisnaningsih et al., 2023) and *Empowerment* in leveraging digital tools for education, employment, and civic engagement such as promoting inclusivity (Tambunan et al., 2024).

Impact of Social Media Influencers on Shaping Representation

India's social media influencers are redefining representation by amplifying marginalised perspectives, challenging established conventions, and increasing awareness about key social concerns. Many Indian influencers have used their platforms to promote diversity and inclusivity. For instance, Kusha Kapila, a popular content creator, has used satire to address issues of classism, sexism, and body image, often questioning societal norms that marginalize women. During the #MeToo movement in India, several influencers used their platforms to share stories of sexual harassment and support survivors. Influencers like Tanmay Bhat and Swara Bhasker amplified these stories, sparking broader conversations on gender justice and workplace harassment. Similarly, Yashica Dutt, an author and anti-caste influencer, has been vocal about Dalit rights, using platforms like Twitter and Instagram to bring visibility to caste-based discrimination. Thenmozhi Soundararajan is a Dalit rights activist, artist, and filmmaker who uses social media to expose caste-based injustice and violence. As the creator of Equality Labs, she works to protect the rights of Dalits and other marginalised groups in India and the Indian diaspora.

Furthermore, Sushant Divgikar (Rani Ko-HE-Nur), is one of India's most famous LGBTQ+ influencers. Divgikar has used their platform to break stereotypes about drag culture and challenge traditional ideas of gender and sexuality. Afreen Fatima is a vocal activist and student leader known for her advocacy of Muslim rights and social justice issues. She has used her platform to speak out against the marginalization of Muslims in India and has been an active voice during protests, such as those against the Citizenship Amendment Act (CAA) and the National Register of Citizens (NRC). Kavita Krishnan, while not Muslim herself is a prominent feminist and human rights activist and has been a strong ally for Muslim rights. Her activism on social media highlights intersectional struggles and supports broader movements for justice and equality for Indian Muslims.

While most of these influencers contribute positively to the efforts to reduce marginalisation and underrepresentation, some others inadvertently reinforce stereotypes, showcasing the complex and dual role influencers play in the digital landscape.

Overcoming Algorithmic Bias for Better Inclusivity

Overcoming algorithmic bias caused by the lack of data access and information available about the under-represented groups necessitates a concentrated effort that addresses basic issues. First and foremost, improving the data collection techniques by ensuring the inclusion of varied corpus of data such as representative datasets of individuals from marginalized populations must be prioritized (Custers & Fosch-Villaronga, 2022). For this, grassroots-level working and research with communities, organizations, and advocacy groups involved in the cause and incorporating their viewpoints and experiences into the data used to train algorithms is necessary. As per the suggestions made by Williams et al. (2018), while developing algorithms independent external experts should review the validity and reliability to reduce people's biases. Another suggestion is to pair a human decision-maker with algorithms, which will ensure fairness in qualitative subject analysis (D. T. Newman et al., 2020).

Ethical data sourcing must also be prioritized (M. Williams et al., 2024), with norms that respect individuals' privacy and rights, such as gaining informed consent and protecting personal information, to foster confidence and engagement. Along with collecting the right representative data, it is important to recognize, understand and address the historical exclusion of marginalized communities from research datasets. This will help in getting broader perspectives about their specific needs and concerns. One way is to actively integrate these marginalized groups in data collection so that their voices can be better represented. The algorithms designed have this tendency to oversimplify the data set and generalize the target group parameters to a large extent which leads to underrepresentation. This should be avoided, and provisions should be made to include a wide variety of features that reflect the diversity within those populations to overcome stereotyping or bias caused by insufficient representation.

Furthermore, incorporating fairness measures while creating algorithms necessitates real-life parallel testing for unequal effects to assure equal performance across demographic groups. Transparency and accountability are also important. FAIR principles, that is, the findability, accessibility, Interoperability and reusability principles (Hanna et al., 2024) used in managing bias in AI models. Calls have been made for the deployment of "fairness-aware" algorithms that take demographic diversity into account and increase transparency in decision-making processes (Selbst et al., 2019). Algorithmic processes should be open to inspection, and companies should encourage the external public to objectively analyze system biases and address them accordingly. Bringing diversity in teams by incorporating people with diverse lived experiences contributes valuable views to the development process, which helps to reduce bias.

Global Policies for Combating Algorithmic Bias

"Ethics Guidelines for Trustworthy AI," released by The European Union delineates seven governance principles: (1) human agency and oversight, (2) technical robustness and safety, (3) privacy and data governance, (4) transparency, (5) diversity, nondiscrimination and fairness, (6) environmental and societal well-being, and (7) accountability (*Ethics Guidelines for Trustworthy AI*, 2019). New York University's AI Now Institute has developed a model framework for government agencies to conduct "Algorithmic Impact Assessments (AIAs)". These assessments evaluate the potential negative consequences of algorithms, similar to how environmental, privacy, data, or human rights impact assessments are carried out. Given the rising impact of algorithms in day-to-day life, equipping common people with 'algorithmic literacy' is a fair and universally accepted method to combat algorithmic bias (Barton et al., 2019). UNESCO's Recommendation on the Ethics of Artificial Intelligence (2021) urges all 194 member nations to "promote the acquisition of "prerequisite skills" for AI education, such as basic literacy, numeracy, coding and digital skills, and media and information literacy, as well as critical and creative thinking, teamwork, communication, socio-emotional and AI ethics skills".

Continuous monitoring and evaluation are required to uncover and correct biases that arise over time, which will ensure that algorithms are working fairly and sensitively to cater to the requirements of individuals from all strata of society. Finally, addressing algorithmic biases necessitates an inclusive, ethical, and collaborative approach, as well as constant evaluation to ensure that algorithms effectively meet the requirements of various groups.

CONCLUSION

India has significant disparities in the representation of marginalized populations on prominent social media platforms, with various underrepresented groups regularly experiencing a lack of visibility and acknowledgement. Furthermore, there is a troubling link between the rise in hate speech instances and a significant drop in the active social media involvement of these marginalized groups in the Indian digital landscape. Addressing these disparities requires a multi-stakeholder approach that includes technology companies, policymakers, educators, and civil society.

For Educators and Institutions: Embracing digital equity and inclusion in higher education not only helps build empowering learning spaces but also equips students with the skills and opportunities needed to meet the challenges and be responsible digital citizens. Moreover, educational institutions should nurture responsible digital citizens who are capable of understanding that their right to freedom of speech and expression on online platforms should not be used to harass, intimidate or bully others. The moral, ethical and responsible use of digital platforms should be incorporated in the curriculums both in schools and universities as we move towards a fully digitalized global society. Teaching intercultural communication and the importance of representation and inclusivity should be given significance. Promoting digital media literacy education which includes a basic understanding of the working of algorithms can reduce the number of individuals falling for misinformation and propaganda content.

For Global Tech Companies: Global tech platforms must take greater responsibility for platform transparency, and content moderation, ensuring the development of bias-free algorithms and AI-driven moderation systems that effectively identify and curb hate speech and discriminatory narratives and promote inclusion in their platforms.

For Policymakers: Addressing the digital divide and its socio-economic causes should be given profound importance. Despite the high internet penetration rates, a significant digital divide remains in India, particularly among rural populations, women, and marginalized groups like Dalits and Adivasis. Promoting digital citizenship involves addressing this divide by enhancing digital literacy and ensuring that all citizens have access to the skills and resources needed to thrive in the digital space. Governments must enforce stricter regulations on digital platforms to hold them accountable for their role in curbing online hate speech and misinformation. Stronger laws and digital rights policies should be implemented to protect marginalized communities from targeted attacks while ensuring that content regulation does not compromise free speech.

For Media: As per the studies conducted by Donati et al.(2022) social media edutainment programs have the potential to change gender norms and attitudes about violence against women online. They effectively raised awareness of gender behaviours and promoted progressive gender standards. In fact, the use of content such as docuseries successfully influenced social media users to take a public stance against violence against women.

For the Common Public: Every individual has a crucial role to play in creating an inclusive digital sphere. Users should engage critically with content, identify and challenge biases, report harmful narratives, and support marginalized voices by sharing and promoting diverse perspectives. The study encourages the common public to have critical online engagement with meaningful sociopolitical and cultural discourse, acknowledging the complex interaction of historical injustices and present power dynamics that contribute to the spread of discriminatory information.

While digital media holds the promise of being an inclusive space where individuals of all backgrounds can express their views freely, this vision will remain unfulfilled without active intervention. Addressing the underrepresentation of marginalized communities is not just about increasing visibility but about creating a responsible and inclusive digital space for everyone.

REFERENCES

Agapoglou, T., Mouratoglou, N., Tsioumis, K., & Bikos, K. (2021). Combating Online Hate Speech through Critical Digital Literacy: Reflections from an Emancipatory Action Research with Roma Youths. *International Journal of Learning and Development, 11*(2), 104. DOI: 10.5296/ijld.v11i2.18524

Ahmed, S., & Matthes, J. (2016). Media representation of Muslims and Islam from 2000 to 2015: A meta-analysis. *The International Communication Gazette, 79*(3), 219–244. DOI: 10.1177/1748048516656305

Arora, P. (2019). Politics of Algorithms, Indian Citizenship, and the Colonial Legacy. In *Global Digital Cultures: Perspectives from South Asia* (p. 37). University of Michigan Press. https://library.oapen.org/bitstream/handle/20.500.12657/23985/1006150.pdf#page=46

Banaji, S. (2018). Vigilante Publics: Orientalism, Modernity and Hindutva Fascism in India. *Javnost - the Public, 25*(4), 333–350. https://doi.org/DOI: 10.1080/13183222.2018.1463349

Banaji, S. (2023, August 2). *A hierarchy of hate*. Media@LSE. Retrieved October 17, 2023, from https://blogs.lse.ac.uk/medialse/2023/08/02/a-hierarchy-of-hate/

Barton, G., Lee, N. T., & Resnick, P. (2019, May 22). Algorithmic bias detection and mitigation: Best practices and policies to reduce consumer harms. *Brookings*. https://www.brookings.edu/articles/algorithmic-bias-detection-and-mitigation-best-practices-and-policies-to-reduce-consumer-harms/

Beck, E., Goin, M. E., Ho, A., Parks, A., & Rowe, S. (2021). Critical digital literacy as method for teaching tactics of response to online surveillance and privacy erosion. *Computers & Composition. Computers and Composition, 61*, 102654. DOI: 10.1016/j.compcom.2021.102654

Bhatia, K. V., & Gajjala, R. (2020, November 28). *Examining Anti-CAA protests at Shaheen Bagh: Muslim Women and politics of the Hindu India*. Bhatia | International Journal of Communication. https://ijoc.org/index.php/ijoc/article/view/16015

Bitman, N. (2021). 'Which part of my group do I represent?': Disability activism and social media users with concealable communicative disabilities. *Information Communication and Society, 26*(3), 619–636. DOI: 10.1080/1369118x.2021.1963463

Botelho, F. H. F. (2021). Accessibility to digital technology: Virtual barriers, real opportunities. *Assistive Technology, 33*(sup1), 27–34. https://doi.org/DOI: 10.1080/10400435.2021.1945705

Chakravarthi, B. R. (2023). Detection of homophobia and transphobia in YouTube comments. *International Journal of Data Science and Analytics, 18*(1), 49–68. DOI: 10.1007/s41060-023-00400-0

Chetty, N., & Alathur, S. (2019). *Racism and social media: A study in Indian context*. https://idr.nitk.ac.in/jspui/handle/123456789/12729

Chopra, D. (2021). The resistance strikes back: Women's protest strategies against backlash in India. *Gender and Development, 29*(2–3), 467–491. DOI: 10.1080/13552074.2021.1981698

Custers, B., & Fosch-Villaronga, E. (2022). *Law and Artificial Intelligence - Regulating AI and Applying AI in Legal Practice - Asser.nl*. Springer. https://www.asser.nl/asserpress/books/?rId=13994

Daga, T., Chandra, V., & Malik, A. (2018). Effect of Reliance Jio on Digital India. *International Journal of Advanced Research, 6*(6), 395–401. DOI: 10.21474/ijar01/7224

Das, S., & Farber, R. (2020). User-generated online queer media and the politics of queer visibility. *Sociology Compass, 14*(9). Advance online publication. DOI: 10.1111/soc4.12824

De Kruijf, J. G. (2015). The DALIT I define: social media and individualized activism in subaltern spheres. *Pacific Asia Journal of the Association for Information Systems*, 11–24. https://doi.org/DOI: 10.17705/1pais.07402

Dey, A. (2019). Sites of exception: Gender violence, digital activism, and Nirbhaya's zone of Anomie in India. *Violence Against Women, 26*(11), 1423–1444. DOI: 10.1177/1077801219862633

Donati, D., Orozco-Olvera, V., & Rao, N. (2022). Using Social Media to Change Gender Norms: An Experiment within Facebook Messenger in India. In *World Bank policy research working paper*. https://doi.org/DOI: 10.1596/1813-9450-10199

Duffy, B. E. (2019, September 22). *Gendered Visibility on social Media: Navigating Instagram's authenticity bind*. Duffy | International Journal of Communication. https://ijoc.org/index.php/ijoc/article/view/11729/2821

Eckert, S., Metzger-Riftkin, J., Kolhoff, S., & O'Shay-Wallace, S. (2019). A*hyper differential counterpublic*: Muslim social media users and Islamophobia during the 2016 US presidential election. *New Media & Society, 23*(1), 78–98. DOI: 10.1177/1461444819892283

Escobar-Viera, C. G., Whitfield, D. L., Wessel, C. B., Shensa, A., Sidani, J. E., Brown, A. L., Chandler, C. J., Hoffman, B. L., Marshal, M. P., & Primack, B. A. (2018). For better or for worse? A systematic review of the evidence on social media use and depression among lesbian, gay, and bisexual minorities. *JMIR Mental Health, 5*(3), e10496. DOI: 10.2196/10496

Ethics guidelines for trustworthy AI. (2019). Shaping Europe's Digital Future. Retrieved February 6, 2025, from https://digital-strategy.ec.europa.eu/en/library/ethics-guidelines-trustworthy-ai

Favaretto, M., De Clercq, E., & Elger, B. S. (2019). Big Data and discrimination: Perils, promises and solutions. A systematic review. *Journal of Big Data, 6*(1). Advance online publication. DOI: 10.1186/s40537-019-0177-4

Fernández-Prados, J. S., Lozano-Díaz, A., & Ainz-Galende, A. (2021). Measuring Digital Citizenship: A comparative analysis. *Informatics (MDPI), 8*(1), 18. DOI: 10.3390/informatics8010018

Fox, J., & Warber, K. M. (2014). Queer Identity Management and Political Self-Expression on Social Networking Sites: A Co-Cultural Approach to the Spiral of Silence. *Journal of Communication, 65*(1), 79–100. DOI: 10.1111/jcom.12137

Fraser, N. (1990). Rethinking the public sphere: A contribution to the critique of actually existing democracy. *Social Text, 25/26*, 56. DOI: 10.2307/466240

Freeman, G., & Acena, D. (2022). "Acting out" Queer Identity: the embodied visibility in social virtual reality. *Proceedings of the ACM on Human-computer Interaction, 6*(CSCW2), 1–32. https://doi.org/DOI: 10.1145/3555153

Gabel, S., Reichert, L., & Reuter, C. (2020). Discussing conflict in social media: The use of Twitter in the Jammu and Kashmir conflict. *Media, War & Conflict, 15*(4), 504–529. DOI: 10.1177/1750635220970997

Hanna, M., Pantanowitz, L., Jackson, B., Palmer, O., Visweswaran, S., Pantanowitz, J., Deebajah, M., & Rashidi, H. (2024). Ethical and bias considerations in artificial intelligence (AI)/Machine learning. *Modern Pathology, 100686*. Advance online publication. DOI: 10.1016/j.modpat.2024.100686

Ippili, T. R., & Venkateswarlu, S. (2017). Social media Gives Voice to Voiceless Dalits in India: An Analysis. *International Journal of Multidisciplinary Educational Research*. https://www.researchgate.net/publication/327011583_Social_Media_Gives_Voice_To_Voiceless_Dalits_In_India_An_Analysis

Isman, A., & Ozlem, C. G. (2014). Digital citizenship. *Turkish Online Journal of Educational Technology*. https://eric.ed.gov/?id=EJ1018088

Issac, A. L. (2023). Understanding Cyber-Feminism and its roles in the digital space. In *Advances in human and social aspects of technology book series* (pp. 321–334). https://doi.org/DOI: 10.4018/978-1-6684-8893-5.ch022

Jain, S. (2020). The Rising Fourth Wave: Feminist activism on digital platforms in India. *ORG Issue Brief, 384*. https://www.orfonline.org/research/the-rising-fourth-wave-feminist-activism-on-digital-platforms-in-india

Jaiswal, Y., & Kapoor, A. R. (2024). Representation of Muslim Women in Social Media: Some Reflections from Western and Indian Media Representations. *jcoma.com*. https://doi.org/DOI: 10.58966/JCM20243215

Jamil, S. (2021). From digital divide to digital inclusion: Challenges for wide-ranging digitalization in Pakistan. *Telecommunications Policy, 45*(8), 102206. DOI: 10.1016/j.telpol.2021.102206

Jonker, A., & Rogers, J. (2024, September 20). Algorithmic bias. *IBM*. Retrieved February 6, 2025, from https://www.ibm.com/think/topics/algorithmic-bias

Jyoti, & Priyanka. (2022). The Jio Effect: Evaluating the impact of Jio on internet usage and digital transformation. *REST Journal on Data Analytics and Artificial Intelligence*. https://doi.org/DOI: 10.46632/jdaai/2/2/16

Kanozia, R., C S, D., & Arya, R. (2022). Critical Media and Information Literacy to Combat Misinformation: Research gaps and future directions. *Journal of Content. Community & Communication, 16*. Advance online publication. DOI: 10.31620/JCCC.12.22/15

Karusala, N., Bhalla, A., & Kumar, N. (2019). Privacy, Patriarchy, and Participation on Social Media. *Proceedings of the 2019 on Designing Interactive Systems Conference,* 2019. https://doi.org/DOI: 10.1145/3322276.3322355

Kemp, S. (2022). Digital 2022: India — DataReportal – Global Digital Insights. *DataReportal – Global Digital Insights*. https://datareportal.com/reports/digital-2022-india

Kemp, S. (2024, February 20). *Digital 2024: India — DataReportal – Global Digital Insights*. DataReportal – Global Digital Insights. Retrieved September 21, 2024, from https://datareportal.com/reports/digital-2024-india#:~:text=More%20broadly%2C%2061.5%20percent%20of,while%2068.6%20percent%20were%20male

Keshari Mishra, B. (2017). Digital media in resisting social inequality The Indian experience. In *Precarity within the Digital Age* (pp. 123–133). Springer Nature., DOI: 10.1007/978-3-658-17678-5

Kharitonova, Y. S., Savina, V. S., & Pagnini, F. (2021). ARTIFICIAL INTELLIGENCE'S ALGORITHMIC BIAS: ETHICAL AND LEGAL ISSUES. *Вестник Пермского Университета Юридические Науки*, *3*(53), 488–515. DOI: 10.17072/1995-4190-2021-53-488-515

Kordzadeh, N., & Ghasemaghaei, M. (2021). Algorithmic bias: Review, synthesis, and future research directions. *European Journal of Information Systems*, *31*(3), 388–409. DOI: 10.1080/0960085x.2021.1927212

Krisnaningsih, E., Dwiyatno, S., Jubaedi, A. D., & Shafitri, A. (2023). Increasing ethical understanding of the use of information technology through digital literacy proficiency training. *Dinamisia Jurnal Pengabdian Kepada Masyarakat*, *7*(3), 789–801. DOI: 10.31849/dinamisia.v7i3.12809

Kumar, S., Sharma, A., & Verma, V. (2021). Social Media A Tool for Social and Political Inclusion among Youth: A Study of Indian University Students. *Current Research Journal of Social Sciences and Humanities*, *4*(1), 82–89. DOI: 10.12944/crjssh.4.1.08

Kumar, V. L., & Goldstein, M. A. (2020). Cyberbullying and adolescents. *Current Pediatrics Reports*, *8*(3), 86–92. DOI: 10.1007/s40124-020-00217-6

Laskar, M. H. (2023). Examining the emergence of digital society and the digital divide in India: A comparative evaluation between urban and rural areas. *Frontiers in Sociology*, *8*. Advance online publication. DOI: 10.3389/fsoc.2023.1145221

Lee, S. (2014). Digital literacy Education for the development of digital literacy. *International Journal of Digital Literacy and Digital Competence*, *5*(3), 29–43. DOI: 10.4018/ijdldc.2014070103

Lewis-Kraus, G. (2022, June 3). How harmful is social media? *The New Yorker*. https://www.newyorker.com/culture/annals-of-inquiry/we-know-less-about-social-media-than-we-think

Majumdar, S., Tewatia, M., Jamkhedkar, D., & Bhatia, K. (2022). "You don't know me so don't try to judge me": Gender and identity performance on social media among young Indian users. *Frontiers in Psychology*, *13*. Advance online publication. DOI: 10.3389/fpsyg.2022.855947

Matamoros-Fernández, A., & Farkas, J. (2021). Racism, Hate Speech, and Social Media: A Systematic Review and critique. *Television & New Media*, *22*(2), 205–224. DOI: 10.1177/1527476420982230

Meherali, S., Rahim, K. A., Campbell, S., & Lassi, Z. S. (2021). Does Digital Literacy Empower adolescent Girls in Low- and Middle-Income Countries: A Systematic review. *Frontiers in Public Health*, *9*. Advance online publication. DOI: 10.3389/fpubh.2021.761394

Mehta, S., & Kaye, D. B. V. (2019). Pushing the next level: Investigating digital content creation in India. *Television & New Media*, *22*(4), 360–378. DOI: 10.1177/1527476419861698

Metherell, T. E., Ghai, S., McCormick, E. M., Ford, T. J., & Orben, A. (2021). Digital exclusion predicts worse mental health among adolescents during COVID-19. medRxiv *(Cold Spring Harbor Laboratory)*. https://doi.org/DOI: 10.1101/2021.11.25.21266853

Metzler, H., & Garcia, D. (2023). Social drivers and algorithmic mechanisms on digital media. *Perspectives on Psychological Science, 19*(5), 735–748. DOI: 10.1177/17456916231185057

Mirchandani, M. (2018). Digital hatred, real violence: majoritarian radicalisation and social media in India. *Ashokauniversity*. https://www.academia.edu/37516844/Digital_Hatred_Real_Violence_Majoritarian_Radicalisation_and_Social_Media_in_India

Mitra, A. (2004). Voices of the Marginalized on the Internet: Examples from a website for women of South Asia. *Journal of Communication, 54*(3), 492–510. DOI: 10.1111/j.1460-2466.2004.tb02641.x

Mkhize, S., Nunlall, R., & Gopal, N. (2020). An examination of social media as a platform for cyber-violence against the LGBT+ population. *Agenda (Durban, South Africa), 34*(1), 23–33. DOI: 10.1080/10130950.2019.1704485

Mukherjee, R. (2018). Jio sparks Disruption 2.0: Infrastructural imaginaries and platform ecosystems in 'Digital India.'. *Media Culture & Society, 41*(2), 175–195. DOI: 10.1177/0163443718818383

Newman, D. T., Fast, N. J., & Harmon, D. J. (2020). When eliminating bias isn't fair: Algorithmic reductionism and procedural justice in human resource decisions. *Organizational Behavior and Human Decision Processes, 160*, 149–167. DOI: 10.1016/j.obhdp.2020.03.008

Newman, L., Browne-Yung, K., Raghavendra, P., Wood, D., & Grace, E. (2016). Applying a critical approach to investigate barriers to digital inclusion and online social networking among young people with disabilities. *Information Systems Journal, 27*(5), 559–588. DOI: 10.1111/isj.12106

Nixon, C. (2014). Current perspectives: The impact of cyberbullying on adolescent health. *Adolescent Health, Medicine and Therapeutics, 143*. Advance online publication. DOI: 10.2147/ahmt.s36456

Pangrazio, L., & Sefton-Green, J. (2021). Digital rights, digital citizenship and digital literacy: What's the difference? *Journal of New Approaches in Educational Research, 10*(1), 15. DOI: 10.7821/naer.2021.1.616

Patel, K., Binjola, H., & Siddiqui, T. (2020). Indian social media politics: New era of election war. *RUDN Journal of Studies in Literature and Journalism, 25*(1), 184–192. DOI: 10.22363/2312-9220-2020-25-1-184-192

Patel, S. (2019). "Brown girls can't be gay": Racism experienced by queer South Asian women in the Toronto LGBTQ community. *Journal of Lesbian Studies, 23*(3), 410–423. DOI: 10.1080/10894160.2019.1585174

Pavez, I., Correa, T., & Farías, C. (2023). The Power of Emotions: The ethics of care in the digital inclusion processes of marginalized communities. *Social Inclusion (Lisboa), 11*(3). Advance online publication. DOI: 10.17645/si.v11i3.6623

Poole, E., & Williamson, M. (2021). Disrupting or reconfiguring racist narratives about Muslims? The representation of British Muslims during the Covid crisis. *Journalism: Theory. Practice & Criticism, 24*(2), 262–279. DOI: 10.1177/14648849211030129

Rao, N., & Lingam, L. (2020). Smartphones, youth and moral panics: Exploring print and online media narratives in India. *Mobile Media & Communication, 9*(1), 128–148. DOI: 10.1177/2050157920922262

Reisman, D., Schultz, J., Crawford, K., & Whittaker, M. (2018). *ALGORITHMIC IMPACT ASSESSMENTS: a PRACTICAL FRAMEWORK FOR PUBLIC AGENCY ACCOUNTABILITY*. https://www.nist.gov/system/files/documents/2021/10/04/aiareport2018.pdf

Ribble, M. (2008). *Passport to Digital Citizenship: Journey toward Appropriate Technology Use at School and at Home*. Retrieved September 21, 2024, from https://eric.ed.gov/?id=EJ904288

Roy, S., Majumder, S., Bose, A., & Chowdhury, I. R. (2024). The rich-poor divide: Unravelling the spatial complexities and determinants of wealth inequality in India. *Applied Geography (Sevenoaks, England), 166*, 103267. DOI: 10.1016/j.apgeog.2024.103267

Sambasivan, N., Arnesen, E., Hutchinson, B., Doshi, T., & Prabhakaran, V. (2021). Re-imagining Algorithmic Fairness in India and Beyond. In *FAccT '21: Proceedings of the 2021 ACM Conference on Fairness, Accountability, and Transparency* (pp. 315–328). https://doi.org/DOI: 10.1145/3442188.3445896

Selbst, A. D., Boyd, D., Friedler, S. A., Venkatasubramanian, S., & Vertesi, J. (2019). Fairness and Abstraction in Sociotechnical Systems. *FAT* '19: Proceedings of the Conference on Fairness, Accountability, and Transparency*, 59–68. https://doi.org/DOI: 10.1145/3287560.3287598

Shafira, V. S., Ramadhani, G., & Rachman, I. F. (2024). How digital literacy can drive inclusive progress towards the 2030 SDGs. *Advances in Economics & Financial Studies, 2*(2). Advance online publication. DOI: 10.60079/aefs.v2i2.260

Shams, R. A., Zowghi, D., & Bano, M. (2023). AI and the quest for diversity and inclusion: A systematic literature review. *AI and Ethics*. Advance online publication. DOI: 10.1007/s43681-023-00362-w

Sharma, D. (2020). Representing pride through hashtags. *Feminist Media Studies, 21*(1), 162–164. DOI: 10.1080/14680777.2021.1864878

Shockley, B., Lari, N., Elmaghraby, E., & Al-Ansari, M. (2020). Social media usage and support for women in community leadership: Evidence from Qatar. *Women's Studies International Forum, 81*, 102374. DOI: 10.1016/j.wsif.2020.102374

Shubham, S. (2022). Caste and the Digital Sphere. *Research Gate*. https://www.researchgate.net/publication/381295637_Caste_and_the_Digital_Sphere

Sinha, S. (2017). Fragile Hegemony: Modi, Social Media and Competitive Electoral Populism in India. *International Journal of Communication*. https://eprints.soas.ac.uk/24533/

Social Media in India - 2023 Stats & Platform Trends - OOSGA. (n.d.). OOSGA. https://oosga.com/social-media/ind/#:~:text=With%20the%20latest%20data%20published,users%20account%20for%20around%2027.01%25

Statista. (2024, May 15). *Internet penetration rate in India 2014-2024*. https://www.statista.com/statistics/792074/india-internet-penetration-rate/

Stinson, C. (2022). Algorithms are not neutral. *AI and Ethics*, *2*(4), 763–770. DOI: 10.1007/s43681-022-00136-w

Subramanian, S. (2015). From the streets to the web: Looking at feminist activism on social media. *Economic and Political Weekly*, *50*(17). https://www.jstor.org/stable/24481828

Tambunan, R., Budimansyah, D., & Darmawan, C. (2024). THE IMPLEMENTATION OF DIGITAL LITERACY TO ENCOURAGE STUDENTS' DEMOCRATIC ENGAGEMENT IN CIVIC EDUCATION LEARNING. *JUPIIS JURNAL PENDIDIKAN ILMU-ILMU SOSIAL, 16*(1), 109–118. DOI: 10.24114/jupiis.v16i1.53503

Tavares, A. O., Areia, N. P., Mellett, S., James, J., Intrigliolo, D. S., Couldrick, L. B., & Berthoumieu, J. (2020). The European Media Portrayal of Climate Change: Implications for the Social Mobilization towards Climate Action. *Sustainability*, *12*(20), 8300. DOI: 10.3390/su12208300

Thakur, A. (2019). New media and the Dalit counter-public sphere. *Television & New Media*, *21*(4), 360–375. DOI: 10.1177/1527476419872133

Titus, D. (2018). Social media as a gateway for young feminists: Lessons from the #IWillGoOut campaign in India. *Gender and Development*, *26*(2), 231–248. DOI: 10.1080/13552074.2018.1473224

UNECSO. (2021). Recommendation on the Ethics of Artificial Intelligence. In *UNESCO Digital Library*. Retrieved February 6, 2025, from https://unesdoc.unesco.org/ark:/48223/pf0000381137

Verma, A., & Misra, S. (2017). Dalits: In the era of Digital media. *Amity Journal of Media & Communication Studies, 7*(2). https://www.academia.edu/43894745/Dalits_In_the_Era_of_Digital_Media

Verma, V. (2019). A crisis of representation: Interests, identities and politics. *Journal of Social Inclusion Studies*, *5*(1), 1–18. DOI: 10.1177/2394481119849289

Williams, B. A., Brooks, C. F., & Shmargad, Y. (2018). How algorithms discriminate based on data they lack: Challenges, solutions, and policy implications. *Journal of Information Policy*, *8*, 78–115. DOI: 10.5325/jinfopoli.8.2018.0078

Williams, M., Karim, W., Gelman, J., & Raza, M. (2024). Ethical data acquisition for LLMs and AI algorithms in healthcare. *NPJ Digital Medicine*, *7*(1). Advance online publication. DOI: 10.1038/s41746-024-01399-9

Yustika, G. P., & Iswati, S. (2020). Digital Literacy in Formal Online Education: A Short review. *Dinamika Pendidikan*, *15*(1), 66–76. DOI: 10.15294/dp.v15i1.23779

KEY TERMS AND DEFINITIONS

Representation Gaps: Representation gaps in the context of social media platforms refers to the disparities in the visibility and portrayal of certain groups which may lead to their marginalization or vice versa.

Counterpublic: A social groups consisting of the marginalised sessions of the society in opposition to the dominant public spheres, elite groups and mainstream discourses. The main motive of such groups is to fight for their equality, collective identity building and advocacy.

Social media: A type of digital media platform that enables users to generate and share content on online networks and communities.

Alternative media: Media platforms that represents marginalized voices, promoting social justice, and offering independent viewpoints on political, social, and cultural issues which are often neglected by the mainstream media.

Homophobia: The irrational fear, dislike and prejudice against homosexual individuals which results in discriminatory attitudes and actions.

Misogyny: It is the manifestation of sexist attitudes, harassment, and discriminatory behavior toward women through social media platforms which may include verbal abuse, trolling, objectification, threats, and actions that reinforces stereotypes.

Islamophobia: Hostility toward Islam and Muslims, often resulting in discrimination and marginalization of individuals based on their religious identity.

Chapter 11
Sustainability and Responsibility in the Digital Era:
Leveraging Green Marketing in Healthcare

M. Vijayalakshmi
https://orcid.org/0000-0002-1324-136X
Saveetha School of Management, Saveetha Institute of Medical and Technical Sciences, India

A. K. Subramani
Saveetha School of Management, Saveetha Institute of Medical and Technical Sciences, India

R. Vettriselvan
Academy of Maritime Education and Training, India

T. C. Catherin
https://orcid.org/0000-0002-2469-5893
Department of Management Studies, Saveetha Engineering College, India

R. Deepika
https://orcid.org/0000-0002-2818-9867
Rajalakshmi School of Business, India

ABSTRACT

This chapter examines the integration of green marketing within the healthcare sector, focusing on how digital platforms can be leveraged to promote sustainability and responsibility. As healthcare organizations increasingly rely on their online presence, they have a unique opportunity to adopt environmentally conscious marketing strategies that align with global sustainability efforts. The chapter highlights the importance of incorporating sustainable practices into healthcare marketing to build trust with eco-conscious consumers and support broader societal objectives, such as environmental preservation and public health. Additionally, it discusses how healthcare institutions can establish a responsible digital presence by advocating for transparency, engaging diverse communities, and fostering a balance between patient well-being and environmental health. The chapter underscores the significance of digital citizenship in promoting sustainability within healthcare.

DOI: 10.4018/979-8-3693-6675-2.ch011

Copyright ©2025, IGI Global Scientific Publishing. Copying or distributing in print or electronic forms without written permission of IGI Global Scientific Publishing is prohibited.

INTRODUCTION

Green marketing, also known as environmental or eco-marketing, is a form of promotion that focuses on the environmental advantages of a product or service. This is an area of special importance in health care, as this industry has a significant impact on the environment due to its high energy consumption, resource utilization, and waste generation. Hospitals and medical centers are among the largest contributors to greenhouse gas emissions, medical waste, and natural resource depletion (Chauhan, 2020). In other words, adding green marketing to healthcare is a strategic step in the direction of addressing environmental issues and ensuring sustainability. Healthcare has seen significant growth in recent years regarding its adoption of sustainable principles. Today, more hospitals and health service providers are opting for energy-efficient technologies, implementing waste reduction practices, and embracing eco-friendly medical supplies. Green marketing acts as a potent tool for these institutions to convey their environmental commitments, thereby making their operations congruent with the increasing consumer demand for sustainable and responsible practices (Singh & Sharma, 2021). Through green marketing strategies, healthcare providers can not only contribute to environmental preservation but also build trust and loyalty among patients who consider sustainability as an important aspect. Green marketing, in the digital age where transparency and accountability are paramount, provides a way for healthcare organizations to demonstrate commitment to both planetary health and patient well-being..

IMPORTANCE OF SUSTAINABILITY AND RESPONSIBILITY IN HEALTHCARE MARKETING

Sustainability in healthcare marketing involves the responsible promotion of services and products that reduce environmental impact. This includes highlighting the use of renewable energy, biodegradable materials, and energy-efficient medical equipment. A cleaner and more sustainable healthcare environment leads to improved public health outcomes, making the integration of green practices both an ethical and strategic imperative (Han & Yoon, 2022). Responsibility in healthcare marketing extends beyond merely promoting green products. It involves ensuring that healthcare organizations are committed to the well-being of their patients, society, and the environment. Increasingly, patients prefer healthcare providers that adopt sustainable practices, as these practices demonstrate a commitment to ethical values and long-term health outcomes (Kumar et al., 2021). For instance, hospitals that implement recycling programs or use renewable energy are more likely to appeal to environmentally conscious consumers. This demonstrates that healthcare organizations engaging in sustainable marketing are not only doing the right thing ethically but are also gaining a competitive advantage. Furthermore, sustainability and responsibility in healthcare marketing are increasingly driven by regulatory pressures. Governments and international bodies are placing stricter regulations on healthcare providers to minimize their carbon footprint and reduce waste. Healthcare organizations are expected to comply with these environmental regulations while meeting consumer demands for greener options (Jackson et al., 2022). These combined factors make sustainable marketing a necessary part of any modern healthcare strategy.

The Role of Digital Platforms in Shaping Consumer Perceptions and Promoting Green Practices

Digital platforms play a central role in shaping consumer perceptions and promoting green practices in healthcare. With the rise of social media, websites, and other digital tools, healthcare providers can now effectively communicate their sustainability initiatives to a broader audience (Rahman, H. 2025). Digital platforms allow organizations to showcase their eco-friendly initiatives, such as reducing waste or transitioning to renewable energy (Brito et al., 2021). Through interactive content, such as blogs, videos, and live webinars, healthcare organizations can educate patients and stakeholders about the importance of sustainability in healthcare. For example, a hospital could use its website to display its environmental performance metrics, highlighting efforts to reduce energy consumption or medical waste. This fosters greater transparency and builds trust with environmentally conscious consumers (D'Souza & Taghian, 2022). Moreover, digital platforms enable healthcare providers to promote green products and services. Through targeted online marketing campaigns, healthcare organizations can showcase eco-friendly medical devices or sustainable treatments. Online reviews and testimonials further enhance the credibility of these claims, allowing satisfied patients to share their experiences with green healthcare practices. Digital platforms thus create a bridge between healthcare organizations and the eco-conscious consumer, allowing for greater engagement and collaboration in promoting sustainability (Zhang & McMullan, 2020).

Objectives of the Chapter and Key Themes Covered

This chapter aims to explore how green marketing can be effectively integrated into the healthcare sector through the use of digital platforms, addressing both the opportunities and challenges associated with promoting sustainability in healthcare. By examining the strategies healthcare providers can adopt to market their green initiatives, the chapter will shed light on the growing influence of consumer behavior on sustainable healthcare practices.

The key themes covered in this chapter include:

- **Green marketing strategies in healthcare**: A review of successful case studies where healthcare providers have implemented eco-friendly practices and effectively marketed them to patients and stakeholders (Han & Yoon, 2022).
- **The role of digital marketing in promoting sustainability**: An exploration of the ways digital tools, such as websites and social media, can enhance public awareness of sustainability efforts in healthcare (Brito et al., 2021).
- **Consumer behavior and preferences**: An analysis of how the environmental consciousness of patients influences their healthcare choices and the potential for healthcare organizations to respond to these preferences (Kumar et al., 2021).
- **Challenges and opportunities in green healthcare marketing**: A discussion of the barriers healthcare providers face in adopting sustainable marketing practices, as well as potential solutions (Jackson et al., 2022).
- **Future trends in green healthcare marketing**: Insights into emerging trends, such as the use of artificial intelligence and blockchain technology, to enhance sustainability in healthcare (Singh & Sharma, 2021).

By addressing these themes, the chapter will provide a comprehensive understanding of how green marketing can drive sustainability in healthcare, particularly through the use of digital platforms.

RESEARCH METHODOLOGY

This study adopts a qualitative research approach to explore the intersection of green marketing and healthcare, particularly focusing on how digital platforms influence consumer behavior and organizational sustainability efforts. Qualitative research was chosen for its ability to provide in-depth insights into complex social phenomena, such as the evolving role of sustainability in healthcare marketing (Creswell, 2014). The research design employs a case study method, where multiple healthcare organizations were selected based on their adoption of green marketing strategies and their use of digital platforms to promote sustainability. Data collection involved semi-structured interviews with marketing professionals and sustainability officers from these organizations, providing firsthand perspectives on the challenges and opportunities they face. Additionally, a content analysis of their digital marketing materials such as websites, social media posts, and sustainability reports was conducted to assess the extent and effectiveness of their green marketing efforts. The data analysis followed a thematic approach, which is ideal for identifying patterns across qualitative data sets (Braun & Clarke, 2006). Through this approach, recurring themes such as consumer engagement, regulatory compliance, and digital innovation were identified. NVivo software was utilized to code and categorize the data, ensuring that the analysis was systematic and consistent. The themes emerging from the interviews and content analysis were then compared to existing literature on green marketing and sustainability in healthcare, allowing for a comprehensive understanding of how digital tools can amplify green marketing strategies in this sector.

Limitations of the Study

While this qualitative approach provides valuable insights into the relationship between green marketing and digital platforms in healthcare, there are inherent limitations to the study. One major limitation is the sample size. Since the study focused on a small number of healthcare organizations, the findings may not be generalizable to the entire healthcare industry (Marshall & Rossman, 2016). The case study method, while useful for in-depth analysis, limits the scope of the research to specific organizations and geographic regions, potentially omitting broader industry trends or global variations in green marketing practices. Additionally, the interviews were conducted with a limited number of key stakeholders, which may not capture the full spectrum of perspectives within the organizations. Another limitation relates to the subjective nature of qualitative research. While thematic analysis provides a structured framework for identifying patterns, the interpretation of qualitative data can be influenced by researcher bias (Maxwell, 2013). Although steps were taken to minimize bias, such as triangulating interview data with content analysis and employing software tools like NVivo, some degree of subjectivity is inevitable. Moreover, the study's focus on digital platforms may overlook other significant factors influencing green marketing in healthcare, such as supply chain logistics or on-site environmental initiatives that are not directly linked to digital communication strategies.

Ethical Considerations

Ethical considerations were paramount in this research, particularly in obtaining consent from participants and ensuring confidentiality. All interviewees were provided with clear information about the purpose of the study, and their consent was secured in writing. To maintain privacy, all data were anonymized, and no identifiable information about the organizations or participants was disclosed (Velmurugan, P. R., Arunkumar, S., Vettriselvan, R., Deepan, A., & Rajesh, D. 2025). Furthermore, the study adhered to ethical guidelines for qualitative research, ensuring that participants could withdraw from the study at any time without repercussions (Orb et al., 2001). By adhering to these ethical standards, the research aimed to foster trust and transparency, allowing for the collection of more authentic and candid data.

Conceptual Framework

Green Marketing: Green marketing refers to the promotion of products or services based on their environmental benefits. It encompasses strategies that highlight eco-friendly attributes such as reduced energy consumption, sustainable materials, and minimal environmental impact (Peattie & Crane, 2005). In healthcare, green marketing involves promoting practices and products that contribute to environmental sustainability, such as energy-efficient medical equipment, waste reduction programs, and eco-friendly packaging.

Digital Citizenship: Digital citizenship involves the responsible use of technology and digital platforms, encompassing aspects such as online behavior, privacy, and digital literacy (Ribble, 2015). It emphasizes the role of individuals and organizations in using digital tools ethically and effectively. In the context of healthcare, digital citizenship includes the responsible management of digital marketing strategies, ensuring that they promote transparency, accuracy, and respect for consumer privacy.

Sustainability: Sustainability refers to the capacity to maintain ecological and social systems over the long term. It involves practices that reduce resource consumption, minimize waste, and promote environmental stewardship (Brundtland Commission, 1987). In healthcare, sustainability focuses on reducing the sector's environmental footprint through practices such as energy conservation, waste management, and the use of sustainable resources.

Eco-Conscious Healthcare Practices: Eco-conscious healthcare practices are initiatives undertaken by healthcare organizations to minimize their environmental impact. This includes adopting green technologies, reducing energy use, and promoting sustainable practices within healthcare facilities (Walton et al., 2020). These practices reflect a commitment to environmental responsibility and align with broader sustainability goals.

INTERSECTION OF CONCEPTS IN HEALTHCARE

The intersection of these concepts in healthcare creates a comprehensive framework for understanding how green marketing and digital citizenship contribute to sustainability and eco-conscious practices. Green marketing in healthcare often leverages digital platforms to communicate and promote sustainable practices, thus integrating digital citizenship into environmental strategies. For instance, healthcare organizations use social media, websites, and digital advertising to showcase their commitment to green practices, such as energy-efficient technologies and waste reduction initiatives (Sweeney et al., 2019 & Sujatha, R., Aarthy, S. L., & Vettriselvan, R. (Eds.). 2021). Digital platforms play a crucial role in this intersection by providing a medium for healthcare organizations to engage with their audience on sustainability issues. Through effective digital marketing, healthcare providers can highlight their eco-conscious practices and educate consumers about the benefits of sustainable healthcare. This digital engagement not only enhances the visibility of green initiatives but also fosters a culture of environmental responsibility among patients and stakeholders (Brito et al., 2021).

Relevance of Digital Platforms in Promoting Green Practices in Healthcare

Digital platforms are increasingly relevant in promoting green practices within healthcare due to their ability to reach a broad audience and facilitate real-time communication. Websites, social media channels, and digital advertising provide healthcare organizations with the tools to effectively communicate their green marketing messages. These platforms allow for the dissemination of information about sustainable practices, such as the adoption of energy-efficient technologies or waste reduction programs, to a global audience (D'Souza & Taghian, 2022). Moreover, digital platforms enable healthcare organizations to engage in interactive communication with consumers, allowing for feedback and dialogue about their sustainability efforts. This engagement helps to build trust and credibility, as consumers can see firsthand the organization's commitment to environmental responsibility (Vettriselvan, R. 2025). Online reviews, social media interactions, and digital content provide avenues for healthcare providers to showcase their green practices and respond to consumer inquiries and concerns (Zhang & McMullan, 2020). In addition, digital tools offer valuable data analytics capabilities that can help healthcare organizations measure the impact of their green marketing efforts. By analyzing digital engagement metrics, organizations can assess the effectiveness of their sustainability messages and make data-driven decisions to enhance their green practices and marketing strategies (Brito et al., 2021). This ability to monitor and adapt strategies based on digital feedback underscores the importance of digital platforms in promoting and advancing green practices in the healthcare sector.

THEORETICAL FRAMEWORK

Ecological Modernization Theory: Ecological modernization theory posits that economic development and environmental protection are not mutually exclusive but can be mutually reinforcing (Hajer, 1995). This theory suggests that technological innovation and institutional reforms can lead to more sustainable practices within industries, including healthcare. In the context of green marketing, ecological modernization emphasizes the role of technological advancements and market mechanisms in promoting environmentally friendly practices (Awawdeh, H. Z., Bani Ahmad, A. Y., Almajali, W. I., Atieh Ali, A. A., & Allahham, M. 2024). For healthcare organizations, this theory supports the idea that investments in green technologies and sustainable practices can lead to both economic and environmental benefits. For instance, the adoption of energy-efficient medical equipment not only reduces operational costs but also minimizes the facility's ecological footprint (Mol, 2001).

Stakeholder Theory: Stakeholder theory, developed by Freeman (1984), asserts that organizations have a responsibility to all stakeholders, not just shareholders. This theory is particularly relevant to green marketing, as it highlights the importance of addressing the interests of various groups affected by the organization's operations, including patients, employees, suppliers, and the community. In healthcare, stakeholder theory underscores the need for organizations to engage with stakeholders on sustainability issues and to communicate their green initiatives transparently. By doing so, healthcare providers can build stronger relationships with stakeholders and enhance their reputation as responsible and ethical organizations (Donaldson & Preston, 1995). This approach is reflected in green marketing strategies that emphasize stakeholder engagement and the promotion of eco-friendly practices.

Theoretical Perspectives on Sustainability in Healthcare

Triple Bottom Line Theory: The Triple Bottom Line (TBL) theory expands the traditional notion of business performance by incorporating three dimensions: economic, environmental, and social (Elkington, 1997). This perspective is particularly pertinent to healthcare, where sustainability efforts must balance financial performance with environmental stewardship and social responsibility. The TBL framework encourages healthcare organizations to assess their performance across these three dimensions, ensuring that their sustainability initiatives contribute to economic viability, environmental protection, and social well-being. For example, a hospital's sustainability program might focus on reducing waste and energy consumption (environmental), improving community health outcomes (social), and maintaining cost-effectiveness (economic) (Norman & MacDonald, 2004).
Resource-Based View (RBV): The Resource-Based View (RBV) theory suggests that organizations can achieve a competitive advantage by leveraging valuable, rare, inimitable, and non-substitutable resources (Barney, 1991). In the context of sustainability in healthcare, RBV emphasizes the importance of integrating unique resources and capabilities, such as advanced green technologies and expertise in sustainable practices, into the organization's strategy. This theoretical perspective supports the idea that healthcare organizations can differentiate themselves and gain a competitive edge by adopting and marketing their green innovations (Teixeira, R. L. P., & Teixeira, C. H. S. B. 2024). For instance, a healthcare provider with cutting-edge energy-efficient systems and a strong commitment to sustainability can attract environmentally conscious patients and enhance its market position (Hart, 1995).

Framework on Consumer Behavior in Relation to Green Healthcare Marketing

Theory of Planned Behavior (TPB): The Theory of Planned Behavior (TPB) explains how individual attitudes, subjective norms, and perceived behavioral control influence behavioral intentions and actions (Ajzen, 1991). In the context of green healthcare marketing, TPB can help explain how patients' attitudes toward environmental sustainability, social norms regarding eco-friendly practices, and their perceived ability to support green initiatives affect their healthcare choices. For example, patients who view sustainable practices positively and believe that their actions can contribute to environmental protection are more likely to choose healthcare providers that prioritize green marketing (White et al., 2019).

Consumer Culture Theory (CCT): Consumer Culture Theory explores how cultural and social factors influence consumer behavior and consumption patterns (Arnould & Thompson, 2005). This theoretical perspective is relevant for understanding how cultural values and societal trends shape consumer preferences for green healthcare services (Darban, K., & Kabbaj, S. 2023). CCT suggests that consumer behavior is influenced by cultural narratives and social identities, which can impact how patients perceive and respond to green marketing efforts. For instance, in cultures where environmental consciousness is a strong value, healthcare providers that emphasize their commitment to sustainability may resonate more effectively with consumers (Sherry & Kozinets, 2008).

Application of Digital Citizenship Theories in Healthcare Marketing

Digital Citizenship Theory: Digital Citizenship Theory encompasses the responsible use of digital tools and technologies, emphasizing ethical behavior, privacy, and effective communication (Ribble, 2015). In healthcare marketing, this theory underscores the importance of leveraging digital platforms responsibly to promote green practices. Healthcare organizations must ensure that their digital marketing strategies align with principles of transparency, accuracy, and respect for consumer privacy. This involves providing clear information about green initiatives, addressing potential misinformation, and engaging with consumers in a respectful and constructive manner (Martin & Grudziecki, 2006).

Technology Acceptance Model (TAM): The Technology Acceptance Model (TAM) explains how users come to accept and use new technologies based on perceived ease of use and perceived usefulness (Davis, 1989). In the realm of digital citizenship and green marketing, TAM can help healthcare organizations understand how patients perceive and interact with digital tools related to sustainability (Sahoo, D. 2024). For example, if patients find a healthcare provider's online platform for tracking eco-friendly practices easy to use and valuable, they are more likely to engage with it and support the provider's green initiatives (Venkatesh & Davis, 2000). The theoretical framework for understanding green marketing in healthcare integrates concepts from ecological modernization, stakeholder theory, and sustainability perspectives, combined with theories of consumer behavior and digital citizenship. By applying these theories, healthcare organizations can effectively navigate the complex interplay between environmental sustainability, consumer expectations, and digital engagement, ultimately fostering more sustainable practices and enhancing their market position.

GREEN MARKETING IN HEALTHCARE: OPPORTUNITIES AND CHALLENGES

Opportunities for Healthcare Organizations to Adopt Green Marketing

Green marketing presents significant opportunities for healthcare organizations to enhance their brand reputation and differentiate themselves in a competitive market. One major opportunity is the ability to attract environmentally conscious consumers who are increasingly prioritizing sustainability in their healthcare choices. By adopting green marketing strategies, healthcare providers can highlight their commitment to environmental stewardship, thereby building trust and loyalty among patients (Peattie & Crane, 2005). For instance, implementing energy-efficient systems and sustainable practices can not only reduce operational costs but also serve as a key marketing point to appeal to eco-conscious patients. Moreover, green marketing offers healthcare organizations a chance to lead by example and influence industry standards (Rosário, A. T., Lopes, P. R., & Rosário, F. S. 2023). By pioneering innovative green practices, such as using eco-friendly materials and reducing waste, healthcare providers can set benchmarks for others in the industry. This leadership can enhance their reputation as pioneers in sustainable healthcare, attracting positive media attention and increasing their visibility (Walton et al., 2020). Additionally, aligning green marketing efforts with corporate social responsibility (CSR) initiatives can further strengthen an organization's commitment to broader societal and environmental goals, enhancing overall organizational credibility (Kotler & Lee, 2005).

The Challenges Healthcare Organizations Face in Promoting Sustainability

Despite the benefits, healthcare organizations encounter several challenges in promoting sustainability. One of the primary challenges is the high initial cost associated with implementing green technologies and practices. Investments in energy-efficient equipment, sustainable building materials, and waste reduction systems can be substantial, posing financial hurdles for many organizations (Brito et al., 2021). Additionally, the complexity of integrating these green practices into existing systems and operations can create logistical and operational challenges, requiring significant adjustments and training. Another challenge is the lack of standardized metrics and frameworks for measuring the effectiveness of green marketing initiatives. Without clear benchmarks and guidelines, it can be difficult for healthcare organizations to assess the impact of their sustainability efforts and communicate their achievements to stakeholders (Sweeney et al., 2019). This lack of standardized reporting can also hinder the ability to compare and benchmark against industry peers, making it challenging to demonstrate leadership in green practices.

Consumer Expectations and the Growing Demand for Eco-Conscious Healthcare

Consumer expectations for eco-conscious healthcare are growing, driven by increasing awareness of environmental issues and a desire for more sustainable choices. Patients are becoming more informed about the environmental impact of healthcare practices and are demanding greater transparency from healthcare providers regarding their sustainability efforts (White et al., 2019). This shift in consumer preferences creates a significant market opportunity for healthcare organizations to differentiate themselves through green marketing.

The growing demand for eco-conscious healthcare is reflected in patient behaviors and preferences. Studies have shown that consumers are willing to choose healthcare providers based on their commitment to sustainability and are increasingly seeking out organizations that demonstrate environmental responsibility (D'Souza & Taghian, 2022). This trend underscores the importance of incorporating green marketing strategies into healthcare organizations' broader marketing efforts to meet evolving consumer expectations and capitalize on the growing demand for sustainable healthcare options.

Case Studies of Healthcare Organizations Leading in Green Marketing

Several healthcare organizations have successfully integrated green marketing into their operations, setting examples for others in the industry. For instance, the Cleveland Clinic has been recognized for its sustainability initiatives, including energy-efficient building designs, waste reduction programs, and the use of sustainable materials in construction (Cleveland Clinic, 2023). The clinic's commitment to environmental sustainability is communicated through its green marketing efforts, enhancing its reputation as a leader in eco-conscious healthcare. Another notable example is the Kaiser Permanente health system, which has implemented a comprehensive sustainability strategy encompassing energy efficiency, waste reduction, and sustainable sourcing (Kaiser Permanente, 2024). Kaiser Permanente's green marketing initiatives highlight its achievements and commitments in sustainability, effectively engaging patients and stakeholders through transparent communication and environmental stewardship. These case studies demonstrate that healthcare organizations can successfully navigate the challenges of green marketing while capitalizing on the opportunities it presents. By adopting and promoting sustainable practices, healthcare providers can enhance their brand reputation, meet growing consumer demands, and contribute to a more sustainable future.

DIGITAL PLATFORMS AND GREEN MARKETING

The Role of Digital Media in Promoting Sustainability in Healthcare

Digital media plays a crucial role in advancing sustainability efforts within the healthcare sector by providing platforms for disseminating information, engaging with consumers, and showcasing eco-friendly practices. Through various digital channels, healthcare organizations can effectively communicate their commitment to sustainability and highlight the environmental benefits of their practices. Digital media allows for the broad dissemination of sustainability initiatives, such as energy-efficient technologies, waste reduction programs, and sustainable sourcing, to a global audience (Sweeney & Johnson, 2019). The use of digital media in green marketing enables healthcare organizations to reach and influence a diverse group of stakeholders, including patients, partners, and the wider community.

How Social Media, Websites, and Online Campaigns Drive Eco-Friendly Practices

Social Media: Social media platforms such as Facebook, Twitter, and Instagram offer powerful tools for promoting green practices in healthcare. These platforms allow organizations to share real-time updates, success stories, and educational content about their sustainability efforts. By leveraging social media, healthcare providers can engage directly with their audience, solicit feedback, and foster a community of supporters for their eco-friendly initiatives (Kaplan & Haenlein, 2010). For example, healthcare organizations can use social media to run campaigns that highlight their recycling programs, energy-saving measures, and participation in environmental events.

Websites: Healthcare organizations' websites serve as central hubs for information on sustainability initiatives. These websites can feature dedicated sections on green practices, detailed case studies, and metrics demonstrating the impact of sustainability efforts. An informative and well-designed website can enhance the credibility of a healthcare organization's green marketing claims and provide a platform for transparency (Iyer, K. A. S., Nagojappa, N. B. S., Kanjanapathy, M., & Kumar, S. H. 2024). Websites also facilitate the integration of interactive tools, such as carbon footprint calculators or sustainability dashboards, that engage users and provide insights into the organization's environmental impact (Smith & Brower, 2012).

Online Campaigns: Online marketing campaigns, including email newsletters, pay-per-click advertising, and content marketing, play a significant role in driving eco-friendly practices. These campaigns can target specific demographics interested in sustainability, promoting green products and services offered by healthcare organizations. Effective online campaigns leverage data analytics to tailor messages and optimize engagement. For instance, targeted email campaigns can inform subscribers about new green initiatives or upcoming events related to sustainability, thereby encouraging participation and support (Cheng et al., 2020).

Engaging with Eco-Conscious Consumers and Building Trust through Transparency and Authenticity

Engaging with eco-conscious consumers requires healthcare organizations to adopt strategies that emphasize transparency and authenticity. Consumers today are increasingly savvy and demand honest communication about environmental practices (Alshatter, Z., & Al Mubarak, M. 2024). To build trust, healthcare providers must clearly communicate their sustainability goals, progress, and achievements. This involves sharing detailed information about the steps taken to reduce environmental impact and addressing any potential challenges or areas for improvement (White et al., 2019). Authenticity is crucial in maintaining consumer trust. Greenwashing, or misleading claims about environmental benefits, can lead to skepticism and damage an organization's reputation. Therefore, healthcare organizations should ensure that their green marketing messages are backed by concrete actions and verifiable data. Engaging in open dialogue with consumers, responding to inquiries, and providing transparent reporting on sustainability metrics are essential practices for building and maintaining trust (Peattie & Crane, 2005).

Examples of Successful Digital Strategies for Green Healthcare Marketing

Several healthcare organizations have successfully implemented digital strategies to promote their green initiatives. For example, the Mayo Clinic has effectively used its website and social media channels to highlight its sustainability efforts, such as energy-efficient building designs and waste reduction programs. The clinic's digital content includes detailed reports on environmental performance and interactive features that engage users in sustainability practices (Mayo Clinic, 2023). Another notable example is the Cleveland Clinic's "Green Team" initiative, which is actively promoted through social media and online campaigns (Baruno, A. D., & Indrasari, M. 2025). The Cleveland Clinic shares updates about its green practices, including renewable energy projects and sustainable healthcare practices, through engaging social media content and interactive web features. This approach has successfully increased awareness and support for the clinic's sustainability efforts (Cleveland Clinic, 2023). These examples demonstrate how healthcare organizations can leverage digital platforms to effectively communicate their commitment to sustainability, engage with eco-conscious consumers, and build a strong reputation as leaders in green healthcare.

BUILDING A RESPONSIBLE ONLINE PRESENCE

Best Practices for Healthcare Organizations to Promote Sustainability Through Digital Platforms

To effectively promote sustainability through digital platforms, healthcare organizations should adopt several best practices. First, creating and maintaining a user-friendly website that showcases sustainability efforts and achievements is crucial. The website should feature dedicated sections for sustainability initiatives, including detailed information on energy-efficient practices, waste management programs, and green certifications. Regularly updated content, such as blog posts, case studies, and annual sustainability reports, can provide transparency and keep stakeholders informed about ongoing efforts and progress (Smith & Brower, 2012). Second, healthcare organizations should leverage social media to amplify their sustainability messages and engage with their audience. This involves crafting compelling content that highlights eco-friendly practices and initiatives, sharing success stories, and promoting interactive campaigns that encourage community participation. Social media platforms such as Facebook, Twitter, and Instagram offer tools for real-time interaction and feedback, enabling organizations to address questions, respond to concerns, and build a community of supporters for their green initiatives (Kaplan & Haenlein, 2010).

Third, incorporating sustainability into digital advertising and email marketing campaigns can help reinforce the organization's commitment to environmental responsibility. Targeted ads and newsletters can feature updates on green practices, upcoming sustainability events, and tips for eco-friendly living. Ensuring that these communications are visually appealing and aligned with the organization's overall brand message will enhance their effectiveness and reach (Cheng et al., 2020).

The Role of Transparency and Corporate Responsibility in Building Trust

Transparency and corporate responsibility are fundamental to building trust with consumers in the digital age. Healthcare organizations must provide clear and honest information about their sustainability practices and goals. This includes sharing detailed reports on environmental impact, disclosing any challenges or setbacks, and openly communicating the steps being taken to address them. Transparency not only builds credibility but also fosters a sense of accountability and trust among stakeholders (Peattie & Crane, 2005). Corporate responsibility extends beyond just environmental issues to encompass ethical practices and community engagement. Demonstrating a commitment to ethical standards, such as fair labor practices and responsible sourcing, complements environmental efforts and enhances overall organizational reputation. By integrating corporate responsibility into their digital presence, healthcare organizations can present a holistic view of their values and priorities, further strengthening their relationship with consumers (Kotler & Lee, 2005).

Engaging Diverse Communities and Fostering a Culture of Sustainability

Engaging diverse communities is essential for fostering a culture of sustainability within the healthcare sector. Healthcare organizations should tailor their digital content and outreach efforts to address the needs and interests of different demographic groups. This involves creating multilingual content, accommodating accessibility needs, and addressing specific community concerns related to sustainability. Furthermore, fostering a culture of sustainability involves promoting inclusivity and encouraging participation from all levels of the organization. Digital platforms can be used to highlight initiatives that support diversity and inclusion, such as partnerships with local organizations, educational programs, and community outreach efforts. By showcasing these efforts online, healthcare organizations can demonstrate their commitment to building a more inclusive and sustainable future (White et al., 2019).

Examples of Healthcare Organizations Promoting Environmental Well-Being Alongside Patient Care

Several healthcare organizations have successfully integrated environmental well-being into their patient care models while maintaining a strong digital presence. The Cleveland Clinic, for example, has actively promoted its green initiatives through its website and social media channels. The Clinic's sustainability efforts include energy-efficient facilities, waste reduction programs, and green building practices, all of which are communicated transparently to the public (Cleveland Clinic, 2023). Similarly, the Mayo Clinic emphasizes its commitment to sustainability on its digital platforms by showcasing its initiatives in renewable energy, waste management, and water conservation. The Clinic's website features detailed reports on these efforts, and social media campaigns highlight the positive impact of their sustainability practices on both the environment and patient care (Mayo Clinic, 2023).

These examples illustrate how healthcare organizations can effectively use digital platforms to promote environmental well-being, engage with diverse communities, and build trust through transparency and corporate responsibility.

SOLUTIONS AND RECOMMENDATIONS

Practical Strategies for Healthcare Organizations to Implement Green Marketing

To successfully implement green marketing, healthcare organizations should adopt practical strategies that effectively communicate their commitment to sustainability while engaging patients and stakeholders. First, organizations should develop a comprehensive green marketing plan that outlines clear sustainability goals and objectives. This plan should include specific initiatives such as reducing energy consumption, minimizing waste, and using eco-friendly materials. The plan should also detail how these initiatives will be communicated to the public through various digital and traditional marketing channels (Peattie & Crane, 2005). Second, healthcare organizations should invest in green certifications and eco-labels to enhance their credibility and marketability. Certifications such as LEED (Leadership in Energy and Environmental Design) for buildings or Energy Star for equipment can serve as tangible proof of an organization's commitment to sustainability. Promoting these certifications through marketing materials, websites, and social media can effectively convey the organization's dedication to green practices (Kotler & Lee, 2005). Third, integrating sustainability into patient engagement strategies is essential. Healthcare providers can use digital platforms to educate patients about the environmental benefits of their practices and encourage them to participate in sustainability efforts. For example, organizations can share tips for reducing personal carbon footprints, promote recycling programs, and highlight the environmental impact of choosing sustainable healthcare options (Cheng et al., 2020).

Steps for Aligning Digital Marketing with Sustainability Goals

Aligning digital marketing efforts with sustainability goals involves several key steps. First, organizations should ensure that their digital content reflects their sustainability commitments. This includes creating engaging and informative content that highlights their green initiatives, shares success stories, and provides updates on sustainability progress. Content should be consistent across all digital platforms, including websites, social media, and email campaigns (Smith & Brower, 2012). Second, organizations should utilize data analytics to track the effectiveness of their green marketing campaigns and adjust strategies as needed. By analyzing metrics such as website traffic, social media engagement, and conversion rates, organizations can identify which aspects of their digital marketing are resonating with their audience and which areas need improvement. This data-driven approach allows for continuous optimization of digital marketing strategies to better align with sustainability goals (Sweeney & Johnson, 2019). Third, integrating sustainability into the organization's overall brand messaging is crucial. All marketing materials, from digital ads to printed brochures, should consistently reflect the organization's commitment to environmental responsibility. This cohesive branding helps reinforce the organization's green message and ensures that sustainability is a core component of its public identity (Kaplan & Haenlein, 2010).

How to Integrate Eco-Friendly Practices into Everyday Healthcare Operations

Integrating eco-friendly practices into everyday healthcare operations requires a systematic approach. First, healthcare organizations should conduct a sustainability audit to assess their current environmental impact and identify areas for improvement. This audit should evaluate aspects such as energy use, waste management, and resource consumption. Based on the findings, organizations can develop an action plan to implement eco-friendly practices, such as energy-efficient lighting, waste segregation, and water-saving technologies (Brito et al., 2021). Second, incorporating sustainability into procurement practices is essential. Organizations should prioritize purchasing eco-friendly products and materials, such as recycled paper, biodegradable cleaning supplies, and energy-efficient medical equipment. By selecting products with lower environmental impacts, healthcare organizations can reduce their overall ecological footprint (White et al., 2019). Third, fostering a culture of sustainability among staff is vital. This can be achieved through training programs that educate employees about sustainable practices and encourage them to adopt eco-friendly behaviors. Creating a sustainability committee or task force can also help drive initiatives and ensure that sustainability remains a priority across all levels of the organization (Peattie & Crane, 2005).

Collaborating with Stakeholders to Enhance Sustainability Efforts

Collaboration with stakeholders is key to enhancing sustainability efforts. Healthcare organizations should engage with a variety of stakeholders, including patients, suppliers, community organizations, and government agencies, to support and advance their green initiatives. Building partnerships with suppliers who share similar sustainability values can help ensure that the organization's procurement practices align with its environmental goals (Kotler & Lee, 2005). Community engagement is also crucial. Organizations can participate in local environmental events, support community-based sustainability programs, and collaborate with non-profit organizations focused on environmental issues. These partnerships can enhance the organization's reputation and demonstrate its commitment to broader sustainability goals

(Cheng et al., 2020). Furthermore, healthcare organizations should work with policymakers and regulatory bodies to advocate for policies and standards that promote sustainability within the healthcare sector. By participating in industry forums and contributing to policy discussions, organizations can help shape the future of sustainable healthcare practices (Smith & Brower, 2012). Implementing green marketing in healthcare requires a strategic approach that includes developing a comprehensive plan, aligning digital marketing with sustainability goals, integrating eco-friendly practices into operations, and collaborating with stakeholders. By following these recommendations, healthcare organizations can effectively promote sustainability, enhance their brand reputation, and contribute to a more environmentally responsible future.

FUTURE RESEARCH DIRECTIONS

Identifying Gaps in Current Research on Green Marketing in Healthcare

Despite growing interest in green marketing within the healthcare sector, several research gaps persist that warrant further exploration. One significant gap is the lack of comprehensive studies on the long-term impact of green marketing initiatives on patient outcomes and organizational performance. Most existing research focuses on immediate consumer perceptions and environmental benefits, leaving a void in understanding how these practices influence patient satisfaction, health outcomes, and overall organizational efficacy over time (Peattie & Crane, 2005). Another gap is the insufficient exploration of the effectiveness of different digital marketing strategies in promoting green practices. While there is considerable research on green marketing in general, specific studies examining how various digital platforms—such as social media, mobile apps, and online patient portals affect the adoption and success of eco-friendly practices in healthcare are limited (Cheng et al., 2020). Addressing these gaps could provide valuable insights into optimizing digital strategies for better sustainability outcomes.

Emerging Trends and Opportunities for Further Exploration

Several emerging trends present opportunities for further research in green marketing within healthcare. One such trend is the increasing focus on circular economy principles, which emphasize recycling, reuse, and the reduction of waste. Research could investigate how healthcare organizations can integrate circular economy practices into their operations and marketing strategies. This includes examining the potential for closed-loop systems in healthcare settings and how these practices are communicated to patients and other stakeholders (Brito et al., 2021). Another emerging trend is the rise of personalized and precision medicine, which could intersect with green marketing efforts. Future research could explore how personalized healthcare solutions can be marketed as environmentally friendly and how these practices align with patients' growing demand for sustainable options. Understanding how precision medicine impacts sustainability and how it can be integrated into green marketing strategies offers a promising area for exploration (Smith & Brower, 2012).

Potential Interdisciplinary Collaborations to Explore the Intersection of Technology, Sustainability, and Healthcare Marketing

Interdisciplinary collaborations are essential for advancing research at the intersection of technology, sustainability, and healthcare marketing. For example, partnerships between environmental scientists, healthcare professionals, and digital marketers can yield innovative solutions for integrating sustainability into healthcare practices. Collaborations with technology experts can also explore how advancements in artificial intelligence (AI) and data analytics can enhance green marketing efforts and optimize sustainability practices (Sweeney & Johnson, 2019). Furthermore, collaborations with behavioral scientists could provide insights into consumer behavior and preferences regarding green healthcare practices. By understanding the psychological and social factors that influence consumer attitudes towards sustainability, researchers can develop more effective marketing strategies that resonate with patients and drive adoption of eco-friendly practices (White et al., 2019).

How Future Research Can Contribute to Shaping Responsible Digital Healthcare

Future research has the potential to significantly shape responsible digital healthcare by providing evidence-based insights into effective green marketing strategies and their impact on patient care and environmental outcomes. By addressing existing research gaps and exploring emerging trends, researchers can develop comprehensive frameworks for integrating sustainability into digital healthcare marketing. This includes identifying best practices for leveraging digital platforms to promote eco-friendly practices and assessing the long-term benefits of these initiatives for both healthcare organizations and patients (Kotler & Lee, 2005). Additionally, research can contribute to developing guidelines and standards for responsible digital healthcare marketing. This involves creating benchmarks for transparency, authenticity, and ethical practices in promoting sustainability. By establishing these guidelines, researchers can help ensure that healthcare organizations engage in responsible marketing that aligns with their sustainability goals and builds trust with patients and other stakeholders (Kaplan & Haenlein, 2010). Future research in green marketing for healthcare should focus on addressing existing gaps, exploring emerging trends, fostering interdisciplinary collaborations, and shaping responsible digital healthcare practices. By doing so, researchers can contribute to a more sustainable and effective approach to healthcare marketing that benefits both the environment and patient care.

CONCLUSION

This chapter has explored the multifaceted intersection of green marketing and healthcare, emphasizing the critical role of sustainability in modern healthcare practices. We began by defining key concepts such as green marketing, digital citizenship, and eco-conscious healthcare practices, highlighting how these elements intersect to influence patient perceptions and organizational behaviors (Kotler & Lee, 2005). We examined the theoretical frameworks underpinning green marketing, including ecological modernization and stakeholder theory, which offer insights into how sustainability can be integrated into healthcare marketing strategies (Peattie & Crane, 2005). Additionally, we discussed the practical strategies for implementing green marketing, aligning digital marketing efforts with sustainability goals, and integrating eco-friendly practices into everyday healthcare operations (Smith & Brower, 2012). The chapter also covered the role of digital platforms in promoting green practices and the need for transparency and corporate responsibility in building trust with consumers (Kaplan & Haenlein, 2010). The importance of sustainability in healthcare marketing has become increasingly pronounced as both consumers and organizations recognize the need for environmental stewardship. Consumers are now more aware of and concerned about the environmental impacts of their choices, and they expect healthcare organizations to reflect these values in their operations and marketing (Cheng et al., 2020). As the demand for eco-conscious healthcare options grows, organizations must adapt their marketing strategies to highlight their sustainability efforts and differentiate themselves in a competitive market. Embracing sustainability not only aligns with consumer expectations but also offers operational benefits such as cost savings through energy efficiency and waste reduction. Healthcare organizations that effectively communicate their commitment to sustainability are likely to build stronger relationships with patients and enhance their reputation within the community (Brito et al., 2021). Digital platforms play a pivotal role in advancing green healthcare marketing by providing channels for organizations to reach and engage with eco-conscious consumers. Social media, websites, and online campaigns enable healthcare providers to share their sustainability initiatives, showcase their achievements, and foster a culture of transparency and accountability (White et al., 2019). These platforms also facilitate real-time interaction with consumers, allowing organizations to address questions, receive feedback, and build a community of supporters for their green practices. The effectiveness of these digital strategies hinges on creating authentic and compelling content that resonates with consumers and reflects the organization's genuine commitment to sustainability (Sweeney & Johnson, 2019). As digital technologies continue to evolve, healthcare organizations must leverage these advancements to enhance their green marketing efforts and drive positive environmental change. In conclusion, healthcare organizations have a significant opportunity and responsibility to lead in sustainable and responsible marketing practices. By integrating green marketing strategies into their operations, aligning digital marketing with sustainability goals, and fostering transparency and accountability, organizations can position themselves as leaders in the movement towards a more environmentally responsible healthcare sector. This involves not only adopting eco-friendly practices but also actively communicating these efforts to patients and stakeholders through effective digital channels. We urge healthcare organizations to commit to sustainability as a core component of their marketing strategies, ensuring that their environmental efforts are both genuine and visible. Leading in sustainable marketing not only benefits the environment but also enhances organizational reputation, builds patient trust, and contributes to a healthier and more sustainable future for all.

REFERENCES

Ajzen, I. (1991). The theory of planned behavior. *Organizational Behavior and Human Decision Processes*, *50*(2), 179–211. DOI: 10.1016/0749-5978(91)90020-T

Alshatter, Z., & Al Mubarak, M. (2024). From Green Tech to Digital Tech: Leveraging Digital Solutions for Sustainable Development (SD) in B2B Markets. In *Equity and Sustainability* (pp. 151–171). Springer Nature Singapore. DOI: 10.1007/978-981-97-4742-9_7

Arnould, E. J., & Thompson, G. J. (2005). Consumer culture theory (CCT): Twenty years of research. *The Journal of Consumer Research*, *31*(4), 868–882. DOI: 10.1086/426626

Awawdeh, H. Z., Bani Ahmad, A. Y., Almajali, W. I., Atieh Ali, A. A., & Allahham, M. (2024). How Does Digital Marketing Influence Consumer Behavior? Examining the Mediating Role of Digital Entrepreneurship in the Healthcare and Pharmaceuticals Sector. Library of Progress-Library Science, Information Technology & Computer, 44(3).

Barney, J. (1991). Firm resources and sustained competitive advantage. *Journal of Management*, *17*(1), 99–120. DOI: 10.1177/014920639101700108

Baruno, A. D., & Indrasari, M. (2025). Leveraging AI To Enhance Green Marketing Strategies. Jurnal Ekonomi, Manajemen, Akuntansi dan Keuangan, 6(1).

Braun, V., & Clarke, V. (2006). Using thematic analysis in psychology. *Qualitative Research in Psychology*, *3*(2), 77–101. DOI: 10.1191/1478088706qp063oa

Brito, M. P., Zhao, Y., & Fan, S. (2021). The role of digital platforms in green marketing. *Sustainable Marketing Journal*, *14*(3), 203–215.

Brundtland Commission. (1987). *Our common future*. Oxford University Press.

Chauhan, A. (2020). Environmental sustainability in healthcare: A green marketing approach. *Journal of Healthcare Management*, *45*(2), 122–131. PMID: 32168188

Cheng, M., Choi, S., & Wang, J. (2020). Online marketing and consumer engagement: Insights from green healthcare marketing. *Journal of Digital Marketing*, *15*(2), 85–99.

Cleveland Clinic. (2023). Sustainability at Cleveland Clinic. Retrieved from https://my.clevelandclinic.org

Creswell, J. W. (2014). *Research design: Qualitative, quantitative, and mixed methods approaches* (4th ed.). SAGE Publications.

D'Souza, C., & Taghian, M. (2022). Green marketing in healthcare: A consumer perspective. *Healthcare Marketing Review*, *8*(4), 451–467.

Darban, K., & Kabbaj, S. (2023). The Transformative Potential of AI in Green Marketing Strategies. *Traditional Journal of Law and Social Sciences*, *2*(02), 14–38.

Davis, F. D. (1989). Perceived usefulness, perceived ease of use, and user acceptance of information technology. *Management Information Systems Quarterly*, *13*(3), 319–340. DOI: 10.2307/249008

Donaldson, T., & Preston, L. E. (1995). The stakeholder theory of the corporation: Concepts, evidence, and implications. *Academy of Management Review*, *20*(1), 65–91. DOI: 10.2307/258887

Elkington, J. (1997). *Cannibals with forks: The triple bottom line of 21st century business*. Capstone Publishing.

Freeman, R. E. (1984). *Strategic management: A stakeholder approach*. Pitman Publishing.

Hajer, M. A. (1995). *The politics of environmental discourse: Ecological modernization and the policy process*. Oxford University Press.

Han, H., & Yoon, H. J. (2022). Sustainable healthcare practices: The role of green marketing. *International Journal of Healthcare Management*, *18*(1), 1–15.

Hart, S. L. (1995). A natural-resource-based view of the firm. *Academy of Management Review*, *20*(4), 986–1014. DOI: 10.2307/258963

Iyer, K. A. S., Nagojappa, N. B. S., Kanjanapathy, M., & Kumar, S. H. (2024). Marketing in the Digital Age: Leveraging Technology for Sustainability. In Utilizing Technology for Sustainable Resource Management Solutions (pp. 194-217). IGI Global.

Jackson, K., Lee, M., & Rogers, D. (2022). Regulatory pressures on sustainable healthcare: The impact of environmental laws. *Journal of Environmental Regulation*, *37*(6), 115–126.

Kaiser Permanente. (2024). Our commitment to sustainability. Retrieved from https://about .kaiserpermanente.org

Kaplan, A. M., & Haenlein, M. (2010). Users of the world, unite! The challenges and opportunities of social media. *Business Horizons*, *53*(1), 59–68. DOI: 10.1016/j.bushor.2009.09.003

Kotler, P., & Lee, N. (2005). *Corporate social responsibility: Doing the most good for your company and your cause*. John Wiley & Sons.

Kumar, P., Jain, R., & Singh, A. (2021). The influence of sustainability on consumer behavior in healthcare. *JMR, Journal of Marketing Research*, *58*(7), 987–1003.

Marshall, C., & Rossman, G. B. (2016). *Designing qualitative research* (6th ed.). SAGE Publications.

Martin, C., & Grudziecki, J. (2006). Digital literacy – What are we really teaching? *European Conference on Information Literacy*, 24(1), 277-289.

Maxwell, J. A. (2013). *Qualitative research design: An interactive approach* (3rd ed.). SAGE Publications.

Mayo Clinic. (2023). Sustainability at Mayo Clinic. Retrieved from https://www.mayoclinic.org

Mol, A. P. J. (2001). Ecological modernization and the global economy. *Environmental Politics*, *10*(1), 1–10. DOI: 10.1080/09644010008414510

Norman, W., & MacDonald, C. (2004). Getting to the bottom of "triple bottom line". *Business Ethics Quarterly*, *14*(2), 243–262. DOI: 10.5840/beq200414211

Orb, A., Eisenhauer, L., & Wynaden, D. (2001). Ethics in qualitative research. *Journal of Nursing Scholarship, 33*(1), 93–96. DOI: 10.1111/j.1547-5069.2001.00093.x PMID: 11253591

Peattie, K., & Crane, A. (2005). Green marketing: Legend, myth, farce or prophecy? *Qualitative Market Research, 8*(4), 357–370. DOI: 10.1108/13522750510619733

Rahman, H. (2025). Blockchain-Driven Knowledge Ecosystems. In Blockchain Technology Applications in Knowledge Management (pp. 29-70). IGI Global Scientific Publishing. DOI: 10.4018/979-8-3693-3956-5.ch002

Ribble, M. (2015). *Digital citizenship in schools: Nine elements all students should know* (6th ed.). International Society for Technology in Education.

Rosário, A. T., Lopes, P. R., & Rosário, F. S. (2023). The Digital Marketing for Sustainable Development.

Sahoo, D. (2024). Chapter-5 Leveraging Digital HR and Marketing Focus for Sustainable Business Practices: A Radical Transformation. Business Management: A New Paradigm, 63.

Sherry, J. L., & Kozinets, R. V. (2008). Special issue: Consumer culture theory. *The Journal of Consumer Research, 35*(1), 151–162.

Singh, R., & Sharma, N. (2021). Green marketing strategies for healthcare organizations: Opportunities and challenges. *Marketing and Sustainability, 29*(3), 343–356.

Smith, A. D., & Brower, T. R. (2012). Longitudinal examination of website sustainability practices: Insights for green marketing. *Sustainable Business Journal, 7*(3), 155–169.

Sujatha, R., Aarthy, S. L., & Vettriselvan, R. (Eds.). (2021). *Integrating Deep Learning Algorithms to Overcome Challenges in Big Data Analytics*. CRC Press. DOI: 10.1201/9781003038450

Sweeney, L., & Johnson, J. (2019). The impact of digital media on green marketing in the healthcare sector. *Journal of Environmental Marketing, 22*(2), 115–132.

Teixeira, R. L. P., & Teixeira, C. H. S. B. (2024). Exploring sustainable human manufacturing: industry 4.0, manufacturing, and green marketing for a balanced future.

Velmurugan, P. R., Arunkumar, S., Vettriselvan, R., Deepan, A., & Rajesh, D. (2025). Strategic Approaches to Corporate Social Responsibility and Sustainable Development: Integrating Leadership, Marketing, and Finance. In Navigating Corporate Social Responsibility Through Leadership and Sustainable Entrepreneurship (pp. 373-406). IGI Global Scientific Publishing.

Venkatesh, V., & Davis, F. D. (2000). A theoretical extension of the technology acceptance model: Four longitudinal field studies. *Management Science, 46*(2), 186–204. DOI: 10.1287/mnsc.46.2.186.11926

Vettriselvan, R. (2025). Harnessing Innovation and Digital Marketing in the Era of Industry 5.0: Resilient Healthcare SMEs. In The Future of Small Business in Industry 5.0 (pp. 163-186). IGI Global Scientific Publishing.

Walton, J., Turner, P., & Simpson, G. (2020). Sustainable healthcare practices: The role of green technologies. *Journal of Sustainable Health Care, 9*(1), 65–77.

White, K., MacDonnell, R., & Ellard, J. H. (2019). Belief in sustainability: The role of consumer attitudes and behavioral intention. *Journal of Marketing*, *83*(3), 78–91.

Zhang, T., & McMullan, R. (2020). Social media and green healthcare: Consumer perceptions and influence. *Journal of Digital Healthcare Management*, *12*(2), 67–81.

KEY TERMS AND DEFINITIONS

Green Marketing: Green marketing involves promoting products or services based on their environmental benefits. This approach emphasizes features such as energy efficiency, use of sustainable materials, or reduced environmental impact to appeal to environmentally conscious consumers. The goal is to differentiate products by highlighting their eco-friendly attributes and contributing to sustainable consumption.

Digital Citizenship: Digital citizenship refers to the responsible use of technology and digital platforms. It involves ethical online behavior, respecting privacy, and engaging positively in digital environments. For healthcare organizations, digital citizenship includes transparent communication and safeguarding patient data in online interactions.

Sustainability: Sustainability is the practice of maintaining or improving environmental, social, and economic conditions over time without compromising the needs of future generations. It involves actions such as conserving resources, reducing waste, and minimizing carbon footprints. In healthcare, sustainability focuses on reducing environmental impact through initiatives like waste management and energy efficiency.

Eco-Conscious Consumer: An eco-conscious consumer prioritizes environmental considerations when making purchasing decisions. This individual seeks out products and services that are environmentally friendly and produced ethically, reflecting a commitment to reducing their personal ecological footprint. Motivations include concerns about climate change, pollution, and resource depletion.

Corporate Social Responsibility (CSR): Corporate Social Responsibility (CSR) refers to a company's commitment to operate ethically and contribute positively to society. It involves addressing the environmental, social, and economic impacts of business activities. In healthcare, CSR includes supporting local health initiatives, improving care access, and adopting sustainable practices.

Transparency in Healthcare: Transparency in healthcare means openly sharing information about medical practices, organizational operations, and patient care. It involves disclosing data on treatment outcomes, pricing, and environmental impacts, which helps build trust and accountability with patients and other stakeholders.

Environmental Preservation: Environmental preservation involves protecting and maintaining natural resources and ecosystems to prevent degradation. It includes efforts to conserve biodiversity, protect habitats, and reduce pollution. In healthcare, this can mean initiatives like waste reduction and resource conservation to support environmental health.

Digital Healthcare Marketing: Digital healthcare marketing uses digital platforms and tools to promote healthcare services and engage with patients. This includes strategies such as social media campaigns, search engine optimization, and email marketing. Effective digital healthcare marketing leverages technology to enhance patient engagement and support organizational objectives.

Chapter 12
Using Protection Motivation Theory to Encourage Healthy Cyber Hygiene Behaviours

Hee Jhee Jiow
https://orcid.org/0000-0002-1125-4036
Singapore Institute of Technology, Singapore

Adeline Tay
https://orcid.org/0009-0001-9490-2253
Singapore Institute of Technology, Singapore

ABSTRACT

In today's connected world, being a responsible digital citizen is extremely important. With our growing dependence on digital platforms for communication, business, and personal use, it is essential to practice healthy cyber hygiene habits to navigate the digital space effectively. This study highlights three key areas for maintaining basic online hygiene: using secure Wi-Fi networks, avoiding unknown links and attachments, and creating strong passwords. Based on a survey of 157 participants, the research emphasises that coping appraisal is vital in encouraging healthy cyber hygiene behaviours. Offering practical and easily doable advice through educational programs can foster more effective and sustainable online hygiene practices. In an increasingly unstable digital landscape, practicing good cyber hygiene is critical to mitigating risks of cybercrimes and their potential impacts. As digital citizens, we must stay informed and adopt safe online behaviours to navigate the digital world securely.

INTRODUCTION

The digital world has increasingly emerged as a parallel reality, standing on equal footing with the physical world we inhabit. Humans operate businesses on the internet, make electronic bank and payment transfers, communicate and facilitate a range of other activities that were once solely constricted to the physical world. While technology has revolutionised the lives of humans, it has also freighted organisations and individuals with a perennial issue; battlegrounds today are no longer constricted to physical spaces but extend to digital realms as well (Doğrul et al., 2023). The International Telecommunications Union

DOI: 10.4018/979-8-3693-6675-2.ch012

Copyright ©2025, IGI Global Scientific Publishing. Copying or distributing in print or electronic forms without written permission of IGI Global Scientific Publishing is prohibited.

(2008) defines cybersecurity as the collection of tools, policies, security concepts, security safeguards, guidelines, risk management approaches, actions, training, best practices, assurance and technologies that can be used to protect the cyber environment and organisation and user's assets. And the National Initiative for Cybersecurity Careers and Studies (2024) defines cybersecurity as the activity or process, ability or capability, or state whereby information and communication systems—and the data contained within—are protected from, or defended against, damage, unauthorised use or modification, or exploitation. Contemporarily, cybersecurity is garnering more attention in light of the growing ubiquity of cyber threats (Arora, 2016; Zhang & Daim, 2023). Ponemon Institute's (2016) Second Annual Cost of Cyber Crime Study unveiled that the average cost of cyberattacks for organisations varies significantly across countries: $17.36 million in the United States, $8.39 million in Japan, $7.84 million in Germany, $7.21 million in the United Kingdom, $5.27 million in Brazil, and $4.3 million in Australia. Fleck (2024) stated that the global cost of cybercrime is expected to rise exponentially from $9.22 trillion in 2024 to $13.82 trillion by 2028. As such, the repercussions of cybercrime can be cataclysmic, making a fortified cybersecurity system more imperative than ever.

No country is immune to cybercrimes, including Singapore, the city-state that has been plagued by a slew of cybercrimes in the yesteryears. Lee & Ng (2023) reported that the overall attack rate in Singapore exceeded the Asia-Pacific average. According to the International Trade Administration (2023), there were nearly 2 million attacks in Singapore in Q2 2022, with ransomware and data theft emerging as the two most prevalent attacks. The financial impact of these attacks can be astronomical, with the average cost of a cybersecurity breach at approximately SGD 1.7 million – the highest in Asia-Pacific. Moreover, at least eight in 10 organisations in Singapore encountered a cybersecurity incident each year (Cyber Security Agency of Singapore, 2023). Of those affected, at least 40% of companies reported that they faced business disruption, reputational damage or data loss due to a cybersecurity incident.

Nonetheless, organisations are not the only entities perniciously impacted by cyberattacks (Cain et al., 2018); individuals are equally susceptible to them. Over the last several years, the personal data of individuals in Singapore was incessantly exploited in a series of data breaches. In 2018, hackers stole the personal particulars of 1.5 million patients in Singapore (Toh, 2018). In 2020, Singapore's homegrown online cashback portal ShopBack was fined S$74,400 following a data leak that affected 1.4 million of its customers (Tang, 2023). In 2022, the personal data of 332,774 Starbucks Singapore customers was put up for sale on a dark web forum (Tang, 2023). Most recently in 2023, the personal data of 665,000 Marina Bay Sands customers was accessed in a data security breach (Hamzah, 2023).

In an increasingly capricious digital climate, it is pivotal that individuals practise exemplary cyber hygiene to inoculate themselves against cybercrimes and the potential pitfalls they may herald on themselves or the companies they are associated with. Kemper (2019) shared that 90% of cybersecurity incidents in organisations are related to human error, corroborating with other studies that identified human factors as a significant contributor to cyber risks. These risks often manifest from the creation and deployment of malicious software, increased use of social engineering by attackers, and a lack of protective behaviours, such as password encryption or the use of antivirus software (Cains et al., 2022). Whether due to intentional or inadvertent actions, many employees are often the root cause of successful cyberattacks (Ponemon Institute & Accenture, 2019), leading to data breaches where attackers can exploit corporate data (Jartelius, 2020).

As such, this study adopts the Protection Motivation Theory to investigate the strategic factors that encourages people to practice safe and hygienic cyber behaviours. The study will first scope out the three domains of cyber behaviours (which are basic) that will be examined and then followed by describing the theory that has been adopted. The research method, findings & discussion and conclusion will ensue.

Three Domains

Adopting Strong Passwords

Password authentication continues to reign as the most prevalently used security mechanism (Grawemeyer & Johnson, 2011), even though many other sophisticated and viable security alternatives have emerged from research and development (Bryant & Campbell, 2006). Despite the looming threat of cybercrimes, studies found that using weak passwords and re-using passwords for multiple accounts is commonplace (Florencio & Herley, 2007; Gaw & Felten, 2006; Grawemeyer & Johnson, 2011; Stobert & Biddle, 2018). This behaviour is not exclusive to individuals who bear little cognisance of password security; instead, studies have uncovered that those who are knowledgeable about password security will, nonetheless, continue to use weak passwords and/or re-use passwords in their daily lives (Riley, 2006; Notoatmodjo & Thomborson, 2009). Shay et al. (2010) and Cain et al. (2018) corroborated this finding, revealing that 80% and 50% of respondents, respectively, reused the same password for multiple accounts.

Such lax security behaviours are engendered by users prioritising the ease of recalling passwords over robust security, leading to the creation of simple, easily memorable passwords rather than secure ones. Grawemeyer & Johnson (2011) noted that many users struggle to balance password memorability with security. According to Bryant & Campbell (2006), robust passwords normally encompassed password lengths of at least 8 characters as well as passwords with mixed case/symbols and non-dictionary words. Given the ostensibly complex requirements for strong passwords—coupled with the myriads of commerce, public access systems and social media accounts that require logins with passwords—it is arduous for individuals to create passwords that are both secure and easy to remember for each account (Chaudhary et al., 2019).

While tools like password managers—a software application that helps a user store and organise passwords (Chaudhary et al., 2019)—are readily available, their adoption is curtailed due to user concerns about both their usability and security. As such, a plethora of users still prefer to depend on their memory to recall and manage their passwords (Madrekar & Jambhale, 2024), often leading to the creation of simple and easily guessable passwords. Additionally, employees exacerbate this issue by displaying their passwords on post-it notes near their office computers, despite warnings from IT departments and directives from senior management (McAlaney & Benson, 2020). Doing so can herald egregious outcomes where their passwords are found by others and used maliciously, consequently debilitating the company's cybersecurity.

Nonetheless, there are pre-emptive measures that individuals can espouse to bolster their password security. For a password to be strong, it has to contain at least 12 characters, and comprise uppercase letters, lowercase letters, numbers or symbols. It is best that it does not contain personal information and easily guessable patterns. Moreover, it is advised that users abstain from using the same password for multiple accounts, as one cyber incident can jeopardise multiple accounts concurrently. Users are also advised to enable a Two-Factor Authentication (2FA) to secure their accounts wherever applicable (Cyber Security Agency of Singapore, 2019).

Avoid Connecting to Unsecure WiFi Networks

In today's world where individuals are in thrall to their devices and possess an implacable desire for constant connectivity, it is unsurprising that they forage for a strong WiFi network wherever they go. According to Norton (2017), more than 55% of consumers globally "wouldn't think twice about exchanging, sharing or even doing something to get a strong WiFi signal". However, public WiFi networks often lack user authentication, rendering them highly insecure, particularly for sensitive data exchange (Wang et al., 2021). Since traffic in public WiFi networks typically passes through a single path, the operator monitoring the WiFi or the captive portal can easily intercept and retrieve users' private information—such as credit card numbers and passwords—and potentially misuse it (Dabrowski et al., 2016). Moreover, attackers can connect to public WiFi networks and leverage network sniffers to intercept users' traffic (Sombatruang et al., 2018).

Despite the inherent dangers, individuals continue to be insouciant in connecting to public WiFi networks, on accounts that public WiFi networks provide the convenience of internet access in various locations such as libraries, airports, cafes, and more (Lotfy et al., 2021; Sombatruang et al., 2018; Sombatruang et al., 2016). This ease of access that public WiFi networks afford to users often supersedes the potential security risks, as numerous individuals prioritise connectivity and immediate access to online resources (Sombatruang et al., 2018). Moreover, some individuals perceive the likelihood of being targeted or affected by a security breach on public WiFi as relatively low (Lotfy et al., 2021; Sombatruang et al., 2016), or they fail to fully grasp the seriousness of their private data being compromised (Sangeen et al., 2023). Users connecting to public WiFi networks are frequently oblivious to the potential risks their data could be exposed to. In most cases, users only realise their private information has been compromised when it is too late to rectify the situation (Sangeen et al., 2023). Interestingly, Maimon et al. (2022) found that individuals who possessed more proficiency with computers reportedly used public WiFi significantly less (50%), likely due to their cognisance about the risks associated with connecting to these networks.

To attenuate the general public's susceptibility to cybercrimes, the Cyber Security Agency of Singapore (n.d.) has enumerated numerous tips for individuals to improve their cyber hygiene – chief among them include advising individuals to refrain from performing online transactions over unsecured WiFi. In a similar vein, Conti et al. (2016) cautioned that users ought to eschew online banking and accessing sensitive data over public WiFi networks, even if these websites are encrypted.

Avoid Opening Links or Attachments from Unknown Senders or Unknown Websites

Phishing – a form of social engineering attack – entails scamming a user into revealing sensitive information such as bank details and credit card numbers, by sending malicious links to the users to direct them to false websites (Alkhalil et al., 2021). Phishing is widely regarded as one of the most pressing cybersecurity threats that affect all internet users – regardless of their technical expertise or level of caution (Althobaiti, 2021) – as cyber criminals have the capacity to engineer these attacks more adroitly, leading to significant losses for the victims.

A survey of working adults and IT professionals worldwide unveiled that bulk phishing attacks were the most prevalent cyber incidents experienced by 85% of organisations in 2022 (Petrosyan, 2024). As phishers deftly engineer their attacks by exploiting humans' psychological triggers, individuals are more vulnerable to phishing scams than they realise. For instance, Furnell (2007) conveyed that individuals are more apt to click on a link within an email when they see authority cues. In a similar vein, Butavicius et al. (2015) expounded that university students lacked the capacity to distinguish between sham and genuine emails, as spear-phishing emails often espoused an authority-based social engineering approach. Students received spam emails from a sender who possessed some level of authority at their school, thereby prompting them to regard the deceptive emails as genuine. This finding aligns with Broadhurst's et al. (2019) research where cyber attackers adroitly craft spam emails that resemble communications from legitimate sources.

Cybercrimes committed through spam emails are considered one of the most prevalently used approaches for distributing malware and executing other established cybercrimes, such as stealing personal information (Alazab & Broadhurst, 2017). Spam can comprise meaningless advertisements through unwanted emails, messages, or social media communications. However, they can also contain malware or malicious bots/viruses that have been created to access the recipient's confidential information from the receivers (Broadhurst et al., 2019). For instance, spam emails may include deceptive attachment buttons that, when clicked, could either install malware such as a virus or worm on the computer, or steal the user's credentials by redirecting them to a fraudulent login page (Alkhalil et al., 2021). Users are highly susceptible to spam emails; Cain et al.'s (2018) study found that 96% of participants clicked on embedded links from unknown senders, and 98% downloaded attachments from unfamiliar senders.

As today's world is rife with calamities, phishers exploit these situations to gain an edge over their victims. For example, individuals were inveigled into opening attachments that falsely claimed to contain information about local Coronavirus cases (Alkhalil et al., 2021). Moreover, phishing attacks have reached unprecedented levels in light of emerging technologies such as mobile and social media (Marforio et al., 2015). Facebook, WhatsApp and Instagram—platforms under social media giant Meta—continue to be the primary avenues for scammers to contact their victims. In 2023, S$14.2 million was lost to phishing scams in Singapore (Ong, 2024).

To safeguard individuals from phishing attacks, The Federal Trade Commission (2022) has doled out several recommendations for users, namely to: protect their devices by leveraging updated security softwares, protect their accounts by using multi-factor authentication, as well as protect their data by backing it up regularly. On the other hand, The Cyber Security Agency of Singapore (n.d.) advised individuals to "think before [they] click on links provided in unsolicited emails and text messages", as well as "always check the authenticity of the email" by calling or checking with the official sources.

LITERATURE REVIEW

Protection Motivation Theory

Protection Motivation Theory (PMT) is a psychological framework that explains how people cope with threats and how motivated they are to protect themselves from such threats. PMT was originally developed by Rogers (1975) to account for the effects of fear appeals on health behaviour (such as disease prevention and health promotion). PMT has been used to explain and predict various health-related behaviours, such as smoking cessation, breast self-examination, condom use, and vaccination.

For example, Floyd et al. (2000) conducted a meta-analysis of 65 relevant studies (with N= approximately 30,000) that covered 20 health issues. In another study, Milne et al. (2000) conducted a meta-analysis of 27 studies that used PMT to design and evaluate interventions for promoting various health behaviours, such as breast self-examination, condom use, tooth-flossing, testicular self-examination, smoking cessation, and alcohol reduction. They found that PMT-based interventions had significant effects on increasing the perceived severity and vulnerability of the health threats, the response efficacy and self-efficacy of the preventive actions, and the intention and behaviour of performing the preventive actions.

PMT is widely used in intervention studies seeking to promote the adoption of hygienic cyber behaviours (see below for the various examples). Moreover, as pointed out earlier in the chapter, human error accounts for a large part of cyber security instances, and addressing the human behaviour then becomes strategic. As such, this study adopted PMT as the theoretical framework to shape hygienic cyber behaviour.

According to PMT, when people encounter a threatening situation, they engage in two types of cognitive appraisal: threat appraisal and coping appraisal. Threat appraisal involves assessing the severity of the threat, as well as the perceived vulnerability to the threat. Coping appraisal involves assessing the available options to cope with the threat, as well as their efficacy and costs. These appraisals lead to the formation of protection motivation, which is the intention to adopt a protective behaviour to reduce or avoid the threat. Protection motivation can be influenced by various factors, such as personality, emotions, social norms, and environmental cues. Lately, PMT has been adopted for use in areas beyond health care. For example, Mou et al.'s (2022) research, based on data sourced from Scopus and Engineering Village, found 164 identified records that satisfies the search parameters using the words "information security" OR "security policy" OR "policy compliance".

Studies involving college student samples have demonstrated that coping self-efficacy and response efficacy, components of coping appraisals, were positively linked with online safety practices (LaRose et al., 2007; Lee et al., 2008) as well as the implementation of security-related software or policies (Herath & Rao, 2009; Ifinedo, 2012; Lee & Larsen, 2009). In the field of online safety, Lee et al. (2008) applied the PMT to the usage of virus protection software and revealed that perceived vulnerability to a threat and both response and coping self-efficacies were significant predictors of the intention to use virus protection. Conversely, perceived severity and response costs did not influence safety intentions. Coping assessment variables, especially coping self-efficacy, were strong indicators of the intention to perform these actions, whereas threat assessment variables were not.

Many websites that provide online safety information often use threat appraisal as a common strategy (Shillair et al., 2014), highlighting susceptibility to online threats (e.g., by indicating that file sharing can lead to virus infections) and their potential severity (e.g., raising concerns about identity theft).

Media reports on online safety issues frequently emphasise threats, often including cautionary tales and warnings that these dangers can affect anyone. Threat-based appeals (e.g., stressing the risks of spyware to elevate perceived threat levels) have been utilized to increase awareness about online security threats (Anderson & Agarwal, 2010; Johnston & Warkentin, 2010; Marett et al., 2011), although with varying degrees of success. Davinson & Sillence (2010) discovered that personalised risk messages boosted participants' intentions to adopt secure behaviours. Awareness of the risks associated with spyware was a strong predictor of intentions to undertake protective actions (Hu & Dinev, 2005). Milne et al. (2009) also found a positive correlation between the perceived likelihood of threats and the adoption of protective security measures (e.g., installing antivirus software) as well as reducing risky behaviours (e.g., not saving passwords). Conversely, Rhee et al. (2009) discovered that merely enumerating the negative outcomes of ignoring security threats had a minimal effect on adopting security practices like password protection. Gurung et al. (2009) noted an influence for severity but found no significant correlation with susceptibility. In general, threats often do not effectively motivate secure behaviours or exhibit only marginal impacts in PMT research (Floyd et al., 2000).

Interventions for online safety information leverage coping appraisals by highlighting the ease and effectiveness of implementing protections. Empirical evidence indicates that self-efficacy is a crucial predictor of the intention to implement online security measures (LaRose et al., 2005a; Lee et al., 2008). Stafford & Poston (2010) discovered that self-efficacy mediated the link between perceptions of a security threat and individuals' motivation to safeguard themselves. Milne et al. (2009) noted that people with high self-efficacy were more prone to engage in protective behaviours (e.g., scanning computers for spyware, deleting information after browsing) and less likely to undertake risky actions such as saving credit card details or setting low browser safety settings. Similarly, Wang (2013) found a positive correlation between users' technical knowledge about online phishing dangers and their intention to adopt appropriate responses. Hu & Dinev (2005) found that perceived controllability and ease of use of online safety solutions, both related to self-efficacy (Ajzen, 1991), predicted intentions to implement protections like anti-spyware programs.

However, the costs associated with adaptive coping behaviours and forgoing the potential rewards of unsafe behaviour pose significant barriers to online safety. Ensuring protection in the digital space often involves investing time and money to acquire and maintain protective software, which must be regularly updated and patched, all while risking that these very protections may cause disruptions (Rifon et al., 2005). Response costs have been found to negatively impact intentions to use anti-spyware software (Chenoweth et al., 2009) and to diminish motivation regarding online safety behaviours (LeFebvre, 2012). Additionally, unsafe behaviours frequently offer immediate rewards, such as free access to the latest movies through file sharing. These rewards are associated with dismissing preventive measures (Howe et al., 2012). Consequently, the refusal to adhere to security advice might be a calculated decision for some individuals when weighing the costs and benefits of security measures (Herley, 2009).

While the above section discusses how PMT has been used for the cyber hygiene domain, this chapter, and specifically the following section will zoom in on 3 domains of cyber hygiene behaviours, namely the adoption of strong passwords, connecting to unsecure (public) WiFi networks, and opening suspicious links and attachments.

In Tsai et al.'s (2016) study, it was found that for threat appraisals, it is the perceived severity of online threats, rather than susceptibility to those threats, that predicts security intentions. Tsai et al.'s (2016) study posture security intentions very generally, but it is aligned to our study's concepts. Examples of common security intentions are "I am likely to take security measures to protect the Internet", "I will change my passwords more often", "I will use passwords that are harder to guess", "I will run protective software regularly to remove spyware from my computer", and "I will update my protective software regularly". However, response efficacy was also found to be a significant predictor. Among coping appraisal variables, response costs, coping self-efficacy, and perceived security support negatively predicted safety intentions, while response efficacy positively predicted them. People who believe in the effectiveness of protective software for detecting malware show greater intention to utilise them. Contrary to PMT expectations, threat severity and coping self-efficacy, were negative predictors in their regression analysis. One plausible explanation, offered by the authors, is that individuals more knowledgeable about online security understand the severity of threats better and have higher coping self-efficacy and perceived security support, yet they may exhibit lower future intentions to take additional security measures due to their confidence in their existing ability to protect themselves online.

Other studies have shown that self-efficacy may either directly or indirectly reduce the tendency to open commercial emails or attachments (Chen et al., 2011; Ng et al., 2009). Moreover, response efficacy was positively connected to the adoption of email authentication measures (Herath et al., 2012). Both response efficacy and coping self-efficacy have been identified as positive predictors for the intention to adopt spyware protection (Johnston & Warkentin, 2010) and fake-website detectors (Zahedi et al., 2015).

In terms of setting strong passwords, Jiow et al. (2021) found their Password Hygiene Training Programme effectively induced positive changes in three PMT factors, with participants demonstrating higher perceived vulnerability, greater perceived effectiveness, and reduced perceived cost following the e-module programme. Considering the significant increase in the time required and number of attempts needed to crack participants' set passwords post-programme found in their study, it is logical to conclude that adolescents are more inclined to follow better password hygiene when they view themselves as more vulnerable to password-related threats and believe that the recommended password strategies are straightforward to implement and effective in mitigating those threats

In Mwagwabi and Jiow's (2021) study, they found, by applying an extended form of PMT (Rogers, 1983), that even when teenage computer users recognise their vulnerability to hacking or the potential harm it could cause, it doesn't influence their choice of passwords. Teenagers' security behaviours are also driven by other motives beyond PMT. Personal norms completely mediate the link between perceived threat severity and intentions to comply, indicating that perceived severity alone isn't adequate to prompt compliance. Teenagers need to feel a sense of obligation to follow through. Though personal norms may encourage compliance, fear of embarrassment or shame if their social media accounts are hacked promotes compliance. Conversely, peer pressure, such as the fear of being mocked if someone hacks their account, discourages compliance. Their study enhances the understanding of early security practices and underscores possible variations between adult and teenage behaviours for future research consideration. For instance, their findings imply that password security guidelines alone may not ensure compliance among teenagers; they might require enforced password rules at the authentication level to prevent any breaches.

Given the diverse conclusions from numerous studies on cyber hygiene regarding which factors most influence intended behaviour, this research will adopt an exploratory approach and will refrain from proposing any hypotheses, in order to achieve the study's objectives of examining factors that motivates a person to practice hygienic cyber behaviours.

RESEARCH METHODOLOGY

An online survey mechanism was developed to capture respondents threat appraisal, coping appraisal and behavioural intentions for the 3 cyber hygiene domains. Data collection was conducted by Kaspersky in accordance with their company policies on research and data protection. The following section elaborates on the research method used.

Instrument

Before conducting the survey, the researchers developed the instruments that would measure the variables of interest. The instruments were carefully developed through consultations with Kaspersky, a global cybersecurity company that provided expertise on cyber hygiene and cyber resilience. Kaspersky also helped to validate the content and design of the instruments, ensuring that they were relevant, clear, and appropriate for the target population.

The instruments consisted of two main parts: the first part collected demographic information about the respondents, such as their age, gender and country; the second part measured the constructs of the Protection Motivation Theory framework, as well as the behavioural intention of performing cyber hygiene practices. The second part largely used a seven-point Likert scale, ranging from 1 (strongly disagree) to 5 (strongly agree), to assess the respondents' agreement with various statements related to the variables.

The instruments were pilot-tested with a small sample of respondents from different countries to check for any errors, ambiguities, or difficulties in understanding the questions. The feedback from the pilot test was used to revise (which resulted in the deletion of Perceived Vulnerability for the domain of Avoid Connecting to Unsecure WiFi Networks) and improve the instruments before launching the survey.

The sections below show the final questions and the scoring system.

Demographic

Demographical data consist of gender (male or female), age (free whole number response option) and country (free response option).

Adopting Strong Passwords

Table 1. Behavioural intention:

I intend to choose a password with a combination of numbers, letters, and symbols	7-point Always to Never scale
I intend to follow recommended security measures	
I intend to protect my digital device(s) from hackers	

Table 2. Perceived vulnerability (Threat appraisal):

It is likely that someone could guess my password	7-point Strongly Agree to Strongly Disagree scale
There is a chance that someone could crack my password	
It is likely that my digital device(s) would be attacked in the future	

Table 3. Perceived severity (Threat appraisal):

The consequences of a password breach would be	7-point Not at all Severe to Extremely Severe scale
Having my digital device(s) accessed by someone without my knowledge would be	
If my personal identifiable information was stolen from my digital device(s) the consequences would be	

Table 4. Perceived response efficacy (Coping appraisal):

Making sure that my passwords contain a combination of numbers, letters and symbols will make it more difficult to guess/crack.	7-point Strongly Agree to Strongly Disagree scale
Making sure that my passwords do not contain any dictionary words will make my passwords difficult to guess/crack	
Making sure that my passwords do not contain any personal information such as my date of birth will make my passwords difficult to guess/crack	

Table 5. Perceived self efficacy (Coping appraisal):

I would be able to use strong passwords if I had a lot of time	7-point Strongly Agree to Strongly Disagree scale
I would be able to remember strong passwords if I had instructions on how to create memorable passwords	
I can protect my digital device(s) without much effort	
For me, creating strong passwords are	7-point Extremely Easy to Extremely Difficult scale

Table 6. Response cost (Coping appraisal):

Strong passwords are difficult to remember	7-point Strongly Agree to Strongly Disagree scale
Strong passwords take too much time to type	
Strong passwords take too much effort to create	
If I change my passwords regularly, it would be difficult for me to remember them	

Avoid Connecting to Unsecure WiFi Networks

Table 7. Behavioural intention:

I will connect to a secure WiFi network	7-point Always to Never scale
I will not connect to any WiFi network if I cannot find a secure one	
I would rather use up my mobile data than take the risk of accessing free public WiFi	

Perceived Vulnerability (Threat Appraisal): Not captured. This was further elaborated in the Instrument section, and will be highlighted in the Limitations.

Table 8. Perceived severity (Threat appraisal):

People can steal personal identifiable information if they have access to my digital devices via my connection with unsecured WiFi networks	7-point Strongly Agree to Strongly Disagree scale
People can perform operations on my digital device(s) if they have access to my digital devices via my connection with unsecured WiFi networks	

Table 9. Perceived response efficacy (Coping appraisal):

Use of secured WiFi networks will stop cyber intrusions into my digital device(s)	7-point Strongly Agree to Strongly Disagree scale

Table 10. Perceived self efficacy (Coping appraisal):

I am confident in identifying which WiFi network is secure	7-point Strongly Agree to Strongly Disagree scale

Table 11. Response cost (Coping appraisal):

Connecting to a secure WiFi network is not complicated	7-point Strongly Agree to Strongly Disagree scale

Avoid Opening Links or Attachments from Unknown Senders or Unknown Websites

Table 12. Behavioural intention:

I scan all my emails sources	7-point Always to Never scale
I delete suspicious emails and its attachments after reporting them to the IT administrator for suspected phishing	
I check all the websites I visit	

Table 13. Perceived vulnerability (Threat appraisal):

It is likely that the links (from unknown senders) I receive are suspicious	7-point Strongly Agree to Strongly Disagree scale
It is likely that the attachments (from unknown senders) I receive are suspicious	
It is likely that the links found on unknown websites are suspicious	

Table 14. Perceived severity (Threat appraisal):

I could introduce viruses and/or malware into my digital device(s) if I open links/attachments from unknown sources	7-point Strongly Agree to Strongly Disagree scale
I could propagate viruses and/or malware to my contacts on my digital device(s) if I open links/attachments from unknown sources	

Table 15. Perceived response efficacy (Coping appraisal):

Not opening links/attachments from unknown sources will prevent all cyber intrusions into my digital device(s)	7-point Strongly Agree to Strongly Disagree scale

Table 16. Perceived self efficacy (Coping appraisal):

I am confident in identifying suspicious sources	7-point Strongly Agree to Strongly Disagree scale
I am capable in dealing with suspicious emails (ie. Report to IT administrator for suspected phishing)	

Table 17. Response cost (Coping appraisal):

It is not time consuming to identify suspicious sources	7-point Strongly Agree to Strongly Disagree scale
It is easy to deal with suspicious emails	

Sampling and Data Collection

An online survey mechanism was deployed by Kaspersky. The survey was distributed as an online questionnaire via email to those registered for Kaspersky Academy's Online Cyber Resilience Workshop for Educators in Asia-Pacific, as well as its face-to-face event in Metro Manila, Philippines. Additionally, it was shared among the network of teachers associated with the Academy's partner, CyberPeace Foundation. The survey ran over a period of five weeks, collecting responses, via convenient sampling, from July through August 2023. The participants included educators, administrative staff, and teaching staff from these three regions.

A total of 157 clean data points were collected and used for this study. This study's sample size is consistent with or larger than those in similar studies (Davinson & Sillence, 2010; Ifinedo, 2012).

Analysis

The study used a multiple regression analysis, which is a statistical technique that allows the examination of the relationship between one dependent variable and several independent variables. In this case, the dependent variable was the behavioural intention, which was calculated based on the responses to questions about the frequency of performing various cyber hygiene practices. The independent variables were the various domains of the Protection Motivation Theory framework, namely Threat Appraisal (perceived severity and perceived vulnerability) and Coping Appraisal (perceived response efficacy, perceived self efficacy and response cost).

The multiple regression analysis enabled the researchers to identify which independent variables had a significant effect on the behavioural intention, and how much variance they explained in the dependent variable. The analysis also provided information about the direction and magnitude of the effects, indicating whether the independent variables had a positive or negative association with the behavioural intention, and how strong this association was.

Similar to Tsai et al.'s (2016) study, reliability testing was not performed.

Findings

The sample (n=157) consisted of 44 males (28%) and 113 females (72%). This section will report the findings by the 3 domain areas. Findings for consolidated behavioural intention (BI) will be highlighted first, broken down into consolidated variables for threat and coping appraisal, their sub-concepts, and then individual variables, followed by individual behavioural intentions. Only the strongest few predictors will be presented. All results are statistically significant ($p < 0.05$).

Adopting Strong Passwords

Table 18. Regression results for "Adoption of Strong Passwords".

Model Statistics	Dependent Variable(s)	Independent Variable(s)
A significant regression was found where $F(2,154) = 28.712$, $p < 0.001$. The R^2 was 0.272 indicating that the independent variables explained approximately **27.2%** of the variance in the dependent variable.	Consolidated BI	**Coping Appraisal** is the strongest predictor with $\beta = 0.497$, $p<0.001$.
A significant regression was found where $F(5,151) = 18.322$, $p < 0.001$. The R^2 was 0.378 indicating that the independent variables explained approximately **37.8%** of the variance in the dependent variable.	Consolidated BI	**Perceived Response Efficacy** is the strongest predictor with $\beta = 0.309$, $p<0.001$.
A significant regression was found where $F(17,139) = 7.750$, $p < 0.001$. The R^2 was 0.487 indicating that the independent variables explained approximately **48.7%** of the variance in the dependent variable.	Consolidated BI	"For me, creating passwords are easy" is the strongest predictor $\beta = 0.174$, $p<0.01$.
A significant regression was found where $F(5,151) = 9.415$, $p < 0.001$. The R^2 was 0.238 indicating that the independent variables explained approximately **23.8%** of the variance in the dependent variable.	I intend to choose a password with a combination of numbers, letters, and symbols	**Perceived Self Efficacy** is the strongest predictor with $\beta = 0.284$, $p<0.01$.
A significant regression was found where $F(17,139) = 4.917$, $p < 0.001$. The R^2 was 0.376 indicating that the independent variables explained approximately **37.6%** of the variance in the dependent variable.	I intend to choose a password with a combination of numbers, letters, and symbols	"I would be able to remember strong passwords if I had instructions on how to create memorable passwords" is the strongest predictor with $\beta = 0.212$, $p<0.01$.
A significant regression was found where $F(5,151) = 16.460$, $p < 0.001$. The R^2 was 0.353 indicating that the independent variables explained approximately **35.3%** of the variance in the dependent variable.	I intend to follow recommended security measures	**Perceived Response Efficacy** is the strongest predictor with $\beta = 0.361$, $p<0.001$.
A significant regression was found where $F(17,139) = 6.544$, $p < 0.001$. The R^2 was 0.445 indicating that the independent variables explained approximately **44.5%** of the variance in the dependent variable.	I intend to follow recommended security measures	"I would be able to remember strong passwords if I had instructions on how to create memorable passwords" is the strongest predictor with $\beta = 0.181$, $p<0.05$.
A significant regression was found where $F(5,151) = 13.350$, $p < 0.001$. The R^2 was 0.307 indicating that the independent variables explained approximately **30.7%** of the variance in the dependent variable.	I intend to protect my digital device(s) from hackers	**Perceived Response Efficacy** is the strongest predictor with $\beta = 0.382$, $p<0.001$.
A significant regression was found where $F(17,139) = 5.810$, $p < 0.001$. The R^2 was 0.415 indicating that the independent variables explained approximately **41.5%** of the variance in the dependent variable.	I intend to protect my digital device(s) from hackers	"Making sure that my passwords do not contain any personal information such as my date of birth will make my passwords difficult to guess/crack" is the strongest predictor with $\beta = 0.221$, $p<0.05$.

Avoid Connecting to Unsecure WiFi Networks

Table 19. Regression results for "Avoid Connecting to Unsecure WiFi Networks".

Model Statistics	Dependent Variable(s)	Independent Variable(s)
A significant regression was found where $F(2,154) = 48.906$, $p < 0.001$. The R^2 was 0.388 indicating that the independent variables explained approximately **38.8%** of the variance in the dependent variable.	Consolidated BI	**Coping Appraisal** is the strongest predictor with $\beta = 0.496$, $p<0.001$.
A significant regression was found where $F(5,151) = 21.764$, $p < 0.001$. The R^2 was 0.419 indicating that the independent variables explained approximately **41.9%** of the variance in the dependent variable.	Consolidated BI	**Perceived Response Efficacy** is the strongest predictor with $\beta = 0.375$, $p<0.001$.
A significant regression was found where $F(6,150) = 18.106$, $p < 0.001$. The R^2 was 0.420 indicating that the independent variables explained approximately **42.0%** of the variance in the dependent variable.	Consolidated BI	"Use of secured WiFi networks will stop cyber intrusions into my digital device(s)" is the strongest predictor $\beta = 0.385$, $p<0.001$.
A significant regression was found where $F(5,151) = 15.794$, $p < 0.001$. The R^2 was 0.343 indicating that the independent variables explained approximately **34.3%** of the variance in the dependent variable.	I will connect to a secure WiFi network	**Perceived Severity** is the strongest predictor with $\beta = 0.285$, $p<0.001$.
A significant regression was found where $F(6,150) = 13.465$, $p < 0.001$. The R^2 was 0.350 indicating that the independent variables explained approximately **35.0%** of the variance in the dependent variable.	I will connect to a secure WiFi network	"People can perform operations on my digital device(s) if they have access to my digital devices via my connection with unsecured WiFi networks" is the strongest predictor with $\beta = 0.289$, $p<0.05$.
A significant regression was found where $F(5,151) = 17.633$, $p < 0.001$. The R^2 was 0.369 indicating that the independent variables explained approximately **36.9%** of the variance in the dependent variable.	I would rather use up my mobile data than take the risk of accessing free public WiFi	**Perceived Response Efficacy** is the strongest predictor with $\beta = 0.427$, $p<0.001$.
A significant regression was found where $F(6,150) = 14.606$, $p < 0.001$. The R^2 was 0.369 indicating that the independent variables explained approximately **36.9%** of the variance in the dependent variable.	I would rather use up my mobile data than take the risk of accessing free public WiFi	"Use of secured WiFi networks will stop cyber intrusions into my digital device(s)" is the strongest predictor with $\beta = 0.424$, $p<0.001$.

Avoid Opening Links or Attachments from Unknown Senders or Unknown Websites

Table 20. Regression results for "Avoid Opening Links or Attachments from Unknown Senders or Unknown Websites".

Model Statistics	Dependent Variable(s)	Independent Variable(s)
A significant regression was found where $F(2,154) = 49.983$, $p < 0.001$. The R^2 was 0.394 indicating that the independent variables explained approximately **39.4%** of the variance in the dependent variable.	Consolidated BI	**Coping Appraisal** is the strongest predictor with $\beta = 0.655$, $p<0.001$.
A significant regression was found where $F(5,151) = 26.420$, $p < 0.001$. The R^2 was 0.467 indicating that the independent variables explained approximately **46.7%** of the variance in the dependent variable.	Consolidated BI	**Perceived Self Efficacy** is the strongest predictor with $\beta = 0.520$, $p<0.001$.
A significant regression was found where $F(10,146) = 14.027$, $p < 0.001$. The R^2 was 0.490 indicating that the independent variables explained approximately **49.0%** of the variance in the dependent variable.	Consolidated BI	"I am confident in identifying suspicious sources" is the strongest predictor $\beta = 0.389$, $p<0.001$.
A significant regression was found where $F(5,151) = 18.637$, $p < 0.001$. The R^2 was 0.382 indicating that the independent variables explained approximately **38.2%** of the variance in the dependent variable.	I scan all my emails sources	**Perceived Self Efficacy** is the strongest predictor with $\beta = 0.549$, $p<0.001$.
A significant regression was found where $F(10,146) = 12.538$, $p < 0.001$. The R^2 was 0.462 indicating that the independent variables explained approximately **46.2%** of the variance in the dependent variable.	I scan all my emails sources	"I am confident in identifying suspicious sources" is the strongest predictor with $\beta = 0.520$, $p<0.001$.
A significant regression was found where $F(5,151) = 15.769$, $p < 0.001$. The R^2 was 0.343 indicating that the independent variables explained approximately **34.3%** of the variance in the dependent variable.	I delete suspicious emails and its attachments after reporting them to the IT administrator for suspected phishing	**Perceived Response Efficacy** is the strongest predictor with $\beta = 0.393$, $p<0.001$.
A significant regression was found where $F(10,146) = 8.046$, $p < 0.001$. The R^2 was 0.355 indicating that the independent variables explained approximately **35.5%** of the variance in the dependent variable.	I delete suspicious emails and its attachments after reporting them to the IT administrator for suspected phishing	"Not opening links/attachments from unknown sources will prevent all cyber intrusions into my digital device(s)" is the strongest predictor with $\beta = 0.365$, $p<0.001$.
A significant regression was found where $F(5,151) = 18.214$, $p < 0.001$. The R^2 was 0.376 indicating that the independent variables explained approximately **37.6%** of the variance in the dependent variable.	I check all the websites I visit	**Perceived Self Efficacy** is the strongest predictor with $\beta = 0.468$, $p<0.001$.
A significant regression was found where $F(10,146) = 10.263$, $p < 0.001$. The R^2 was 0.413 indicating that the independent variables explained approximately **41.3%** of the variance in the dependent variable.	I check all the websites I visit	"I am capable in dealing with suspicious emails (ie. Report to the IT administrator for suspected phishing" is the strongest predictor with $\beta = 0.351$, $p<0.01$.

DISCUSSION AND IMPLICATIONS

The findings show that manipulating the coping appraisal would have the largest influence on healthy cyber hygiene behavioural motivation for the 3 domains, namely avoid connecting to unsecure WiFi networks, avoid opening links or attachments from unknown senders or unknown websites, and adopting strong passwords. Coping appraisal involves assessing the effectiveness of a behavioural response to a threat, the difficulty in implementing the response, and the associated costs. However, there are some nuanced findings which will be elaborated further below, together with its implications.

In the domain of adopting strong passwords, being provided with instructions on how to create memorable passwords that are strong, would greatly motivate a person to follow the recommended security measures and choosing a password with a difficult combination of numbers, letters and symbols. This has valuable educational implications for educators, school administrators, parents, guardians and advocates of responsible digital citizenry. Instead of focussing on the threat appraisal – teaching and warning people of the possible dangers (perceived severity) associated with a weak password and/or the susceptibility (perceived vulnerability) of being hacked because of the weak password – one should focus educational efforts on how to set a strong password that is easy to remember or recall, as fear-based messaging have often proved insufficient to change behaviour (Ahluwalia, 2000). For example, one of the authors' experience was to teach teenagers to choose a favourite sentence or phrase in their favourite song, and then change some of the letters in that sentence or phrase into numbers and symbols. This was done to make it easier for them to remember their passwords. One should not despise the old-school method of writing down passwords in a secure physical notebook (of course, one needs to ensure the notebook is not made publicly available). Alternatively, one can also rely on technology for help. For example, the promotion of password manager applications would help one's perceived lack of self-efficacy, although that would put reliance on technology which may be susceptible to hacking too. Another example would be the use of biometrics (in replacement of password verification) afforded by digital devices.

In the domain of avoiding connection to unsecure WiFi networks, people's fear that their digital devices may be compromised as a result of connecting to unsecure WiFi networks will strongly motivate them to connect to a more secure one. This suggests that focus should be put on warning digital citizens of the severity of consequences for connecting to an unsecure WiFi network. While this has significant educational implications, it may help push for policy and technological implications requiring companies to have warning labels on WiFi connection applications. Moreover, establishments such as airports, hotels, malls and other public places, can also put up notices informing its patrons of legitimate, secure and free WiFi networks that patrons can log on to. Related to the preceding point, this study also found that belief in connecting to secure WiFi networks will strongly motivate a person to use their mobile data instead of taking a risk to connect to free public WiFi networks. This finding also reinforces the point that warning labels should be put up, and an associated message explaining the various protection one receives when using one's own mobile data plan.

In the domain of avoiding the opening of unknown and/or suspicious links or attachments, confidence in one's ability to identify and detect, as well as appropriately deal with suspicious emails and/or sources, strongly predicts one's motivation to scan all email sources and websites visited. The educational implications would be for educators to continually focus on informing users of what tell-tale signs to look out for, such as hovering over the link with the cursor to ensure that it leads to a legitimate website and looking carefully more at the URL and/or email sender. Other flags include observing if the email was written in proper English and if the sender is someone known to the user. Other measures to undertake would be to seek help from friends, family members and experts on how genuine such sources or attachments are. In this regard, technology can help too. With the advancement of artificial intelligence, such technologies can be deployed to alert the user to the possibility of suspicious emails and/or attachments for the user to further examine the contents. Additionally, plug-ins on web browsers could also help provide gatekeeping facilities to alert the users.

In light of these findings, several policy recommendations can be made to enhance cybersecurity awareness in individuals. First, governments can mandate that individuals enable multi-factor authentication or use password managers for to strengthen digital security practices. Furthermore, providing subsidies or financial incentives for individuals to enrol in cybersecurity courses can help raise awareness and improve digital literacy.

Limitations

Due to the concern for respondent fatigue, a decision was undertaken to reduce the number of questions. This resulted in the removal of one of the conceptual variables (perceived vulnerability) for "Avoid Connecting to Unsecure WiFi Networks". This is not ideal, and future studies could afford to balance this out further to ensure that all conceptual variables are represented. Alternatively, a case can be made to reduce the concepts in coping appraisal by combing the response cost and perceived self-efficacy which are highly similar (Milne et al., 2000; Sommestad et al., 2015), thereby cutting down on the questions asked. Along those lines, future studies may also explore capturing more demographic variables such as prior experiences with consequences of poor cyber hygiene behaviours (Tsai et al., 2019). Further research may also investigate other common and basic cyber behaviours such as installing anti-virus software (or other technological safety tools) and use of 2-Factor-Authentication.

While statistical significance was attained for all the results listed, and the study's sample size was largely consistent with similar studies, future studies should explore larger sample sizes in order to make stronger generalisable claims.

CONCLUSION

In today's interconnected world, the role of a digital citizen has gained immense importance. As we increasingly rely on digital platforms for communication, business, and personal activities, it is crucial to exhibit healthy online behaviours to navigate this digital realm effectively. This study has identified three essential domains for maintaining basic online hygiene: connecting to secure WiFi networks, avoiding unknown links and attachments, and adopting strong passwords.

Overall, the study underscores the importance of coping appraisal in motivating healthy cyber hygiene behaviours. By focusing on practical, easily actionable advice and emphasizing the effectiveness of protective behaviours, educational initiatives can lead to more effective and sustainable online hygienic practices. In an increasingly volatile digital environment, practicing exemplary cyber hygiene is crucial to protect against cybercrimes and their potential consequences. As digital citizens, it is our responsibility to stay informed and adopt healthy online behaviours to navigate the digital world safely and securely.

ACKNOWLEDGEMENT

This study was done in collaboration with Kaspersky. We would like to acknowledge and thank Kaspersky for providing the anonymized data for our use in this chapter. The views and opinions expressed in this publication are those of the authors and do not necessarily reflect or represent the views and opinions of Kaspersky.

REFERENCES

Ahluwalia, R. (2000). An Examination of Psychological Processes Underlying Resistance to Persuasion. *The Journal of Consumer Research*, *27*(2), 217–232. DOI: 10.1086/314321

Ajzen, I. (1991). The theory of planned behavior. *Organizational Behavior and Human Decision Processes*, *50*(2), 179–211. DOI: 10.1016/0749-5978(91)90020-T

Alazab, M., & Broadhurst, R. (2017). *An analysis of the nature of spam as cybercrime. Cyber-Physical Security*. Springer International Publishing.

Alkhalil, Z., Hewage, C., Nawaf, L., & Khan, I. (2021). Phishing Attacks: A Recent Comprehensive Study and a New Anatomy. *Frontiers of Computer Science*, *3*(563060), 1–23.

Althobaiti, M. M. (2021). Assessing User's Susceptibility and Awareness of Cybersecurity Threats. *Intelligent Automation & Soft Computing*, *28*(1), 167–177. DOI: 10.32604/iasc.2021.016660

Anderson, C. L., & Agarwal, R. (2010). Practicing safe computing: A multimedia empirical examination of home computer user security behavioral intentions. *Management Information Systems Quarterly*, *34*(3), 613–643. DOI: 10.2307/25750694

Arora, B. (2016). Exploring and analysing Internet Crimes and their behaviours. *Perspectives in Science*, *8*, 540–542. DOI: 10.1016/j.pisc.2016.06.014

Broadhurst, R., Skinner, K., Sifniotis, B., Matamoros-Macias, B., & Ipsen, Y. (2019). Phishing and cybercrime risks in a university student community. *International Journal of Cybersecurity Intelligence & Cybercrime*, *2*(1), 4–23. DOI: 10.52306/02010219RZEX445

Bryant, K., & Campbell, J. (2006). User Behaviours Associated With Password Security And Management. *AJIS. Australasian Journal of Information Systems*, *14*(1), 81–100. DOI: 10.3127/ajis.v14i1.9

Butavicius, M., Parsons, K., Pattinson, M., & McCormac, A. (2015). Breaching the human firewall: Social engineering in phishing and spear phishing emails. *Australasian Conference on Information Systems (pp. 12-23)*. AISeL. https://aisel.aisnet.org/acis2015/98/

Cain, A. A., Edwards, M. E., & Still, J. D. (2018). An exploratory study of cyber hygiene behaviours and knowledge. *Journal of Information Security and Applications*, *42*, 36–45. DOI: 10.1016/j.jisa.2018.08.002

Cains, M. G., Flora, L., Taber, D., King, Z., & Henshel, D. S. (2022). Defining Cyber Security and Cyber Security Risk within a Multidisciplinary Context using Expert Elicitation. *Risk Analysis*, *42*(8), 1643–1669. DOI: 10.1111/risa.13687 PMID: 33586204

Chaudhary, S., Schafeitel-Tähtinen, T., Helenius, M., & Berki, E. (2019). Usability, security and trust in password managers: A quest for user-centric properties and features. *Computer Science Review*, *33*, 69–90. DOI: 10.1016/j.cosrev.2019.03.002

Chen, R., Wang, J., Herath, T., & Rao, H. R. (2011). An investigation of email processing from a risky decision making perspective. *Decision Support Systems*, *52*(1), 73–81. DOI: 10.1016/j.dss.2011.05.005

Chenoweth, T., Minch, R., & Gattiker, T. (2009, January 5–8). Application of protection motivation theory to adoption of protective technologies. Paper presented at the 42nd Hawaii International Conference on System Sciences, HICSS'09, Honolulu, HI.

Conti, M., Dragoni, N., & Lesyk, V. (2016). A Survey of Man In The Middle Attacks. *IEEE Communications Surveys and Tutorials*, *18*(3), 2027–2051. DOI: 10.1109/COMST.2016.2548426

Cyber Security Agency of Singapore. (2019). *Secure Your Password.* Retrieved from https://www.csa.gov.sg/Tips-Resource/Resources/gosafeonline/2019/Secure-Your-Password

Cyber Security Agency of Singapore. (2023). *Singapore Cybersecurity Health Report.* Retrieved from https://www.csa.gov.sg/docs/default-source/our-programmes/support-for-enterprises/sg-cyber-safe-programme/csa-singapore-cybersecurity-health-report-2023.pdf?sfvrsn=c611e1b_1

Cyber Security Agency of Singapore. (n.d.). *Live Savvy Tips.* Retrieved from https://www.csa.gov.sg/docs/default-source/gso/file/resources/live-savvy-tips.pdf?sfvrsn=e4d879b8_0

Cyber Security Agency of Singapore. (n.d.). *Beware of Phishing Scams.* Retrieved from https://www.csa.gov.sg/our-programmes/cybersecurity-outreach/cybersecurity-campaigns/the-unseen-enemy-campaign/beware-of-phishing-scams

Dabrowski, A., Merzdovnik, G., Kommenda, N., & Weippl, E. (2016). *Browser History Stealing with Captive WiFi Portals. IEEE Security and Privacy Workshops.* IEEE.

Davinson, N., & Sillence, E. (2010). It won't happen to me: Promoting secure behaviour among internet users. *Computers in Human Behavior*, *26*(6), 1739–1747. DOI: 10.1016/j.chb.2010.06.023

Doğrul, M., Yalçın, H., & Daim, T. U. (2023). *Cybersecurity: A Technology Landscape Analysis. Cybersecurity Technology: A Landscape Analysis.* Springer Nature Switzerland.

Federal Trade Commission. (2022). *How to Recognize and Avoid Phishing Scams.* Retrieved from https://consumer.ftc.gov/articles/how-recognize-and-avoid-phishing-scams

Fleck, A. (2024). Cybercrime Expected To Skyrocket in Coming Years. *Statista.* Retrieved from https://www.statista.com/chart/28878/expected-cost-of-cybercrime-until-2027

Florencio, D., & Herley, C. (2007). A large-scale study of web password habits. *Proceedings of the 16th International Conference on World Wide Web* (pp.657-666). Association for Computing Machinery. DOI: 10.1145/1242572.1242661

Floyd, D. L., Prentice-Dunn, S., & Rogers, R. W. (2000). A Meta-Analysis of Research on Protection Motivation Theory. *Journal of Applied Social Psychology*, *30*(2), 407–429. DOI: 10.1111/j.1559-1816.2000.tb02323.x

Furnell, S. (2007). An assessment of website password practices. *Computers & Security, 26*(7), 445–451. DOI: 10.1016/j.cose.2007.09.001

Gaw, S., & Felton, E. W. (2006). Password Management Strategies for Online Accounts. *Proceedings of the Second Symposium on Usable Privacy and Security*. Association for Computing Machinery. DOI: 10.1145/1143120.1143127

Grawemeyer, B., & Johnson, H. (2011). Using and managing multiple passwords: A week to a view. *Interacting with Computers, 23*(3), 256–267. DOI: 10.1016/j.intcom.2011.03.007

Gurung, A., Luo, X., & Liao, Q. (2009). Consumer motivations in taking action against spyware: an empirical investigation. *Information Management & Computer Security, 17*(3), 276–289. https://doi .org/, F. (2023). Personal data of 665,000 Marina Bay Sands lifestyle rewards members accessed in data security breach. *CNA*. Retrieved from https://www.channelnewsasia.com/singapore/marina-bay-sands -mbs-customers-personal-data-security-breach-3902491DOI: 10.1108/09685220910978112Hamzah

Herath, T., Chen, R., Wang, J., Banjara, K., Wilbur, J., & Rao, H. R. (2012). Security services as coping mechanisms: An investigation into user intention to adopt an email authentication service. *Information Systems Journal*. Advance online publication. DOI: 10.1111/j.1365-2575.2012.00420.x

Herath, T., & Rao, H. R. (2009). Protection motivation and deterrence: A framework for security policy compliance in organisations. *European Journal of Information Systems, 18*(2), 106–125. DOI: 10.1057/ejis.2009.6

Herley, C. (2009, September 8–11). So long, and no thanks for the externalities: The rational rejection of security advice by users. Paper presented at the 2009 Workshop on New Security Paradigms, Oxford, UK. DOI: 10.1145/1719030.1719050

Howe, A. E., Ray, I., Roberts, M., Urbanska, M., & Byrne, Z. (2012, May 20–23). The psychology of security for the home computer user. Paper presented at the 2012 IEEE Symposium on Security and Privacy (SP), San Francisco, CA. DOI: 10.1109/SP.2012.23

Hu, Q., & Dinev, T. (2005). Is spyware an internet nuisance or public menace? *Communications of the ACM, 48*(8), 61–65. DOI: 10.1145/1076211.1076241

Ifinedo, P. (2012). Understanding information systems security policy compliance: An integration of the theory of planned behavior and the protection motivation theory. *Computers & Security, 31*(1), 83–95. DOI: 10.1016/j.cose.2011.10.007

International Telecommunications Union (ITU). (2008). *X.1205 Overview of Cybersecurity*. Series X: Data Networks, Open System Communications and Security – Telecommunication Security. Retrieved from https://www.itu.int/rec/T-REC-X.1205-200804-I

International Trade Administration. (2023). *Singapore Cybersecurity Market*. Retrieved from https://www.trade.gov/market-intelligence/singapore-cybersecurity-market

Jartelius, M. (2020). The 2020 Data Breach Investigations Report – a CSO's Perspective. *Network Security, 2020*(7), 9–12. DOI: 10.1016/S1353-4858(20)30079-9

Johnston, A. C., & Warkentin, M. (2010). Fear appeals and information security behaviors: An empirical study. *Management Information Systems Quarterly*, *34*(3), 549–566. DOI: 10.2307/25750691

Jiow, H. J., Mwagwabi, F., & Low-Lim, A. (2021). Effectiveness of protection motivation theory based: Password hygiene training programme for youth media literacy education. *The Journal of Media Literacy Education*, *13*(1), 67–78. DOI: 10.23860/JMLE-2021-13-1-6

Kemper, G. (2019). Improving employees' cyber security awareness. *Computer Fraud & Security*, *2019*(8), 11–14. DOI: 10.1016/S1361-3723(19)30085-5

LaRose, R., Rifon, N. J., Liu, X., & Lee, D. (2005, May 26–30). Understanding online safety behavior: A multivariate model. Paper presented at the International Communication Association Conference, New York, NY.

LaRose, R., Rifon, N. J., & Wirth, C. (2007). Online safety begins with you and me: Getting Internet users to protect themselves. *Paper presented at the57th International Communication Association Conference*, San Francisco, CA.

Lee, C., & Ng, D. (2023). Digital fraud attack rate in Singapore higher than APAC average: Cybercrime report. *CNA*. Retrieved from https://www.channelnewsasia.com/singapore/cyber-crime-digital-fraud-attack-bots-lexisnexis-report-3992166

Lee, D., LaRose, R., & Rifon, N. J. (2008). Keeping our network safe: A model of online protection behaviour. *Behaviour & Information Technology*, *27*(5), 445–454. DOI: 10.1080/01449290600879344

Lee, Y., & Larsen, K. R. (2009). Threat or coping appraisal: Determinants of SMB executives' decision to adopt anti-malware software. *European Journal of Information Systems*, *18*(2), 177–187. DOI: 10.1057/ejis.2009.11

LeFebvre, R. (2012, October 12–13). The human element in cyber security: A study on student motivation to act. Paper presented at the 2012 Information Security Curriculum Development Conference, Kennesaw, GA. DOI: 10.1145/2390317.2390318

Lotfy, Y., Zaki, A., Abd El-Hafeez, T., & Mahmoud, T. M. (2021). Privacy Issues of Public WiFi Networks. *Proceedings of the International Conference on Artificial Intelligence and Computer Vision (AICV2021)* (pp.656-665). Springer.

Madrekar, A., & Jambhale, R. (2024). Password Manager Statistics By Reasons, Usage and Subscription Rate. *Coolest Gadgets*. Retrieved from https://www.coolest-gadgets.com/password-manager-statistics

Maimon, D., Jordan Howell, C., Jacques, S., & Perkins, R. C. (2022). Situational awareness and public Wi-Fi users' self-protective behaviors. *Security Journal*, *35*(1), 154–174. DOI: 10.1057/s41284-020-00270-2

Marett, K., McNab, A. L., & Harris, R. B. (2011). Social networking websites and posting personal information: An evaluation of protection motivation theory. *AIS Transactions on Human-Computer Interaction*, *3*(3), 170–188. DOI: 10.17705/1thci.00032

Marforio, C., Masti, R. J., Soriente, C., Kostiainen, K., & Capkun, S. (2015). Personalized security indicators to detect application phishing attacks in mobile platforms. Retrieved from https://arxiv.org/abs/1502.06824

McAlaney, J., & Benson, V. (2020). *Cyber Influence and Cognitive Threats*. Elsevier Inc.

Milne, G. R., Labrecque, L. I., & Cromer, C. (2009). Toward an understanding of the online consumer's risky behavior and protection practices. *The Journal of Consumer Affairs*, *43*(3), 449–473. DOI: 10.1111/j.1745-6606.2009.01148.x

Milne, S., Sheeran, P., & Orbell, S. (2000). Prediction and Intervention in Health-Related Behavior: A Meta-Analytic Review of Protection Motivation Theory. *Journal of Applied Social Psychology*, *30*(1), 106–143. DOI: 10.1111/j.1559-1816.2000.tb02308.x

Mou, J., Cohen, J. F., Bhattacherjee, A., & Kim, J. (2022). A Test of Protection Motivation Theory in the Information Security Literature: A Meta-Analytic Structural Equation Modeling Approach. *Journal of the Association for Information Systems*, *23*(1), 196–236. DOI: 10.17705/1jais.00723

Mwagwabi, F., & Jiow, H. J. (2021). Compliance with security guidelines in teenagers: The conflicting role of peer influence and personal norms. *AJIS. Australasian Journal of Information Systems*, *25*. Advance online publication. DOI: 10.3127/ajis.v25i0.2953

Ng, B. Y., Kankanhalli, A., & Xu, Y. C. (2009). Studying users' computer security behavior: A health belief perspective. *Decision Support Systems*, *46*(4), 815–825. DOI: 10.1016/j.dss.2008.11.010

Norton. (2017). *Norton WiFi Risk Report: Report of Online Survey Results in 15 Global Markets*. Retrieved from https://www.nortonlifelock.com/us/en/newsroom/press-kits/norton-WiFi-risk-report-2017

Notoatmodjo, G., & Thomborson, C. D. (2009). Passwords and Perceptions. *Proceedings of the 7th Australasian Information Security Conference (AISC 2009)* (pp.71-78). Australian Computer Society, Inc.

Ong, J. (2024). More scammers exploited social media in 2023; Facebook, WhatsApp and Instagram of 'particular concern': Police. *Today*. Retrieved from https://www.todayonline.com/singapore/more-scammers-exploited-social-media-2023-facebook-whatsapp-instagram-2363481

Petrosyan, A. (2024). Percentage of organizations worldwide experiencing cyberattacks from 2021 to 2023, by attack type. *Statista*. Retrieved from https://www.statista.com/statistics/1376249/cyber-attack-global-firms-by-type/

Ponemon Institute. (2016). *2016 Cost of Cyber Crime Study & the Risk of Business Innovation*. Retrieved from https://www.ponemon.org/local/upload/file/2016%20HPE%20CCC%20GLOBAL%20REPORT%20FINAL%203.pdf

Ponemon Institute & Accenture. (2019). *The Cost of Cybercrime: Ninth Annual Cost of Cybercrime Study*. Retrieved from https://iapp.org/media/pdf/resource_center/accenture_cost_of_cybercrime_study_2019.pdf

Rhee, H. S., Kim, C., & Ryu, Y. U. (2009). Self-efficacy in information security: Its influence on end users' information security practice behavior. *Computers & Security*, *28*(8), 816–826. DOI: 10.1016/j.cose.2009.05.008

Rifon, N. J., Quilliam, E., & LaRose, R. (2005, May 26–30). Consumer perceptions of online safety. Paper presented at the 2005 Annual Meeting of the International Communication Association, New York, NY.

Riley, S. (2006). Password Security: What Users Know and What They Actually Do. *Usability News*, *8*(1), 1–5.

Rogers, R. (1983). Cognitive and physiological processes in fear appeals and attitude change: A revised theory of protection motivation. In Cacioppo, J. T., & Petty, R. E. (Eds.), *Social Psychophysiology* (pp. 153–176). Guilford Press.

Sangeen, M., Bhatti, N. A., Kifayat, K., Alsadhan, A. A., & Wang, H. (2023). Blind-trust: Raising awareness of the dangers of using unsecured public Wi-Fi networks. *Computer Communications*, *209*, 359–367. DOI: 10.1016/j.comcom.2023.07.011

Shay, R., Komanduri, S., Kelley, P. G., Bauer, L., Leon, P. G., Christin, N., Mazurek, M. L., & Cranor, L. F. (2010). Encountering stronger password requirements: User attitudes and behaviors. *Symposium on Usable Privacy and Security (SOUPS)*. Redmond, WA, USA. DOI: 10.1145/1837110.1837113

Shillair, S., Boehmer, J., Cotten, S. R., LaRose, R., & Rifon, N. (2014, May 25). Communicating online safety: A critical content analysis of websites providing online safety information to consumers over time. Paper presented at the International Communication Association, Seattle, WA.

Sombatruang, N., Kadobayashi, Y., Sasse, A. M., Baddeley, M., & Miyamoto, D. (2018). The continued risks of unsecured public WiFi and why users keep using it: Evidence from Japan. *2018 16th Annual Conference on Privacy, Security and Trust (PST)* (pp. 1-11). IEEE.

Sombatruang, N., Sasse, A. M., & Baddeley, M. (2016). Why Do People Use Unsecure Public WiFi: An Investigation of Behaviour and Factors Driving Decisions. *Proceedings of the 6th Workshop on Socio-Technical Aspects in Security and Trust* (pp. 61-72). Association for Computing Machinery. DOI: 10.1145/3046055.3046058

Sommestad, T., Karlzén, H., & Hallberg, J. (2015). A Meta-Analysis of Studies on Protection Motivation Theory and Information Security Behaviour. *International Journal of Information Security and Privacy*, *9*(1), 26–46. DOI: 10.4018/IJISP.2015010102

Stafford, T. F., & Poston, R. (2010). Online security threats and computer user intentions. *Computer*, *43*(1), 58–64. DOI: 10.1109/MC.2010.18

Stobert, E., & Biddle, R. (2018). The Password Life Cycle. *ACM Transactions on Privacy and Security*, *21*(3), 1–32. DOI: 10.1145/3183341

Tang, L. (2023). ShopBack fined S$74,400 over leak of more than 1.4 million customers' personal data. *CNA*. Retrieved from https://www.channelnewsasia.com/singapore/shopback-fined-data-leak-million-customers-personal-3705006

Tang, L. (2023). Loyalty marketing agency fined S$10,000 over data leak of Starbucks Singapore customers. *CNA*. Retrieved from https://www.channelnewsasia.com/singapore/starbucks-singapore-330000-customers-data-leak-ascentis-fined-3925281

Toh, E. (2018). SingHealth cyber attack: PM Lee and other ministers among 1.5 million patients whose records were stolen. *The Straits Times.* Retrieved from https://www.straitstimes.com/singapore/pm-lee-and-other-ministers-among-15-million-patients-whose-records-were-stolen

Tsai, H. S., Jiang, M., Alhabash, S., LaRose, R., Rifon, N. J., & Cotten, S. R. (2016). Understanding online safety behaviors: A protection motivation theory perspective. *Computers & Security*, *59*, 138–150. DOI: 10.1016/j.cose.2016.02.009

Wang, P. A. (2013, June 23–28). Assessment of cybersecurity knowledge and behavior: An anti-phishing scenario. Paper presented at the eighth International Conference on Internet Monitoring and Protection (ICIMP 2013), Rome, Italy.

Wang, W., Huang, H., Zhang, L., & Su, C. (2021). Secure and efficient mutual authentication protocol for smart grid under blockchain. *Peer-to-Peer Networking and Applications*, *14*(5), 2681–2693. DOI: 10.1007/s12083-020-01020-2

Zahedi, F. M., Abbasi, A., & Chen, Y. (2015). Fake-website detection tools: Identifying elements that promote individuals' use and enhance their performance. *Journal of the Association for Information Systems*, *16*(6), 448–484. DOI: 10.17705/1jais.00399

Zhang, H., & Daim, T. U. (2023). *Cybersecurity: A Technology Landscape Analysis. Technology Assessment of Cybersecurity*. Springer Nature Switzerland.

Chapter 13
Media Manipulation and Disinformation:
The Role of AI in Fake News Creation and Its Impact on Trust in Media

Mayuree Pal
https://orcid.org/0009-0001-7583-3889
Presidency University, India

C. P. Rashmi
Presidency University, India

ABSTRACT

The emergence of the digital era has brought about a fundamental change in how information is generated, distributed, and consumed. Artificial intelligence (AI), previously celebrated as a champion of pursuing truth, now assumes a dual function in this intricate environment. This chapter examines the complicated relationship between AI and media manipulation, specifically focusing on its capacity to generate convincing deepfakes that control public sentiment and disseminate false information via social media bots. Therefore, a comprehensive literature review was undertaken by gathering many publications to enhance understanding and obtain clarification on the subject matter. To cultivate a knowledgeable and resilient information environment, it is crucial to provide individuals with the necessary skills to assess sources critically, recognize content generated by artificial intelligence, and acknowledge their own biases. This chapter intends to illuminate AI systems in moderating misinformation and determining the most effective strategy for addressing this significant issue.

INTRODUCTION

In the news industry and society, there is significant emphasis on identifying fake news. This arises from the ongoing issue of misleading information being widely spread, creating a need for continuous assessment of the accuracy of digital content. The validity of information has long-term implications for both print and digital media in society. The practice of using attention-grabbing and sensational headlines, which are not always accurate, to engage audiences and market information has persisted throughout

DOI: 10.4018/979-8-3693-6675-2.ch013

Copyright ©2025, IGI Global Scientific Publishing. Copying or distributing in print or electronic forms without written permission of IGI Global Scientific Publishing is prohibited.

the dissemination of information. The spread and impact of information on social media platforms are greatly amplified and occur rapidly. This means that distorted, erroneous, or fraudulent information can lead to enormous consequences within minutes, affecting millions of users (Zafarani et al., 2019).

The rapid emergence of social media platforms like Facebook, Twitter, YouTube, and WhatsApp has significantly influenced our lifestyle and information acquisition methods. Approximately a decade ago, the principal sources of news were traditional journalism and reputable media organizations that followed established ethical standards on reporting and news dissemination (Hunt & Matthew, 2017). Conversely, much of the information we receive now originates from social media platforms, with brief interactions predominantly consisting of tweets, which, although appearing credible at first glance, lack substantial verification. Yingxuan Yang (2024) asserts that disseminating deceptive information aims to mislead individuals.

In recent years, there has been an increasing prevalence of disseminating misinformation using various digital media platforms, including YouTube, LinkedIn, Facebook, Instagram, WhatsApp, and Twitter. How individuals communicate has significantly evolved due to the emergence of social media platforms, which serve as primary channels for disseminating deliberate misinformation rooted in meticulously orchestrated conspiracies and responses to global crises such as pandemics. These media have often been misused to disseminate misinformation (Zafarani et al., 2019; Agarwal et al., 2022; Zhang et al., 2019). The pervasive utilization of social media for disseminating incorrect information and sensational content has engendered concerns that underscore the pressing necessity to implement systems capable of identifying and addressing disinformation, a significant issue in contemporary society.

Over the past decade, significant transformations have occurred in information transmission and consumption methods, mostly due to the increasing prevalence of social media. Despite the considerable growth of these platforms, substantial issues about deception necessitate reevaluating our media consumption practices. Enhanced literacy about online material is critical (Yang et al., 2023).

This alteration reflects a broader cultural phenomenon in which the traditional media's perceived authority has been scrutinized, and the distinction between genuine information and deception has become more nebulous. All of these consequences affect several aspects of our living conditions. It is vital to possess information grounded in critical thinking and discernment among consumers within this contemporary milieu, including political discourse, social interactions, and personal convictions (Hunt & Matthew, 2017).

The proliferation of disinformation poses a challenge to individuals seeking to remain informed and seems to erode public faith in media institutions and political entities. In navigating this intricate digital landscape, we must establish efficient methods for identifying and mitigating the dissemination of disinformation while concurrently fostering a more informed and discerning populace. Due to these developments, educational institutions, media entities, and civil society must collaborate to enhance the quality of information disseminated and consumed. This will facilitate the discernment of reality from falsehood during an era of rapid information dissemination via social media (Keith, 2023). This necessitates the creation of frameworks and tools that enable individuals to critically evaluate the information they encounter critically, so enhancing overall media literacy and fostering a culture of clarity and truth in communication.

As we go toward a future where social media will increasingly shape public debate and personal relationships, it is imperative to confront the issues of misinformation and its societal impact. To cultivate a better-educated citizenry, it is essential to adopt proactive strategies that foster responsible information dissemination, ensure accountability among platform providers, and advance literacy efforts that enable consumers to make informed decisions regarding their information intake. This would foster a healthier, more educated society (Zhang et al., 2019).

The spread of fake news threatens journalism, business, democracy, and all other fields (Nasir et al., 2021). It engenders violence and fosters bullying within society. The content misleads and manipulates individuals (Setiawan et al., 2021; Garg and Sharma, 2022). The phenomenon leads to perplexity and exasperation among individuals who engage in social media platforms (Khan et al., 2022). The spread of false information causes individuals to experience turmoil, unease, trepidation, and unfounded judgments (Yun and Ahn, 2018; Zervopoulos et al., 2022). Disinformation propagates illogical goals across society to serve the self-interested objectives of particular factions and entities. The financial status of a country may be adversely impacted (Mishra et al., 2022; Kishwar and Zafar, 2022). In their study, Stieglitz and Dang-Xuan (2013) determined that disseminating unverified information on social media platforms, commonly known as fake news, had detrimental effects on social institutions. Sivek (2018) contended that emotions significantly influenced the consumption and regulation of fake news on digital media since individuals' emotions motivated them to disseminate information to the online community. Bakir and Andrew (2018) argued that individual emotions and sentiments compelled Internet users to disseminate false information on digital platforms. Tsipursky et al. (2018) argued that psychological factors contributed to the spread of false information on social media. The sentiments expressed by internet users played a significant role in producing and disseminating biased content on social media platforms (Lunenborg and Maier, 2018).

The distortion of truth is a repeating occurrence throughout history. *Damnatio memoriae*, often known as the deliberate eradication of individuals from historical records, illustrates the intentional manipulation of reality previously employed in Ancient Egypt. However, due to the swift progress in information and communication technologies (ICT) and their growing ubiquity, deceptive information can now be easily created realistically. It can also be rapidly and extensively spread to a specific audience through artificial intelligence (AI) techniques. The repercussions are significant, with wide-ranging ramifications. For example, the media ecosystem has influenced citizens' viewpoints and voting choices regarding the 2016 US presidential election and the 2016 UK referendum on exiting the European Union (EU) (Howard and Kollanyi, 2016).

Facebook has served as a valuable tool in Myanmar for individuals aiming to propagate hatred against Rohingya Muslims (Human Rights Council, 2018, para 74). Several murders in India were caused by falsehoods spread on WhatsApp (Dixit and Mac, 2018). In France, a professor was tragically assassinated as a result of an intense and malicious internet campaign directed towards him on social media (Bindner and Gluck, 2020). Conspiracy theories are currently thriving. During the Covid-19 pandemic, we are experiencing what the World Health Organization (WHO) refers to as an infodemic. This refers to the overwhelming abundance of accurate and inaccurate information, which harms individuals and society.

In April 2018, a video became widely shared on WhatsApp, the globally renowned mobile instant messaging platform. The video, apparently captured by a closed-circuit television (CCTV) camera, depicts a gathering of children engaged in a game of cricket on the street. Two individuals on a motorcycle approach and seize one of the youngest children before swiftly departing. The dissemination of this film depicting a purported kidnapping leads to extensive bewilderment and alarm, inciting a period of mob violence lasting eight weeks, resulting in the tragic deaths of at least nine individuals who were not involved in the incident (BBC News, 2018).

The inciting clip that triggered these vendettas was a skillfully fabricated rendition—an alteration of a video from a public educational initiative in Pakistan explicitly crafted to enhance consciousness regarding child abductions. The educational film commences with a depiction of a kidnapping. However, shortly thereafter, a contracted performer dismounts from the motorbike and displays a notice advising viewers to exercise vigilance in safeguarding their children. The video that gained widespread popularity in India was manipulated to remove the gadget used for the "big reveal," leaving only highly realistic footage depicting the abduction of a child.

In the same month, BuzzFeed releases a video featuring former US President Barack Obama addressing the camera directly, appearing to be in the Oval Office. During the initial 35 seconds, the camera focuses exclusively on Obama's face. After making a few slightly unusual remarks, Obama shocks everyone by stating, "President Trump is a complete and utter imbecile." Following a momentary pause, he proceeds, "Presently... it is important to note that I would never express these sentiments, particularly in a public speech, but there are individuals who would... individuals akin to Jordan Peele." At this juncture, the genuine purpose of BuzzFeed's film becomes apparent. The speaker is not, in fact, President Obama. A split screen displays former US President Barack Obama on the left and the acclaimed American actor, comedian, and director Jordan Peele on the right. The facial expressions and mouth motions of Obama and Peele are perfectly synchronized. Peele's production team has utilized artificial intelligence (AI) to digitally accurately recreate Obama's face. Using AI to combine Peele's facial features and impersonation of Obama's speech serve as a clever public service message highlighting the potential for manipulation in online videos (Facebook, 2018; Twitter, 2018; YouTube, 2018).

Online social networks (OSNs) are seeing tremendous growth and have emerged as a significant platform for millions of users to access a wide range of global and local news. Online social networks (OSNs) have both positive and negative aspects. While they provide significant benefits, such as unrestricted and convenient communication and immediate access to news and information, they can also offer several drawbacks and challenges. One of their primary problematic challenges is the propagation of fake news. The identification of counterfeit news remains a complex and unresolved matter. In addition, identifying false information on online social networks (OSNs) possesses specific attributes and difficulties that render the search for a resolution far from simple.

The increasing intricacy of artificial intelligence (AI) is a challenge that humanity now confronts as it evolves. The advancement of artificial intelligence has accelerated at an alarming pace in recent years, surpassing the projections given by experts a few years prior. The exponential tendency has already led to several popular applications in industries often characterized by flexible or absent regulations. This unites two domains: first, the rules and frameworks that guarantee the proper functioning of ethical AI, and second, the increasing necessity for implementation. In response to artificial intelligence's augmented capabilities and implications, industry leaders have recently advocated for institutional regulation of AI research. A compilation has been established to disseminate concepts that foster ethical considerations in developing AI systems (Candelon et al., 2021; Gibbs, 2017). The proposed regulations tackle the urgent responsibility of practising Registered Investment Advisors to ensure that technological advancements do not surpass the established regulatory and ethical frameworks designed to protect society.

The ongoing discourse over artificial intelligence extends beyond technical specialists; it encompasses a broader spectrum. It includes a varied cohort of participants, including those lacking technical expertise yet cognizant of the implications of artificial intelligence on society and business. Using these systems in creative domains like graphic design and music composition suggests that these technologies may do jobs hitherto limited to humans. These improvements prompt significant concerns over the future of human professionals in fields traditionally characterized by creativity and skill. ChatGPT was launched in November 2022, representing a notable progression in conversational AI. The newly developed language model acquired an impressive user base of 1 million persons within just one week of its launch (Buchholz, 2023). ChatGPT was constructed on the foundation of its predecessor, GPT-3, which initiated a new era in which technology could participate in dynamic conversations with users. This capability significantly alters human-computer interaction by enabling artificial intelligence to write text and replicate certain conversational behaviours (Rossomakha et al., 2023).

Despite much discourse on the impact of these technologies, mainly on image and text generation, their implications have elicited considerable intrigue and concern. Economists and business professionals are increasingly scrutinizing the ramifications of artificial intelligence capabilities on labour markets and the overall economy. Researchers like Malone (2018) and Mayor (2019) have articulated concerns over the potential for artificial intelligence (AI) to usurp employment presently held by humans. This fear is particularly pertinent now, as AI technologies are advancing and proliferating, with discussions around the possibility of significant job upheaval. (Kaplan, J., 2016).

The emergence of artificial intelligence and its swift advancement present several challenges that mankind must address. As we embark on this new epoch of technological progress, it is imperative to establish ethical guidelines, engage in informed dialogues with diverse stakeholders, and conduct critical evaluations of the impact of artificial intelligence on labor markets. As artificial intelligence advances and permeates various facets of human existence, it is imperative to engage in continuous discourse regarding the implications of AI on society, the economy, and our comprehension of humanity in an era dominated by intelligent machines (Nguyen & Van Dinh, 2024).

Considering the increasing utilization of advanced AI algorithms for creating and detecting fabricated content, it is imperative to examine concerns such as disinformation within the framework of generative AI. Amaro et al. (2023) and Dadkhah et al. (2023). To develop more effective strategies against generative AI and investigate the ethical ramifications of AI-generated misinformation, it is essential to understand the capabilities of generative AI in creating credible false narratives. Artificial intelligence algorithms accelerate the dissemination of misinformation online via two distinct channels. Artificial intelligence (AI) methodologies facilitate generating and altering many forms of content, including text, images, audio, and video. They generate new opportunities. Secondly, internet platforms employ artificial intelligence algorithms to enhance user engagement markedly. This, thus, facilitates the rapid and extensive dissemination of inaccurate information online. The issue is based on these strategies. This issue presents several ethical dilemmas that need a thorough examination.

This substantial challenge remains unresolved by artificial intelligence techniques. Machine learning and deep learning are two artificial intelligence techniques employed to create impressionistic forgeries by generating and disseminating counterfeit content. The automated detection of bogus news remains challenging. The primary reason is that the deliberately fabricated material closely resembles reality, making it difficult for artificial intelligence to assess correctness using supplementary data from other sources (Westerlund 2019).

This chapter also explores the influence of Generative AI on people's digital content consumption. It is crucial to acknowledge that AI can create content that was previously non-existent. It can adopt a creative approach that involves generating abstract images. A few years ago, the technology known as deepfake was introduced. It involved replacing the faces of regular individuals in any video with those of well-known individuals, resulting in placing them in troublesome circumstances. Present-day technology has far surpassed previous levels of advancement (Westerlund, 2019).

WHAT IS AN AI?

Artificial intelligence (AI) is a primary field of research within computer science. It pertains to developing systems and robots engineered to do tasks that often need human-like intelligence. Artificial intelligence (AI) encompasses several methodologies and technological frameworks that enable computers to do activities, including cognition, problem-solving, natural language comprehension, and environmental interpretation (Thakre et al., 2024). Artificial intelligence systems are designed to replicate specific traits of human intellect, like learning from experience and adapting to new information or unfamiliar environments, among other capabilities. The domain is vast since it includes disciplines like computer science, cognitive psychology, neurology, and linguistics, all of which aid in the objective of developing robots that can replicate human thought processes (Liang et al., 2023).

The concept of an intelligent agent is essential to artificial intelligence. An intelligent agent is a system that actively observes its environment and executes activities designed to attain certain objectives or enhance the probability of favorable results. Intelligent agents operate semi-autonomously, allowing them to ignore numerous inputs and make decisions solely based on their assessments. Such agents will likely be identified in the domains of robotics, autonomous vehicles, and electronic personal assistants. An intelligent agent must continuously learn and adapt to environmental changes in all these applications. Self-driving vehicles utilize sensors to evaluate their surroundings, avoid obstacles, and make instantaneous judgments based on real-time data. These exemplify the dynamic capabilities of intelligent agents in practical scenarios (Becerra-Bonache et al., 2014).

The preponderance of contemporary computer advancements is predicated on artificial intelligence, conceived in the pursuit of systems capable of executing specific tasks while exhibiting attributes akin to human reasoning and learning. The advent of artificial intelligence (AI) technologies has transformed perceptions and applications of technology. This has influenced user engagement with technology, decision-making processes, and the resolution of complex problems across several sectors (Sage et al., 2024).

Artificial Intelligence has many uses, such as chatbots and automated complex data processing, enhancing service delivery in the financial and healthcare sectors. Its application across several sectors has shown its capacity to improve efficiency, minimize human error, and support data-driven decision-making (Saini, 2023).

AI Methods

- **Machine Learning:** This is an example of an AI application where robots do not follow explicit instructions to complete specific jobs. Instead, they learn and enhance their performance via experience autonomously. Deep Learning is a branch of machine learning that uses artificial neural networks to do predictive analysis. Machine learning encompasses several algorithms, including Unsupervised, Supervised, and Reinforcement Learning. In Unsupervised Learning, the algorithm operates without using labelled data or instruction. In Supervised Learning, a function is derived from the training data, which comprises an input object and its corresponding desired output. Machines employ reinforcement learning to determine optimal actions that maximize rewards (Saini, 2023).
- **Natural Language Processing(NLP):** The interactions between computers and human language occur when computers are programmed to process natural languages. Machine Learning is a dependable technique used in Natural Language Processing to extract semantic information from

human languages. In Natural Language Processing (NLP), the machine records the spoken words of a human. The conversion from audio to text takes place, followed by the processing of the text, where the data is transformed into audio. The algorithm utilizes the auditory input to generate responses directed at human individuals. Natural Language Processing is applied in various domains, including IVR programs used in contact centres, language translation applications like Google Translate, and word processors such as Microsoft Word for grammar accuracy verification in text.

The intricacies of human languages pose a challenge for Natural Language Processing due to the complex rules governing the transmission of information through natural language, which computers struggle to comprehend. Natural Language Processing (NLP) employs algorithms to identify and generalize the patterns inherent in human languages. This converts unstructured data from human languages into a computer-readable format (Saini, 2023).

- **Automation & Robotics:** Automation aims to delegate monotonous and repetitive jobs to machines, enhancing productivity and achieving cost-effective and more efficient outcomes. Several firms employ machine learning, neural networks, and graphs for automated purposes. Implementing CAPTCHA technology in online financial transactions can effectively mitigate fraud risks. Robotic process automation is designed to execute large quantities of repetitive activities and can easily adjust to varying conditions (Saini, 2023).
- **Machine Vision:** Visual information can be captured by machines and subsequently analyzed. In this context, cameras are utilized to record the visual information, the process of converting the image into digital data is achieved by analogue to digital conversion and digital signal processing is implemented to manipulate the data. Subsequently, the acquired data is inputted into a computer. In machine vision, two crucial factors are sensitivity, which refers to the machine's capacity to detect faint signals, and resolution, which pertains to the system's ability to differentiate between objects at varying levels of detail. Machine vision is utilized in several applications, such as signature detection, pattern recognition, and medical picture analysis (Saini, 2023).
- **Knowledge-Based Systems(KBS):** A Knowledge-Based System (KBS) is a computer system that utilizes expertise from a human expert to deliver guidance on a specific topic. KBS is characterized by the clear distinction between the knowledge base, expressed through various means like rules, frames, or cases, and the inference engine or algorithm, which utilizes the knowledge base to conclude (Saini, 2023).
- **Neural Networks:** Neural networks (NNs) are systems that take inspiration from biology and are composed of a highly interconnected network of computational "neurons" arranged in layers. By manipulating the network weights, neural networks can be "trained" to approximate almost any nonlinear function with a desired level of precision. Neural networks are commonly equipped with a collection of input and output examples. A learning algorithm, such as backpropagation, would be employed to modify the weights in the network, ensuring that the network produces the intended output. This form of learning is generally called supervised learning (Saini, 2023).

WHAT IS FAKE NEWS?

The phenomenon of "fake news" has emerged as a significant concern in contemporary discourse, particularly about its impact on information consumption and creation in a digital society. The Collins English Dictionary defines "fake news" as incorrect material, often sensationalized, presented in a manner that mimics authentic news reporting. This definition exposes the inaccuracy of such information while emphasizing an effort to deceive the audience by portraying it as fact-based journalism.

Allcott and Gentzkow's 2017 definition of fake news refers to news stories deliberately created to be untrue and explicitly deceive the viewer. This definition necessitates a conscious effort to mislead, distinguishing phoney news from erroneous or misleading journalism that may arise from inadvertent errors or genuine misunderstandings. The authors indicated that such news narratives are inherently dangerous as they might influence public perception and direct actions based on inaccuracies (Baptista et al., 2022).

Academic literature has contested many attributes that delineate false news. The talks primarily focus on the absence of authenticity; in other words, false information represents a divergence from truth. This feature has sparked academic discourse over the conceptual limits of false news. They have explicitly questioned whether the elements of satire, rumours, conspiracy theories, disinformation, and hoaxes should be incorporated or omitted from the definition of fake news. Each of them possesses distinct characteristics, hence complicating the task of formulating a singular description for false news.

Nakov (2020, p. 2) observes that "fake news" interpretations are increasingly diverging among various groups and people. In politics, some politicians have adopted an expansive meaning of "fake news," leading to phrases such as "news I do not like," reflecting their contempt for any reporting that opposes their ideas or narratives. This illustrates the term's capacity for weaponization, functioning as a rhetorical tactic while undermining media organizations regarded as antagonistic or unfriendly (Weiss et al., 2020).

The notion of fake news includes several components, such as deliberate deception, a departure from accurate reporting, and larger interpretations that arise in public debate. The focus has transitioned from a specific denunciation of poor journalism to a broader reproach applicable to any news considered objectionable, frequently indicating increased scepticism towards the media by both governmental and social entities. The ongoing discourse on fake news necessitates inquiries into the interplay between journalism, information credibility, and public perceptions of truth (Baade 2020).

According to Melissa Zimdars at Merrimack College, there are numerous categories of fake news. Here are some to look out for:

- **Satire:** It is a highly effective method of commenting on political or cultural events. Satirical pieces aim to deceive readers into criticizing or mocking a particular circumstance momentarily. However, they can be detrimental when disseminated as "New," so it is crucial to ensure that they are explicitly identified and recognized as opinions rather than objective truth. Satire websites should incorporate disclaimers, typically in the header or "About Us" section, on their web pages.
- **Extreme Bias:** Certain websites selectively utilize correct information while disregarding other accurate information that contradicts their argument or presents an alternative perspective. Despite incorporating certain factual information, these sources remain unreliable due to their restricted scope.
- **Clickbait:** Articles that use sensational headlines to attract traffic to their websites are categorized as clickbait. These headlines frequently lack substantiation from the corresponding article and

may pertain to another topic. The website generates substantial advertising revenue from the large number of clicks generated by clickbait.

- **Conspiracy:** Websites that include fake news may also try to link unrelated events together to prove that there is a conspiracy. They often rely on the argument that "It hasn't been disproved, so it must be true." Find corroborated evidence before believing these types of claims.
- **State News (Propaganda):** States may disseminate their disinformation to preserve control over a nation or consolidate authority in a leader. An impartial, external source should corroborate news sources from the government or politicians.

FAKE NEWS: A MODERN DAY PROBLEM

The propagation of disinformation began before the printing machine was invented, making it easier to share misinformation. For instance, more than two and a half thousand years ago, Socrates was sentenced to death based on false information that said he had shown disrespect to the gods of Athens and had corrupted the youth of the day. An examination of Google Trends data on the phrase "fake news" shows that it became much more prevalent around the 2016 US presidential election. Recently, organizations like the European Union and NATO have addressed the problem of recognizing fraudulent information (Bontridder, N., and Poullet, Y., 2021).

Artificial Intelligence (AI) Methods to Enhance the Production and Spread of False Information

AI algorithms enhance the disinformation phenomena on the internet through two distinct methods. AI approaches enable creating and manipulating many forms of information, such as texts, images, audio, and video, opening up new prospects. Artificial intelligence technologies created and implemented by online platforms play a crucial role in promoting the efficient and swift spread of false material online. These tactics are the primary cause of the problem. This situation has Several ethical concerns that should be carefully analysed (Bontridder, N., and Poullet, Y., 2021).

Artificial Intelligence Approaches Enable the Production of Fake Content

Applying AI algorithms to create deceptive content results in deepfakes. Deepfakes, a blend of deep learning and fake, arise from two artificial intelligence algorithms working together within a Generative Adversarial Network (GAN). GANs are defined as a computational method for generating novel forms of data by utilizing existing datasets. For example, a GAN can analyze numerous images of Donald Trump and produce a new image that resembles the analyzed pictures without replicating any specific image. This technology can be applied across various content formats, including photos, videos, audio, and text. The term "deepfake" is primarily used to refer to audio and video content (Walorska, 2020, p. 9).

Although it was always possible to manipulate words and images using less advanced technologies, the emergence of deepfakes has made it much easier to create fake films and audio. They propagate misleading information and erode the trustworthiness of genuine information, casting suspicion on any information encountered, whether from the mainstream media or government authorities. In the context of disinformation, those who deceive or mislead others can alter the truth in two efficient ways: they may present phoney content as genuine, and they can show legitimate information as fake NATO (Bontridder, N., and Poullet, Y., 2021).

Artificial Intelligence Techniques Available Online Can Enhance the Spread of False Information

Micro-targeting is an effective method for disseminating misinformation to particular demographics. The creation of AI-driven monitoring systems by information and communication platforms, integral to the web's advertising-centric economic model, accounts for this accomplishment. These technologies enable firms to direct advertisements to specific clients, thus enhancing their efficacy. This tactic has historically employed traditional tracking techniques, such as browser cookies and third-party tracking. Companies include many types of content, including advertising networks, social media widgets, and website analytics scripts, into the first-party websites accessed by users. This linkage allows third-party entities to track users across other websites, facilitating the development of comprehensive profiles derived from an individual's browsing history (Akers et al., 2018, p. 4).

Browser fingerprinting is a prevalent technique for tracking users. It recognises individuals through their browsers and operating systems' distinctive characteristics and specific configurations. This creates a unique fingerprint that may link a user's activities across many websites. These tactics were first designed to engage prospective purchasers explicitly for advertising purposes. Advertisers often select targeting parameters encompassing demographics and inferred interests in this context. The platform's algorithms determine which individuals will view advertisements and designate the locations for these advertisements on the platform (Maréchal and Biddle, 2020, p. 13).

Tracking technologies provide an effective means of delivering personalized content to each user. Funding is derived not from user fees but from advertising, as per the prevailing economic model employed by major internet platforms, which relies on "the economics of attention" (Festré & Garrouste, 2015).

As micro-targeting evolves, it is crucial to understand the intricate mechanisms of these monitoring systems and their impact on user engagement and information dissemination, particularly concerning the spread of misinformation. The Council of the EU's new findings underscore algorithms' significant role in facilitating users' search for desired content (findability) and in presenting media material without explicit queries (discoverability). This underscores the substantial influence of algorithms on consumers' perceptions of online settings, which sometimes occur without a complete understanding of the underlying business considerations (Bontridder, N., and Poullet, Y., 2021).

OBJECTIVE

This chapter's fundamental objective is to examine the role of artificial intelligence in propagating false information within society and to ascertain methods for mitigating the spread of misinformation and disinformation among individuals. It mainly analyzes the role of artificial intelligence technology in generating, disseminating, and enhancing misinformation. The research will demonstrate how artificial intelligence may exacerbate and alleviate deception-related challenges. By evaluating the role of artificial intelligence in information processing, this study will examine interventions designed to mitigate the dissemination of false narratives and promote factual information among individuals.

METHODOLOGY

Misinformation is commonly defined as information that is erroneous or misleading. Misinformation engenders several societal problems, including diminished trust in the media, disturbances in political contexts, and misdirection of public debate. As artificial intelligence technologies progress, the capacity to generate and manipulate information has heightened concerns over the proliferation of misinformation. The beginning of artificial intelligence, particularly in natural language processing and machine learning, has fundamentally altered the creation and dissemination of knowledge. Data-driven systems can analyze extensive datasets, produce content at an unmatched speed, and tailor information dissemination to specific audience segments. These abilities confer significant advantages in some aspects, yet they also facilitate the intentional dissemination of misinformation. Artificial intelligence (AI) use across several industries has transformed perceptions regarding its possible societal impacts, particularly concerning the spread of misinformation. Literature review studies have been undertaken to investigate the effects of artificial intelligence on the propagation of misinformation. This study collects information from several sources, including academic publications, published books, and newspaper articles. This chapter seeks to illustrate the role of artificial intelligence (AI) in the extensive propagation of misinformation. The digital era has experienced a swift increase in awards for creating AI-generated content.

CONCEPTUAL FRAMEWORK

- **Digital Citizenship:** The idea of being a responsible, ethical, and courteous member of an online community. It emphasizes the need to understand digital rights and responsibilities and the repercussions of online behaviour. The deliberate effort to maintain a personal online identity and reputation in a favourable light ensures that one's digital footprints reflect the values and ideals that one intends to portray in both personal and professional situations.
- **Digital Footprint:** All online activities impact a person's digital reputation and appearance. Understanding digital footprints is crucial for being a good citizen in cyberspace, as they may affect one's privacy and chances of getting a job.
- **Cyber Safety:** It is a broad term that refers to actions taken to prevent online dangers. It covers topics such as understanding and managing privacy settings, recognizing phishing schemes, and safely navigating the Internet.
- **Digital Literacy:** The capacity to locate, evaluate, and utilize information effectively across various digital media. Digital literacy helps people behave responsibly online and make educated decisions about the information they come across, especially when telling the difference between trustworthy sources and disinformation.
- **Netiquette:** It refers to the rules of politeness and respect that should be followed when communicating online. Good netiquette helps create a positive online community and ensures that discussions are productive and courteous.
- **Media literacy:** It is the ability to critically analyze media messages and comprehend the function of media in society. This includes recognizing prejudice, understanding other views, and spotting disinformation. Media literacy is essential to digital citizenship since it enables individuals to engage with material thoughtfully.

- **Data Privacy:** The ability to decide and manage how personal information is gathered, used, and shared on the internet. Advocacy for data privacy is important because it protects individual rights and encourages responsible behaviour online.
- **Digital Rights:** The rights individuals possess in the digital realm, such as access to information, express ideas, and retain privacy. To be considered a successful digital citizen, it is essential to recognize and advocate for digital rights.
- **Content moderation:** It refers to the guidelines and procedures online platforms employ to regulate the material users create and uphold community standards. It is an effective way to establish a secure and courteous online environment and encourage responsible online behaviour.
- **False News:** It refers to information that is wholly fabricated yet presented as if it were legitimate news. This material is often exaggerated to attract more clicks, shares, or views, which can mislead the audience. There are many different types of fake news, including entirely fabricated stories, headlines that do not accurately reflect the content, and misrepresentations of information. Fake news undermines the public's trust in legitimate news sources and can significantly influence their thoughts and actions.
- **Disinformation:** Disinformation is false information distributed to deceive others. The term emphasizes the intent behind the distribution; it is created and shared with full awareness of its falsity. Disinformation campaigns frequently target specific individuals, groups, or entire populations to manipulate perceptions, create division, or achieve political, economic, or ideological goals. Some tactics involve creating fake websites or social media accounts to spread inaccurate information.
- **Misinformation:** Unlike disinformation, this refers to false information released without the intention to cause harm. Misinformation can manifest as errors, misinterpretations, or outdated information shared unintentionally. Misinformation, however, can still have negative consequences, such as causing confusion or leading to false beliefs, particularly in critical situations like health-related emergencies (for instance, misinformation regarding vaccinations).
- **Deepfakes:** Deepfakes are videos or audio recordings that appear highly realistic but are fabricated. They are created using artificial intelligence, which can alter a person's appearance or voice. The term combines "deep learning," artificial intelligence technology, with "fake." Deepfakes can convincingly make it seem like individuals are saying or doing things they never did, raising significant concerns. This can erode trust in digital media, potentially resulting in reputational damage or misinformation.
- **Micro-targeting:** This method uses data analytics to categorize viewers into groups and craft messages tailored to them based on their online behaviour, interests, and demographics. Micro-targeting is especially effective in misinformation, as it can disseminate misinformation or propaganda to narrow audiences. This increases success since it closely aligns with the beliefs and preferences of the targeted groups.
- **AI-Generated Content:** It refers to any material (text, images, videos) produced using artificial intelligence. AI systems can generate large volumes of misleading articles or media that appear realistic. While artificial intelligence can positively impact content creation, it also poses the risk of developing and spreading fake news, as it can mislead a vast audience quickly and easily.
- **Information Literacy:** It refers to recognising when information is needed and finding, evaluating, and effectively using that information. An information-literate person can identify trustworthy sources, helping to mitigate the impact of disinformation or fake news. This skill is vital

for navigating today's information landscape, which is overloaded with various sources that often provide conflicting information.

- **Algorithms:** These are specific rules or computations used by social media platforms and search engines to determine which content is displayed to users. Algorithms can promote sensational or entertaining content, often leading to the disproportionate spread of disinformation. Understanding algorithms can help you understand why specific news stories appear in your feeds and how they may contribute to disseminating false information.
- **Psychographics:** This area of research focuses on consumers and their psychological traits, including values, interests, and lifestyles. Psychographic data enables businesses to create messages that deeply resonate with specific target audiences. In terms of misinformation, understanding the psychographics of a group assists bad actors in tailoring their disinformation campaigns to align with the beliefs and emotions of those targeted audiences.

THEORETICAL FRAMEWORK

- **Social Cognitive Theory:** One of the most significant frameworks for understanding social learning is Social Cognitive Theory, established by Albert Bandura. Bandura states that learning is not just about incentives or punishments that come from direct experiences. It may also happen through observational learning, which is when you observe other people learn. This hypothesis about misinformation on social media will explain how individuals connect and communicate with one another to propagate misleading information. Users will mimic the attitudes and actions of their friends, influencers, or content providers and make decisions based on the tales they come across on the internet. This mimicry opens the door for various narratives, including disinformation, which helps to disseminate it even further among users.

Social Cognitive Theory helps us understand how social media can raise awareness and promote positive role models. The detrimental impact of disinformation can be minimized in an environment where accurate and truthful information is valued and shared. By actively practising ethical online conduct, people may substantially contribute to establishing a more educated and responsible digital realm. This will assist in flattening the curve of misinformation through improved vigilance and informed distribution.

- **Media Literacy Theory:** In general, media literacy is the ability to access, comprehend, analyze, evaluate, and produce media material. The chapter emphasizes the need to improve these media literacy abilities at a time when difficulties are mostly driven by material created by artificial intelligence and the rapid spread of misinformation. As a result of technology such as deepfakes, it is becoming more and more difficult to trust information that is given online. This makes it essential for individuals to improve their critical thinking skills.

Initiatives and officially organized programs to improve media literacy can empower people to evaluate the reliability of diverse sources. As their capacity to judge whether the information is accurate and understand the implications of their consumption increases, they can spot harmful misinformation more quickly. This extra-critical discernment helps people make decisions while improving society's ability to deal with disinformation.

- **Confirmation Bias Hypothesis:** According to the Confirmation Bias Hypothesis, people tend to seek out and assign greater importance to information that supports their current ideas while ignoring information that contradicts them. This section illustrates how social media creates echo chambers, where users hear only ideas and perspectives that are similar to their own. This reinforcing nature of material might further consolidate users' beliefs, making them more susceptible to disinformation.

To reduce the effect of confirmation bias, it is necessary to have tactics that can break down echo chambers and force people to face differing viewpoints. Suppose social media is advertised as a venue for open and critical discourse and exposes users to a broader diversity of perspectives. In that case, it can contribute to healthy information sharing instead of perpetuating a single narrative. Recognizing confirmation bias can build more inclusive environments where various perspectives can be heard.

- **Framing Theory:** The Framing Theory states that how information is presented affects how a person understands or interprets it. As a result, the chapter argues that misinformation is carefully framed within specific settings to trigger emotional responses or create a feeling of urgency, which in turn influences public opinion. Sensationalism may hide the truth about reality because the way information is presented can generate biases and harm the audience's feelings.

One of the most essential tools in the fight against disinformation is neutralizing the influence of framing. Advocates can establish a narrative space where truthful representations are more prominent by disputing biased assertions and providing facts neutrally and objectively. This will help limit the spread of false narratives while also boosting the public's comprehension of complex topics based on facts rather than sensationalist framing.

- **Disruptive Innovation Theory:** This theory states that emerging technologies that disrupt current market systems can have a significant effect. This chapter finds that the disruptive nature of artificial intelligence systems that create deepfakes profoundly changes how information is distributed and consumed. The difficulty these technologies provide to traditional journalism is that we are now in a new paradigm where information can be easily generated or changed substantially.

Media organisations and content curators will need to adapt to limit the spread of disinformation about these breakthroughs. Media stakeholders should reestablish themselves as trustworthy sources of information by creating strong frameworks for accountability and developing new tactics to combat misinformation. Media firms that take a proactive approach to disruptive technology can more effectively navigate the complicated disinformation landscape.

- **Networked Publics Theory:** The Networked Publics Theory examines how digital technologies impact social structures and how individuals interact with one another. Social media is a contradictory instrument in the fight against false information. It provides opportunities for people to work together but also allows for the rapid spread of falsehoods. Networked publics can inspire coordinated actions against disinformation by utilizing the power of togetherness to activate structured campaigns.

Groups can utilize networked publics to gather resources and collective intelligence in order to combat the spread of false information. Citizens can be encouraged to actively participate in participatory projects with a sense of responsibility. This may help galvanize efforts against the spread of misinformation and ultimately cultivate an educated public that engages actively in the conversation.

- **Social Identity Theory:** Henri Tajfel and John Turner established the Social Identity Theory. This theory explains how people form their self-concept based on their organisational membership. The hypothesis is critical because it helps us understand how group memberships affect how people respond to information from outside their organizations. The chapter discusses the phenomenon in which members of some social groups are hesitant to accept information from sources that are not part of their in-group. This, in turn, makes them more likely to be misled or misinformed.

Intergroup communication provides insight into how individualized interventions help overcome gaps in understanding and support different groups. Interacting with people from other groups can enhance one's capacity to critically analyze information and lessen the chance of becoming locked in an echo chamber that promotes false information.

- **Actor-Network Theory (ANT):** Actor-Network Theory states that social realities are created by the relationships between a diverse range of players, including both human and non-human actors. This is particularly relevant when observing how users engage with technical actors, especially artificial intelligence, in discussions about disinformation. Algorithms that govern platforms driven by artificial intelligence impact how users experience the transmission of information, including spreading fake news.

Researchers can better understand how digital tools affect public communication by closely examining the relationship between human activities and technical infrastructures. This will also help them understand how these interactions contribute to spreading misinformation, leading to better tactics to combat its reach and effect. Thus, people will have the knowledge they need to navigate the digital information world.

- **The Ethics of Technology Theory:** The Ethics of Technology The theory investigates the ethical consequences that are believed to be connected to developing and using new technologies. This chapter discusses the moral problems that arise when artificial intelligence is used to create content. It also addresses technology corporations' responsibilities to avoid the misuse of artificial intelligence technologies. It is stated that ethical principles are essential in developing and applying artificial intelligence to ensure that new technologies do not cause issues with the spread of false pieces of information.

A framework of ethical accountability in technology allows parties to address the social and legal consequences of regulating technology. As the world talks about the consequences of fast technological advancement, it is crucial to prioritise responsible innovation, with ethics being the most critical factor in creating new laws.

- **Constructivist Learning Theory:** According to Constructivist Learning Theory, knowledge is built based on personal experiences and social interactions. As a result, this chapter calls for educational approaches that actively promote digital citizenship among students. Modules of critical thinking and collaborative learning are important considerations because they can help students to discuss themes like disinformation, ethical AI usage, and responsible sharing behaviours. This, in turn, empowers individuals to navigate the digital world effectively.

Education systems that teach students to think critically and collaborate may create a generation that is more engaged with knowledge. As students become accustomed to being critical and introspective about the digital content they consume, their ability to combat disinformation through politically informed behaviours improves. This, in turn, results in a better media environment.

FINDINGS

Upon through the articles, specific findings have been discovered and categorized into distinct headings:

- **Special Bot Spreaders**

Multiple authors (Shu et al. 2018b, 2017; Shi et al. 2019; Bessi and Ferrara 2016; Shao et al. 2018a) have demonstrated that fake news is often generated and disseminated by non-human accounts with comparable characteristics and network structure, such as social bots (Ferrara et al. 2016). Bots, abbreviated forms of software robots, have existed since the inception of computers. A social bot is a computer algorithm that autonomously generates material and engages with individuals on social media platforms to imitate and influence their actions. While intended to be beneficial, they can have detrimental effects, such as facilitating the dissemination of unconfirmed information or rumors. It is crucial to acknowledge that bots are merely instruments devised and backed by humans to serve certain undisclosed motives (Ferrara et al., 2016).

According to Ismailov et al. (2020), the methods employed to identify bots vary based on the specific social media platform and the intended goal. It is observed that distinguishing between a bad bot that aims to establish connections with several accounts and a bot that repeatedly shares links to harmful websites necessitates employing distinct detection methods. They propose two models for identifying rogue accounts, each utilizing a different set of characteristics. Detection in social context models is accomplished by analyzing features associated with an account's social presence, such as its relationships with other accounts, similarity to other users' activities, and various graph-based features. User behaviour models typically concentrate on characteristics associated with an individual user's behaviour, such as the frequency of actions (e.g., the number of tweets or posts during a specific time), patterns of activity, and sequences of clicks.

Using social bots by malevolent actors also enhances the efficient propagation of false information. These bots, short for software robots, are user accounts on social media platforms that are either totally or partially automated. They are created to communicate and mimic human behaviour online. Various categories of social bots exist, including basic bots that automate specific aspects of communication (such as liking or sharing), partially human-controlled accounts with automated features (known as hybrid bots or 'cyborgs'), and autonomous agents with AI and learning capabilities like Microsoft's Zo1 or Replika.ai (Assenmacher et al., 2020).

- **The impact of Artificial Intelligence on human sensitivity**

Bots acquire knowledge of users' social media usage, frequency of visits to specific websites, and preferences for particular information patterns. The dissemination of false information by bots is contingent upon the actions of individuals on social media platforms (Vosoughi et al., 2018). Human behavior on social media platforms endorses a specific message to be displayed in other users' news feeds. When a Facebook post garners more likes, regardless of its topic, individuals are more inclined to shape their perspectives based on the content.

Walther and Jang (2012) state that the number of likes a video receives is directly related to the number of comparable videos people will see on YouTube. The more agreement there is in a blog post after a dispute, the more likely it is that someone's opinion will be changed. Hashtags help organize news feeds by showing information that matches a person's interests while reducing exposure to differing or opposing viewpoints or news. A person's statement can influence how other people feel about the topic. This comparison is relevant when discussing the spread of incorrect information. According to Walther et al. (2018), the perceived veracity of information is directly affected by the amount of agreement and exposure to a misleading news piece.

- **The role of cookies and analytics**

Social media analytics helps political actors sustain fake news. Websites, Facebook pages, and blogs that contain false information target their audience. Analytics enable them to ensure that what they send is what people are likely looking for (Emplifi, n.d.).

According to Gayatri et al. (2017), analytics gives the audience profile, which may be used to adapt messaging to convince each individual. Political actors make use of the chances that cookies present to build an audience that is open to their messaging. Cookies on computers or mobile devices keep track of and document what users do on the internet. They keep track of and document individuals' browsing histories. When users choose to save cookies, these cookies can also save information from websites that users have not logged into, called third-party cookies. This creates a problem. The browser has accessed these cookies, which are retrieved from domains the user has never visited. Some outside organizations have collected information about people without getting their permission.

Facebook, Google, and other prominent data platforms provide trackers to analyze the websites users visit. These trackers profile users' actions while on websites, gathering information about what other pages were opened, what was clicked on, and how much time was spent there. They can analyse this data by running scripts on users' web browsers, and users may not even be aware of it (Tiku, 2017).

- **The role of echo – chambers**

"Echo Chambers" refers to exposure only to a single type of content. Essentially, the greater the level of engagement with the content, the higher the likelihood of encountering similar material. If a child interacts with an influencer who promotes misogyny, they will be exposed to a more significant amount of related content. As individuals engage with the material, they encounter more content, perpetuating this cycle. This continues until all they see revolves around misogyny. An algorithm creates an echo chamber by selectively presenting content that aligns with the user's views, limiting exposure to opposing perspectives. As a result, understanding and accepting alternative viewpoints becomes exceedingly difficult, hindering the expansion of one's worldview. When confronted with dissent, individuals often adopt a defensive posture and are likely to spread hostility (Online Safety Concerns, n.d.).

Echo chambers significantly contribute to the spread of misinformation. Within these restricted environments, false or deceptive information can rapidly be seen as truth since individuals tend to disseminate and reinforce notions that align with their beliefs. This places individuals in a reinforcing environment since they are surrounded by like-minded others who tend to bolster their beliefs. Inaccurate narratives originating from echo chambers can proliferate rapidly and uncontrolled, resulting in a distorted perception of reality. The insularity reinforces existing preconceptions and inhibits the development of critical thinking since individuals are less likely to question the dominant narrative or seek information that contradicts their beliefs. (GCF Global, n.d.).

Future Recommendations

Following a thorough study of articles, the following specific proposals have been identified to combat the spread of fake news effectively:

- Invest in research and development to create more advanced algorithms and detection systems that can recognize increasingly lifelike deepfakes. Discover artificial intelligence (AI) solutions that can effectively adjust to the ever-changing tactics of deepfakes.
- Facilitate cooperation among technology corporations, researchers, governments, and civil society to establish uniform procedures for identifying and countering deepfakes. Promote the exchange of data and collaborative efforts to address the issue collaboratively.
- Establish inclusive educational initiatives in schools, workplaces, and communities to augment media literacy, analytical reasoning, and digital proficiency. Provide individuals with the necessary abilities to identify and react to deepfakes.
- Establish resilient and flexible legal structures that effectively tackle the production, dissemination, and abuse of deepfakes. Propose revisions to privacy, defamation, and intellectual property legislation to address infringements associated to deepfake technology.
- Ensure that social media platforms and technology corporations are held responsible for actively monitoring and regulating deepfake content present on their platforms. Promote the adoption of clear and open policies and mechanisms that allow users to report and authenticate material.
- Develop and implement ethical principles and industry norms to ensure the appropriate utilization of AI technology, encompassing the production and distribution of deepfake content. Advocate for integrating ethical considerations in the entire process of researching, developing, and implementing AI systems.
- Promote global cooperation to foster worldwide collaboration and joint efforts in tackling the cross-border dissemination of deepfakes. Promote the exchange of information and implementation of effective strategies amongst nations to counteract the worldwide dissemination of false information.
- Facilitate collaborations across governments, academia, and corporate sectors to finance and bolster research endeavors focused on comprehending and mitigating the repercussions of deepfakes.

- Provide users with tools and resources to authenticate the genuineness of media material. Create accessible verification tools and platforms that empower individuals to evaluate the reliability of information.
- Continuously monitor and adapt to evolving deepfake techniques to remain aware and responsive. Consistently oversee and revise tactics, technology, and policies to proactively address emerging risks that distorted media presents.

CONCLUSION

We need to work together to address the problem of disinformation driven by artificial intelligence. By empowering individuals, promoting ethical AI development, fostering cooperation, and supporting independent media, we may build a more resilient statistical landscape in which the truth triumphs. The benefits are even more significant than the considerable trials. Navigating this complicated environment is essential for the future of informed discourse and democratic activities.

Big internet corporations and platforms also have an essential role and cannot sit idly by. They are in charge of overseeing content moderation and making sure that algorithms are transparent. They also work to stop the spread of incorrect information through their systems. It is important to create valuable partnerships with groups that examine facts, support various perspectives, and prioritise the safety of user data.

To combat the spread of false information, it is crucial for us to unite as a community. Citizens, media organizations, technology companies, and policymakers can collaborate to find effective solutions by fostering open communication and teamwork. In a society significantly influenced by technology, it is vital to take steps to restore trust. Some of these actions include promoting the ethical advancement of AI, enhancing digital literacy, and ensuring accountability on platforms.

REFERENCES

Aimeur, E., Amri, S., & Brassard, G. (2023). Fake news, disinformation and misinformation in social media: A review. *Social Network Analysis and Mining*, *13*(1), 1–36. DOI: 10.1007/s13278-023-01028-5 PMID: 36789378

Akhtar, P., Ghouri, A. M., & Khan, H. U. R., Amin ul Haq, M., Awan, U., Zahoor, N., Khan, Z., & Ashraf, A. (2022). Detecting fake news and disinformation using artificial intelligence and machine learning to avoid supply chain disruptions. *Annals of Operations Research*, *327*. Advance online publication. DOI: 10.1007/s10479-022-05015 PMID: 36338350

Akkor Gül, A., Ertürk, Y., & Elmer, P. (2020). *Digital Transformation in Media & Society*. Istanbul University Press., DOI: 10.26650/B/SS07.2020.013

Al Tamimi, H. A. H., Duqi, A., Kanas, A., & Zervopoulos, P. D. (2021). Directional distance function DEA estimators for evaluating efficiency gains from possible mergers and acquisitions. *The Journal of the Operational Research Society*, *73*(6), 1240–1257. DOI: 10.1080/01605682.2021.1907243

Ali Adeeb, R., & Mirhoseini, M. (2023). The Impact of Affect on the Perception of Fake News on Social Media: A Systematic Review. *Social Sciences (Basel, Switzerland)*, *12*(12), 674. DOI: 10.3390/socsci12120674

Allcott, H., & Gentzkow, M. (2017). Social Media and Fake News in the 2016 Election. *The Journal of Economic Perspectives*, *31*(2), 211–236. DOI: 10.1257/jep.31.2.211

Bakir, V., & McStay, A. (2018). Fake News and the Economy of Emotions: Problems, Causes, Solutions. *Digital Journalism (Abingdon, England)*, *6*(2), 154–175. DOI: 10.1080/21670811.2017.1345645

Becerra-Bonache, L., & Jiménez-López, M. D. (2014). Linguistic Models at the Crossroads of Agents, Learning and Formal Languages. *ADCAIJ: Advances In Distributed Computing And Artificial Intelligence Journal.*, *3*(4), 67–87. DOI: 10.14201/ADCAIJ2014346787

Biju, P. R., & Gayathri, O. (2023). Self-Breeding Fake News. The International Journal of Information, Diversity, &. *Inclusion (Washington, D.C.)*, *7*(1/2), 1–25.

Bindner, L., & Gluck, R. (2017). Wilayat Internet: ISIS' Resilience across the Internet and Social Media. Bellingcat. Retrieved from - https://www.researchgate.net/publication/323005907_Wilayat_Internet_ISIS'_Resilience_across_the_Internet_and_Social_Media

Bonnet, J. (n.d.). Subject Guides: Hot Topics: Fake News and Misinformation: Identifying Fake News. Libguides.library.umaine.edu. https://libguides.library.umaine.edu/fakenews/identification

Bontridder, N., & Poullet, Y. (2021). The role of artificial intelligence in disinformation. *Data & Policy*, *3*, e32. DOI: 10.1017/dap.2021.20

Buchholz, R. R., D'Attilo, G., Deeter, M., Edwards, D., Emmons, L., Francis, G., Gaubert, B., Gille, J., Mao, D., Martinez-Alonso, S., Worden, H., & Ziskin, D. (2023). MOPITT CO monthly climatology 2002-2021. Version 1.0. UCAR/NCAR - GDEX. DOI: 10.5065/s9fy-9w35

Chesney, R., & Citron, D. K. (2019). Deep Fakes: A Looming Challenge for Privacy, Democracy, and National Security. SSRN Electronic Journal, 107(1753). https://doi.org/DOI: 10.2139/ssrn.3213954

Dame Adjin-Tettey, T. (2022). Combating fake news, disinformation, and misinformation: Experimental evidence for media literacy education. *Cogent Arts & Humanities*, *9*(1), 2037229. Advance online publication. DOI: 10.1080/23311983.2022.2037229

Dixit, A., & Mac, D. (2017). Global Marketing Strategies of McDonald's Corporation (with Reference to India and Russia). IJAR, 3, 870-875. Scientific Research Publishing. Retrieved from - https://www.scirp.org/reference/referencespapers?referenceid=2807253

Emplifi. (n.d.). Social media analytics for political campaigns. Emplifi White Paper. Retrieved from - https://emplifi.io/resources/blog/social-media-analytics-the-complete-guide

Endres, D., Hedler, L., & Wodajo, K. (2023). Bias in Social Media Content Management: What Do Human Rights Have to Do with It? *AJIL Unbound.*, *117*, 139–144. DOI: 10.1017/aju.2023.23

Ferrara, E., Varol, O., Davis, C., Menczer, F., & Flammini, A. (2016). The rise of social bots. *Communications of the ACM*, *59*(7), 96–104. DOI: 10.1145/2818717

Festré, A., & Garrouste, P. (2015). The 'economics of attention': A history of economic thought perspective. Œconomia. History, Methodology, Philosophy, (5-1), 3-36.

Figueira, Á., & Oliveira, L. (2017). The current state of fake news: Challenges and opportunities. *Procedia Computer Science*, *121*, 817–825. DOI: 10.1016/j.procs.2017.11.106

Franquet, M., & Eldridge, S. A. (2020). The Ethics of Photojournalism in the Digital Age. Routledge. Retrieved from - https://www.routledge.com/The-Ethics-of-Photojournalism-in-the-Digital-Age/SantosSilva-EldridgeII/p/book/9781138586307?srsltid=AfmBOoosnTC3CRnUVXECRSFHPjMNpNs4-bL6EBfWS4k4o-lHWDF-wA5j

Gayatri, P., & Ajitha, M. & Kumar, Kiran & Kumar, P.P. & Yamsani, Madhusudan. (2017). Validated high performance liquid chromatographic method for estimation of nisoldipine from matrix type transdermal films: Application to ex vivo skin permeability studies. *Asian Journal of Pharmaceutics*, *11*, S407–S411.

Gibbs, J. L., Sivunen, A., & Boyraz, M. (2017). Investigating the impacts of team type and design on virtual team processes. *Human Resource Management Review*, *27*(4), 590–603. DOI: 10.1016/j.hrmr.2016.12.006

Gordon, R., Spotswood, F., & Dibb, S. (2022). Critical social marketing: Towards emancipation? *Journal of Marketing Management*, *38*(11–12), 1043–1071. DOI: 10.1080/0267257X.2022.2131058

Gupta, A., Kumar, N., Prabhat, P., Gupta, R., Tanwar, S., Sharma, G., Bokoro, P. N., & Sharma, R. (2022). Combating Fake News: Stakeholder Interventions and Potential Solutions. *IEEE Access : Practical Innovations, Open Solutions*, *10*, 78268–78289. DOI: 10.1109/ACCESS.2022.3193670

Gupta, M., Dennehy, D., Parra, C. M., Mäntymäki, M., & Dwivedi, Y. K. (2023). Fake news believability: The effects of political beliefs and espoused cultural values. *Information & Management*, *60*(2), 103745. DOI: 10.1016/j.im.2022.103745

Hölbl, Marko & Zadorozhny, Vladimir & Družovec, Tatjana & Komapara, Marko & Nemec Zlatolas, Lili. (2024). Browser Fingerprinting: Overview and Open Challenges. .DOI: 10.3233/FAIA231163

Howard, P. N., & Kollanyi, B. (2016, June 1). Bots, #StrongerIn, and #Brexit: Computational Propaganda during the UK-EU Referendum. NASA ADS. Retrieved from - https://doi.org//arXiv.1606.06356DOI: 10.48550

Human Rights Council. (2018). Report on Myanmar and Facebook. United Nations Human Rights Council, para 74. Retrieved from - https://www.ohchr.org/sites/default/files/Documents/Publications/OHCHRreport2018.pdf

Hussein, F., & Hejase, H. (2023). Artificial Intelligence and Fake News. Žurnalistikos tyrimai. Vol. 16. pp. 39-71. DOI: 10.15388/ZT/JR.2022.2

India lynchings: WhatsApp sets new rules after mob killings. (2018). BBC News. Retrieved from - https://www.bbc.com/news/world-asia-india-44897714 (Visited on 5th June, 2024)

Iqbal, A., Shahzad, K., Khan, S. A., & Chaudhry, M. S. (2023, October 03). Shakeel Ahmad Khan, & Muhammad Shahzad Chaudhry. (2023). The relationship of artificial intelligence (AI) with fake news detection (FND): A systematic literature review. *Global Knowledge, Memory and Communication.* Advance online publication. DOI: 10.1108/GKMC-07-2023-0264

Ismailov, R., Simoens, S., & Khasanova, Z. (2020). Greater awareness of biosimilars and shared decision-making among patients attending rheumatology practices in Colorado, United States: Real-world data. *World Journal of Rheumatology*, *10*(1), 1–10. DOI: 10.5499/wjr.v10.i1.1

Jacobides, M., Brusoni, S., & Candelon, F. (2021). The Evolutionary Dynamics of the Artificial Intelligence Ecosystem. *Strategy Science*, *6*(4), 412–435. Advance online publication. DOI: 10.1287/stsc.2021.0148

Karnouskos, S. (2020). Artificial Intelligence in Digital Media: The Era of Deepfakes. *IEEE Transactions on Technology and Society*, *1*(3), 138–147. DOI: 10.1109/TTS.2020.3001312

Khan, R., Kulasiri, D., & Samarasinghe, S. (2023). A multifarious exploration of synaptic tagging and capture hypothesis in synaptic plasticity: Development of an integrated mathematical model and computational experiments. *Journal of Theoretical Biology*, *556*, 111326. DOI: 10.1016/j.jtbi.2022.111326 PMID: 36279957

Kim, A., & Dennis, A. R. (2019). Says who? The effects of presentation format and source rating on fake news in social media. *Management Information Systems Quarterly*, *43*(3), 1025–1039. DOI: 10.25300/MISQ/2019/15188

Kim, B., Xiong, A., Lee, D., & Han, K. (2021). A systematic review on fake news research through the lens of news creation and consumption: Research efforts, challenges, and future directions. *PLoS One*, *16*(12), e0260080. DOI: 10.1371/journal.pone.0260080 PMID: 34882703

Kochan, T. A. (2021). Artificial Intelligence and the Future of Work: A Proactive Strategy. *AI Magazine*, *42*(1), 16–24. DOI: 10.1002/j.2371-9621.2021.tb00006.x

Kolo, C., Joschka Mütterlein, & Sarah Anna Schmid. (2022). Believing Journalists, AI, or Fake News: The Role of Trust in Media. Proceedings of the ... Annual Hawaii International Conference on System Sciences. https://doi.org/DOI: 10.24251/HICSS.2022.393

Kumar, A., & Taylor, J. W. (2023). Feature importance in the age of explainable AI: Case study of detecting fake news & misinformation via a multi-modal framework. *European Journal of Operational Research*, *317*(2), 401–413. Advance online publication. DOI: 10.1016/j.ejor.2023.10.003

Lunenborg, J., & Maier, R. (2018). Bias in social media content. *International Journal of Communication*, *12*(1), 1–18.

Martens, B., & Aguiar, L. GGmez, E., & Mueller-Langer, F. (2018). The Digital Transformation of News Media and the Rise of Disinformation and Fake News. SSRN Electronic Journal. https://doi.org/DOI: 10.2139/ssrn.3164170

Matters, I. (2024). Learn about fake news to support children. Internet Matters. https://www.internetmatters.org/issues/fake-news-and-misinformation-advice-hub/learn-about-fake-news-to-support-children/

Mishra, S., Ewing, M. T., & Cooper, H. B. (2022). Artificial intelligence focus and firm performance. *Journal of the Academy of Marketing Science*, *50*(6), 1176–1197. Advance online publication. DOI: 10.1007/s11747-022-00876-5

Msughter, A. E., Perpetua, A. O., & Itiafa, A. L. (2023). Artificial Intelligence and the Media: Revisiting Digital Dichotomy Theory. In Information Systems Management. IntechOpen.

Nakova, P. (2020). Fake news: Definitions and debates. *Mediaeval Studies*, *45*(1), 2–15.

Nguyen, T., & Van Dinh, L. (2024). AI ethics in the age of intelligent machines. *AI and Ethics*, *4*(2), 1–12.

Pang, A., Shin, W., Lew, Z., & Walther, J. B. (2016). Building relationships through dialogic communication: Organizations, stakeholders, and computer-mediated communication. *Journal of Marketing Communications*, *24*(1), 68–82. DOI: 10.1080/13527266.2016.1269019

Patil, M., Yadav, H., Gawali, M., Suryawanshi, J., Patil, J., Yeole, A., Shetty, P., & Potlabattini, J. (2023). A Novel Approach to Fake News Detection Using Generative AI. *International Journal of Intelligent Systems and Applications in Engineering*, *12*(4s), 343–354. https://www.ijisae.org/index.php/IJISAE/article/view/3797

Raman, R., Kumar Nair, V., Nedungadi, P., Kumar Sahu, A., Kowalski, R., Ramanathan, S., & Achuthan, K. (2024). Fake news research trends, linkages to generative artificial intelligence and sustainable development goals. *Heliyon*, *10*(3), e24727. DOI: 10.1016/j.heliyon.2024.e24727 PMID: 38322879

Rossomakha, A.. (2023). ChatGPT and conversational AI. *Journal of Natural Language Processing*, *30*(1), 1–15.

Safadi, H., Li, W., Rahmati, P., & Soleymani, S. (2020). Curtailing Fake News Propagation with Psychographics. SSRN Electronic Journal. https://doi.org/DOI: 10.2139/ssrn.3558236

Sage, A. Bagheri, Maryam & Bagheritaba, Mohsen & Alizadeh, Sohila & Salemizadeh Parizi, Sam & Matoufinia, Parisa & Luo, Yang. (2024). AI-Driven Decision-Making in Healthcare Information Systems: A Comprehensive Review. DOI: 10.20944/preprints202406.0790.v1

Saini, R. (2023). Applications of AI in healthcare and finance. *Tech Innovations Journal, 15*(2), 45–60.

Sandrini, L., & Somogyi, R. (2023). Generative AI and deceptive news consumption. *Economics Letters, 232*, 111317–111317. DOI: 10.1016/j.econlet.2023.111317

Setiawan, B., Nugraha, D. P., Irawan, A., Nathan, R. J., & Zoltan, Z. (2021). User Innovativeness and Fintech Adoption in Indonesia. *Journal of Open Innovation, 7*(3), 188. DOI: 10.3390/joitmc7030188

Shi, S., Liang, Y., Ang, E. L., & Zhao, H. (2019). Delta Integration CRISPR-Cas (Di-CRISPR) in Saccharomyces cerevisiae. *Methods in Molecular Biology (Clifton, N.J.), 1927*, 73–91. DOI: 10.1007/978-1-4939-9142-6_6 PMID: 30788786

Shu, K., Mahudeswaran, D., & Liu, H. (2019). FakeNewsTracker: A tool for fake news collection, detection, and visualization. *Computational & Mathematical Organization Theory, 25*(1), 60–71. Advance online publication. DOI: 10.1007/s10588-018-09280-3

Singh, M. K., Ahmed, J., Raghuvanshi, K. K., & Alam, M. A.Manish Kumar Singh. (2023). Bharat-FakeNewsTracker: An Ensemble Learning Tool to Spot Fake News in India. *International Journal of Engineering Trends and Technology, 71*(9), 76–91. DOI: 10.14445/22315381/IJETT-V71I9P208

Singh, P., & Dhiman, D. B. (2023). Exploding AI-Generated Deepfakes and misinformation: A threat to global concern in the 21st century. Available at *SSRN* 4651093.

Sivek, S. D. (2018). Both Facts and Feelings: Emotion and News Literacy. *The Journal of Media Literacy Education, 10*(2), 123–138. DOI: 10.23860/JMLE-2018-10-2-7

Stieglitz, S., & Dang-Xuan, L. (2013). Social media and political communication. *Journal of Computer-Mediated Communication, 18*(3), 379–395.

Thakare, A., Sedani, S., Kriplani, S., Patel, A., & Umre, U. (2024). Chitosan: A Versatile Biomaterial Revolutionizing Endodontic Therapy. *Cureus, 16*(6), e62506. DOI: 10.7759/cureus.62506 PMID: 39022517

Tiku, V. N., Jain, C., Raz, Y., Nakamura, S., Heestand, B., Liu, W., Späth, M., Suchiman, H. E. D., Müller, R. U., Slagboom, P. E., Partridge, L., & Antebi, A. (2017). Small nucleoli are a cellular hallmark of longevity. *Nature Communications, 8*(1), 16083. DOI: 10.1038/ncomms16083 PMID: 28853436

Trandabă , D., & Gifu, D. (2023). Discriminating AI-generated Fake News. *Procedia Computer Science, 225*, 3822–3831. DOI: 10.1016/j.procs.2023.10.378

Tsipursky, G., Votta, F., & Roose, K. M. (2018). Fighting fake news and post-truth politics with behavioral science: The Pro-Truth Pledge. *Behavior and Social Issues, 27*(1), 47–70. DOI: 10.5210/bsi.v27i0.9127

Vaccari, C., & Chadwick, A. (2020). Deepfakes and Disinformation: Exploring the Impact of Synthetic Political Video on Deception, Uncertainty, and Trust in News. *Social Media + Society, 6*(1), 2056305120903408. Advance online publication. DOI: 10.1177/2056305120903408

Walorska, K. (2020). Deepfakes: Technology and implications. *Journal of Digital Forensics, 12*(3), 1–10.

Walther, J., & Jang, S. (2012). Social media likes and information exposure. *Journal of Computer-Mediated Communication*, *17*(3), 249–265.

Węcel, K., Sawiński, M., Stróżyna, M., Lewoniewski, W., Księżniak, E., Stolarski, P., & Abramowicz, W. (2023). Artificial intelligence—Friend or foe in fake news campaigns. *Economics and Business Review*, *9*(2). Advance online publication. DOI: 10.18559/ebr.2023.2.736

Yang, L., Wang, H., & Li, J. (2023). Digital Literacy and Misinformation: A Global Perspective. *International Journal of Communication*, *17*, 1234–1256.

Yingxuan, Y. (2024). Deceptive information dissemination. *Media Studies Quarterly*, *46*(1), 56–70.

Yun, J., & Ahn, J. (2018). Fake news and psychological effects. *Journal of Media Psychology*, *30*(2), 78–90.

Zafarani, R., Abbasi, M. A., & Liu, H. (2014). *Social media mining: an introduction*. Cambridge University Press.

Zhang, X., Liu, Y., & Chen, Z. (2019). The Role of Social Media in Spreading Fake News. *Journal of Information Science*, *45*(4), 567–582.

Zhang, Y., McAreavey, K., & Liu, W. (2022). Developing andExperimenting on Approaches to Explainability in AI Systems. In A. P.Rocha, L. Steels, & J. van den Herik (Eds.), Proceedings of the 14thInternational Conference on Agents and Artificial Intelligence (ICAART'22). Vol. 2, pp. 518-527. DOI: 10.5220/0010900300003116

Chapter 14
Gen Z's Code:
Cracking the Authenticity Formula in Influencer-Driven Social Media Advertising

Ana Venes
https://orcid.org/0009-0002-4501-5938
FCH, Catholic University of Portugal, Portugal

Alexandre Duarte
https://orcid.org/0000-0002-2665-864X
ICNOVA / FCSH-Universidade Nova de Lisboa, Portugal

ABSTRACT

This chapter explores Gen Z's authenticity code regarding influencer-driven social media advertising. In a context where social media is cluttered with sponsored content, Gen Z seeks for authenticity, contributing to an influencer marketing landscape where brands must crack a unique formula to engage this tech-savvy and socially conscious generation. By analyzing the data gathered from an online survey with a non-probabilistic sample of 105 Gen Z members, this research reveals details regarding transparency, influencer partnerships, and the delicate balance between creativity and authenticity. Our findings show that Gen Z prefers real content instead of amazing but hollow messages, values simplicity instead of megalomaniac advertising productions, prefers the truth over spectacularity, prefers to understand the logic and involvement of influencers with products or services instead of advertisements with scripts read sounding fake, and values unique, original and creative ideas to showcase the products, that inform the audience at the same time as it also entertain and engage them.

INTRODUCTION

Social media has become a crucial part of our lives in today's technology-driven world. It has completely transformed how we interact, form connections, and access information (Ortiz-Ospina & Roser, 2023; Lee & Cho, 2020). In this new social landscape, new stars have emerged: digital influencers (Gaenssle & Budzinski, 2021; Dias & Duarte, 2022a). However, every coin has two sides. As this phenomenon gains popularity, social media platforms have become increasingly cluttered with sponsored content and advertisements (Jung & Heo, 2021), making it challenging for brands to stand out and engage with

DOI: 10.4018/979-8-3693-6675-2.ch014

Copyright ©2025, IGI Global Scientific Publishing. Copying or distributing in print or electronic forms without written permission of IGI Global Scientific Publishing is prohibited.

their target audience. In this context, and due to today's consumers' "tendency to minimize ad exposure" (Jung & Heo, 2021, p. 592), Generation Z (Gen Z) specifically has developed an authenticity formula in influencer-driven social media advertising that brands need to "crack" to capture their attention. This article aims to explore the impact of authenticity in social influencer marketing on Gen Z.

Theoretical Background

Gen Z, the generation that is currently in their early twenties and late teenage years (Dimock, 2019), has never experienced life without the internet (Wilson, 2019) and is considered to be the most tech-savvy (Sachdeva, 2016) and socially conscious (Singh, 2014) generation. Growing up in a reality where there is no clear boundary between offline and online realities, highly increased Gen Z's awareness of the most common marketing tactics used to catch consumers' attention and for inauthenticity and decreased their chances of actually believing them. Thus, Gen Zers need marketing tactics to be customized to their unique needs and preferences (Dias & Duarte, 2022b) and are, therefore, skeptical towards traditional advertising, as they "feel it is invasive or forced upon them" (Wilson, 2019, p. 228). As in any communication strategy, the first step that brands need to do to be effective - in this specific case, to catch Generation Z's attention and crack their authenticity formula - is getting to know them by understanding their passions, values, and preferences. We are talking about a generation that values authenticity, transparency (Blakley, 2016; Read & Fromm, 2018), and social responsibility (Singh, 2014). Fancy advertisements or celebrity endorsements might work to persuade older generations, but not Gen Z. Instead, this generation seeks real people with relatable life stories and experiences supporting brands that align with their values and make a positive impact in the world (The Shelf, 2023). This is where influencer marketing (IM) comes into play. According to Vrontis et al. (2021), influencer marketing is a form of advertising in which brands stimulate popularity and influence individuals on social media (digital/social media influencers – SMIs), either by paying them or by offering some kind of advantage, to promote their products and services. IM is believed to be a crucial contributor to shaping social media's future and evolution (Ge & Gretzel, 2018; Appel et al., 2020). This tactic can be advantageous for brands because by partnering with SMIs, they can reach an already-built audience that trusts their opinions (Nieto, 2018; Breves et al., 2021; Duarte & Albuquerque, 2024). During this partnering process, both the SMIs and their followers contribute to the co-creation of the brand's image on social media (Martínez-López et al., 2020). Considering this, digital influencers become opinion leaders crucial in Generation Z's decision-making. However, their role extends beyond product or service promotion – genuineness, transparency, authenticity, and alignment with brands are essential (Mendes et al., 2020; Kay et al., 2020). This matters because financial motivations can be perceived as intrusive, especially in product-only posts, as posts that showcase the influencer with the product generate a higher credibility sense (Jin & Muqaddam, 2019). Furthermore, maintaining a healthy balance between sponsored and non-sponsored content is vital, as an excess of advertisements may lead to unfollowing (Djafarova & Trofimenko, 2019).

Authenticity, nevertheless, is not a straightforward concept. It has been used to refer to the reality or truth (Kennick, 1985), the fact of being natural, honest, or simple (Boyle, 2006), and it has also been used to refer to sincerity, innocence, and originality (Fine, 2003). In fact, authenticity is defined by the experience of diverse audiences, and it is therefore largely influenced by their interests and/or knowledge towards a specific topic (Grazian, 2005).

Regarding social influence marketing, some scholars believe that authenticity means way more than the influencer's endorsement, as it is also directly connected to the way they do it. Generation Z is a sucker for honest, relatable, and informative content that can be shared through different types of content, such as behind-the-scenes footage, product reviews, and story times of real-life experiences (Ki et al., 2020; Ki & Kim, 2019). In other words, this means that instead of simply sharing overly scripted promotions, SMIs should find unique and original ways to showcase the product or service (Casaló et al., 2020) while entertaining, informing, and engaging with their audience.

Lastly, some studies collide regarding sponsorship disclosures, which are directly connected to the sense of transparency in the influencer marketing field. De Jans et al. (2020) agree that by disclosing a sponsorship, the SMI enhances the brand image and increases the advertisement's acceptance. In contrast, a study conducted by De Veirman and Hudders (2020) shows that a sponsorship disclosure has negative effects on both the influencer's credibility and the user's attitude towards the brand since there is an increased ad recognition, which leads to ad skepticism.

METHODOLOGY

Research Questions

Our study addresses the following research questions:

RQ1—What are the practices of the Portuguese Gen Z (16–27-year-olds) on social media?
RQ2—What is and how important is authenticity of the digital influencers?
RQ3—Which type of content publish by digital influencers does Gen Z relate more?

Research Design and Data Collection Techniques

This study uses a quantitative research design to enrich the existing literature and discussion on Generation Z's authenticity code in influencer-driven social media advertising. The research is exploratory, and data were collected at a single point in time (Creswell, 2003). An online questionnaire was conducted via Google Forms as the data collection technique. This method was chosen due to its suitability for reaching the sample, which consisted of 105 respondents belonging to Generation Z, aged between 16 and 27, who are familiar with digital environments and follow influencers on social media. The survey questions can be consulted in Table 1. The insights derived from the questionnaire are used in the discussion section to either support or contradict the opinions of various authors cited in the literature review.

Table 1. Structure of the survey.

Parts	**Questions**	**Theoretical Grounding**
Informed Consent (eliminatory)	Informed Consent	General Regulation of Data Protection - Law nº 58/2019
Part 1- Sociodemographic data	*Q1:* Age (eliminatory, only those aged between 16 and 27 proceed)	Bossen and Kottasz (2020)
	Q2: Gender	Kennedy (2020)
Part 2- Social Media practices	*Q3:* Which social media platforms do you use most frequently?	Alhabash and Ma (2017)
	Q4: How much time, on average, do you spend on social media daily?	Kuss and Griffiths (2011)
	Q5: Do you feel like social media is cluttered with advertisements?	Hudson et al. (2015)
	Q6: Do you consider that, because of this ad cluttering, you have the tendency to minimize ad exposure when scrolling through social media?	Cho and as- (2004)
	Q7: Do you feel like the rise of influencer marketing is contributing to social media's ad cluttering?	De Veirman et al. (2017)
Part 3- Digital Influencers relationship	*Q8:* Do you follow influencers on social media?	Djafarova and Rushworth (2017)
	Q9: What types of influencers do you follow?	Abidin (2015)
	Q10: How often do you engage with influencer content? (e.g., liking, commenting, sharing)	Casaló et al. (2020)
Part 4- Digital Influencers Authenticity	*Q11:* What does authenticity mean to you in the context of influencer content?	(Audrezet et al., 2018)
	Q12: How important is authenticity when you engage with influencer content?	Djafarova and Trofimenko's (2019)
	Q13: How do you identify if an influencer's promotion is authentic or inauthentic?	Schouten et al. (2019)
	Q14: Have you ever felt that an influencer was not genuine in their endorsement?	Campbell and Farrell (2020)
	Q15: If you feel like an influencer's promotion is inauthentic, how does it affect your willingness to buy the product/service they are promoting?	Lou and Yuan (2018)

continued on following page

Table 1. Continued

Parts	Questions	Theoretical Grounding
Part 5- Digital Influencers Content	*Q16:* Do you prefer influencers who partner with brands that align with their personal values?	Hughes et al. (2019)
	Q17: How does an influencer's partnership with a brand impact your perception of both the influencer and the brand?	Sokolova and Kefi (2019)
	Q18: What type of content do you find most engaging from influencers?	Breves et al. (2019)
	Q19: How much do you value creativity and innovation in influencer content?	Rijitha (2021)
	Q20: How important is it for influencers to be transparent about the products or services they promote?	Evans et al. (2017)
	Q21: Do you appreciate influencers who actively support or promote social responsibility initiatives?	Ki et al. (2020)
	Q22: Do you appreciate influencers partnering with brands that take a stand on important social issues?	Kian Yeik et al. (2022)
	Q23: Can you recall any example where an influencer partnered with a brand that took a stand on an important social issue? If so, briefly explain how it affected your perception of both the brand and the influencer.	Audrezet et al. (2018)
Part 6- Best practices' Suggestions	*Q24:* What actions can brands and influencers take to build trust with Gen Z?	Djafarova and Bowes (2020)
	Q25: Are there specific changes or improvements you'd like to see in influencer-driven social media advertising?	Ashley and Tuten (2015)

To make the questionnaire accessible to everybody who is not familiar with technical language, the authors opted to use direct questions to avoid misunderstandings and biased results. Thus, we used mainly closed questions but also some Likert scales (1 - completely disagree to 5 - completely agree) in specific items designed to measure the importance or value of some phrases, and finally, two open questions.

The survey was online in the first week of the year, between 3rd and 7th of January 2024.

Data Analysis Techniques

The data were analyzed through descriptive statistics performed using Excel and IBM SPSS, version 29.0.0.0 (de Vaus, 2013).

Sampling and Sample

To reach a sample of Social Media users for this study, the authors started by recruiting volunteers on their networks by sharing the link of the online survey on their social media platforms such as Facebook, Instagram, TikTok, and LinkedIn, as well as sending some WhatsApp messages, asking for participation, thus using a convenience sample (Kalton, 2020). In total, 105 answers were obtained in the online survey, as presented in Table 2, with a brief characterization of the sample (n = 105).

Table 2. Brief characterization of the sample.

Parts	Questions
	16-18 years old: 19%
Age (Q1)	19-23 years old: 71,4%
	24-27 years old: 9,5%
Gender (Q2)	Female: 74,3%
	Male: 25,7%

FINDINGS AND DISCUSSION

Ortiz-Ospina and Roser (2023) and Lee and Cho (2020) believe social media completely transformed the way we communicate with each other. Although there's no way to deny this, the authors believe that influencers play a key part in this digital transformation. In Figure 1, to answer RQ1, we can see the more used platforms by our respondents from Gen Z (Q3), where Instagram dominates with 95,2% of the respondents assuming the use of this platform, followed by TikTok (71,4%) and YouTube (49,5%). A note for the fact that just one respondent referred to Facebook and Be Real. In Figure 2, we can see the daily time spent on social media (Q4), with more than 3/4 answering that they spent 2 or more hours a day.

Figure 1. Main used social media platforms by Gen Z

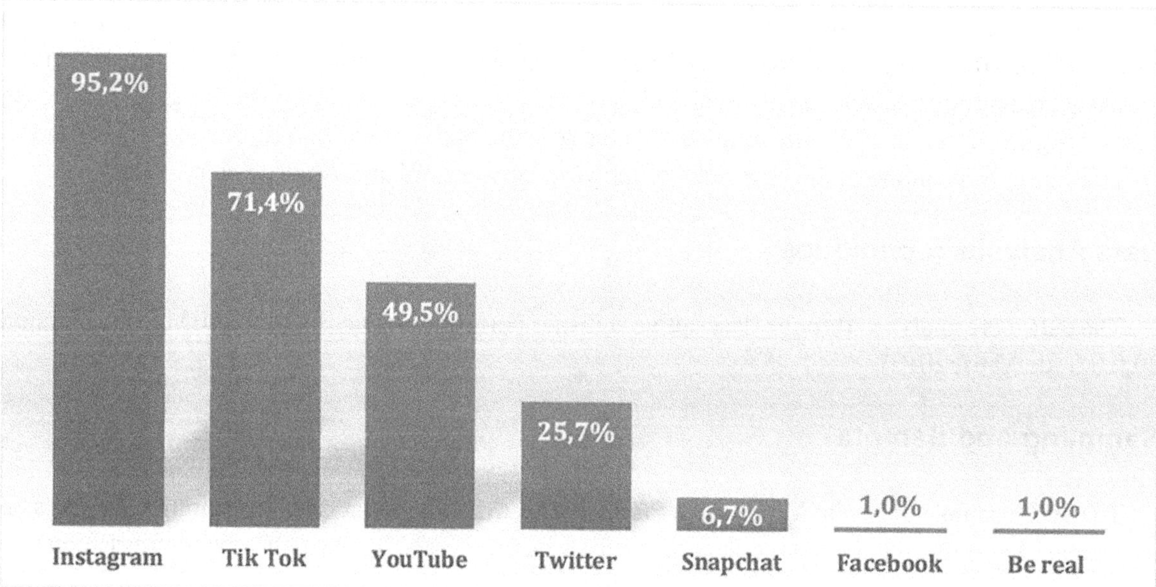

Figure 2. Daily time used on social media platforms

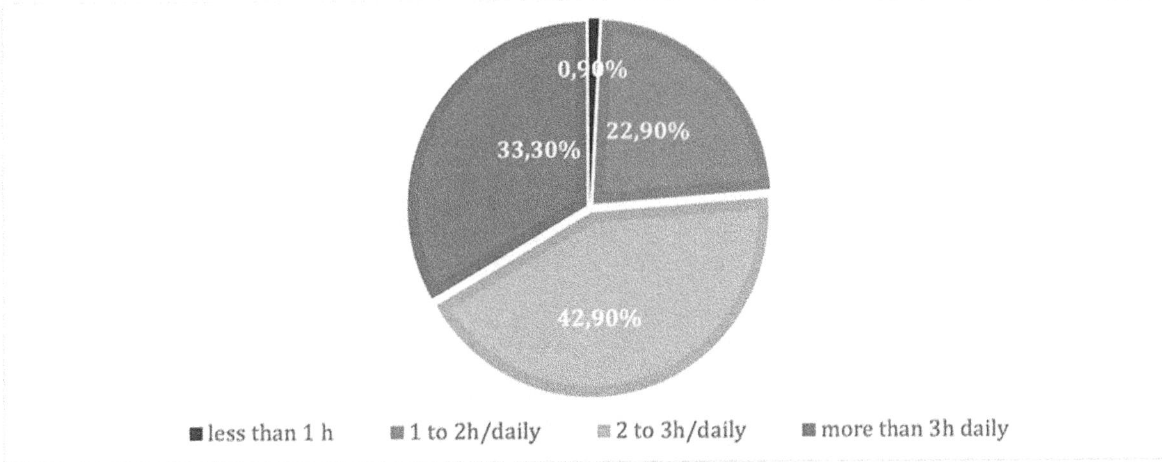

In this daily time consumed by social media, digital Influencers occupy a big part of Gen Z's screen time, working as opinion leaders (Mendes et al., 2020; Pinto et al., 2022), which makes influencer marketing extra relevant in several areas, namely in commercial communication.

88,6% of our sample follows digital influencers (Q8) in many different areas, from Beauty to Lifestyle, from Fashion to travel, and from Gaming to Fitness (Q9). Regarding the frequency of interaction (liking, commenting, or sharing) with the digital influencers' posts (Q10), as shown in Figure 3, 76,1% of the respondents rarely or only occasionally interact (37,1% and 39% respectively), 17,1% interacts frequently and 6,8% always interact.

Figure 3. Frequent interaction with digital influencers' posts

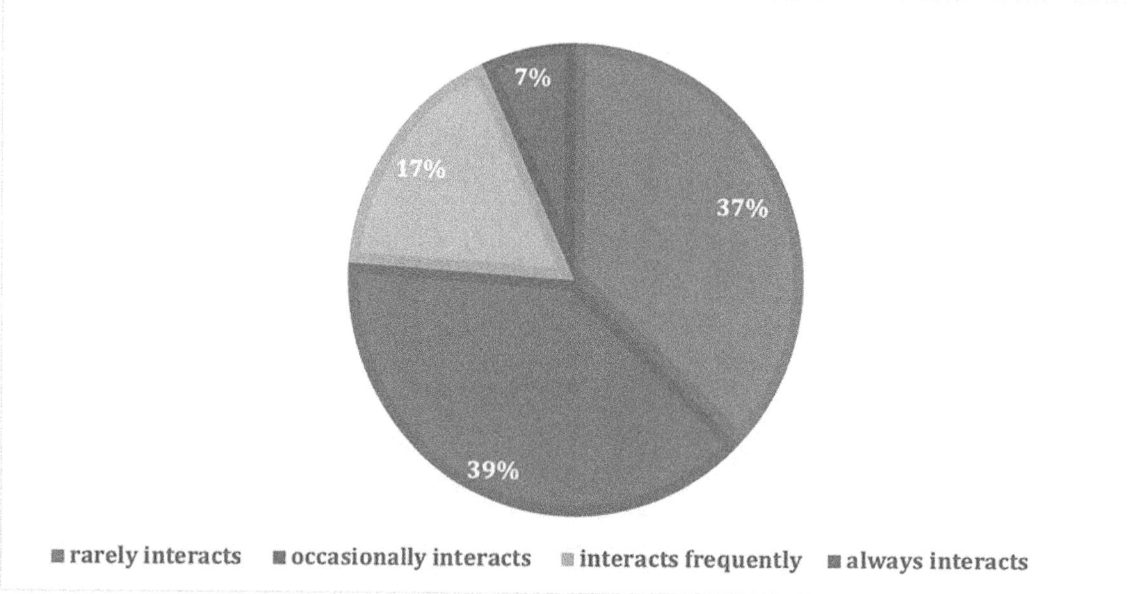

Nonetheless, 96,6% of the respondents in our study, consider that social media is cluttered with advertisements (Q5) and 91,5% feel like the rise of influencer marketing is the main reason for this (Q7). This fact could be a problem for brands, but especially for influencers, as warned by Djafarova and Trofimenko (2019), once the excess of advertisements frequently leads to unfollowing. Along with Jung and Heo's (2021) belief, 86,4% stated that they tend to minimize ad exposure when scrolling through social media (Q6) precisely due to this fact.

Analyzing Gen Z's preference regarding the emphasis on transparency and authentic content is another important for our discussion. Read and Fromm (2018), as well as Blakley (2016) highlighted the importance of transparency, which, paired with the questionnaire data, reveals a relevant perspective. While 68,8% of the respondents accentuated the importance of transparency, the challenge for brands and influencers lies in defining what transparency means. The respondents, as highlighted by the data (Q11), gravitate towards "genuine and honest representation" (48,4%), "relatable and unfiltered content" (39,8%), and "Transparency about sponsorships (11,8%).

One of the first steps for organizations and brands to achieve transparency is to be clear about their manufacturing processes and material sourcing. This way, partnering with digital influencers can also be fully transparent when promoting their products or services, and brands will be able to meet Gen Z's values while also enhancing the authenticity of their influencer marketing campaigns.

The existing literature also introduces conflicting views on sponsorship disclosures. While De Jans et al. (2020) believe that disclosure enhances brand image, De Veirman and Hudders (2020) suggest some negative effects. This dichotomy underlines the difficulties brands and influencers face in balancing the disclosure of sponsorships with maintaining credibility.

To answer the RQ2, the questionnaire findings provide valuable insights regarding authenticity preferences among Gen Zers, with a substantial 88.1% rating authenticity as important (33,3%) or highly important (54,8%) when engaging with influencer content (Q12). This is in line with the findings of Kay et al. (2020) and Mendes et al. (2020), which refer to the importance of genuineness, transparency, authenticity, and alignment with brands. However, achieving this level of authenticity requires a symbiotic relationship between brands and influencers. Martínez-López et al. (2020) emphasize the critical role of influencers choosing brand partnerships judiciously, advocating for collaboration only with brands they genuinely believe in and intend to use because otherwise, the audience is not going to feel like the content is genuine and relatable. This feeling of authenticity is essential, as identified by the respondents (figure 4), who are skeptical towards influencer promotions (Q13) that lack personal experience with the product (54.8%), overly scripted endorsements (51.6%), and frequently sponsored content (46.2%).

Figure 4. How do you identify if an influencer's promotion is authentic or inauthentic?

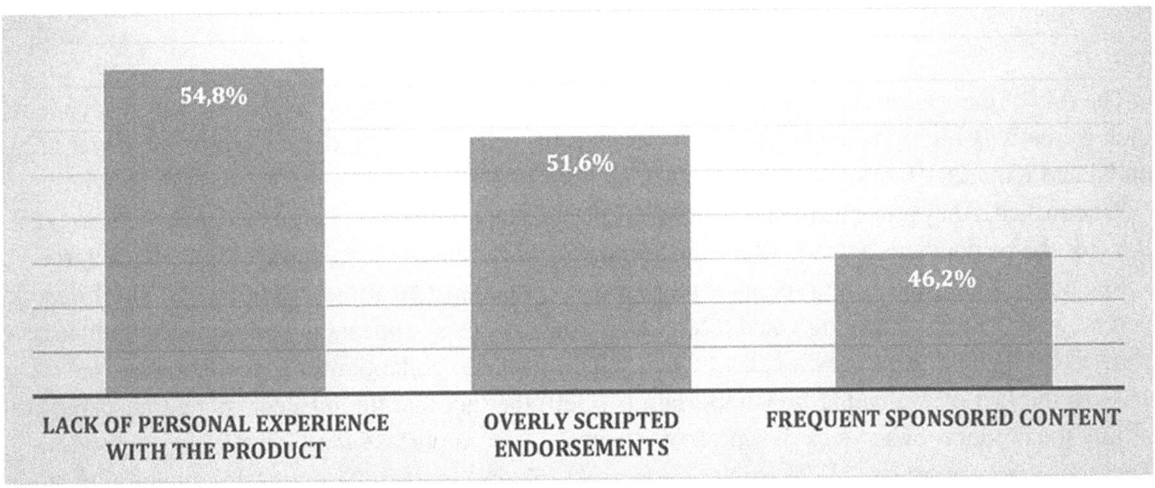

These findings, specifically the ones regarding the frequency of sharing sponsored content, are supported by Djafarova and Trofimenko's (2019) research, which underlines the potential consequences of inauthentic promotion on follower engagement and loyalty.

92,4% of the respondents assumed that they already felt that a digital influencer wasn't being authentic in their sponsored content (Q14). When asked how that feeling of inauthenticity affects the buying intention of that product, more than 60% said agree (24,7%) or totally agree (36,6%), as shown in Figure 5.

Figure 5. How does the feeling of unauthenticity affect your purchase intention?

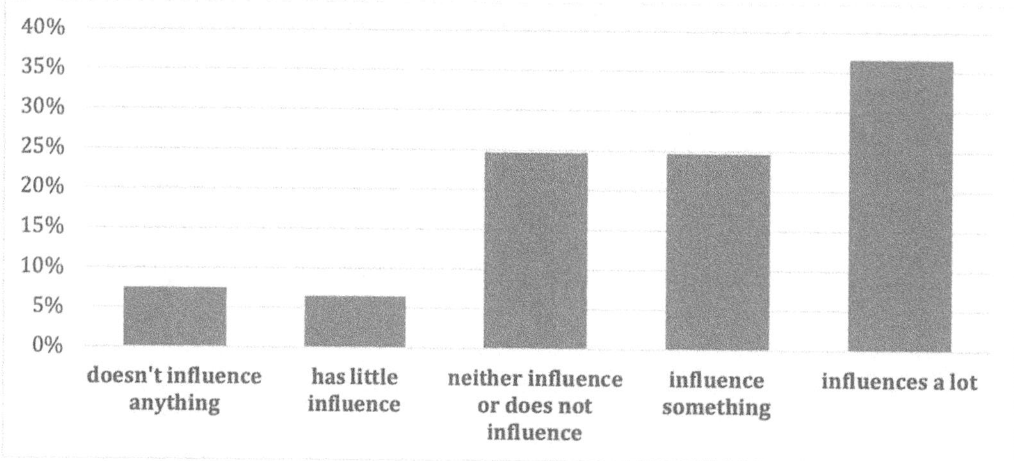

The role of innovation also reinforces the importance of authenticity, with 87.1% of our respondents valuing creativity highly, being aligned with Casaló et al.'s (2020) perspective. However, Ki et al. (2020) and Ki and Kim (2019) bring attention to the challenge of achieving authenticity through creativity.

When asked if they prefer influencers that make partnerships with brands align with their values, (Q16), 79,6% of the respondents agreed, 18,3% answered it's indifferent to them, while only 2,1% disagreed.

Regarding RQ3, about the type of content respondents consider more engaging (Q18), "Lifestyle content" was the most engaging voted (65,6%), "product reviews" appears in second place with 46,5%, followed by "behind the scenes" with 43% and "challenges and collaborations" with 23,7%.

As to the fact of evaluating how important is (Q20) the fact that the influencers be transparent regarding the products or services they promote (from 1 - "not important at all" to 5 - "very important"), 92,3% consider important (23,7%) or very important (68,6%), with 7,7% opting for a neutral position - "neither important nor not important" and none of the respondents considering low or zero important.

Almost all of the respondents (96,8%) assumed they appreciate influencers that support or promote actively social responsibility initiatives (Q21) and 92,5% appreciate those who personally position themselves towards important social issues (Q22).

Finally, in the first of the last two open questions, we asked for suggestions about what brands and influencers can do to build trust with Gen Zers (Q24). Some of the answers (for brands) were: " Learn what their audience values and try to align themselves with said values", or "Partner with influencers that were using the product before or were brand lovers". As suggestions, for influencers, and supporting Jin and Muqaddam (2019) research proved that posts that showcase the influencer with the product generate a higher credibility sense, we highlight "Try to be more transparent with what you share and don't get involved with dubious brands" and "Only promote products that you've already tried and actually liked." This is in line with several authors (e.g. Blakley, 2016; Read & Fromm, 2018) that already explained how Gen Z values authenticity and transparency.

In the last open question, the authors aimed to know what improvements they would like to see in influencer-driven social media advertising (Q25), and some interesting insights were also pointed out as "Authenticity should be the priority. I don't want to see scripted ads. Makes social media and advertising boring." or "More transparency in what they are really advertising and what they just like. Sometimes can't tell the difference." Aligned with the writings of Singh (2014) that proved how this generation is attached to social responsibility, two other relevant testimonials are: "Promote ideas/products that bring rewards/benefits in the future/long term and not so much ideas that satisfy just pleasures or momentary ideas that end up fading" and "I believe that if the brands can show to the public the impact of their brand, or just like a real example where we can use their product and impact our environment would be interesting." As shown in The Shelf (2023) article, Ki et al. (2020), Ki and Kim (2019), and others, contrary to older generations that were easily stimulated by huge fancy advertisements or celebrity endorsements, Gen Z is looking for real stories, from real people, with which they can relate to, and therefore preferring brands aligned with their values, but especially with those who aim to cause a positive impact in the community, society, and the world.

CONCLUSION

Behavioral differences between generations are generally accepted and are well-documented academically (De Pelsmacker et al., 2005). According to Pueschel et al. (2020), Generation Z is generally very entrepreneurial, responsible, loyal, very technology-oriented, and is more motivated by challenging projects than by money. They are individuals who, for example, more than being happy about getting a job and are concerned with understanding the purpose that awaits them in this job (Seifert et al., 2023). All this fits into an attitude of authenticity, highly valued by this generation, whose need for understanding is fundamental to establishing bonds of communication and relationships.

Due to its own subjectivity and personal interpretation, there is no absolute consensus on the authenticity definition in the social influence marketing context. Even so, most of the respondents in our study highlighted the honest and genuine representation of the promoted products by the influencers, relatable and non-filtered content, and honesty and transparency regarding sponsorships as the main factors to be considered.

This generation's skepticism and concerns regarding advertising, as pointed out by Wilson (2019), raise the question of how to gain attention and ensure the trust organizations need when reaching these individuals with commercial messages. Knowing that digital communication, particularly through social media platforms, is a very important channel for them, both brands and influencers need to understand what their passions, values, preferences, and motivations are, but mainly what disturbs them and can harm this relationship.

Our findings show that in line with the writings of Casaló et al. (2020), this generation prefers real content instead of amazing but hollow messages, values simplicity instead of megalomaniac advertising productions, prefers the truth over spectacularity, prefers to understand the logic and involvement of influencers with products or services instead of advertisements with scripts read sounding fake, and values unique, original and creative ideas to showcase the products, that inform the audience at the same time as it also entertain and engage them.

Nevertheless, this may raise another question: how can brands and influencers create and deliver constantly innovative, creative content, while remaining authentic? The respondent's request for "genuine and creative marketing" brings us to a very subjective context in which authenticity intersects with content originality, providing a rich field for further investigation.

REFERENCES

Abidin, C. (2015). Communicative intimacies: Influencers and Perceived Interconnectedness. Ada: *A Journal of Gender, New Media, and Technology,* 8, 1–16. DOI: 10.7264/N3MW2FFG

Alhabash, S., & Ma, M. (2017). A Tale of Four Platforms: Motivations and Uses of Facebook, Twitter, Instagram, and Snapchat Among College Students? *Social Media + Society*, *3*(1), 205630511769154. DOI: 10.1177/2056305117691544

Appel, G., Grewal, L., Hadi, R., & Stephen, A. T. (2020). The future of social media in marketing. *Journal of the Academy of Marketing Science*, *48*(1), 79–95. DOI: 10.1007/s11747-019-00695-1 PMID: 32431463

Audrezet, A., Kerviler, G., & Moulard, J. (2018). Authenticity under threat: When social media influencers need to go beyond self-presentation. *Journal of Business Research*, *117*, 557–569. Advance online publication. DOI: 10.1016/j.jbusres.2018.07.008

Blakley, C. (2016, December 5). How to Build a Marketing Campaign that Appeals to Generation Z. Fortune. https://fortune.com/2016/12/05/marketing-campaign-generation-z/

Bossen, C. B., & Kottasz, R. (2020). Uses and gratifications sought by pre-adolescent and adolescent TikTok consumers. *Young Consumers*, *21*(4), 463–478. DOI: 10.1108/YC-07-2020-1186

Boyle, D. (2006). *Authenticity: Brands, fakes, spin and the lust for real life*. Harper Perennial.

Breves, P., Amrehn, J., Heidenreich, A., Liebers, N., & Schramm, H. (2021). Blind trust? The importance and interplay of parasocial relationships and advertising disclosures in explaining influencers' persuasive effects on their followers. *International Journal of Advertising*, *40*(7), 1209–1229. DOI: 10.1080/02650487.2021.1881237

Breves, P., Liebers, N., Abt, M., & Kunze, A. (2019). The Perceived Fit between Instagram Influencers and the Endorsed Brand: How Influencer–Brand Fit Affects Source Credibility and Persuasive Effectiveness. *Journal of Advertising Research*, *59*(4), 440–454. DOI: 10.2501/JAR-2019-030

Campbell, C., & Farrell, J. (2020). More than meets the eye: The functional components underlying influencer marketing. *Business Horizons*, *63*(4), 469–479. Advance online publication. DOI: 10.1016/j.bushor.2020.03.003

Casaló, L. V., Flavián, C., & Ibáñez-Sánchez, S. (2020). Influencers on Instagram: Antecedents and consequences of opinion leadership. *Journal of Business Research*, *117*, 510–519. DOI: 10.1016/j.jbusres.2018.07.005

Cho, C. H., & as-, U. T. A. (2004). Why do people avoid advertising on the internet? *Journal of Advertising*, *33*(4), 89–97. DOI: 10.1080/00913367.2004.10639175

Creswell, J. (2003). *Research Design, Qualitative, Quantitative, and Mixed Methods Approaches*. Sage Publications.

De Pelsmacker, P., Geuens, M., & Van den Bergh, J. (2005). *Marketingcommunicatie*. Pearson Education Benelux.

de Vaus, D. (2013). *Surveys in Social Research*. Routledge.

De Veirman, M., Cauberghe, V., & Hudders, L. (2017). Marketing through Instagram influencers: The impact of number of followers and product divergence on brand attitude. *International Journal of Advertising*, *36*(5), 798–828. DOI: 10.1080/02650487.2017.1348035

De Veirman, M., & Hudders, L. (2020). Disclosing sponsored Instagram posts: The role of material connection with the brand and message-sidedness when disclosing covert advertising. *International Journal of Advertising*, *39*(1), 94–130. DOI: 10.1080/02650487.2019.1575108

Dias, P., & Duarte, A. (2022a). How Portuguese adolescents relate to influencers and brands on TikTok. *Journal of Digital & Social Media Marketing*, *10*(1), 82–95. DOI: 10.69554/VSHW2788

Dias, P., & Duarte, A. (2022b). TikTok Practices among Teenagers in Portugal: A Uses & Gratifications Approach. *Journalism and Media*, *3*(4), 615–632. DOI: 10.3390/journalmedia3040041

Dimock, M. (2019, January 17). *Defining generations: Where Millennials end and Generation Z begins*. Pew Research Center. https://www.pewresearch.org/short-reads/2019/01/17/where-millennials-end-and-generation-z-begins/

Djafarova, E., & Bowes, T. (2020). 'Instagram made Me buy it': Generation Z impulse purchases in fashion industry. *Journal of Retailing and Consumer Services*, *59*, 102345. DOI: 10.1016/j.jretconser.2020.102345

Djafarova, E., & Rushworth, C. (2017). Exploring the credibility of online celebrities' Instagram profiles in influencing the purchase decisions of young female users. *Computers in Human Behavior*, *68*, 1–7. DOI: 10.1016/j.chb.2016.11.009

Djafarova, E., & Trofimenko, O. (2019). 'Instafamous' – credibility and self-presentation of micro-celebrities on social media. *Information Communication and Society*, *22*(10), 1432–1446. DOI: 10.1080/1369118X.2018.1438491

Duarte, A., & Albuquerque, L. (2024). The Influence of TikTok in Portuguese Millennials' Footwear Consumer Behaviour. In: Rocha, Á., Adeli, H., Dzemyda, G., Moreira, F., Poniszewska-Marańda, A. (eds) Good Practices and New Perspectives in Information Systems and Technologies. WorldCIST 2024. *Lecture Notes in Networks and Systems, vol 990,* pp. 117–126. Springer https://doi.org/DOI: 10.1007/978-3-031-60328-0_12

Evans, N., Phua, J., Lim, J., & Jun, H. (2017). Disclosing Instagram Influencer Advertising: The Effects of Disclosure Language on Advertising Recognition, Attitudes, and Behavioral Intent. *Journal of Interactive Advertising,* 17, 00–00. DOI: 10.1080/15252019.2017.1366885

Fine, G. A. (2003). Crafting authenticity: The validation of identity in self-taught art. *Theory and Society*, *32*(2), 153–180. DOI: 10.1023/A:1023943503531

Ge, J., & Gretzel, U. (2018). Emoji rhetoric: A social media influencer perspective. *Journal of Marketing Management*, *34*(15-16), 1272–1295. Advance online publication. DOI: 10.1080/0267257X.2018.1483960

Grazian, D. (2005). *Blue Chicago: The search for authenticity in urban blues clubs*. University of Chicago Press.

Hudson, S., Huang, L., Roth, M., & Madden, T. (2015). The influence of social media interactions on consumer–brand relationships: A three-country study of brand perceptions and marketing behaviors. *International Journal of Research in Marketing*, *33*(1), 27–41. Advance online publication. DOI: 10.1016/j.ijresmar.2015.06.004

Hughes, C., Swaminathan, V., & Brooks, G. (2019). Driving Brand Engagement Through Online Social Influencers: An Empirical Investigation of Sponsored Blogging Campaigns. *Journal of Marketing*, *83*(5), 78–96. DOI: 10.1177/0022242919854374

Jin, S. V., & Muqaddam, A. (2019). Product placement 2.0: "Do Brands Need Influencers, or Do Influencers Need Brands?". *Journal of Brand Management*, *26*(5), 522–537. DOI: 10.1057/s41262-019-00151-z

Jung, A.-R., & Heo, J. (2021). Does cluttered social media environment hurt advertising effectiveness? The moderation of ad types and personalization. *Journal of Research in Interactive Marketing*, *15*(4), 592–606. DOI: 10.1108/JRIM-11-2020-0238

Kalton, G. (2020). *Introduction to Survey Sampling*. Sage.

Kay, S., Mulcahy, R., & Parkinson, J. (2020). When less is more: The impact of macro and micro social media influencers' disclosure. *Journal of Marketing Management*, *36*(3–4), 248–278. DOI: 10.1080/0267257X.2020.1718740

Kennedy, M. (2020). 'If the Rise of the TikTok Dance and e-Girl Aesthetic has taught us Anything, it's that Teenage Girls Rule the Internet Right Now': TikTok Celebrity, Girls and the Coronavirus Crisis. *European Journal of Cultural Studies*, *23*(6), 1069–1076. DOI: 10.1177/1367549420945341

Kennick, W. E. (1985). Art and Inauthenticity. *The Journal of Aesthetics and Art Criticism*, *44*(1), 3–12. DOI: 10.1111/1540_6245.jaac44.1.0003

Ki, C.-W., Cuevas, L. M., Chong, S. M., & Lim, H. (2020). Influencer marketing: Social media influencers as human brands attaching to followers and yielding positive marketing results by fulfilling needs. *Journal of Retailing and Consumer Services*, *55*, 102133. DOI: 10.1016/j.jretconser.2020.102133

Ki, C.-W., & Kim, Y.-K. (2019). The mechanism by which social media influencers persuade consumers: The role of consumers' desire to mimic. *Psychology and Marketing*, *36*(10), 905–922. DOI: 10.1002/mar.21244

Kian Yeik, K., Cheung, M.-L., Soh, P., & Teoh, C. W. (2022). Social media influencer marketing: The moderating role of materialism. *European Business Review*, *34*(2), 224–243. DOI: 10.1108/EBR-02-2021-0032

Kuss, D., & Griffiths, M. (2011). Online Social Networking and Addiction—A Review of the Psychological Literature. *International Journal of Environmental Research and Public Health*, *8*(9), 3528–3552. DOI: 10.3390/ijerph8093528 PMID: 22016701

Lee, H., & Cho, C.-H. (2020). Digital advertising: Present and future prospects. *International Journal of Advertising*, *39*(3), 332–341. DOI: 10.1080/02650487.2019.1642015

Lou, C., & Yuan, S. (2018). Influencer Marketing: How Message Value and Credibility Affect Consumer Trust of Branded Content on Social Media. *Journal of Interactive Advertising*, *19*(1), 1–45. DOI: 10.1080/15252019.2018.1533501

Martínez-López, F. J., Anaya-Sánchez, R., Fernández Giordano, M., & Lopez-Lopez, D. (2020). Behind influencer marketing: Key marketing decisions and their effects on followers' responses. *Journal of Marketing Management*, *36*(7–8), 579–607. DOI: 10.1080/0267257X.2020.1738525

Mendes, M., Magano, J., & Silva, C. S. e. (2020). Marketing de Influência e Autenticidade: Impacto no Processo de Compra da Geração Z. *Percursos & Ideias*, *10*, 16–27. DOI: 10.56123/percursos.2020.n10.16

Nieto, B. G. (2018). El influencer: Herramienta clave en el contexto digital de la publicidad engañosa. *Methaodos.Revista de Ciencias Sociales*, *6*(1), 1. Advance online publication. DOI: 10.17502/m.rcs.v6i1.212

Ortiz-Ospina, E., & Roser, M. (2023). *The rise of social media. Our World in Data.* https://ourworldindata.org/rise-of-social-media

Pinto, B., Duarte, A., & Dias, P. (2022). The influence of YouTubers on children (8-12 years old): Current issues and brands. *Doxa Comunicación. Revista Interdisciplinar De Estudios De Comunicación Y. Ciencia y Sociedad*, (36), 321–340. DOI: 10.31921/doxacom.n36a1638

Pueschel, A., Johnson, R. C., & Dhanani, L. Y. (2020). Putting Gen Z first: Educating with a generational mind-set. *Industrial and Organizational Psychology: Perspectives on Science and Practice*, *13*(4), 594–598. DOI: 10.1017/iop.2020.103

Read, A., & Fromm, J. (2018). *Marketing to Gen Z: The Rules for Reaching This Vast and Very Different Generation of Influencers*. AMACOM.

Rijitha, R. (2021). The Impact of Social Media Marketing on Consumer Purchase Intention. In *International Conference on Contemporary Media Persuasive Technology and Psychology of Visual Communication* (pp. 100-119).

Sachdeva, D. (2016). Gen Z – THE TECHILICIOUS GENERATION Case Study of a Digitally Hooked Gen Zee. TIJ's *Research Journal of Social Science & Management - RJSSM,* 6. https://api.semanticscholar.org/CorpusID:151938534

Schouten, A., Janssen, L., & Verspaget, M. (2019). Celebrity vs. Influencer endorsements in advertising: The role of identification, credibility, and Product-Endorser fit. *International Journal of Advertising*, *39*(2), 1–24. DOI: 10.1080/02650487.2019.1634898

Seifert, C. F., Van Ness, R. K., Eddy, E. R., Buff, C., & D'Abate, C. P. (2023). Generational Work Ethic Differences: From Baby Boomers to Gen Z. *Journal of Managerial Issues*, *35*(4), 401–422.

Singh, A. (2014). Challenges and Issues of Generation Z. *IOSR Journal of Business and Management*, *16*(7), 59–63. DOI: 10.9790/487X-16715963

Sokolova, K., & Kefi, H. (2019). Instagram and YouTube bloggers promote it, why should I buy? How credibility and parasocial interaction influence purchase intentions. *Journal of Retailing and Consumer Services*, *53*, 101742. Advance online publication. DOI: 10.1016/j.jretconser.2019.01.011

The Shelf. (2023, May 15). *Gen Z Social Media: How They Use It and the Content They Want to See* [Website]. The Shelf Full-Service Influencer Marketing. https://www.theshelf.com/insights/gen-z-social-media/

Vrontis, D., Makrides, A., Christofi, M., & Thrassou, A. (2021). Social media influencer marketing: A systematic review, integrative framework and future research agenda. *International Journal of Consumer Studies*, *45*(4), 617–644. DOI: 10.1111/ijcs.12647

Wilson, B. (2019). Book Review: Marketing to Gen Z: The Rules for Reaching This Vast and Very Different Generation of Influencers. *Journal of Macromarketing*, *39*(2), 228–230. DOI: 10.1177/0276146719830156

Compilation of References

Abdikarimov, K. (2024). Media literacy as a means of protecting Uzbekistan's media environment from disinformation. *International Journal of Media and Communications in Central Asia*. Tashkent. *UzJOKU*, (Special issue), 47–55.

Abhishek, G. (2023). Digital citizenship: empowering responsible online engagement. https://www.linkedin.com/pulse/digital-citizenship-empowering-responsible-online-abhishek-guguloth-/

Abidin, C. (2015). Communicative intimacies: Influencers and Perceived Interconnectedness. Ada: *A Journal of Gender, New Media, and Technology,* 8, 1–16. DOI: 10.7264/N3MW2FFG

Acquisti, A., Brandimarte, L., & Loewenstein, G. (2015). Privacy and human behavior in the age of information. *Science*, *347*(6221), 509–514. DOI: 10.1126/science.aaa1465 PMID: 25635091

Adyin, A., Arda, B., Güneş, B., & Erbaş, O. (2021). Psychopathology of cyberbullying and internet trolling. *Journal of Experimental and Basic Medical Sciences.*, *2*(3), 380–391. DOI: 10.5606/jebms.2021.75680

Agapoglou, T., Mouratoglou, N., Tsioumis, K., & Bikos, K. (2021). Combating Online Hate Speech through Critical Digital Literacy: Reflections from an Emancipatory Action Research with Roma Youths. *International Journal of Learning and Development*, *11*(2), 104. DOI: 10.5296/ijld.v11i2.18524

Agger, B. (2012). *Oversharing: Presentations of self in the internet age*. Routledge. DOI: 10.4324/9780203125090

Ahluwalia, R. (2000). An Examination of Psychological Processes Underlying Resistance to Persuasion. *The Journal of Consumer Research*, *27*(2), 217–232. DOI: 10.1086/314321

Ahmed, S., & Matthes, J. (2016). Media representation of Muslims and Islam from 2000 to 2015: A meta-analysis. *The International Communication Gazette*, *79*(3), 219–244. DOI: 10.1177/1748048516656305

AIESEC. (n.d.). Leadership is for every young person. https://aiesec.org/

Aimeur, E., Amri, S., & Brassard, G. (2023). Fake news, disinformation and misinformation in social media: A review. *Social Network Analysis and Mining*, *13*(1), 1–36. DOI: 10.1007/s13278-023-01028-5 PMID: 36789378

Ainley, K. (2017). *Virtue ethics.* Oxford University Press and the International Studies Association.

Ajzen, I. (1991). The theory of planned behavior. *Organizational Behavior and Human Decision Processes*, *50*(2), 179–211. DOI: 10.1016/0749-5978(91)90020-T

Akhmetova, L.S. & Shorokhov, D.P. (2018). Media and Information Literacy: The solution of questions of formation of a civic stand in Kazakhstan. *Публічне урядування, 3*(13), 14-27.

Akhtar, P., Ghouri, A. M., & Khan, H. U. R., Amin ul Haq, M., Awan, U., Zahoor, N., Khan, Z., & Ashraf, A. (2022). Detecting fake news and disinformation using artificial intelligence and machine learning to avoid supply chain disruptions. *Annals of Operations Research, 327*. Advance online publication. DOI: 10.1007/s10479-022-05015 PMID: 36338350

Akkor Gül, A., Ertürk, Y., & Elmer, P. (2020). *Digital Transformation in Media & Society*. Istanbul University Press., DOI: 10.26650/B/SS07.2020.013

Al Tamimi, H. A. H., Duqi, A., Kanas, A., & Zervopoulos, P. D. (2021). Directional distance function DEA estimators for evaluating efficiency gains from possible mergers and acquisitions. *The Journal of the Operational Research Society*, *73*(6), 1240–1257. DOI: 10.1080/01605682.2021.1907243

Alazab, M., & Broadhurst, R. (2017). *An analysis of the nature of spam as cybercrime. Cyber-Physical Security*. Springer International Publishing.

Aldahshan. (2016). Digital citizenship as an introduction to Arab education in the digital age. *Journal of Criticism and Enlightenment*, 5/(2), 72-104.

Alhabash, S., & Ma, M. (2017). A Tale of Four Platforms: Motivations and Uses of Facebook, Twitter, Instagram, and Snapchat Among College Students? *Social Media + Society*, *3*(1), 205630511769154. DOI: 10.1177/2056305117691544

Ali Adeeb, R., & Mirhoseini, M. (2023). The Impact of Affect on the Perception of Fake News on Social Media: A Systematic Review. *Social Sciences (Basel, Switzerland)*, *12*(12), 674. DOI: 10.3390/socsci12120674

Ali, S., Wang, D., Hussain, T., & Li, B. (2023). The impact of virtual society on social capital formation: A comparative analysis of Facebook and WhatsApp. *Sage Journals*https://journals.sagepub.com/doi/full/10.1177/21582440231210077

Ali, H., Hussain, T., Ali, S., Jumani, N. M., Li, B., & Zhang, G. (2017). Coverage of educational issues in Pakistan by Daily Dawn and the news: September 2015-December 2015. *Science International (Lahore)*, *29*(5), 1069–1072.

Al-Jenaibi, B. (2020). Digital ethics and social norms: A Middle Eastern perspective. *Journal of Media Studies*, *12*(3), 45–63.

Alkhalil, Z., Hewage, C., Nawaf, L., & Khan, I. (2021). Phishing Attacks: A Recent Comprehensive Study and a New Anatomy. *Frontiers of Computer Science*, *3*(563060), 1–23.

Allahrakha, N. (2023). Balancing cyber-security and privacy: Legal and ethical considerations in the digital age. *Legal Issues in the digital. The Age (Melbourne, Vic.)*, (2), 78–121.

Allcott, H., & Gentzkow, M. (2017). Social Media and Fake News in the 2016 Election. *The Journal of Economic Perspectives*, *31*(2), 211–236. DOI: 10.1257/jep.31.2.211

Allen, C., Berg, C., & Lane, A. M. (2016). *The Path to Self-Sovereign Identity*. Retrieved December 22, 2025 from https://ssimeetup.org/

Allen, J., Howland, B., Mobius, M., Rothschild, D., & Watts, D. J. (2020). Evaluating the fake news problem at the scale of the information ecosystem. *Science Advances*, *6*(14), eaay3539. Advance online publication. DOI: 10.1126/sciadv.aay3539 PMID: 32284969

Almasry, M. W., & Shaath, A. H. (2013). The level of digital citizenship of a sample of Palestine university learners from their point of view. *Palestine University Journal for Research and Studies*, 7/(2), 171-203.

Alqayed, M. (2014). The concept of digital citizenship. Retrieved from https://www.neweduc.com/definition-of-digital-citizenship

Alselehat, Y. R., Alfalloh, F. R., & Alsarhan, A. K. (2018). The degree of awareness of the concept of digital citizenship for undergraduate learners in the Faculty of Educational Sciences at the University of Jordan. *Dirasat: Educational Sciences*, 45/(3), 19-33.

Alshatter, Z., & Al Mubarak, M. (2024). From Green Tech to Digital Tech: Leveraging Digital Solutions for Sustainable Development (SD) in B2B Markets. In *Equity and Sustainability* (pp. 151–171). Springer Nature Singapore. DOI: 10.1007/978-981-97-4742-9_7

Alsuhaim, A. A., & Alibrahim, A. A. (2019). The extent to which computer teachers activate the digital citizenship standards in the secondary school. *Specialized Educational Journal*, 8/(4), 1-12.

Alter, A. (2017). *Irresistible: The rise of addictive technology and the business of keeping us hooked.* Penguin.

Althobaiti, M. M. (2021). Assessing User's Susceptibility and Awareness of Cybersecurity Threats. *Intelligent Automation & Soft Computing*, *28*(1), 167–177. DOI: 10.32604/iasc.2021.016660

Ameen, K., & Gorman, G. E. (2009). Information and digital literacy: A stumbling block to development? A Pakistan perspective. *Library Management*, *30*(1/2), 99–112. DOI: 10.1108/01435120910927565

Amin, F., & Khan, M. F. (2021). Online reputation and stress: Discovering the dark side of social media. *FIIB Business Review*, *10*(2), 181–192. DOI: 10.1177/2319714520948171

Andema, S. (2014). *Promoting digital literacy in African education: ICT innovations in a Ugandan primary teachers' college* (Doctoral dissertation, University of British Columbia).

Anderson, C. L., & Agarwal, R. (2010). Practicing safe computing: A multimedia empirical examination of home computer user security behavioral intentions. *Management Information Systems Quarterly*, *34*(3), 613–643. DOI: 10.2307/25750694

Andrews, M. (2010). Good government means different things in different countries. *Governance: An International Journal of Policy, Administration and Institutions*, *23*(1), 7–35. DOI: 10.1111/j.1468-0491.2009.01465.x

Andriushchenko, K., Rozhko, O., Tepliuk, M., Semenyshyna, I., Kartashov, E., & Liezina, A. (2020). Digital literacy development trends in the professional environment. *International Journal of Learning. Teaching and Educational Research*, *19*(7), 55–79.

Anglen, J. (2024). AI & Blockchain: digital identity 2024. *Rapid Innovation*. https://www.rapidinnovation.io/post/ai-and-blockchain-fusion-advancing-digital-identity-in-2024

Ante, L., Fischer, C., & Strehle, E. (2022). A bibliometric review of research on digital identity: Research streams, influential works and future research paths. *Journal of Manufacturing Systems*, *62*, 523–538. DOI: 10.1016/j.jmsy.2022.01.005

Anthonysamy, L., & Sivakumar, P. (2024). A new digital literacy framework to mitigate misinformation in social media infodemic. *Global Knowledge, Memory and Communication*, *73*(6/7), 809–827. DOI: 10.1108/GKMC-06-2022-0142

Appel, G., Grewal, L., Hadi, R., & Stephen, A. T. (2020). The future of social media in marketing. *Journal of the Academy of Marketing Science*, *48*(1), 79–95. DOI: 10.1007/s11747-019-00695-1 PMID: 32431463

Arnould, E. J., & Thompson, G. J. (2005). Consumer culture theory (CCT): Twenty years of research. *The Journal of Consumer Research*, *31*(4), 868–882. DOI: 10.1086/426626

Arora, P. (2019). Politics of Algorithms, Indian Citizenship, and the Colonial Legacy. In *Global Digital Cultures: Perspectives from South Asia* (p. 37). University of Michigan Press. https://library.oapen.org/bitstream/handle/20.500.12657/23985/1006150.pdf#page=46

Arora, B. (2016). Exploring and analysing Internet Crimes and their behaviours. *Perspectives in Science*, *8*, 540–542. DOI: 10.1016/j.pisc.2016.06.014

Arriagada, A., & Ibáñez, F. (2020). "You need at least one picture daily, if not, you're dead": Content creators and platform evolution in the social media ecology. *Social Media + Society*, *6*(3), 2056305120944624. DOI: 10.1177/2056305120944624

Arrieta, A. B., Díaz-Rodríguez, N., Del Ser, J., Bennetot, A., Tabik, S., Barbado, A., & Herrera, F. (2020). Explainable artificial intelligence (XAI): Concepts, taxonomies, opportunities, and challenges toward responsible AI. *Information Fusion*, *58*, 82–115. DOI: 10.1016/j.inffus.2019.12.012

Ashdown v. Telegraph Group Ltd, [2001] EWCA Civ 1142 (UK).

Astana, I. T. University (2024a). *About us*. Retrieved from: https://astanait.edu.kz/en/about-2/

Astana, I. T. University (2024b). *Digital Journalism*. Retrieved from: https://astanait.edu.kz/en/digital-journalism-2/

Astana, I. T. University (2024c). *Digital Public Administration and Services*. Retrieved from: https://astanait.edu.kz/en/digital-public-administration-and-services-2/

Audrezet, A., Kerviler, G., & Moulard, J. (2018). Authenticity under threat: When social media influencers need to go beyond self-presentation. *Journal of Business Research*, *117*, 557–569. Advance online publication. DOI: 10.1016/j.jbusres.2018.07.008

Audrin, C., & Audrin, B. (2023). More than just emotional intelligence online: Introducing "digital emotional intelligence". *Frontiers in Psychology*, *14*, 1154355. DOI: 10.3389/fpsyg.2023.1154355 PMID: 37205063

Aufderheide, P., & Jaszi, P. (2018). *Reclaiming Fair Use: How to Put Balance Back in Copyright*. University of Chicago Press., DOI: 10.7208/chicago/9780226374222.001.0001

Australian Law Reform Commission (ALRC). (2018). *Fair Use and Copyright Reform in Australia: A Legal Perspective*. ALRC.

Aviram, A., & Eshet-Alkalai, Y. (2006). *Towards a theory of digital literacy: Three scenarios for the next steps.* http://www.eurodl.org/index.php?p=archives&year=2006&halfyear=1&article=223

Awawdeh, H. Z., Bani Ahmad, A. Y., Almajali, W. I., Atieh Ali, A. A., & Allahham, M. (2024). How Does Digital Marketing Influence Consumer Behavior? Examining the Mediating Role of Digital Entrepreneurship in the Healthcare and Pharmaceuticals Sector. Library of Progress-Library Science, Information Technology & Computer, 44(3).

Ayres, R., & Williams, E. (2004). The digital economy: Where do we stand? *Technological Forecasting and Social Change*, *71*(4), 315–339. DOI: 10.1016/j.techfore.2003.11.001

Badenes-Ribera, L., Fabris, M. A., Gastaldi, F. G. M., Prino, L. E., & Longobardi, C. (2019). Parent and peer attachment as predictors of Facebook addiction symptoms in different developmental stages (early adolescents and adolescents). *Addictive Behaviors*, *95*, 226–232. DOI: 10.1016/j.addbeh.2019.05.009 PMID: 31103243

Bailey, L., Chan, H., & Williams, M. (2022). Social media and mental health advocacy: The role of empathy in online communities. *Journal of Digital Psychology*, *8*(3), 112–129.

Bailey, R. A. (2011). The legal, financial, and ethical implications of online medical consultations. *J. Tech. L. & Pol'y*, *16*, 53.

Bakir, V., & McStay, A. (2018). Fake News and the Economy of Emotions: Problems, Causes, Solutions. *Digital Journalism (Abingdon, England)*, *6*(2), 154–175. DOI: 10.1080/21670811.2017.1345645

Balgabayeva, A.E., Karataeva, T.O, Karabulatova, I.S. et al. (2024). Digital Literacy as a Meta-Cognitive Component of Younger Students' Intellectual and Creative Potential in Foreign Language Lessons. *Rupkatha Journal, 16*(1).

Balkin, J. (2017). The three laws of robotics in the age of big data. *Faculty Scholarship Series.*http://hdl.handle.net/20.500.13051/4697

Banaji, S. (2018). Vigilante Publics: Orientalism, Modernity and Hindutva Fascism in India. *Javnost - the Public, 25*(4), 333–350. https://doi.org/DOI: 10.1080/13183222.2018.1463349

Banaji, S. (2023, August 2). *A hierarchy of hate*. Media@LSE. Retrieved October 17, 2023, from https://blogs.lse.ac.uk/medialse/2023/08/02/a-hierarchy-of-hate/

Banerjee, S. (2020). Aadhaar: Digital identity as a developmental tool in India. *Information Systems Frontiers*, *22*(3), 557–569. DOI: 10.1007/s10796-019-09931-5

Banić, L., & Orehovački, T. (2024). A Comparison of Parenting Strategies in a Digital Environment: A Systematic Literature Review. *Multimodal Technologies and Interaction*, *8*(4), 32. DOI: 10.3390/mti8040032

Barbu, M., Zamfiroiu, A., Marinescu, I. A., Iordache, D., & Bumbac, R. (2023). Empowering digital education: Understanding students' perceptions about risks and threats in the shifting educational paradigm. *Studies in Informatics and Control, 32*(4), 105–114. DOI: 10.24846/v32i4y202310

Barney, J. (1991). Firm resources and sustained competitive advantage. *Journal of Management, 17*(1), 99–120. DOI: 10.1177/014920639101700108

Barton, G., Lee, N. T., & Resnick, P. (2019, May 22). Algorithmic bias detection and mitigation: Best practices and policies to reduce consumer harms. *Brookings.* https://www.brookings.edu/articles/algorithmic-bias-detection-and-mitigation-best-practices-and-policies-to-reduce-consumer-harms/

Baruno, A. D., & Indrasari, M. (2025). Leveraging AI To Enhance Green Marketing Strategies. Jurnal Ekonomi, Manajemen, Akuntansi dan Keuangan, 6(1).

Başarmak, U., Yakar, H., Güneş, E., & Kuş, Z. (2019). Analysis of digital citizenship subject contents of secondary education curricula. *Turkish Online Journal of Qualitative Inquiry, 10*(1), 26–51. DOI: 10.17569/tojqi.438333

Basyah, N., & Hayati, E. (2020, March). Strengthening Integration and Diversity Through Digital Ethics. In *2nd Annual Civic Education Conference (ACEC 2019)* (pp. 558-562). Atlantis Press.

Bauernschuster, S., Falck, O., & Woessmann, L. (2014). Surfing alone? The internet and social capital: Evidence from an unforeseeable technological mistake. *Journal of Public Economics, 117*, 73–89. DOI: 10.1016/j.jpubeco.2014.05.007

Bawden, D. (2001). Information and digital literacies: A review of concepts. *The Journal of Documentation, 57*(2), 218–259. DOI: 10.1108/EUM0000000007083

Bawden, D. (2008). Origins and concepts of digital literacy. In Lankshear, C., & Knobel, M. (Eds.), *Digital literacies: Concepts, policies and practices* (1st ed., pp. 17–30). Peter Lang Publishing.

Bazarhanova, A., & Smolander, K. (2020). The review of non-technical assumptions in digital identity architectures.

Beauchamp, T. L., & Childress, J. F. (1983). *Principles of biomedical ethics* (2nd ed.). Oxford University Press.

Becerra-Bonache, L., & Jiménez-López, M. D. (2014). Linguistic Models at the Crossroads of Agents, Learning and Formal Languages. *ADCAIJ: Advances In Distributed Computing And Artificial Intelligence Journal., 3*(4), 67–87. DOI: 10.14201/ADCAIJ2014346787

Beck, E., Goin, M. E., Ho, A., Parks, A., & Rowe, S. (2021). Critical digital literacy as method for teaching tactics of response to online surveillance and privacy erosion. *Computers & Composition. Computers and Composition, 61*, 102654. DOI: 10.1016/j.compcom.2021.102654

Bejaković, P., & Mrnjavac, Ž. (2020). The importance of digital literacy on the labour market. *Employee Relations, 42*(4), 921–932. DOI: 10.1108/ER-07-2019-0274

Bélanger, F., & Crossler, R. E. (2011). Privacy in the digital age: A review of information privacy research in information systems. *Management Information Systems Quarterly, 35*(4), 1017–1041. DOI: 10.2307/41409971

Benkler, Y. (2000). Internet regulation: A case study in the problem of unilateralism. *European Journal of International Law, 11*(1), 171–185. DOI: 10.1093/ejil/11.1.171

Benlahcene, A., Zainuddin, R. B., Syakiran, N., & Ismail, A. B. (2018). A narrative review of ethics theories: Teleological & deontological Ethics. *Journal of Humanities and Social Science (IOSR-JHSS), 23*(1), 31-32.

Bennett, W. L. (2008). Changing citizenship in the digital age. In Bennett, W. L. (Ed.), *Civic life online: Learning how digital media can engage youth.* The MIT Press.

Bently, L., & Sherman, B. (2014). *Intellectual Property Law* (4th ed.). Oxford University Press., DOI: 10.1093/he/9780199645558.001.0001

Bhatia, K. V., & Gajjala, R. (2020, November 28). *Examining Anti-CAA protests at Shaheen Bagh: Muslim Women and politics of the Hindu India.* Bhatia | International Journal of Communication. https://ijoc.org/index.php/ijoc/article/view/16015

Bhatia, R. (2020). Digital empathy in crisis response: Social media and refugee support networks. *Global Media Journal, 18*(2), 55–73.

Bhuvaneswari, G., & Vijayakumar, S. (2021). Emotional intelligence and values in digital world through emoticons among indian students and faculty. *International Journal of Asian Education, 2*(2), 267–276. DOI: 10.46966/ijae.v2i2.142

Bicher, L. A., & Fathy, S. A. (2020). Ethical Considerations for Using Social Media:"Is it Okay to Share? [AJMCR]. *The Arab Journal of Media and Communication Research, 2020*(31), 172–187. DOI: 10.21608/jkom.2020.156744

Biju, P. R., & Gayathri, O. (2023). Self-Breeding Fake News. The International Journal of Information, Diversity, &. *Inclusion (Washington, D.C.), 7*(1/2), 1–25.

Bindner, L., & Gluck, R. (2017). Wilayat Internet: ISIS' Resilience across the Internet and Social Media. Bellingcat. Retrieved from - https://www.researchgate.net/publication/323005907_Wilayat_Internet_ISIS'_Resilience_across_the_Internet_and_Social_Media

Binns, R. (2018). Fairness in machine learning: Lessons from political philosophy. *Proceedings of the 2018 Conference on Fairness, Accountability, and Transparency*, 149-159.

Bin, Y., Khan, W., & Mastafa, E., Hadi, Rooth, S., & Khan, R. A. (2021). Social media: A source of bridging and bonding social capital. *Journal of Media and Communication Studies, 1*(1), •••.

Bitman, N. (2021). 'Which part of my group do I represent?': Disability activism and social media users with concealable communicative disabilities. *Information Communication and Society, 26*(3), 619–636. DOI: 10.1080/1369118x.2021.1963463

Blackshaw, P. (2009). A pocket guide to social media and kids. [Web log comment]. Retrieved from http://blog.nielsen.com/nielsenwire/consumer/a-pocket-guide-to-social-media-and-kids/

Blakley, C. (2016, December 5). How to Build a Marketing Campaign that Appeals to Generation Z. Fortune. https://fortune.com/2016/12/05/marketing-campaign-generation-z/

Bleazby, J. (2006). Autonomy, democratic community, and citizenship in philosophy for children: Dewey and philosophy for children's rejection of the individual/community dualism. Analytic Teaching, 26(1), 30-52. https://journal.viterbo.edu/index.php/at/article/view/832

Blue, J., Condell, J., & Lunney, T. (2018, July). Digital footprints: your unique identity. In *Proceedings of the 32nd International BCS Human Computer Interaction Conference*. BCS Learning & Development.

Bode, L., & Jones, M. L. (2018). Do Americans want a right to be forgotten? Estimating public support for digital erasure legislation. *Policy and Internet, 10*(3), 244–263. DOI: 10.1002/poi3.174

Bolter, J. (2000). Identity. In Swiss, T. (Ed.), *Unspun: Key Concepts for Understanding the World Wide Web* (pp. 17–29). New York University Press.

Bonnet, J. (n.d.). Subject Guides: Hot Topics: Fake News and Misinformation: Identifying Fake News. Libguides.library.umaine.edu. https://libguides.library.umaine.edu/fakenews/identification

Bontridder, N., & Poullet, Y. (2021). The role of artificial intelligence in disinformation. *Data & Policy, 3*, e32. DOI: 10.1017/dap.2021.20

Bossen, C. B., & Kottasz, R. (2020). Uses and gratifications sought by pre-adolescent and adolescent TikTok consumers. *Young Consumers, 21*(4), 463–478. DOI: 10.1108/YC-07-2020-1186

Botelho, F. H. F. (2021). Accessibility to digital technology: Virtual barriers, real opportunities. *Assistive Technology, 33*(sup1), 27–34. https://doi.org/DOI: 10.1080/10400435.2021.1945705

Boucher, D. (2022). *The Ethics of Artificial Intelligence in Content Creation: Fair Use in the Age of Deepfakes*. MIT Press.

Bourhis, P., Demartini, G., Elbassuoni, S., Hoareau, E., & Rao, H. R. (2019). Ethical challenges in the future of work. *A Quarterly Bulletin of the Computer Society of the IEEE Technical Committee on Data Engineering*.

Boyle, D. (2006). *Authenticity: Brands, fakes, spin and the lust for real life*. Harper Perennial.

Bradshaw, P. (2018). Assessment of Computing and ICT at 16: What are the perceptions of learners? In Younie, S., & Bradshaw, P. (Eds.), *Debates in Computing and ICT Education* (1st ed., pp. 43–52). Routledge.

Branch, J. (2021). What's in a Name? Metaphors and cybersecurity. *International Organization, 75*(1), 39–70. DOI: 10.1017/S002081832000051X

Braun, V., & Clarke, V. (2006). Using thematic analysis in psychology. *Qualitative Research in Psychology, 3*(2), 77–101. DOI: 10.1191/1478088706qp063oa

Brenner, S. W. (2010). *Cybercrime: Criminal threats from cyberspace.* ABC-CLIO. DOI: 10.5040/9798400636554

Breves, P., Amrehn, J., Heidenreich, A., Liebers, N., & Schramm, H. (2021). Blind trust? The importance and interplay of parasocial relationships and advertising disclosures in explaining influencers' persuasive effects on their followers. *International Journal of Advertising, 40*(7), 1209–1229. DOI: 10.1080/02650487.2021.1881237

Breves, P., Liebers, N., Abt, M., & Kunze, A. (2019). The Perceived Fit between Instagram Influencers and the Endorsed Brand: How Influencer–Brand Fit Affects Source Credibility and Persuasive Effectiveness. *Journal of Advertising Research, 59*(4), 440–454. DOI: 10.2501/JAR-2019-030

Brey, P. (2017). Ethics of Emerging Technology. In *The Ethics of Technology Methods and Approaches.* Rowan & Littlefield International.

Brinda, T., Napierala, S., & Behler, G. (2018). What do secondary school students associate with the digital world? *Proceedings Of The 13Th Workshop In Primary And Secondary Computing Education, Postdam, Germany.*https://doi.org/ [Accessed 18 February 2021].DOI: 10.1145/3265757.3265763

Britannica. (n. d.). Social capital. https://www.britannica.com/topic/social-capital

Brito, M. P., Zhao, Y., & Fan, S. (2021). The role of digital platforms in green marketing. *Sustainable Marketing Journal, 14*(3), 203–215.

Broadhurst, R., Skinner, K., Sifniotis, B., Matamoros-Macias, B., & Ipsen, Y. (2019). Phishing and cybercrime risks in a university student community. *International Journal of Cybersecurity Intelligence & Cybercrime, 2*(1), 4–23. DOI: 10.52306/02010219RZEX445

Brundtland Commission. (1987). *Our common future.* Oxford University Press.

Bryant, K., & Campbell, J. (2006). User Behaviours Associated With Password Security And Management. *AJIS. Australasian Journal of Information Systems, 14*(1), 81–100. DOI: 10.3127/ajis.v14i1.9

Bryman, A. (2004). *Social research methods* (2nd ed.). Oxford University Press.

Brynjolfsson, E., & Kahin, B. (2002). *Understanding the digital economy.* The MIT Press.

Brynjolfsson, E., & McAfee, A. (2016). *The second machine age.* W. W. Norton.

Brynjolfsson, E., & Saunders, A. (2013). *Wired for innovation.* MIT Press.

Buchanan, E. A. (2020). Digital ethics: A framework for online conduct. *Ethics and Information Technology, 22*(1), 23–37.

Buchholz, R. R., D'Attilo, G., Deeter, M., Edwards, D., Emmons, L., Francis, G., Gaubert, B., Gille, J., Mao, D., Martinez-Alonso, S., Worden, H., & Ziskin, D. (2023). MOPITT CO monthly climatology 2002-2021. Version 1.0. UCAR/NCAR - GDEX. DOI: 10.5065/s9fy-9w35

Buckingham, D. (2008). Defining digital literacy - What do young people need to know about media? In Lankshear, C., & Knobel, M. (Eds.), *Digital literacies: Concepts, policies, and practices* (1st ed., pp. 73–90). Peter Lang Publishing.

Buolamwini, J., & Gebru, T. (2018). Gender shades: Intersectional accuracy disparities in commercial gender classification. *Proceedings of the 1st Conference on Fairness, Accountability, and Transparency*, 81, 77–91. Retrieved December 24, 2024 from https://proceedings.mlr.press/v81/buolamwini18a.html

Burgess, S., Sellitto, C., Cox, C., & Buultjens, J. (2011). Trust perceptions of online travel information by different content creators: Some social and legal implications. *Information Systems Frontiers*, *13*(2), 221–235. DOI: 10.1007/s10796-009-9192-x

Burton, K. L. (2015). Eternal or ephemera? The myth of permanence in online writing. *Computers and Composition*, *38*, 57–67. DOI: 10.1016/j.compcom.2015.09.010

Butavicius, M., Parsons, K., Pattinson, M., & McCormac, A. (2015). Breaching the human firewall: Social engineering in phishing and spear phishing emails. *Australasian Conference on Information Systems (pp. 12-23).* AISeL. https://aisel.aisnet.org/acis2015/98/

Bykov, I., Gladchenko, I. A., Ibrayeva, G. Zh., & Myssayeva, K. N. (2019). *Digital Literacy and Strategy of PR-education in Russia and Kazakhstan: Comparative analysis. In 2019 Communication Strategies in Digital Society Workshop (ComSDS).* IEEE.

Bylieva, D., Lobatyuk, V., & Safonova, A. (2019). Online forums: Communication model, categories of online communication regulation and norms of behavior. *Humanities & Social Sciences Reviews*, *7*(1), 332–340. DOI: 10.18510/hssr.2019.7138

Caballero, B. (2002). Ethical issues for collaborative research in developing countries12. *The American Journal of Clinical Nutrition*, *76*(4), 717–720. DOI: 10.1093/ajcn/76.4.717 PMID: 12324282

Cadwalladr, C., & Graham-Harrison, E. (2018). Revealed: 50 million Facebook profiles harvested for Cambridge Analytica in major data breach. *The Guardian*. Retrieved December 15, 2024 from https://www.theguardian.com/

Cain, A. A., Edwards, M. E., & Still, J. D. (2018). An exploratory study of cyber hygiene behaviours and knowledge. *Journal of Information Security and Applications*, *42*, 36–45. DOI: 10.1016/j.jisa.2018.08.002

Cains, M. G., Flora, L., Taber, D., King, Z., & Henshel, D. S. (2022). Defining Cyber Security and Cyber Security Risk within a Multidisciplinary Context using Expert Elicitation. *Risk Analysis*, *42*(8), 1643–1669. DOI: 10.1111/risa.13687 PMID: 33586204

Cameron, K. (2005). *The Laws of Identity*. Retrieved December 24, 2025 from https://www.identityblog.com/?p=354

Campbell v. Acuff-Rose Music, Inc., 510 U.S. 569 (1994).

Campbell, C., & Farrell, J. (2020). More than meets the eye: The functional components underlying influencer marketing. *Business Horizons*, *63*(4), 469–479. Advance online publication. DOI: 10.1016/j.bushor.2020.03.003

Capisizu, L. A. (2020). Digital identity. *Conferin a Interna ională de Drept, Studii Europene i Rela ii Interna ionale,* *8*(VIII), 256-263.

Carlsson, B. (2004). The digital economy: What is new and what is not? *Structural Change and Economic Dynamics*, *15*(3), 245–264. DOI: 10.1016/j.strueco.2004.02.001

Casaló, L. V., Flavián, C., & Ibáñez-Sánchez, S. (2020). Influencers on Instagram: Antecedents and consequences of opinion leadership. *Journal of Business Research*, *117*, 510–519. DOI: 10.1016/j.jbusres.2018.07.005

Castells, M. (2012). Networks of outrage and hope: Social movements in the Internet age. *Polity*.

Chaiken, S. (1987). The heuristic model of persuasion. In *Social influence: The Ontario symposium* (Vol. 5, pp. 3–39). Lawrence Erlbaum Associates, Inc.

Chakravarthi, B. R. (2023). Detection of homophobia and transphobia in YouTube comments. *International Journal of Data Science and Analytics*, *18*(1), 49–68. DOI: 10.1007/s41060-023-00400-0

Chan, C. K., Zhao, M. M., & Lee, P. S. N. (2022). Determinants of escape from echo chambers: The predictive power of political orientation, social media use, and demographics. In *Global Media and China*. https://journals.sagepub.com/doi/10.1177/20594364221140820

Chancellor Masters and Scholars of the University of Oxford v. Rameshwari Photocopy Services, (2016) 160 DRJ 679 (Del).

Chase, Z., & Laufenberg, D. (2011). Embracing the squishiness of digital literacy. *Journal of Adolescent & Adult Literacy*, *54*(7), 535–537. DOI: 10.1598/JAAL.54.7.7

Chatterjee, A. (2022). The Role of Blockchain and NFTs in Copyright Protection. *Journal of Digital Economy*, *15*(3), 87–102.

Chaudhary, S., Schafeitel-Tähtinen, T., Helenius, M., & Berki, E. (2019). Usability, security and trust in password managers: A quest for user-centric properties and features. *Computer Science Review*, *33*, 69–90. DOI: 10.1016/j.cosrev.2019.03.002

Chauhan, A. (2020). Environmental sustainability in healthcare: A green marketing approach. *Journal of Healthcare Management*, *45*(2), 122–131. PMID: 32168188

Chelysheva, I. (2019). Media literacy education in Uzbekistan, Kyrgyzstan and Turkmenistan. *International Journal of Media and Information Literacy*, *4*(1), 3–10.

Chen, L. (2020). Data connectivity in the Greater Mekong Subregion. *Subregional Development Strategy in ASEAN after COVID-19: Inclusiveness and Sustainability in the Mekong Subregion (Mekong 2030)*, 42.

Cheng, M., Choi, S., & Wang, J. (2020). Online marketing and consumer engagement: Insights from green healthcare marketing. *Journal of Digital Marketing*, *15*(2), 85–99.

Chen, K., & Yen, D. C. (2004). Improving the quality of online presence through interactivity. *Information & Management*, *42*(1), 217–226. DOI: 10.1016/j.im.2004.01.005

Chen, L. L., Mirpuri, S., Rao, N., & Law, N. (2021). Conceptualization and measurement of digital citizenship across disciplines. *Educational Research Review*, *33*, 100379. DOI: 10.1016/j.edurev.2021.100379

Chenoweth, T., Minch, R., & Gattiker, T. (2009, January 5–8). Application of protection motivation theory to adoption of protective technologies. Paper presented at the 42nd Hawaii International Conference on System Sciences, HICSS'09, Honolulu, HI.

Chen, R. (2013). Living a private life in public social networks: An exploration of member self-disclosure. *Decision Support Systems*, *55*(3), 661–668. DOI: 10.1016/j.dss.2012.12.003

Chen, R., Wang, J., Herath, T., & Rao, H. R. (2011). An investigation of email processing from a risky decision making perspective. *Decision Support Systems*, *52*(1), 73–81. DOI: 10.1016/j.dss.2011.05.005

Chen, Y., & Xu, Y. (2021). Exploring the effect of social support and empathy on user engagement in online mental health communities. *International Journal of Environmental Research and Public Health*, *18*(13), 6855. DOI: 10.3390/ijerph18136855 PMID: 34206719

Chesney, R., & Citron, D. K. (2019). Deep Fakes: A Looming Challenge for Privacy, Democracy, and National Security. SSRN Electronic Journal, 107(1753). https://doi.org/DOI: 10.2139/ssrn.3213954

Chetty, N., & Alathur, S. (2019). *Racism and social media: A study in Indian context*. https://idr.nitk.ac.in/jspui/handle/123456789/12729

Cho, C. H., & as-, U. T. A. (2004). Why do people avoid advertising on the internet? *Journal of Advertising*, *33*(4), 89–97. DOI: 10.1080/00913367.2004.10639175

Cho, H., Li, W., Cannon, J., Lopez, R., & Song, C. (2020). Testing three explanations for stigmatization of people of Asian descent during COVID-19: Maladaptive coping, biased media use, or racial prejudice? *Ethnicity & Health*, *0*(0), 1–16. PMID: 33059486

Choi, M. (2016). A concept analysis of digital citizenship for democratic citizenship education in the internet age. *Theory and Research in Social Education*, *44*(4), 565–607. DOI: 10.1080/00933104.2016.1210549

Chopra, D. (2021). The resistance strikes back: Women's protest strategies against backlash in India. *Gender and Development*, *29*(2–3), 467–491. DOI: 10.1080/13552074.2021.1981698

Choraś, M., Pawlicki, M., & Kozik, R. (2019). The feasibility of deep learning use for adversarial model extraction in the cybersecurity domain. In Yin, H., Camacho, D., Tino, P., Tallón-Ballesteros, A., Menezes, R., & Allmendinger, R. (Eds.), Lecture Notes in Computer Science: Vol. 11872. *Intelligent Data Engineering and Automated Learning–IDEAL 2019*. Springer. DOI: 10.1007/978-3-030-33617-2_36

Choy, Y. N., Lau, E. Y. H., & Wu, D. (2024). Digital parenting and its impact on early childhood development: A scoping review. *Education and Information Technologies*, *29*(16), 1–41. DOI: 10.1007/s10639-024-12643-w

Chyung, S. Y. Y. (2007). Age and gender differences in online behavior, self-efficacy, and academic performance. *Quarterly Review of Distance Education*, *8*(3).

Cialdini, R. B., & Fincham, R. L. (1990). Trust and the development of interpersonal relationships. *Psychological Bulletin*, •••, 1990.

Citron, D. K. (2014). *Hate crimes in cyberspace*. Harvard University Press. DOI: 10.4159/harvard.9780674735613

Cleveland Clinic. (2023). Sustainability at Cleveland Clinic. Retrieved from https://my.clevelandclinic.org

Cohen, C. (2023). Becoming a good digital citizen: navigating the online world ethically. https://www.cyberwise.org/post/becoming-a-good-digital-citizen-navigating-the-online-world-ethically#:~:text=1.%20Respect,for%20Others

Cohen, L., Manion, L., & Morrison, K. (2011). *Research methods in education* (7th ed.). Routledge.

Coklar, A. N., & Tatli, A. (2020). Evaluation of digital citizenship levels of teachers in the context of information literacy and internet and computer use self-efficacy. *Asian Journal of Contemporary Education, 4*(2), 80–90. DOI: 10.18488/journal.137.2020.42.80.90

Coleman, J. (1988). Social capital in the creation of human capital. *American Journal of Sociology, 94*, 95–120. DOI: 10.1086/228943

Colley, S. (2014). Ethics and digital heritage. In *The ethics of cultural heritage* (pp. 13–32). Springer New York.

Colliander, J. (2019). “This is fake news”: Investigating the role of conformity to other users’ views when commenting on and spreading disinformation in social media. *Computers in Human Behavior, 97*, 202–215. DOI: 10.1016/j.chb.2019.03.032

Common Sense Media. (2009Pru). Digital citizenship. Retrieved from www.common sense.org/education/digital-citizenship

Computing at School (CAS). (n.d.). Community for computing teachers. https://www.computingatschool.org.uk/

Conti, M., Dragoni, N., & Lesyk, V. (2016). A Survey of Man In The Middle Attacks. *IEEE Communications Surveys and Tutorials, 18*(3), 2027–2051. DOI: 10.1109/COMST.2016.2548426

Corti, L., & Bishop, L. (2020). Ethical issues in data sharing and archiving. *Handbook of research ethics and scientific integrity*, 403-426.

Court of Justice of the European Union. (2014). Deckmyn v. Vandersteen. CASE (Philadelphia, Pa.), C-201(13).

Court of Justice of the European Union. (2019). Pelham GmbH v. Hütter. *CASE (Philadelphia, Pa.), C-476*(17).

Court, D. H. (2011). Super Cassettes Industries Ltd. v. [Del]. *MySpace Inc., 48*(PTC), 49.

Couto, R. (2024). How to build an inclusive online community (and why it matters). Groundwork. https://groundwork.space/building-an-inclusive-online-community/Cyber

Crawford, K. (2021). *Atlas of AI: Power, politics, and the planetary costs of artificial intelligence*. Yale University Press.

Creswell, J. (2003). *Research Design, Qualitative, Quantitative, and Mixed Methods Approaches*. Sage Publications.

Creswell, J. W. (2014). *Research design: Qualitative, quantitative, and mixed methods approaches* (4th ed.). SAGE Publications.

Cross, D., Lester, L., & Barnes, A. (2019). Digital interventions in cyberbullying prevention: Evaluating the "Reword" initiative. *Australian Journal of Educational Technology, 35*(1), 65–80.

CRS - The Congressional Research Service. (2023). S.394 - Digital Citizenship and Media Literacy Act. https://www.congress.gov/bill/118th-congress/senate-bill/394

Cruz, J., & Plaisance, P. L. (2021). Virtue ethics and a technomoral framework for online activism. *International Journal of Communication, 15*, 19.

Cruz, T., & Simões, P. (2021). Down the rabbit hole: Fostering active learning through guided exploration of a scada cyber range. *Applied Sciences (Basel, Switzerland), 11*(20), 9509. DOI: 10.3390/app11209509

Cuomo, M. T., & Foroudi, P. (2023). Digital transformation: Corporate branding, identity, and reputation. In *Digital Transformation and Corporate Branding* (pp. 1–6). Routledge. DOI: 10.4324/9781003263784-1

Custers, B., & Fosch-Villaronga, E. (2022). *Law and Artificial Intelligence - Regulating AI and Applying AI in Legal Practice - Asser.nl*. Springer. https://www.asser.nl/asserpress/books/?rId=13994

Cyber Security Agency of Singapore. (2019). *Secure Your Password.* Retrieved from https://www.csa.gov.sg/Tips-Resource/Resources/gosafeonline/2019/Secure-Your-Password

Cyber Security Agency of Singapore. (2023). *Singapore Cybersecurity Health Report.* Retrieved from https://www.csa.gov.sg/docs/default-source/our-programmes/support-for-enterprises/sg-cyber-safe-programme/csa-singapore-cybersecurity-health-report-2023.pdf?sfvrsn=c611e1b_1

Cyber Security Agency of Singapore. (n.d.). *Beware of Phishing Scams.* Retrieved from https://www.csa.gov.sg/our-programmes/cybersecurity-outreach/cybersecurity-campaigns/the-unseen-enemy-campaign/beware-of-phishing-scams

Cyber Security Agency of Singapore. (n.d.). *Live Savvy Tips.* Retrieved from https://www.csa.gov.sg/docs/default-source/gso/file/resources/live-savvy-tips.pdf?sfvrsn=e4d879b8_0

Cyber Wellness Program (updated 2024). Practising cyber wellness. https://www.moe.gov.sg/education-in-sg/our-programmes/cyber-wellness

D'Errico, F., & Paciello, M. (2018). Online moral disengagement and hostile emotions in discussions on hosting immigrants. *Internet Research, 28*(5), 1313–1335. DOI: 10.1108/IntR-03-2017-0119

D'Souza, C., & Taghian, M. (2022). Green marketing in healthcare: A consumer perspective. *Healthcare Marketing Review, 8*(4), 451–467.

Dabrowski, A., Merzdovnik, G., Kommenda, N., & Weippl, E. (2016). *Browser History Stealing with Captive WiFi Portals. IEEE Security and Privacy Workshops*. IEEE.

Dadakhonov, A. O. (2023). Theoretical approaches to Media and Information Literacy in Foreign Aid Projects (Model for Central Asian countries). *Communication Papers.Media Literacy and Gender Studies, 12*(25), 23–46.

Dadakhonov, A. O. (2024a). Sustaining foreign aid initiatives in the field of media and information literacy within the Central Asian context. *Medical Education*, *15*(1), 57–70. DOI: 10.36253/me-15703 PMID: 38212802

Dadakhonov, A. O. (2024b). Main Issues of Implementing Media and Information Literacy in School Education in Central Asian Countries (The Case of Kazakhstan, Kyrgyzstan, and Uzbekistan). *Media Literacy and Academic Research*, *7*(1), 51–71. DOI: 10.34135/mlar-24-01-03

Daga, T., Chandra, V., & Malik, A. (2018). Effect of Reliance Jio on Digital India. *International Journal of Advanced Research*, *6*(6), 395–401. DOI: 10.21474/ijar01/7224

Dahal, N. (2023). Digital citizenship and digital ethics. *Advances in Social Networking and Online Communities*, 249–257. Advance online publication. DOI: 10.4018/978-1-6684-8934-5.ch014

Dame Adjin-Tettey, T. (2022). Combating fake news, disinformation, and misinformation: Experimental evidence for media literacy education. *Cogent Arts & Humanities*, *9*(1), 2037229. Advance online publication. DOI: 10.1080/23311983.2022.2037229

Daniels, J. (2018). The algorithmic rise of the "alt-right.". *Contexts*, *17*(1), 60–65. DOI: 10.1177/1536504218766547

Darban, K., & Kabbaj, S. (2023). The Transformative Potential of AI in Green Marketing Strategies. *Traditional Journal of Law and Social Sciences*, *2*(02), 14–38.

Das, S., & Farber, R. (2020). User-generated online queer media and the politics of queer visibility. *Sociology Compass*, *14*(9). Advance online publication. DOI: 10.1111/soc4.12824

Davinson, N., & Sillence, E. (2010). It won't happen to me: Promoting secure behaviour among internet users. *Computers in Human Behavior*, *26*(6), 1739–1747. DOI: 10.1016/j.chb.2010.06.023

Davis, F. D. (1989). Perceived usefulness, perceived ease of use, and user acceptance of information technology. *Management Information Systems Quarterly*, *13*(3), 319–340. DOI: 10.2307/249008

Davison, M., & Hugenholtz, P. B. (2017). *Research Handbook on Copyright Law*. Edward Elgar Publishing.

De George, R. T. (1999). *Business ethics*. Prentice Hall.

de Kerckhove, D., & De Almeida, C. M. (2013). What is a digital persona? *Technoetic Arts : a Journal of Speculative Research*, *11*(3), 277–287. DOI: 10.1386/tear.11.3.277_1

De Kruijf, J. G. (2015). The DALIT I define: social media and individualized activism in subaltern spheres. *Pacific Asia Journal of the Association for Information Systems*, 11–24. https://doi.org/DOI: 10.17705/1pais.07402

De Pelsmacker, P., Geuens, M., & Van den Bergh, J. (2005). *Marketingcommunicatie*. Pearson Education Benelux.

de Terwangne, C. (2014). The right to be forgotten and informational autonomy in the digital environment. In *The ethics of memory in a digital age: Interrogating the right to be forgotten* (pp. 82–101). Palgrave Macmillan UK.

de Vaus, D. (2013). *Surveys in Social Research.* Routledge.

De Veirman, M., Cauberghe, V., & Hudders, L. (2017). Marketing through Instagram influencers: The impact of number of followers and product divergence on brand attitude. *International Journal of Advertising, 36*(5), 798–828. DOI: 10.1080/02650487.2017.1348035

De Veirman, M., & Hudders, L. (2020). Disclosing sponsored Instagram posts: The role of material connection with the brand and message-sidedness when disclosing covert advertising. *International Journal of Advertising, 39*(1), 94–130. DOI: 10.1080/02650487.2019.1575108

Delhi High Court (2013). India TV v. Yashraj Films, (2013) SCC Online Del 500.

Dewey, J. (2008). *The later works of John Dewey, volume 7, 1925-1953: Ethics.* J. Boydston (Ed.). Carbondale, IL: Southern Illinois University Press.

Dey, A. (2019). Sites of exception: Gender violence, digital activism, and Nirbhaya's zone of Anomie in India. *Violence Against Women, 26*(11), 1423–1444. DOI: 10.1177/1077801219862633

Dias, P., & Duarte, A. (2022a). How Portuguese adolescents relate to influencers and brands on TikTok. *Journal of Digital & Social Media Marketing, 10*(1), 82–95. DOI: 10.69554/VSHW2788

Dias, P., & Duarte, A. (2022b). TikTok Practices among Teenagers in Portugal: A Uses & Gratifications Approach. *Journalism and Media, 3*(4), 615–632. DOI: 10.3390/journalmedia3040041

Diepeveen, S., & Pinet, M. (2022). User perspectives on digital literacy as a response to misinformation. *Development Policy Review, 40*(S2), e12671. DOI: 10.1111/dpr.12671

Digital Citizenship. (2024). Education Technology (CA Department of Education). ca.gov. https://www.cde.ca.gov/ls/et/dc/index.asp

Dimock, M. (2019, January 17). *Defining generations: Where Millennials end and Generation Z begins.* Pew Research Center. https://www.pewresearch.org/short-reads/2019/01/17/where-millennials-end-and-generation-z-begins/

Dinner, I., & Polovinko, V. (2018). Dinner, Igor, and Vladimir *Polovinko. Features of Professional Self-determination and Professional Orientation of Young People in the Digital Economy.* In The 12th International Days of Statistics and Economics (pp. pp. 335-343). Prague. Retrieved 18 February 2021, from https://msed.vse.cz/msed_2018/article/204-Dinner-Igor-paper.pdf. [Accessed 18 February 2021].

Disco. (2024). 10 effective community rules examples and how to implement them. https://www.disco.co/blog/10-effective-community-rules-examples-and-how-to-implement-them#:~:text=Notably%2C%20the%20language%20and%20content,members%2C%20including%20other%20community%20members

Dixit, A., & Mac, D. (2017). Global Marketing Strategies of McDonald's Corporation (with Reference to India and Russia). IJAR, 3, 870-875. Scientific Research Publishing. Retrieved from - https://www.scirp.org/reference/referencespapers?referenceid=2807253

Djafarova, E., & Bowes, T. (2020). 'Instagram made Me buy it': Generation Z impulse purchases in fashion industry. *Journal of Retailing and Consumer Services, 59*, 102345. DOI: 10.1016/j.jretconser.2020.102345

Djafarova, E., & Rushworth, C. (2017). Exploring the credibility of online celebrities' Instagram profiles in influencing the purchase decisions of young female users. *Computers in Human Behavior, 68*, 1–7. DOI: 10.1016/j.chb.2016.11.009

Djafarova, E., & Trofimenko, O. (2019). 'Instafamous' – credibility and self-presentation of micro-celebrities on social media. *Information Communication and Society, 22*(10), 1432–1446. DOI: 10.1080/1369118X.2018.1438491

Doğrul, M., Yalçın, H., & Daim, T. U. (2023). *Cybersecurity: A Technology Landscape Analysis. Cybersecurity Technology: A Landscape Analysis.* Springer Nature Switzerland.

Donaldson, T., & Preston, L. E. (1995). The stakeholder theory of the corporation: Concepts, evidence, and implications. *Academy of Management Review, 20*(1), 65–91. DOI: 10.2307/258887

Donati, D., Orozco-Olvera, V., & Rao, N. (2022). Using Social Media to Change Gender Norms: An Experiment within Facebook Messenger in India. In *World Bank policy research working paper*. https://doi.org/DOI: 10.1596/1813-9450-10199

Donoso, V., Pyzalski, J., Walter, N., Retzmann, N., Iwanicka, A., d'Haenens, L., & Bartkowiak, K. (2021). *Report on Interviews with Experts on Digital Skills in Schools and on the Labour Market*. Zenodo: European Commission. Retrieved from http://doi.org/ [Accessed 18 February 2021].DOI: 10.5281/zenodo.4274613

Dow, M. J. (2008). Implications of social presence for online learning: A case study of MLS students. *Journal of Education for Library and Information Science*, •••, 231–242.

Dredge, R., Gleeson, J., & de la Piedad Garcia, X. (2020). Social responsibility in online interactions: The impact of the #IAmAWitness campaign. *Cyberpsychology & Behavior, 23*(2), 78–95.

Duarte, A., & Albuquerque, L. (2024). The Influence of TikTok in Portuguese Millennials' Footwear Consumer Behaviour. In: Rocha, Á., Adeli, H., Dzemyda, G., Moreira, F., Poniszewska-Marańda, A. (eds) Good Practices and New Perspectives in Information Systems and Technologies. WorldCIST 2024. *Lecture Notes in Networks and Systems, vol 990,* pp. 117–126. Springer https://doi.org/DOI: 10.1007/978-3-031-60328-0_12

Duffy, B. E. (2019, September 22). *Gendered Visibility on social Media: Navigating Instagram's authenticity bind.* Duffy | International Journal of Communication. https://ijoc.org/index.php/ijoc/article/view/11729/2821

Dufva, T., & Dufva, M. (2019). Grasping the future of the digital society. *Futures, 107*, 17–28. DOI: 10.1016/j.futures.2018.11.001

ECJ. (2019). Pelham GmbH v. Hütter. *CASE (Philadelphia, Pa.), C-476*(17).

Eckert, S., Metzger-Riftkin, J., Kolhoff, S., & O'Shay-Wallace, S. (2019). A*hyper differential counterpublic*: Muslim social media users and Islamophobia during the 2016 US presidential election. *New Media & Society, 23*(1), 78–98. DOI: 10.1177/1461444819892283

Edelman (2022). 2022 Edelman trust barometer. https://www.edelman.com/trust/2022-trust-barometer

Education Uplifted. (2024). Enhancing civic engagement through technology in education. https://educationuplifted.com/civic-engagement-through-technology-2/

Edwards, R., & Weller, S. (2016). Ethical dilemmas around anonymity and confidentiality in longitudinal research data sharing: the death of Dan. In *Qualitative ethics in practice* (pp. 97–108). Routledge.

Elkington, J. (1997). *Cannibals with forks: The triple bottom line of 21st century business.* Capstone Publishing.

Elliott, D., & Spence, E. H. (2017). *Ethics for a digital era.* John Wiley & Sons. DOI: 10.1002/9781118968888

Ellison, N. B., Steinfield, C., & Lampe, C. (2007, July 1). The benefits of Facebook "Friends:" social capital and college students' use of online social network sites. *Journal of Computer-Mediated Communication*, *12*(4), 1143–1168. DOI: 10.1111/j.1083-6101.2007.00367.x

Emplifi. (n.d.). Social media analytics for political campaigns. Emplifi White Paper. Retrieved from - https://emplifi.io/resources/blog/social-media-analytics-the-complete-guide

Endres, D., Hedler, L., & Wodajo, K. (2023). Bias in Social Media Content Management: What Do Human Rights Have to Do with It? *AJIL Unbound.*, *117*, 139–144. DOI: 10.1017/aju.2023.23

Erlanitasari, Y., Rahmanto, A., & Wijaya, M. (2020). Digital economic literacy micro, small and medium enterprises (SMES) go online. *Informasi*, *49*(2), 145–156. DOI: 10.21831/informasi.v49i2.27827

Escobar-Viera, C. G., Whitfield, D. L., Wessel, C. B., Shensa, A., Sidani, J. E., Brown, A. L., Chandler, C. J., Hoffman, B. L., Marshal, M. P., & Primack, B. A. (2018). For better or for worse? A systematic review of the evidence on social media use and depression among lesbian, gay, and bisexual minorities. *JMIR Mental Health*, *5*(3), e10496. DOI: 10.2196/10496

Eshet-Alkali, Y. (2004). 'Digital literacy: A conceptual framework for survival skills in the digital era. *Journal of Educational Multimedia and Hypermedia*, *13*(1), 93–106.

Eshet-Alkali, Y., & Amichai-Hamburger, Y. (2004). Experiments in Digital Literacy. *Cyberpsychology & Behavior*, *7*(4), 421–429. DOI: 10.1089/cpb.2004.7.421 PMID: 15331029

Ethics guidelines for trustworthy AI. (2019). Shaping Europe's Digital Future. Retrieved February 6, 2025, from https://digital-strategy.ec.europa.eu/en/library/ethics-guidelines-trustworthy-ai

Etim, A. E., Uzonna, I., & Worgu Steve, C. (2018). Social media usage and firm performance: Reflections from the Nigerian telecommunication sector. *International Journal of Management Science and Business Administration*, *4*(6), 7–16. DOI: 10.18775/ijmsba.1849-5664-5419.2014.46.1001

eTwinning – European Commission. (n.d.). European School Education Platform. https://school-education.ec.europa.eu/en/etwinning

European Commission. (2019). Copyright in the Digital Single Market. Retrieved from https://ec.europa.eu/digital-strategy

European Commission. (2020, September 9). *Communication from the Commission to the European Parliament, the Council, the European Economic and Social Committee and the Committee of the Regions on the Digital Education Action Plan. COM(2020) 624 final*. https://eur-lex.europa.eu/legal-content/EN/TXT/?uri=celex:52020DC0624

European Parliament. (2016). *General Data Protection Regulation (GDPR)*. Retrieved December 20, 2024 from https://eur-lex.europa.eu/legal-content/EN/TXT/?uri=CELEX%3A32016R0679

European Union (EU). (2014). *Regulation (EU) No 910/2014 on electronic identification and trust services for electronic transactions in the internal market (eIDAS)*. Retrieved December 20, 2025 from https://eur-lex.europa.eu

Evans, N., Phua, J., Lim, J., & Jun, H. (2017). Disclosing Instagram Influencer Advertising: The Effects of Disclosure Language on Advertising Recognition, Attitudes, and Behavioral Intent. *Journal of Interactive Advertising*, 17, 00–00. DOI: 10.1080/15252019.2017.1366885

Evans, M. (2019). International policy transfer: Between the global and sovereign and between the global and local. In Stone, D., & Moloney, K. (Eds.), *The Oxford Handbook of Global Policy and Transnational Administration* (pp. 94–110). Oxford University Press., DOI: 10.1093/oxfordhb/9780198758648.013.3

Fabbri, M. (2019). Digital citizenship skills as an essential factor in Lifelong-Lifewide-Lifedeep Education. *Pedagogia Oggi*, *17*(2), 285–299.

Facer, K. (2011). *Learning futures: Education, technology and social change*. Taylor & Francis. DOI: 10.4324/9780203817308

Favaretto, M., De Clercq, E., & Elger, B. S. (2019). Big Data and discrimination: Perils, promises and solutions. A systematic review. *Journal of Big Data*, *6*(1). Advance online publication. DOI: 10.1186/s40537-019-0177-4

FCC - the U.S. Federal Communications Commission. (n.d.) Emergency broadband benefit. https://www.fcc.gov/broadbandbenefit

Federal Trade Commission. (2022). *How to Recognize and Avoid Phishing Scams*. Retrieved from https://consumer.ftc.gov/articles/how-recognize-and-avoid-phishing-scams

Feng, S., & Tan, C. Y. (2024). Toward conceptual clarity for digital cultural and social capital in student learning: Insights from a systematic literature review. *Humanities & Social Sciences Communications*, *11*(1), 68. DOI: 10.1057/s41599-023-02519-8

Ferdinand, P. (2000). The internet, democracy and democratization. *Democratization*, *7*(1), 1–17. DOI: 10.1080/13510340008403642

Fernández-Prados, J. S., Lozano-Díaz, A., & Ainz-Galende, A. (2021). Measuring Digital Citizenship: A Comparative Analysis. *Informatics (MDPI)*, *8*(1), 18. DOI: 10.3390/informatics8010018

Ferrara, E., Varol, O., Davis, C., Menczer, F., & Flammini, A. (2016). The rise of social bots. *Communications of the ACM*, *59*(7), 96–104. DOI: 10.1145/2818717

Festré, A., & Garrouste, P. (2015). The 'economics of attention': A history of economic thought perspective. Œconomia. History, Methodology, Philosophy, (5-1), 3-36.

Fidel, R. (1984). The case study method: A case study. *Library & Information Science Research*, *6*(3), 273–288.

Figueira, Á., & Oliveira, L. (2017). The current state of fake news: Challenges and opportunities. *Procedia Computer Science*, *121*, 817–825. DOI: 10.1016/j.procs.2017.11.106

Fine, G. A. (2003). Crafting authenticity: The validation of identity in self-taught art. *Theory and Society*, *32*(2), 153–180. DOI: 10.1023/A:1023943503531

Finland Education Hub. (2023). Digital literacy in Finnish education: A model for the world. https://finlandeducationhub.com/digital-literacy-in-finnish-education-a-model-for-the-world/

Fleck, A. (2024). Cybercrime Expected To Skyrocket in Coming Years. *Statista.* Retrieved from https://www.statista.com/chart/28878/expected-cost-of-cybercrime-until-2027

Fletcher, R., Cornia, A., Graves, L., & Kleis Nielsen, R. (2018, February). *Measuring the reach of "fake news" and online disinformation in Europe*. Reuters Institute Factsheet. In *Reuters Institute for the Study of Journalism.* University of Oxford. https://reutersinstitute.politics.ox.ac.uk/our-research/measuring-reach-fake-news-and-online-disinformation-europe

Florencio, D., & Herley, C. (2007). A large-scale study of web password habits. *Proceedings of the 16th International Conference on World Wide Web* (pp.657-666). Association for Computing Machinery. DOI: 10.1145/1242572.1242661

Floridi, L. (2019). *The ethics of information.* Oxford University Press.

Floridi, L., & Cowls, J. (2022). Ethical guidelines for trustworthy AI: Principles, practices, and challenges. *AI & Society*, *37*(1), 1–18.

Floridi, L., & Taddeo, M. (2020). *The ethics of artificial intelligence: Principles, challenges, and opportunities.* Oxford University Press.

Floyd, D. L., Prentice-Dunn, S., & Rogers, R. W. (2000). A Meta-Analysis of Research on Protection Motivation Theory. *Journal of Applied Social Psychology*, *30*(2), 407–429. DOI: 10.1111/j.1559-1816.2000.tb02323.x

Formosa, P. (2023). A principlist-based study of the ethical design and acceptability of artificial social agents. *International Journal of Human-Computer Studies*, *2023*(172), 102980.

Fox, J., & Warber, K. M. (2014). Queer Identity Management and Political Self-Expression on Social Networking Sites: A Co-Cultural Approach to the Spiral of Silence. *Journal of Communication*, *65*(1), 79–100. DOI: 10.1111/jcom.12137

Franquet, M., & Eldridge, S. A. (2020). The Ethics of Photojournalism in the Digital Age. Routledge. Retrieved from - https://www.routledge.com/The-Ethics-of-Photojournalism-in-the-Digital-Age/SantosSilva-EldridgeII/p/book/9781138586307?srsltid=AfmBOoosnTC3CRnUVXECRSFHPjMNpNs4-bL6EBfWS4k4o-lHWDF-wA5j

Fraser, N. (1990). Rethinking the public sphere: A contribution to the critique of actually existing democracy. *Social Text, 25/26*, 56. DOI: 10.2307/466240

Frau-Meigs, D., O'Neill, B., Soriani, A., & Tomé, V. (2017). Digital citizenship education: Volume 1: Overview and new perspectives.

Freeman, G., & Acena, D. (2022). "Acting out" Queer Identity: the embodied visibility in social virtual reality. *Proceedings of the ACM on Human-computer Interaction, 6*(CSCW2), 1–32. https://doi.org/ DOI: 10.1145/3555153

Freeman, R. E. (1984). *Strategic management: A stakeholder approach*. Pitman Publishing.

Friedman, T. L. (2017). *'Thank you for being late: An optimist's guide to thriving in the age of acceleration*. Picador/Farrar Straus and Giroux.

Friesem, Y. (2016). Empathy for the digital age: Using video production to enhance social, emotional, and cognitive skills. In *Emotions, technology, and behaviors* (pp. 21–45). Academic Press. DOI: 10.1016/ B978-0-12-801873-6.00002-9

Fumasoli, T., & Rossi, F. (2021). The role of higher education institutions in transnational networks for teaching and learning innovation: The case of the Erasmus+ programme. *European Journal of Education, 56*(2), 200–218. DOI: 10.1111/ejed.12454

Furnell, S. (2007). An assessment of website password practices. *Computers & Security, 26*(7), 445–451. DOI: 10.1016/j.cose.2007.09.001

Gabel, S., Reichert, L., & Reuter, C. (2020). Discussing conflict in social media: The use of Twitter in the Jammu and Kashmir conflict. *Media, War & Conflict, 15*(4), 504–529. DOI: 10.1177/1750635220970997

Gagliardone, I., Diepeveen, S., Findlay, K., Olaniran, S., Pohjonen, M., & Tallam, E. (2021). Demystifying the COVID-19 infodemic: Conspiracies, context, and the agency of users. *Social Media + Society, 7*(3), 20563051211044233. DOI: 10.1177/20563051211044233

Gallagher, B. (2016), "What impact does fake news have on the real world?", Mary Jane. December 5th, available at: https://merryjane.com/culture/fake-news-internetimpact

Gangneux, J. (2019). Digital citizenship in a datafied society: By Arne Hintz, Lina Dencik and Karin Wahl-Jorgensen, Cambridge, UK, Polity Press, 2019,180 pp. *Information Communication and Society, 22*(14), 2211–2213. DOI: 10.1080/1369118X.2019.1635186

Gaw, S., & Felton, E. W. (2006). Password Management Strategies for Online Accounts. *Proceedings of the Second Symposium on Usable Privacy and Security*. Association for Computing Machinery. DOI: 10.1145/1143120.1143127

Gayatri, P., & Ajitha, M. & Kumar, Kiran & Kumar, P.P. & Yamsani, Madhusudan. (2017). Validated high performance liquid chromatographic method for estimation of nisoldipine from matrix type transdermal films: Application to ex vivo skin permeability studies. *Asian Journal of Pharmaceutics, 11*, S407–S411.

Ge, X. (2020). Social media reduce users' moral sensitivity: Online shaming as a possible consequence. *Aggressive Behavior, 46*(5), 359–369. . 21904.DOI: 10. 1002/ ab

Geiger, C., Hilty, R., Griffiths, J., & Sganga, C. (2020). Fair Use in the European Union: Inconsistencies and Legal Uncertainties. *European Intellectual Property Review, 42*(2), 97–112.

Ge, J., & Gretzel, U. (2018). Emoji rhetoric: A social media influencer perspective. *Journal of Marketing Management, 34*(15-16), 1272–1295. Advance online publication. DOI: 10.1080/0267257X.2018.1483960

Giannopoulou, A. (2023). Digital identity infrastructures: A critical approach of self-sovereign identity. *Digital Society : Ethics, Socio-Legal and Governance of Digital Technology, 2*(2), 18. DOI: 10.1007/s44206-023-00049-z PMID: 37200582

Gibbs, J. L., Sivunen, A., & Boyraz, M. (2017). Investigating the impacts of team type and design on virtual team processes. *Human Resource Management Review, 27*(4), 590–603. DOI: 10.1016/j.hrmr.2016.12.006

Gillespie, T. (2018). *Custodians of the internet: Platforms, content moderation, and the hidden decisions that shape social media*. Yale University Press.

Gilster, P. (1997). *Digital literacy*. John Wiley & Sons.

Global Educators Network. https://www.gentefl.org/

Global Nomads Group – USA. (Since 1998). Designed by global youth for global youth. https://gng.org/

Goffman, E. (1959). *The Presentation of Self in Everyday Life*. Doubleday.

Goldman, E. (2020). Content ID and the Fair Use Dilemma in Digital Platforms. *Journal of Internet Law, 23*(4), 12–18.

Goldman, E. (2020). Content moderation remedies. *Michigan Technology Law Review, 26*(2), 215–267.

Goldman, E. (2020). Copyright in the Age of Automated Content Moderation. *Stanford Law Review, 72*(4).

Goldman, E. (2020). *Internet Law: Cases and Problems* (9th ed.). Semaphore Press.

Gómez-Galán, J., Martínez-López, J. Á., Lázaro-Pérez, C., & Fernández-Martínez, M. D. M. (2021). Usage of Internet by University Students of Hispanic Countries: Analysis Aimed at Digital Literacy Processes in Higher Education. *European Journal of Contemporary Education, 10*(1), 53–65.

Gong, Y.-C., Wang, M., Liang, W., Hu, F., & Zhang, Z.-K. (2023). UHIR: An effective information dissemination model of online social hypernetworks based on user and information attributes. *Information Sciences, 644*, 119284. DOI: 10.1016/j.ins.2023.119284

Gonzalez-Mohino, M., Rodriguez-Domenech, M., Callejas-Albiñana, A. I., & Castillo-Canalejo, A. (2023). Empowering critical thinking: The role of digital tools in citizen participation. *Journal of New Approaches in Educational Research, 12*(2), 258–275. DOI: 10.7821/naer.2023.7.1385

Google LLC v. Oracle America, Inc., 593 U.S. ___ (2021).

Gordon, R., Spotswood, F., & Dibb, S. (2022). Critical social marketing: Towards emancipation? *Journal of Marketing Management, 38*(11–12), 1043–1071. DOI: 10.1080/0267257X.2022.2131058

Gotterbarn, D., Bruckman, A., Flick, C., Miller, K. W., & Wolf, M. J. (2018). ACM Code of Ethics and Professional Conduct. *Communications of the ACM, 61*(1), 19–26.

Graham, C. M. (2009). Communications technology, emotional intelligence, and impact on performance: A conceptual exploration of connections. *Advances in Developing Human Resources, 11*(6), 773–783. DOI: 10.1177/1523422309360695

Grawemeyer, B., & Johnson, H. (2011). Using and managing multiple passwords: A week to a view. *Interacting with Computers, 23*(3), 256–267. DOI: 10.1016/j.intcom.2011.03.007

Grazian, D. (2005). *Blue Chicago: The search for authenticity in urban blues clubs*. University of Chicago Press.

Greengard, S. (2021). *The internet of things*. MIT Press. DOI: 10.7551/mitpress/13937.001.0001

Grindle, M. S. (2007). Good enough governance revisited. *Development Policy Review, 25*(5), 533–574. DOI: 10.1111/j.1467-7679.2007.00385.x

Grondin, F., Lomanowska, A. M., & Jackson, P. L. (2019). Empathy in computer-mediated interactions: A conceptual framework for research and clinical practice. *Clinical Psychology : a Publication of the Division of Clinical Psychology of the American Psychological Association, 26*(4), e12298. DOI: 10.1111/cpsp.12298

Gruszczynska, A., Merchant, G., & Pountney, R. (2013). "Digital futures in teacher education": Exploring open approaches towards digital literacy. *Electronic Journal of e-Learning, 11*(3), 193–206.

Gruszczynska, A., & Pountney, R. (2013). Developing the concept of digital literacy in the context of schools and teacher education. *Enhancing Learning in the Social Sciences, 5*(1), 25–36. DOI: 10.11120/elss.2013.05010025

Guedes, I., Martins, M., & Cardoso, C. S. (2022). Exploring the determinants of victimization and fear of online identity theft: An empirical study. *Security Journal, 21*(1), 1–26. DOI: 10.1057/s41284-022-00350-5

Guess, A. M., Lerner, M., Lyons, B., Montgomery, J. M., Nyhan, B., Reifler, J., & Sircar, N. (2020). A digital media literacy intervention increases discernment between mainstream and false news in the United States and India. *Proceedings of the National Academy of Sciences of the United States of America, 2020*(27), 15536–15545. DOI: 10.1073/pnas.1920498117 PMID: 32571950

Gui, M., & Argentin, G. (2011). Digital skills of Internet natives: Different forms of digital literacy in a random sample of northern Italian high school students. *New Media & Society, 13*(6), 963–980. DOI: 10.1177/1461444810389751

Gupta, A., Kumar, N., Prabhat, P., Gupta, R., Tanwar, S., Sharma, G., Bokoro, P. N., & Sharma, R. (2022). Combating Fake News: Stakeholder Interventions and Potential Solutions. *IEEE Access : Practical Innovations, Open Solutions, 10*, 78268–78289. DOI: 10.1109/ACCESS.2022.3193670

Gupta, M., Dennehy, D., Parra, C. M., Mäntymäki, M., & Dwivedi, Y. K. (2023). Fake news believability: The effects of political beliefs and espoused cultural values. *Information & Management, 60*(2), 103745. DOI: 10.1016/j.im.2022.103745

Gurung, A., Luo, X., & Liao, Q. (2009). Consumer motivations in taking action against spyware: an empirical investigation. *Information Management & Computer Security, 17*(3), 276–289. https://doi .org/, F. (2023). Personal data of 665,000 Marina Bay Sands lifestyle rewards members accessed in data security breach. *CNA*. Retrieved from https://www.channelnewsasia.com/singapore/marina-bay-sands -mbs-customers-personal-data-security-breach-3902491DOI: 10.1108/09685220910978112Hamzah

Gurung, A., & Raja, M. K. (2016). Online privacy and security concerns of consumers. *Information and Computer Security, 24*(4), 348–371. DOI: 10.1108/ICS-05-2015-0020

Hajer, M. A. (1995). *The politics of environmental discourse: Ecological modernization and the policy process*. Oxford University Press.

Haleem, A., Javaid, M., Qadri, M. A., & Suman, R. (2022). Understanding the role of digital technologies in education: a review. *Sustainable Operations and Computers*. V. 3, 2022, Pages 275-285https:// DOI: 10.1016/j.susoc.2022.05.004

Hamid, N., & Khalid, F. (2016). Entrepreneurship and innovation in the digital economy. *The Lahore Journal Of Economics, 21*(Special Edition), 273–312. DOI: 10.35536/lje.2016.v21.isp.a12

Han, H., & Yoon, H. J. (2022). Sustainable healthcare practices: The role of green marketing. *International Journal of Healthcare Management, 18*(1), 1–15.

Hanna, M., Pantanowitz, L., Jackson, B., Palmer, O., Visweswaran, S., Pantanowitz, J., Deebajah, M., & Rashidi, H. (2024). Ethical and bias considerations in artificial intelligence (AI)/Machine learning. *Modern Pathology, 100686*. Advance online publication. DOI: 10.1016/j.modpat.2024.100686

Harasymiw, T. (2012). *Cyberbullying and the law*. The Rosen Publishing Group.

Harris, T. (2019). *The Center for Humane Technology: Ethical tech design principles.*

Harris, B., & Birnbaum, R. (2015). Ethical and legal implications on the use of technology in counselling. *Clinical Social Work Journal, 43*(2), 133–141. DOI: 10.1007/s10615-014-0515-0

Harrison, T., & Polizzi, G. (2021). drawing on moral theory to advance digital citizenship education.

Harrison, J. (2023). The role of virtue ethics in digital citizenship education. *Journal of Digital Ethics, 15*(2), 112–127.

Harrison, T. (2022). A new educational model for online flourishing: A pragmatic approach to integrating moral theory for cyber-flourishing. *Pastoral Care in Education, 40*(2), 128–151. DOI: 10.1080/02643944.2021.1898665

Hart, S. L. (1995). A natural-resource-based view of the firm. *Academy of Management Review, 20*(4), 986–1014. DOI: 10.2307/258963

Haseeb, M., Hussain, H. I., Ślusarczyk, B., & Jermsittiparsert, K. (2019). Industry 4.0: A solution towards technology challenges of sustainable business performance. *Social Sciences (Basel, Switzerland), 8*(5), 154. DOI: 10.3390/socsci8050154

Hengstler, J. (2017). Managing your digital footprint: Ostriches v. eagles. Education for a digital world, 2(1), 89-139.

He, Q., Hong, Y., & Raghu, T. S. (2024). Platform Governance with Algorithm-Based Content Moderation: An Empirical Study on Reddit. *Information Systems Research*, isre.2021.0036. DOI: 10.1287/isre.2021.0036

Herath, T., Chen, R., Wang, J., Banjara, K., Wilbur, J., & Rao, H. R. (2012). Security services as coping mechanisms: An investigation into user intention to adopt an email authentication service. *Information Systems Journal*. Advance online publication. DOI: 10.1111/j.1365-2575.2012.00420.x

Herath, T., & Rao, H. R. (2009). Protection motivation and deterrence: A framework for security policy compliance in organisations. *European Journal of Information Systems*, *18*(2), 106–125. DOI: 10.1057/ejis.2009.6

Herley, C. (2009, September 8–11). So long, and no thanks for the externalities: The rational rejection of security advice by users. Paper presented at the 2009 Workshop on New Security Paradigms, Oxford, UK. DOI: 10.1145/1719030.1719050

High Court of Australia (1975). University of New South Wales v. Moorhouse, (1975) 133 CLR 1.

High Court of Australia (2012). Roadshow Films Pty Ltd. v. iiNet Ltd., (2012) HCA 16.

Hinds, J., & Joinson, A. (2019). Human and computer personality prediction from digital footprints. *Current Directions in Psychological Science*, *28*(2), 204–211. DOI: 10.1177/0963721419827849

Hintz, A., Dencik, L., & Wahl-Jorgensen, K. (2019). Digital citizenship in a datafied society. *Polity Press, Cambridge*, UK, 2019

Hölbl, Marko & Zadorozhny, Vladimir & Družovec, Tatjana & Komapara, Marko & Nemec Zlatolas, Lili. (2024). Browser Fingerprinting: Overview and Open Challenges. .DOI: 10.3233/FAIA231163

Hollandsworth, R., Dowdy, L., & Donovan, J. (2011). Digital Citizenship in K-12: It Takes a Village. *TechTrends*, *55*(4), 37–47. DOI: 10.1007/s11528-011-0510-z

Howard, P. N., & Kollanyi, B. (2016, June 1). Bots, #StrongerIn, and #Brexit: Computational Propaganda during the UK-EU Referendum. NASA ADS. Retrieved from - https://doi.org//arXiv.1606.06356DOI: 10.48550

Howe, A. E., Ray, I., Roberts, M., Urbanska, M., & Byrne, Z. (2012, May 20–23). The psychology of security for the home computer user. Paper presented at the 2012 IEEE Symposium on Security and Privacy (SP), San Francisco, CA. DOI: 10.1109/SP.2012.23

Huang, C., & Chou, C. (2021). Cyberbullying and cultural values: A cross-national comparison. *International Journal of Cyber Behavior, Psychology and Learning*, *11*(4), 1–15.

Hudson, S., Huang, L., Roth, M., & Madden, T. (2015). The influence of social media interactions on consumer–brand relationships: A three-country study of brand perceptions and marketing behaviors. *International Journal of Research in Marketing*, *33*(1), 27–41. Advance online publication. DOI: 10.1016/j.ijresmar.2015.06.004

Huerta, E., & Sandoval-Almazán, R. (2007). Digital literacy: Problems faced by telecenter users in Mexico. *Information Technology for Development*, *13*(3), 217–232. DOI: 10.1002/itdj.20071

Hughes, C., Swaminathan, V., & Brooks, G. (2019). Driving Brand Engagement Through Online Social Influencers: An Empirical Investigation of Sponsored Blogging Campaigns. *Journal of Marketing*, *83*(5), 78–96. DOI: 10.1177/0022242919854374

Hughes, J. (2019). *Copyright and Its Limits: Privacy, Publicity, and the Legal Uses of Fair Use*. Cambridge University Press.

Hughes, J. (2019). The role of transformative use in copyright law. *Harvard Journal of Law & Technology*, *33*(1), 45–79.

Hughes, J. (2019). *Transformative Use and Digital Copyright: Navigating User-Generated Content*. Oxford University Press.

Hughes, J. (2019). Transformative Use in the Digital Age: The Expansion of Fair Use and Its Implications. *Columbia Journal of Law & Arts*, *42*(3), 245–267.

Hultenius, J. (2023). The challenges of digital citizenship. https://www.linkedin.com/pulse/challenges-digital-citizenship-jonas-hultenius-g2ccf

Human Rights Council. (2018). Report on Myanmar and Facebook. United Nations Human Rights Council, para 74. Retrieved from - https://www.ohchr.org/sites/default/files/Documents/Publications/OHCHRreport2018.pdf

Hu, Q., & Dinev, T. (2005). Is spyware an internet nuisance or public menace? *Communications of the ACM*, *48*(8), 61–65. DOI: 10.1145/1076211.1076241

Hussein, F., & Hejase, H. (2023). Artificial Intelligence and Fake News. Žurnalistikos tyrimai. Vol. 16. pp. 39-71. DOI: 10.15388/ZT/JR.2022.2

Hutt, R. (2015). What are your digital rights? https://www.weforum.org/agenda/2015/11/what-are-your-digital-rights-explainer/

Idowu, A., & Elbanna, A. (2022). Digital platforms of work and the crafting of career path: The crowdworkers' perspective. *Information Systems Frontiers*, *24*(2), 441–457. DOI: 10.1007/s10796-020-10036-1

IEEE. (2024). Consequences of the digital divide in education. https://ctu.ieee.org/consequences-of-the-digital-divide-in-education/

Ifinedo, P. (2012). Understanding information systems security policy compliance: An integration of the theory of planned behavior and the protection motivation theory. *Computers & Security*, *31*(1), 83–95. DOI: 10.1016/j.cose.2011.10.007

Immanuel Kant Baltic Federal University. (2024). *Russian Dictation*. Retrieved from: https://kantiana.ru/en/international/russian-dictation/

India lynchings: WhatsApp sets new rules after mob killings. (2018). BBC News. Retrieved from - https://www.bbc.com/news/world-asia-india-44897714 (Visited on 5th June, 2024)

International Society for Technology in Education. (2007). *National Educational Technology Standards for Students (NETSS)*. https://iste.org/standards/students

International Society for Technology in Education. (2008). *National Educational Technology Standards for Teachers (NETSS)*. https://iste.org/standards/educators

International Society for Technology in Education. (2016a). *ISTE Standards for Students (NETSS)*. https://iste.org/standards/students

International Society for Technology in Education. (2016b). *ISTE Standards for Educators (NETSS)*. https://iste.org/standards/educators

International Society for Technology in Education. (2021). *ISTE Standards adoption by state (NETSS)*. https://iste.org/adopting-the-standards

International Telecommunications Union (ITU). (2008). *X.1205 Overview of Cybersecurity*. Series X: Data Networks, Open System Communications and Security – Telecommunication Security. Retrieved from https://www.itu.int/rec/T-REC-X.1205-200804-I

International Trade Administration. (2023). *Singapore Cybersecurity Market.* Retrieved from https://www.trade.gov/market-intelligence/singapore-cybersecurity-market

Ippili, T. R., & Venkateswarlu, S. (2017). Social media Gives Voice to Voiceless Dalits in India: An Analysis. *International Journal of Multidisciplinary Educational Research*. https://www.researchgate.net/publication/327011583_Social_Media_Gives_Voice_To_Voiceless_Dalits_In_India_An_Analysis

Iqbal, A., Shahzad, K., Khan, S. A., & Chaudhry, M. S. (2023, October 03). Shakeel Ahmad Khan, & Muhammad Shahzad Chaudhry. (2023). The relationship of artificial intelligence (AI) with fake news detection (FND): A systematic literature review. *Global Knowledge, Memory and Communication*. Advance online publication. DOI: 10.1108/GKMC-07-2023-0264

Irkinovich, N. R. (2022). THE DIGITAL ECONOMY TODAY. Academicia Globe. *Inderscience Research, 3*(10), 198–203.

Isin, E. F. (2020). *Being digital citizens*. Evelyn Sharon Ruppert (2nd ed.). London, United Kingdom: Rowman & Littlefield Publishing Group. p. 17. ISBN 978-1-78661-447-6. OCLC 1133661431.

Ismailov, R., Simoens, S., & Khasanova, Z. (2020). Greater awareness of biosimilars and shared decision-making among patients attending rheumatology practices in Colorado, United States: Real-world data. *World Journal of Rheumatology, 10*(1), 1–10. DOI: 10.5499/wjr.v10.i1.1

Isman, A., & Ozlem, C. G. (2014). Digital citizenship. *Turkish Online Journal of Educational Technology*. https://eric.ed.gov/?id=EJ1018088

Isono, I., & Prilliadi, H. (2023). *Accelerating Artificial Intelligence Discussions in ASEAN: Addressing Disparities, Challenges, and Regional Policy Imperatives*. Economic Research Institute for ASEAN and East Asia.

Israel, M. J., & Amer, A. (2021). Ethical implications of digital infrastructures for pluralistic perspectives. *Ethics and Information Technology, 23*(3), 399–417. DOI: 10.1007/s10676-021-09582-2

Issac, A. L. (2023). Understanding Cyber-Feminism and its roles in the digital space. In *Advances in human and social aspects of technology book series* (pp. 321–334). https://doi.org/DOI: 10.4018/978-1-6684-8893-5.ch022

ISTE - International Society for Technology in Education. (n.d.). Dream big. Transfer teaching. Empower learners. https://iste.org/

Iyer, K. A. S., Nagojappa, N. B. S., Kanjanapathy, M., & Kumar, S. H. (2024). Marketing in the Digital Age: Leveraging Technology for Sustainability. In Utilizing Technology for Sustainable Resource Management Solutions (pp. 194-217). IGI Global.

Jackson, K., Lee, M., & Rogers, D. (2022). Regulatory pressures on sustainable healthcare: The impact of environmental laws. *Journal of Environmental Regulation*, *37*(6), 115–126.

Jacobides, M., Brusoni, S., & Candelon, F. (2021). The Evolutionary Dynamics of the Artificial Intelligence Ecosystem. *Strategy Science*, *6*(4), 412–435. Advance online publication. DOI: 10.1287/stsc.2021.0148

Jafari, S. M., & Moharrani, R. (2019). Conceptualizing digital social capital. *International Seminar on New Topics in Business Management*. https://www.researchgate.net/publication/340262485_Conceptualizing_Digital_Social_Capital

Jain, S. (2020). The Rising Fourth Wave: Feminist activism on digital platforms in India. *ORG Issue Brief, 384*. https://www.orfonline.org/research/the-rising-fourth-wave-feminist-activism-on-digital-platforms-in-india

Jain, S. L. (2020). Sham Lal Jain v. Union of India. Delhi High Court.

Jain, A. K., Ross, A., & Nandakumar, K. (2016). *Introduction to Biometrics*. Springer., DOI: 10.1007/978-0-387-77326-1

Jaiswal, Y., & Kapoor, A. R. (2024). Representation of Muslim Women in Social Media: Some Reflections from Western and Indian Media Representations. *jcoma.com*. https://doi.org/DOI: 10.58966/JCM20243215

James, C., Weinstein, E., & Mendoza, K. (2019). Teaching digital citizens in today's world: Research and insights behind the Common Sense K–12 digital citizenship curriculum. Project Zero. http://www.pz.harvard.edu/resources/teaching-digital-citizens-in-todays-world

James, L. (2005). *Phishing exposed*. Elsevier.

Jamil, S. (2021). From digital divide to digital inclusion: Challenges for wide-ranging digitalization in Pakistan. *Telecommunications Policy*, *45*(8), 102206. DOI: 10.1016/j.telpol.2021.102206

Jankowicz, N., Hunchak, J., Pavluic, A., Davies, C., Pierson, S., & Kaufmann, Z. (2021). *Malign creativity: How gender, sex and lies are weaponized against women online*. Wilson Centre. https://www.wilsoncenter.org/publication/malign-creativity-how-gender-sex-and-lies-are-weaponized-against-women-online

Jartelius, M. (2020). The 2020 Data Breach Investigations Report – a CSO's Perspective. *Network Security*, *2020*(7), 9–12. DOI: 10.1016/S1353-4858(20)30079-9

Javerian, N. (2023). Digital literacy and online social capital. *Iranian Journal of Educational Sociology, 6*(4), 246–255. DOI: 10.61186/ijes.6.4.246

Jennings, B., & Finkelstein, A. (2009). Digital identity and reputation in the context of a bounded social ecosystem. In *Business Process Management Workshops: BPM 2008 International Workshops, Milano, Italy, September 1-4, 2008. Revised Papers 6* (pp. 687-697). Springer Berlin Heidelberg. DOI: 10.1007/978-3-642-00328-8_69

Jin, S. V., & Muqaddam, A. (2019). Product placement 2.0: "Do Brands Need Influencers, or Do Influencers Need Brands?". *Journal of Brand Management, 26*(5), 522–537. DOI: 10.1057/s41262-019-00151-z

Jiow, H. J., Mwagwabi, F., & Low-Lim, A. (2021). Effectiveness of protection motivation theory based: Password hygiene training programme for youth media literacy education. *The Journal of Media Literacy Education, 13*(1), 67–78. DOI: 10.23860/JMLE-2021-13-1-6

Johansson, S., & Felmlee, D. (2022). Digital responsibility in Swedish education: A virtue ethics approach. *Scandinavian Journal of Educational Research, 66*(2), 189–205.

Johnson, D. G., Powers, T., & Turner, J. (2021). Ethical considerations in online community management: Fostering inclusivity and respect in digital spaces. *Journal of Digital Ethics, 13*(3), 145–158.

Johnson, T. J., & Kaye, B. K. (2004). Wag the blog: How reliance on traditional media and the Internet influence credibility perceptions of Weblogs among blog users. *Journalism & Mass Communication Quarterly, 81*(3), 622–642. DOI: 10.1177/107769900408100310

Johnston, A. C., & Warkentin, M. (2010). Fear appeals and information security behaviors: An empirical study. *Management Information Systems Quarterly, 34*(3), 549–566. DOI: 10.2307/25750691

Jones-Jang, S. M., Mortensen, T., & Liu, J. (2021). Does media literacy help identification of fake news? Information literacy helps, but other literacies don't. *The American Behavioral Scientist, 65*(2), 371–388. DOI: 10.1177/0002764219869406

Jones, L. M., & Mitchell, K. J. (2016). Defining and measuring youth digital citizenship. New Media &amp. *New Media & Society, 18*(9), 2063–2079. DOI: 10.1177/1461444815577797

Jonker, A., & Rogers, J. (2024, September 20). Algorithmic bias. *IBM*. Retrieved February 6, 2025, from https://www.ibm.com/think/topics/algorithmic-bias

Jonsson, P. O. (2011). On utilitarianism vs virtue ethics as foundations of economic choice theory. *Humanomics, 27*(1), 24–40. DOI: 10.1108/08288661111110150

Jung, A.-R., & Heo, J. (2021). Does cluttered social media environment hurt advertising effectiveness? The moderation of ad types and personalization. *Journal of Research in Interactive Marketing, 15*(4), 592–606. DOI: 10.1108/JRIM-11-2020-0238

Jyoti, & Priyanka. (2022). The Jio Effect: Evaluating the impact of Jio on internet usage and digital transformation. *REST Journal on Data Analytics and Artificial Intelligence*. https://doi.org/DOI: 10.46632/jdaai/2/2/16

Kairat, A., Sultanbayeva, G. S., & Kuash, R. (2024). Ways to solve the problems of information and digital literacy within the framework of the "One Belt, One Road" initiative. *Bulletin of LN Gumilyov Eurasian National University.JOURNALISM Series*, *146*(1), 57–73.

Kaiser Permanente. (2024). Our commitment to sustainability. Retrieved from https://about.kaiserpermanente.org

Kalton, G. (2020). *Introduction to Survey Sampling*. Sage.

Kanozia, R., C S, D., & Arya, R. (2022). Critical Media and Information Literacy to Combat Misinformation: Research gaps and future directions. *Journal of Content. Community & Communication*, *16*. Advance online publication. DOI: 10.31620/JCCC.12.22/15

Kantanen, H., & Manninen, J. (2016). Hazy boundaries: Virtual communities and research ethics. *Media and Communication*, *4*(4), 86–96. DOI: 10.17645/mac.v4i4.576

Kaplan, A. M., & Haenlein, M. (2010). Users of the world, unite! The challenges and opportunities of social media. *Business Horizons*, *53*(1), 59–68. DOI: 10.1016/j.bushor.2009.09.003

Kapoun, J. (2020). Teaching undergrads WEB evaluation: A guide for library instruction | Kapoun | College & Research Libraries News. *College & Research Libraries News*, *59*(7), 522–523. DOI: 10.5860/crln.59.7.522

Kara, N. (2018). Understanding university learners' thoughts and practices about digital citizenship: A mixed methods study. *Journal of Educational Technology & Society*, 21/(1), 172-185.

Karnouskos, S. (2020). Artificial Intelligence in Digital Media: The Era of Deepfakes. *IEEE Transactions on Technology and Society*, *1*(3), 138–147. DOI: 10.1109/TTS.2020.3001312

Karusala, N., Bhalla, A., & Kumar, N. (2019). Privacy, Patriarchy, and Participation on Social Media. *Proceedings of the 2019 on Designing Interactive Systems Conference,* 2019. https://doi.org/DOI: 10.1145/3322276.3322355

Katz, A. (2020). Artificial Intelligence and the Future of Copyright: Who Owns AI-Generated Works? *Journal of Intellectual Property Law*, *28*(3).

Katz, S. (2020). AI and Copyright: Challenges and Opportunities. *Journal of Intellectual Property*, *57*(2), 115–132.

Kaye, J., Whitley, E. A., Lund, D., Morrison, M., Teare, H., & Melham, K. (2015). Dynamic consent: A patient interface for twenty-first-century research networks. *European Journal of Human Genetics : EJHG*, *23*(2), 141–146. DOI: 10.1038/ejhg.2014.71 PMID: 24801761

Kayes, I., & Iamnitchi, A. (2017). Privacy and security in online social networks: A survey. *Online Social Networks and Media*, *3*, 1–21. DOI: 10.1016/j.osnem.2017.09.001

Kay, S., Mulcahy, R., & Parkinson, J. (2020). When less is more: The impact of macro and micro social media influencers' disclosure. *Journal of Marketing Management*, *36*(3–4), 248–278. DOI: 10.1080/0267257X.2020.1718740

Keen, A. (2015). *The Internet is not the answer*. Atlantic Books.

Ke, F., & Kwak, D. (2013). Online learning across ethnicity and age: A study on learning interaction participation, perception, and learning satisfaction. *Computers & Education, 61*, 43–51. DOI: 10.1016/j.compedu.2012.09.003

Keller, J. (2021). A practical guide to the NIST Cybersecurity Framework. *Journal of Information Security, 10*(2), 45–61.

Kemp, S. (2022). Digital 2022: India — DataReportal – Global Digital Insights. *DataReportal – Global Digital Insights*. https://datareportal.com/reports/digital-2022-india

Kemp, S. (2024, February 20). *Digital 2024: India — DataReportal – Global Digital Insights*. DataReportal – Global Digital Insights. Retrieved September 21, 2024, from https://datareportal.com/reports/digital-2024-india#:~:text=More%20broadly%2C%2061.5%20percent%20of,while%2068.6%20percent%20were%20male

Kemper, G. (2019). Improving employees' cyber security awareness. *Computer Fraud & Security, 2019*(8), 11–14. DOI: 10.1016/S1361-3723(19)30085-5

Kemp, S. (2017). *Digital in 2017: Global Overview. A collection of internet, social media, and mobile data from around the world*. Wearesocial & Hootsuite.

Kende, A., Becker, J. C., & Hegedűs, P. (2022). Cross-cultural approaches to digital ethics: The impact of culture on online behavior. *Computers in Human Behavior, 127*, 107060.

Kennedy, M. (2020). 'If the Rise of the TikTok Dance and e-Girl Aesthetic has taught us Anything, it's that Teenage Girls Rule the Internet Right Now': TikTok Celebrity, Girls and the Coronavirus Crisis. *European Journal of Cultural Studies, 23*(6), 1069–1076. DOI: 10.1177/1367549420945341

Kenner, S., & Lange, D. (2019). Digital citizenship education. Challenge and opportunity. *Scuola Democratica, 10*(4), 47–55.

Kennick, W. E. (1985). Art and Inauthenticity. *The Journal of Aesthetics and Art Criticism, 44*(1), 3–12. DOI: 10.1111/1540_6245.jaac44.1.0003

Keselman, D., & Saxe-Braithwaite, M. (2021, May). Authentic and ethical leadership during a crisis. []. Sage CA: Los Angeles, CA: SAGE Publications.]. *Healthcare Management Forum, 34*(3), 154–157. DOI: 10.1177/0840470420973051 PMID: 33327784

Keshari Mishra, B. (2017). Digital media in resisting social inequality The Indian experience. In *Precarity within the Digital Age* (pp. 123–133). Springer Nature., DOI: 10.1007/978-3-658-17678-5

Khamdamov, N. N. (2024). Strategies to Improve Students' Digital Literacy in Higher Education. *Educational Research in Universal Sciences, 3*(6), 243–248.

Khan, R., Kulasiri, D., & Samarasinghe, S. (2023). A multifarious exploration of synaptic tagging and capture hypothesis in synaptic plasticity: Development of an integrated mathematical model and computational experiments. *Journal of Theoretical Biology, 556*, 111326. DOI: 10.1016/j.jtbi.2022.111326 PMID: 36279957

Kharitonova, Y. S., Savina, V. S., & Pagnini, F. (2021). ARTIFICIAL INTELLIGENCE'S ALGORITHMIC BIAS: ETHICAL AND LEGAL ISSUES. *Вестник Пермского Университета Юридические Науки*, *3*(53), 488–515. DOI: 10.17072/1995-4190-2021-53-488-515

Kian Yeik, K., Cheung, M.-L., Soh, P., & Teoh, C. W. (2022). Social media influencer marketing: The moderating role of materialism. *European Business Review*, *34*(2), 224–243. DOI: 10.1108/EBR-02-2021-0032

Ki, C.-W., Cuevas, L. M., Chong, S. M., & Lim, H. (2020). Influencer marketing: Social media influencers as human brands attaching to followers and yielding positive marketing results by fulfilling needs. *Journal of Retailing and Consumer Services*, *55*, 102133. DOI: 10.1016/j.jretconser.2020.102133

Ki, C.-W., & Kim, Y.-K. (2019). The mechanism by which social media influencers persuade consumers: The role of consumers' desire to mimic. *Psychology and Marketing*, *36*(10), 905–922. DOI: 10.1002/mar.21244

Kim, A., & Dennis, A. R. (2019). Says who? The effects of presentation format and source rating on fake news in social media. *Management Information Systems Quarterly*, *43*(3), 1025–1039. DOI: 10.25300/MISQ/2019/15188

Kim, B., Xiong, A., Lee, D., & Han, K. (2021). A systematic review on fake news research through the lens of news creation and consumption: Research efforts, challenges, and future directions. *PLoS One*, *16*(12), e0260080. DOI: 10.1371/journal.pone.0260080 PMID: 34882703

Kim, J. J., & Hong, S. P. (2011). A method of risk assessment for multi-factor authentication. *Journal of Information Processing Systems*, *7*(1), 187–198. DOI: 10.3745/JIPS.2011.7.1.187

Kim, J., & Park, J. (2021). The impact of ease of use and feature relevance on social media use: A cross-platform study. *Information Systems Research*, *32*(2), 436–454.

Kindervag, J. (2020). Zero Trust architecture: Rethinking network security. *Cybersecurity Journal*, *15*(3), 34–49.

Kloza, B. (2023). Consequences of the digital divide in education - connecting the unconnected. *Connecting the Unconnected.* https://ctu.ieee.org/consequences-of-the-digital-divide-in-education/

Kochan, T. A. (2021). Artificial Intelligence and the Future of Work: A Proactive Strategy. *AI Magazine*, *42*(1), 16–24. DOI: 10.1002/j.2371-9621.2021.tb00006.x

Kolo, C., Joschka Mütterlein, & Sarah Anna Schmid. (2022). Believing Journalists, AI, or Fake News: The Role of Trust in Media. Proceedings of the ... Annual Hawaii International Conference on System Sciences. https://doi.org/DOI: 10.24251/HICSS.2022.393

Koltay, T. (2011). The media and the literacies: Media literacy, information literacy, digital literacy. *Media Culture & Society*, *33*(2), 211–221. DOI: 10.1177/0163443710393382

Kordzadeh, N., & Ghasemaghaei, M. (2021). Algorithmic bias: Review, synthesis, and future research directions. *European Journal of Information Systems*, *31*(3), 388–409. DOI: 10.1080/0960085x.2021.1927212

Kotler, P., & Lee, N. (2005). *Corporate social responsibility: Doing the most good for your company and your cause*. John Wiley & Sons.

Kowalski, R. M., Giumetti, G. W., Schroeder, A. N., & Lattanner, M. R. (2019). Bullying in the digital age: A critical review. *Psychological Review*, *126*(2), 347–383.

Krisnaningsih, E., Dwiyatno, S., Jubaedi, A. D., & Shafitri, A. (2023). Increasing ethical understanding of the use of information technology through digital literacy proficiency training. *Dinamisia Jurnal Pengabdian Kepada Masyarakat*, *7*(3), 789–801. DOI: 10.31849/dinamisia.v7i3.12809

Kumar, A., & Taylor, J. W. (2023). Feature importance in the age of explainable AI: Case study of detecting fake news & misinformation via a multi-modal framework. *European Journal of Operational Research*, *317*(2), 401–413. Advance online publication. DOI: 10.1016/j.ejor.2023.10.003

Kumar, D., & Vyas, S. (2023). *The Role of Social Media in Shaping Public Perception of Legal Cases and Its Influence on the Justice Delivery System*. Rawat Prakashan.

Kumar, P. (2021). Content ID and Fair Use: A Case Study of YouTube's Monetization Model. *Indian Journal of Media Studies*, *27*(1), 54–71.

Kumar, P., Jain, R., & Singh, A. (2021). The influence of sustainability on consumer behavior in health-care. *JMR, Journal of Marketing Research*, *58*(7), 987–1003.

Kumar, S., Sharma, A., & Verma, V. (2021). Social Media A Tool for Social and Political Inclusion among Youth: A Study of Indian University Students. *Current Research Journal of Social Sciences and Humanities*, *4*(1), 82–89. DOI: 10.12944/crjssh.4.1.08

Kumar, V. L., & Goldstein, M. A. (2020). Cyberbullying and adolescents. *Current Pediatrics Reports*, *8*(3), 86–92. DOI: 10.1007/s40124-020-00217-6

Kundu, K. (2024). The #MeToo Movement : Investigating the lasting international impacts. https://hir .harvard.edu/metoo-movement-international/

Kuss, D., & Griffiths, M. (2011). Online Social Networking and Addiction—A Review of the Psychological Literature. *International Journal of Environmental Research and Public Health*, *8*(9), 3528–3552. DOI: 10.3390/ijerph8093528 PMID: 22016701

Kusumastuti, A., & Nuryani, A. (2020, March). Digital literacy levels in ASEAN (comparative study on ASEAN countries). In *IISS 2019: Proceedings of the 13th International Interdisciplinary Studies Seminar, IISS 2019,30-31 October 2019,Malang, Indonesia* (p. 269). European Alliance for Innovation. DOI: 10.4108/eai.23-10-2019.2293047

Kwak, K. T., Lee, S. Y., Ham, M., & Lee, S. W. (2021). The effects of internet proliferation on search engine and over-the-top service markets. *Telecommunications Policy*, *45*(8), 102146. DOI: 10.1016/j.telpol.2021.102146

Laak, K. J., & Aru, J. (2024). AI and Personalized Learning: Bridging the gap with modern educational goals. *arXiv preprint arXiv:2404.02798*.

Lambiotte, R., & Kosinski, M. (2014). Tracking the digital footprints of personality. *Proceedings of the IEEE, 102*(12), 1934–1939. DOI: 10.1109/JPROC.2014.2359054

Land, M. (2013). Toward an international law of the internet. *Harv. Int'l LJ, 54*, 393.

Lanham, R. (1995). Digital literacy. *Scientific American, 273*(3), 160–161.

Lankshear, C. J., & Knobel, M. (2008). *Introduction: digital literacies: concepts, policies and practices*. Peter Lang Publishing.

LaRose, R., Rifon, N. J., & Wirth, C. (2007). Online safety begins with you and me: Getting Internet users to protect themselves. *Paper presented at the57th International Communication Association Conference*, San Francisco, CA.

LaRose, R., Rifon, N. J., Liu, X., & Lee, D. (2005, May 26–30). Understanding online safety behavior: A multivariate model. Paper presented at the International Communication Association Conference, New York, NY.

Laskar, M. H. (2023). Examining the emergence of digital society and the digital divide in India: A comparative evaluation between urban and rural areas. *Frontiers in Sociology, 8*. Advance online publication. DOI: 10.3389/fsoc.2023.1145221

Laurent, M., Denouël, J., Levallois-Barth, C., & Waelbroeck, P. (2015). Digital identity. In *Digital identity management* (pp. 1–45). Elsevier. DOI: 10.1016/B978-1-78548-004-1.50001-8

Law, N. W. Y., Chow, S. L., & Fu, K. W. (2018). Digital citizenship and social media: A curriculum perspective. *Second handbook of information technology in primary and secondary education.*

Lcom Team. (2024). The importance of digital citizenship: a 2024 perspective. https://www.learning.com/blog/the-importance-of-digital-citizenship-a-2024-perspective/#:~:text=Digital%20citizenship%20is%20an%20increasingly,Read%20the%20full%20article%20here)

Leary, M. G. (2010). Reasonable expectations of privacy for youth in a digital age. *Mississippi Law Journal, 80*, 1035.

Lee, C., & Ng, D. (2023). Digital fraud attack rate in Singapore higher than APAC average: Cybercrime report. *CNA*. Retrieved from https://www.channelnewsasia.com/singapore/cyber-crime-digital-fraud-attack-bots-lexisnexis-report-3992166

Lee, D., & Rubin, T. (2021). Digital citizenship instruction: challenges and opportunities. https://infobase.com/blog/digital-citizenship-instruction-challenges-and-opportunities/

Lee, D., LaRose, R., & Rifon, N. J. (2008). Keeping our network safe: A model of online protection behaviour. *Behaviour & Information Technology, 27*(5), 445–454. DOI: 10.1080/01449290600879344

Lee, H., & Cho, C.-H. (2020). Digital advertising: Present and future prospects. *International Journal of Advertising, 39*(3), 332–341. DOI: 10.1080/02650487.2019.1642015

Lee, M. R. (2009). E-ethical leadership for virtual project teams. *International Journal of Project Management, 27*(5), 456–463. DOI: 10.1016/j.ijproman.2008.05.012

Lee, S. (2014). Digital literacy Education for the development of digital literacy. *International Journal of Digital Literacy and Digital Competence*, *5*(3), 29–43. DOI: 10.4018/ijdldc.2014070103

Lee, Y., & Larsen, K. R. (2009). Threat or coping appraisal: Determinants of SMB executives' decision to adopt anti-malware software. *European Journal of Information Systems*, *18*(2), 177–187. DOI: 10.1057/ejis.2009.11

LeFebvre, R. (2012, October 12–13). The human element in cyber security: A study on student motivation to act. Paper presented at the 2012 Information Security Curriculum Development Conference, Kennesaw, GA. DOI: 10.1145/2390317.2390318

Lemphane, P., & Prinsloo, M. (2014). Children's digital literacy practices in unequal South African settings. *Journal of Multilingual and Multicultural Development*, *35*(7), 738–753. DOI: 10.1080/01434632.2014.908894

Lenkaitis, C. A. (2019). Valuing technology in the L2 classroom: Learner and teacher perceptions, preferences, and digital identity. In Montebello, M. (Ed.), *Embracing Digital Learners in an Age of Global Educational Change and Rapid Technological Advancements* (pp. 173–189). IGI Global., DOI: 10.4018/978-1-5225-9304-1.ch011

Lessig, L. (2015). *Free Culture: The Nature and Future of Creativity*. Penguin Books.

Let's Talk Science. (2020). Digital citizenship & ethics. *Let's Talk Science*. https://letstalkscience.ca/educational-resources/backgrounders/digital-citizenship-ethics

Lewandowsky, S., Ecker, U. K., & Cook, J. (2020). Beyond misinformation: Understanding and coping with the "post-truth" era. *Journal of Applied Research in Memory and Cognition*, *6*(4), 353–369. DOI: 10.1016/j.jarmac.2017.07.008

Lewis-Kraus, G. (2022, June 3). How harmful is social media? *The New Yorker*. https://www.newyorker.com/culture/annals-of-inquiry/we-know-less-about-social-media-than-we-think

Leyden, K. M. (2003). Social capital and the built environment - PubMed Central. In *American Journal of Public Health*. https://pmc.ncbi.nlm.nih.gov/articles/PMC1448008/

Lievens, E. (2011). Risk-reducing regulatory strategies for protecting minors in social networks. *Strategy for Telecommunications. Information and Media*, *13*(6), 43–54. DOI: 10.1108/14636691111174252

Litman, J. (2020). Harper & Row and the Limits of Fair Use: Analyzing Copyright Protection in Journalism. *Columbia Law Review*, *38*(2).

Litman, J. (2021). *Digital Copyright: Law and Policy*. Wolters Kluwer.

Littlejohn, A., Beetham, H., & McGill, L. (2012). Learning at the digital frontier: A review of digital literacies in theory and practice. *Journal of Computer Assisted Learning*, *28*(6), 547–556. DOI: 10.1111/j.1365-2729.2011.00474.x

Liu, D., Ainsworth, S. E., & Baumeister, R. F. (2016). A meta-analysis of social networking online and social capital. *Review of General Psychology*, *20*(4), 369–391. DOI: 10.1037/gpr0000091

Liu, Y. (2023). Investigating the dynamics of trust in social capital on social media platforms. In Hoda, N., & Arabia, S. (Eds.), *Social Capital in the Age of Online Networking: Genesis, Manifestations, and Implications*. IGI Global., DOI: 10.4018/978-1-6684-8953-6.ch005

Livingstone, I., & Hope, A. (2011). Next Gen [online]. Nesta., Available at https://www.nesta.org.uk/report/next-gen/, Retrieved May 28, 2019, from.

Livingstone, S., Stoilova, M., & Nandagiri, R. (2018). Children's data and privacy online: Growing up in a digital age. *London School of Economics and Political Science*. Retrieved December 18, 2024 from https://www.lse.ac.uk/

Livingstone, S., & Sefton-Green, J. (2016). *The class*. New York University Press.

Livingstone, S., Stoilova, M., & Nandagiri, R. (2020). Children's data and privacy online: Growing up in a digital age. *New Media & Society*, *22*(8), 1365–1382.

Livingstone, S., & Third, A. (2017). Children and young people's rights in the digital age: An emerging agenda. *New Media & Society*, *19*(5), 657–670. DOI: 10.1177/1461444816686318

Lombana-Bermudez, A., Cortesi, S., Fieseler, C., Gasser, U., Hasse, A., Newlands, G., & Wu, S. (2020). Youth and the Digital Economy: Exploring Youth Practices, Motivations, Skills, Pathways, and Value Creation. Berkman Klein Center for Internet & Society., Retrieved from https://cyber.harvard.edu/ publication/2020/youth-and-digital-economy, Retrieved February 18, 2021, from.

Lotfy, Y., Zaki, A., Abd El-Hafeez, T., & Mahmoud, T. M. (2021). Privacy Issues of Public WiFi Networks. *Proceedings of the International Conference on Artificial Intelligence and Computer Vision (AICV2021)* (pp.656-665). Springer.

Lou, C., & Yuan, S. (2018). Influencer Marketing: How Message Value and Credibility Affect Consumer Trust of Branded Content on Social Media. *Journal of Interactive Advertising*, *19*(1), 1–45. DOI: 10.1080/15252019.2018.1533501

Lozano-Díaz, A., & Fernández-Prados, J. S. (2020). Educating digital citizens: An opportunity to critical and activist perspective of sustainable development goals. *Sustainability (Basel)*, *12*(18), 7260. Advance online publication. DOI: 10.3390/su12187260

Lu, H., Tong, P., & Zhu, R. (2020). Does internet use affect netizens' trust in government? Empirical evidence from China. *Social Indicators Research*, *149*(1), 167–185. DOI: 10.1007/s11205-019-02247-0

Luke, A. (2018). Digital ethics now. *Language and Literature*, *20*(3), 185–198.

Lunenborg, J., & Maier, R. (2018). Bias in social media content. *International Journal of Communication*, *12*(1), 1–18.

Lyon, D. (1991). The information society: Ideology or utopia. In Mackay, H., Young, M., & Beynon, J. (Eds.), *Understanding Technology in Education* (1st ed., pp. 93–108). The Falmer Press.

Lyu, H. (2012). Internet policy in Korea: A preliminary framework for assigning moral and legal responsibility to agents in internet activities. *Government Information Quarterly*, *29*(3), 394–402. . giq. 2011. 12. 008.DOI: 10. 1016/j

Madden, M., Lenhart, A., Cortesi, S., & Gasser, U. (2013). Teens, social media, and privacy. *Pew Research Center*. Retrieved December 21, 2024 from https://www.pewresearch.org/

Madrekar, A., & Jambhale, R. (2024). Password Manager Statistics By Reasons, Usage and Subscription Rate. *Coolest Gadgets*. Retrieved from https://www.coolest-gadgets.com/password-manager-statistics

Mahdi, H. R. (2018). The awareness of the digital citizenship among the users of social networks and its relation to some variables. *International Journal of Learning Management Systems*, 6/(1), 11-25. https://doi.org/DOI: 10.18576/ijlms/060102

Maimon, D., Jordan Howell, C., Jacques, S., & Perkins, R. C. (2022). Situational awareness and public Wi-Fi users' self-protective behaviors. *Security Journal*, *35*(1), 154–174. DOI: 10.1057/s41284-020-00270-2

Majumdar, S., Tewatia, M., Jamkhedkar, D., & Bhatia, K. (2022). "You don't know me so don't try to judge me": Gender and identity performance on social media among young Indian users. *Frontiers in Psychology*, *13*. Advance online publication. DOI: 10.3389/fpsyg.2022.855947

Mandarano, L., Meenar, M., & Steins, C. (2010). Building social capital in the digital age of civic engagement. *Journal of Planning Literature*, *25*(2), 123–135. DOI: 10.1177/0885412210394102

Marav, D. (2016). Mongolian students' digital literacy practices: The interface between English and the internet. *Trabalhos em Lingüística Aplicada*, *55*(02), 293–318. DOI: 10.1590/010318134962176441

Marengo, D., Longobardi, C., Fabris, M. A., & Settanni, M. (2018). Highly-visual social media and internalizing symptoms in adolescence: The mediating role of body image concerns. *Computers in Human Behavior*, *82*, 63–69. DOI: 10.1016/j.chb.2018.01.003

Marett, K., McNab, A. L., & Harris, R. B. (2011). Social networking websites and posting personal information: An evaluation of protection motivation theory. *AIS Transactions on Human-Computer Interaction*, *3*(3), 170–188. DOI: 10.17705/1thci.00032

Marforio, C., Masti, R. J., Soriente, C., Kostiainen, K., & Capkun, S. (2015). Personalized security indicators to detect application phishing attacks in mobile platforms. Retrieved from https://arxiv.org/abs/1502.06824

Marin, L., & Copeland, S. M. (2022). Self-Trust and Critical Thinking Online: A Relational Account. *Social Epistemology*, 1–13. DOI: 10.1080/02691728.2022.2151330

Marín-López, I., Zych, I., Monks, C. P., & Ortega-Ruiz, R. (2019). Empathy, morality and social and emotional competencies in interpersonal interactions online. *Thriving in Digital Workspaces: Emerging Issues for Research and Practice*, 217-233.

Marshall, C., & Rossman, G. B. (2016). *Designing qualitative research* (6th ed.). SAGE Publications.

Marshall, P. D., Kim, B., & Christopher, M. (2017). Academic persona: The construction of online reputation in the modern academy. In *The digital academic* (pp. 47–62). Routledge. DOI: 10.4324/9781315473611-4

Martens, B., & Aguiar, L. GGmez, E., & Mueller-Langer, F. (2018). The Digital Transformation of News Media and the Rise of Disinformation and Fake News. SSRN Electronic Journal. https://doi.org/ DOI: 10.2139/ssrn.3164170

Martin, A., & Grudziecki, J. (2006). DigEuLit: Concepts and tools for digital literacy development. *Innovation in teaching and learning in information and computer sciences, 5*(4), 249-267.

Martin, A. (2005). DigEuLit – A European framework for digital literacy: A progress report. *Journal Of Eliteracy*, *2*(2), 130–136.

Martin, C., & Grudziecki, J. (2006). Digital literacy – What are we really teaching? *European Conference on Information Literacy*, 24(1), 277-289.

Martínez-López, F. J., Anaya-Sánchez, R., Fernández Giordano, M., & Lopez-Lopez, D. (2020). Behind influencer marketing: Key marketing decisions and their effects on followers' responses. *Journal of Marketing Management*, *36*(7–8), 579–607. DOI: 10.1080/0267257X.2020.1738525

Matamoros-Fernández, A., & Farkas, J. (2021). Racism, Hate Speech, and Social Media: A Systematic Review and critique. *Television & New Media*, *22*(2), 205–224. DOI: 10.1177/1527476420982230

Matters, I. (2024). Learn about fake news to support children. Internet Matters. https://www.internetmatters.org/issues/fake-news-and-misinformation-advice-hub/learn-about-fake-news-to-support-children/

Maurushat, A. (2019). *Ethical hacking* (p. 370). University of Ottawa Press/Les Presses de l'Université d'Ottawa.

Maxwell, J. (2010). Validity: How might you be wrong? In Luttrell, W. (Ed.), *Qualitative Educational Research* (1st ed., pp. 279–287). Routledge.

Maxwell, J. A. (2013). *Qualitative research design: An interactive approach* (3rd ed.). SAGE Publications.

Mayer-Schönberger, V., & Cukier, K. (2013). *Big data: A revolution that will transform how we live, work, and think*. Houghton Mifflin Harcourt.

Mayo Clinic. (2023). Sustainability at Mayo Clinic. Retrieved from https://www.mayoclinic.org

McAlaney, J., & Benson, V. (2020). *Cyber Influence and Cognitive Threats*. Elsevier Inc.

McCutcheon, D. (2021). *AI-Generated Works and the Future of Fair Use: Challenges and Opportunities*. MIT Press.

McCutcheon, J. (2021). *AI and Intellectual Property: New Horizons*. Oxford University Press.

McCutcheon, J. (2021). Artificial intelligence and the evolving boundaries of fair use. *Journal of Intellectual Property Law & Practice*, *16*(4), 239–258.

McDermot, M. (2018). Digital footprints: Creation, implication, and higher education. *FDLA Journal*, *3*(1), 11.

McGucken v. Newsweek LLC, 464 F. Supp. 3d 594 (2022).

McGucken v. Newsweek LLC, U.S. District Court for the Southern District of New York (2022) Aufderheide, P., & Jaszi, P. (2018). Reclaiming Fair Use: How to Put Balance Back in Copyright. University of Chicago Press.

MediaSmarts – Canada's Center for Digital Media Literacy. (n.d.). Media literacy week. https://mediasmarts.ca/?https://mediasmarts.ca/break-fake&gad_source=1&gclid=CjwKCAjw68K4BhAuEiwAylp3klut5HIt4l5pFNp9Ph-f_Bl8kerdbDHL10_Aj_-9j-OKiTuuubDG_BoCoVEQAvD_BwE

Meherali, S., Rahim, K. A., Campbell, S., & Lassi, Z. S. (2021). Does Digital Literacy Empower adolescent Girls in Low- and Middle-Income Countries: A Systematic review. *Frontiers in Public Health*, *9*. Advance online publication. DOI: 10.3389/fpubh.2021.761394

Mehta, S., & Kaye, D. B. V. (2019). Pushing the next level: Investigating digital content creation in India. *Television & New Media*, *22*(4), 360–378. DOI: 10.1177/1527476419861698

Meissner, M. (2017). China's social credit system: A big-data enabled approach to market regulation with broad implications for doing business in China. *Mercator Institute for China Studies (MERICS). Papers on China*, *16*, 1–24. Retrieved December 20, 2024, from https://merics.org

Mendes, M., Magano, J., & Silva, C. S. e. (2020). Marketing de Influência e Autenticidade: Impacto no Processo de Compra da Geração Z. *Percursos & Ideias*, *10*, 16–27. DOI: 10.56123/percursos.2020.n10.16

Metherell, T. E., Ghai, S., McCormick, E. M., Ford, T. J., & Orben, A. (2021). Digital exclusion predicts worse mental health among adolescents during COVID-19. medRxiv *(Cold Spring Harbor Laboratory)*. https://doi.org/DOI: 10.1101/2021.11.25.21266853

Metzger, M. J., & Flanagin, A. J. (2013). Credibility and trust of information in online environments: The use of cognitive heuristics. *Journal of Pragmatics*, *59*, 210–220. DOI: 10.1016/j.pragma.2013.07.012

Metzger, M. J., Flanagin, A. J., & Medders, R. B. (2010). Social and Heuristic Approaches to Credibility Evaluation Online. *Journal of Communication*, *60*(3), 413–439. DOI: 10.1111/j.1460-2466.2010.01488.x

Metzler, H., & Garcia, D. (2023). Social drivers and algorithmic mechanisms on digital media. *Perspectives on Psychological Science*, *19*(5), 735–748. DOI: 10.1177/17456916231185057

Meyer, M. N. (2018). Practical tips for ethical data sharing. *Advances in Methods and Practices in Psychological Science*, *1*(1), 131–144. DOI: 10.1177/2515245917747656

Meza, M. B., Luengo-Charath, X., Arancibia, M., & Madrid, E. (2018). Council for International Organizations of Medical Sciences (CIOMS) Ethical Guidelines: Advancements and unsolved topics in 2016 upgrade. *Medwave*, *18*(2), e7208. DOI: 10.5867/medwave.2018.02.7208 PMID: 29701729

Micheli, M., Lutz, C., & Büchi, M. (2018). Digital footprints: An emerging dimension of digital inequality. *Journal of Information. Communication and Ethics in Society*, *16*(3), 242–251. DOI: 10.1108/JICES-02-2018-0014

Milne, G. R., Labrecque, L. I., & Cromer, C. (2009). Toward an understanding of the online consumer's risky behavior and protection practices. *The Journal of Consumer Affairs*, *43*(3), 449–473. DOI: 10.1111/j.1745-6606.2009.01148.x

Milne, S., Sheeran, P., & Orbell, S. (2000). Prediction and Intervention in Health-Related Behavior: A Meta-Analytic Review of Protection Motivation Theory. *Journal of Applied Social Psychology, 30*(1), 106–143. DOI: 10.1111/j.1559-1816.2000.tb02308.x

Mirchandani, M. (2018). Digital hatred, real violence: majoritarian radicalisation and social media in India. *Ashokauniversity.* https://www.academia.edu/37516844/Digital_Hatred_Real_Violence_Majoritarian_Radicalisation_and_Social_Media_in_India

Misha, A. (2024), The growing importance of digital citizenship. https://www.edsurge.com/news/2024-02-19-the-growing-importance-of-digital-citizenship

Mishra, S., Ewing, M. T., & Cooper, H. B. (2022). Artificial intelligence focus and firm performance. *Journal of the Academy of Marketing Science, 50*(6), 1176–1197. Advance online publication. DOI: 10.1007/s11747-022-00876-5

Mitra, A. (2004). Voices of the Marginalized on the Internet: Examples from a website for women of South Asia. *Journal of Communication, 54*(3), 492–510. DOI: 10.1111/j.1460-2466.2004.tb02641.x

Mittelstadt, B. D., Allo, P., Taddeo, M., Wachter, S., & Floridi, L. (2016). The ethics of algorithms: Mapping the debate. *Big Data & Society, 3*(2), 1–21. DOI: 10.1177/2053951716679679

Mkhize, S., Nunlall, R., & Gopal, N. (2020). An examination of social media as a platform for cyber-violence against the LGBT+ population. *Agenda (Durban, South Africa), 34*(1), 23–33. DOI: 10.1080/10130950.2019.1704485

Mohammadyari, S., & Singh, H. (2015). Understanding the effect of e-learning on individual performance: The role of digital literacy. *Computers & Education, 82*, 11–25. DOI: 10.1016/j.compedu.2014.10.025

Mol, A. P. J. (2001). Ecological modernization and the global economy. *Environmental Politics, 10*(1), 1–10. DOI: 10.1080/09644010008414510

Mont'Alerne, C., Arguedas, A. R., Toff, B., Fletcher, R., & Nielsen, R. K. (2022). The trust gap: how and why news on digital platforms is viewed more skeptically versus news in general. https://reutersinstitute.politics.ox.ac.uk/trust-gap-how-and-why-news-digital-platforms-viewed-more-sceptically-versus-news-general#header--0

Moore, C., Barbour, K., & Lee, K. (2017). Five dimensions of online persona. *Persona Studies, 3*(1), 1–11. DOI: 10.21153/ps2017vol3no1art658

Moore, E. J., Price, D. L., Van Abel, K. M., Janus, J. R., Moore, E. T., Martin, E., Morris, J. M., & Alexander, A. E. (2021). Association of Virtual Surgical Planning with external incisions in complex maxillectomy reconstruction. *JAMA Otolaryngology-Head & Neck Surgery, 147*(6), 526–531. DOI: 10.1001/jamaoto.2021.0251 PMID: 33792635

Moore, S. C. (2012). Digital Footprints on the Internet. *The International Journal of Childbirth Education, 27*(3).

Mossberger, K., Tolbert, C. J., & McNeal, R. S. (2008). *Digital citizenship: the internet, society, and participation.* MIT Press., DOI: 10.7551/mitpress/7428.001.0001

Mou, J., Cohen, J. F., Bhattacherjee, A., & Kim, J. (2022). A Test of Protection Motivation Theory in the Information Security Literature: A Meta-Analytic Structural Equation Modeling Approach. *Journal of the Association for Information Systems*, *23*(1), 196–236. DOI: 10.17705/1jais.00723

Msughter, A. E., Perpetua, A. O., & Itiafa, A. L. (2023). Artificial Intelligence and the Media: Revisiting Digital Dichotomy Theory. In Information Systems Management. IntechOpen.

Mueller, M. L., Park, Y., Lee, J., & Kim, T. Y. (2006). Digital identity: How users value the attributes of online identifiers. *Information Economics and Policy*, *18*(4), 405–422. DOI: 10.1016/j.infoecopol.2006.04.002

Mukherjee, R. (2018). Jio sparks Disruption 2.0: Infrastructural imaginaries and platform ecosystems in 'Digital India.'. *Media Culture & Society*, *41*(2), 175–195. DOI: 10.1177/0163443718818383

Muratshina, K. G., & Valeeva, M. V. (2020). Russia—Post-Soviet Central Asia Cooperation in Information and Media. *Communication trends in the post-literacy era: polylingualism, multimodality and multiculturalism as prerequisites for new creativity. Ekaterinburg, 2020*, 476-495

Mwagwabi, F., & Jiow, H. J. (2021). Compliance with security guidelines in teenagers: The conflicting role of peer influence and personal norms. *AJIS. Australasian Journal of Information Systems*, *25*. Advance online publication. DOI: 10.3127/ajis.v25i0.2953

Nakamura, Y. (2021). Fostering digital empathy: The effectiveness of Japan's Kind Speech Movement. *Asian Journal of Communication*, *31*(4), 327–342.

Nakova, P. (2020). Fake news: Definitions and debates. *Mediaeval Studies*, *45*(1), 2–15.

Nambisan, P. (2011). Information seeking and social support in online health communities: Impact on patients' perceived empathy. *Journal of the American Medical Informatics Association : JAMIA*, *18*(3), 298–304. DOI: 10.1136/amiajnl-2010-000058 PMID: 21486888

National Council for the Social Studies. (2013). *The college, career, and civic life (C3) framework for social studies state standards: Guidance for enhancing the rigor of K-12 civics, economics, geography, and history*. Author.

National Institute of Standards and Technology (NIST). (2017). *Digital Identity Guidelines*. Special Publication 800-63-3. Retrieved from DOI: 10.6028/NIST.SP.800-63-3

Nawi, A., Hussin, Z., Ren, C. C., Norsaidi, N. S., & Mohd Pozi, M. S. (2020). Identifying the types of digital footprint data used to predict psychographic and human behaviour. In *Digital Libraries at Times of Massive Societal Transition: 22nd International Conference on Asia-Pacific Digital Libraries, ICADL 2020, Kyoto, Japan, November 30–December 1, 2020* [Springer International Publishing.]. *Proceedings*, *22*, 287–296.

Newman, D. T., Fast, N. J., & Harmon, D. J. (2020). When eliminating bias isn't fair: Algorithmic reductionism and procedural justice in human resource decisions. *Organizational Behavior and Human Decision Processes*, *160*, 149–167. DOI: 10.1016/j.obhdp.2020.03.008

Newman, L., Browne-Yung, K., Raghavendra, P., Wood, D., & Grace, E. (2016). Applying a critical approach to investigate barriers to digital inclusion and online social networking among young people with disabilities. *Information Systems Journal, 27*(5), 559–588. DOI: 10.1111/isj.12106

Newman, N., Fletcher, R., Robertson, C. T., Eddy, K., & Nielsen, R. K. (2022). *Digital news report 2022*. Reuters Institute for the Study of Journalism.

Ng, C. K., Yusof, Y., & Ab Aziz, N. S. N. (2021, December). DFBC Recon Tool: Digital Footprint and Breach Check Reconnaissance Tool. In *2021 14th International Conference on Developments in eSystems Engineering (DeSE)* (pp. 526-530). IEEE.

Ng, B. Y., Kankanhalli, A., & Xu, Y. C. (2009). Studying users' computer security behavior: A health belief perspective. *Decision Support Systems, 46*(4), 815–825. DOI: 10.1016/j.dss.2008.11.010

Nguyen, T., & Van Dinh, L. (2024). AI ethics in the age of intelligent machines. *AI and Ethics, 4*(2), 1–12.

Ng, W. (2012). Can we teach digital natives digital literacy? *Computers & Education, 59*(3), 1065–1078. DOI: 10.1016/j.compedu.2012.04.016

Nieto, B. G. (2018). El influencer: Herramienta clave en el contexto digital de la publicidad engañosa. *Methaodos.Revista de Ciencias Sociales, 6*(1), 1. Advance online publication. DOI: 10.17502/m.rcs.v6i1.212

Nissenbaum, H. (2019). *Privacy in context: Technology, policy, and the integrity of social life*. Stanford University Press.

Nithiananthan, N. (2021). Identify User behavior on digital devices by using digital footprints (Doctoral dissertation).

Nixon, C. (2014). Current perspectives: The impact of cyberbullying on adolescent health. *Adolescent Health, Medicine and Therapeutics, 143*. Advance online publication. DOI: 10.2147/ahmt.s36456

Noakes, T., Walton, M., & Cronjé, J. (2018) Capital meets capabilities: negotiating cultural exclusion in participatory culture. Proceedings of the *International Conference on e-Learning*, ICEL

Noble, S. U. (2018). *Algorithms of oppression: How search engines reinforce racism*. NYU Press. DOI: 10.18574/nyu/9781479833641.001.0001

Norman, W., & MacDonald, C. (2004). Getting to the bottom of "triple bottom line". *Business Ethics Quarterly, 14*(2), 243–262. DOI: 10.5840/beq200414211

Norton. (2017). *Norton WiFi Risk Report: Report of Online Survey Results in 15 Global Markets*. Retrieved from https://www.nortonlifelock.com/us/en/newsroom/press-kits/norton-WiFi-risk-report-2017

Notoatmodjo, G., & Thomborson, C. D. (2009). Passwords and Perceptions. *Proceedings of the 7th Australasian Information Security Conference (AISC 2009)* (pp.71-78). Australian Computer Society, Inc.

Nyst, C., Makin, P., Pannifer, S., & Whitley, E. (2016). *Digital identity: Issue analysis: executive summary*. Consult Hyperion.

O'Reilly, M. (2020). Social media and adolescent mental health: The good, the bad and the ugly. *Journal of Mental Health (Abingdon, England)*, *29*(2), 200–206. DOI: 10.1080/09638237.2020.1714007 PMID: 31989847

Obar, J. A., & Oeldorf-Hirsch, A. (2020). The biggest lie on the internet: Ignoring the privacy policies and terms of service policies of social networking services. *Information Communication and Society*, *23*(1), 128–147. DOI: 10.1080/1369118X.2018.1486870

OECD. (2013). *Guidelines on the protection of privacy and transborder flows of personal data*. OECD Publishing.

OECD. (2021). *21st-century readers: Developing literacy skills in a digital world*. Organisation for Economic Co-operation and Development., DOI: 10.1787/a83d84cb-

OECDa. (2016). Policy Brief on the Future of Work- Skills for a Digital World. OECD. Retrieved from https://www.oecd.org/employment/future-of-work.htm [Accessed 18 February 2021].

OECDb. (2016). Skills for a Digital World. OECD Publishing. htm., Retrieved February 18, 2021, from.

Okechukwu, A. A. (2011). PEOPLE-CENTERED GOVERNANCE AS THE BED-ROCK OF SUSTAINABLE DEVELOPMENT. *ISIG JOURNAL*, 108.

Ong, J. (2024). More scammers exploited social media in 2023; Facebook, WhatsApp and Instagram of 'particular concern': Police. *Today*. Retrieved from https://www.todayonline.com/singapore/more-scammers-exploited-social-media-2023-facebook-whatsapp-instagram-2363481

Orb, A., Eisenhauer, L., & Wynaden, D. (2001). Ethics in qualitative research. *Journal of Nursing Scholarship*, *33*(1), 93–96. DOI: 10.1111/j.1547-5069.2001.00093.x PMID: 11253591

Ortiz-Ospina, E., & Roser, M. (2023). *The rise of social media. Our World in Data*.https://ourworldindata.org/rise-of-social-media

Oscar.howell/Social Capital and Digital Citizenship - Technologies and Politics of Control. (2024). harvard.edu. https://cyber.harvard.edu/is2011/Oscar.howell/Social_Capital_and_Digital_Citizenship

Ostanina, E., & Titova, E. (2023). Legitimate Expectations of Privacy in the Era of Digitalization. *BRICS LJ*, *10*(1), 109–125. DOI: 10.21684/2412-2343-2023-10-1-109-125

Oxford dictionary - Lexicon.com (n.d.) Social capital. https://www.lexicon.com/search?q=social+capital

Oxley, L., Walker, P., Thorns, D., & Wang, H. (2008). The knowledge economy/society: the latest example of" Measurement without theory"?. Journal of Philosophical Economics, 2(Articles).

Oxley, C. (2011). Digital citizenship: Developing an ethical and responsible online culture. *Access : the Newsmagazine of the American Dental Hygienists' Association*, *25*(3), 5–9.

Öztürk, G. (2021). Digital citizenship and its teaching: A literature review. *Journal of Educational Technology & Online Learning*, *4*(1), 31–45.

Pang, A., Shin, W., Lew, Z., & Walther, J. B. (2016). Building relationships through dialogic communication: Organizations, stakeholders, and computer-mediated communication. *Journal of Marketing Communications, 24*(1), 68–82. DOI: 10.1080/13527266.2016.1269019

Pangrazio, L. (2019). *Young People's Literacies in the Digital Age: Continuities, Conflicts and Contradictions.* Routledge.

Pangrazio, L., Godhe, A. L., & Ledesma, A. G. L. (2020). What is digital literacy? A comparative review of publications across three language contexts. *E-Learning and Digital Media, 17*(6), 442–459. DOI: 10.1177/2042753020946291

Pangrazio, L., & Sefton-Green, J. (2021). Digital rights, digital citizenship and digital literacy: What's the Difference? *Journal of New Approaches in Educational Research, 10*(1), 15–27. DOI: 10.7821/naer.2021.1.616

Papacharissi, Z. (2015). A private sphere: Democracy in a digital age. *Polity.*

Parate, S., Josyula, H. P., & Reddi, L. T. (2023). Digital identity verification: Transforming KYC processes in banking through advanced technology and enhanced security measures. *International Research Journal of Modernization in Engineering Technology and Science, 5*(9), 128–137.

Paris, C., Colineau, N., Nepal, S., Bista, S. K., & Beschorner, G. (2013). Ethical considerations in an online community: The balancing act. *Ethics and Information Technology, 15*(4), 301–316. DOI: 10.1007/s10676-013-9315-4

Pariser, E. (2011). *The filter bubble: What the Internet is hiding from you.* Penguin.

Park, R. (2022). Forgiving Without Forgetting? Privacy in an Age of Digital Permanence.

Parthasarathy, D. (2005). *Taking responsibilities and participation seriously: A Critique of 'Good Governance. Droit, gouvernance et développement durable, numéro spécial des Cahiers d'Anthropologie du Droit.* Éditions Karthala.

Patel, K., Binjola, H., & Siddiqui, T. (2020). Indian social media politics: New era of election war. *RUDN Journal of Studies in Literature and Journalism, 25*(1), 184–192. DOI: 10.22363/2312-9220-2020-25-1-184-192

Patel, S. (2019). "Brown girls can't be gay": Racism experienced by queer South Asian women in the Toronto LGBTQ community. *Journal of Lesbian Studies, 23*(3), 410–423. DOI: 10.1080/10894160.2019.1585174

Patil, M., Yadav, H., Gawali, M., Suryawanshi, J., Patil, J., Yeole, A., Shetty, P., & Potlabattini, J. (2023). A Novel Approach to Fake News Detection Using Generative AI. *International Journal of Intelligent Systems and Applications in Engineering, 12*(4s), 343–354. https://www.ijisae.org/index.php/IJISAE/article/view/3797

Patkos, T., Flouris, G., Papadakos, P., Bikakis, A., Casanovas, P., González-Conejero, J., & Stan, A. (2015, September). Privacy-by-norms privacy expectations in online interactions. In *2015 IEEE International Conference on Self-Adaptive and Self-Organizing Systems Workshops* (pp. 1-6). IEEE.

Paul, R., & Elder, L. (2014). *The Miniature Guide to Critical Thinking Concepts & Tools* (Elder, L., Ed.). The Foundation for Critical Thinking.

Pavez, I., Correa, T., & Farías, C. (2023). The Power of Emotions: The ethics of care in the digital inclusion processes of marginalized communities. *Social Inclusion (Lisboa), 11*(3). Advance online publication. DOI: 10.17645/si.v11i3.6623

Pawlicka, A., Choraś, M., Kozik, R., & Pawlicki, M. (2021). First broad and systematic horizon scanning campaign and study to detect societal and ethical dilemmas and emerging issues spanning over cybersecurity solutions. *Personal and Ubiquitous Computing*, •••, 2021.

Pawlicki, M., Choraś, M., & Kozik, R. (2020). Defending network intrusion detection systems against adversarial evasion attacks. *Future Generation Computer Systems, 2020*(110), 148–154. DOI: 10.1016/j.future.2020.04.013

Peart, M., Cubo-Delgado, S. C., & Gutiérrez-Esteban, P. (2022). Exploring the role of digital and socio-civic skills for promoting youth participation and digital citizenship. *European Journal of Educational Research, 1*(2), 697–709. DOI: 10.12973/eu-jer.11.2.697

Peattie, K., & Crane, A. (2005). Green marketing: Legend, myth, farce or prophecy? *Qualitative Market Research, 8*(4), 357–370. DOI: 10.1108/13522750510619733

Pengl, J., & Li, C. W. (2022). Security breaches and modifications on cybersecurity disclosures. *Accounting and Management Information Systems, 21*(3), 452–470.

Perrin, A. (2015). *Social media usage: 2005–2015*. Pew Research Trust.

Peterson, H. L., & Johnston, M. A. (2015). Digital footprints: Creating responsible digital citizens. *Journal of Digital and Media Literacy, 3*(2), 101–118.

Petrosyan, A. (2024). Internet penetration rate worldwide 2024, by region. https://www.statista.com/statistiques/269329/penetration-rate-of-the-internet-by-region/

Petrosyan, A. (2024). Percentage of organizations worldwide experiencing cyberattacks from 2021 to 2023, by attack type. *Statista.* Retrieved from https://www.statista.com/statistics/1376249/cyber-attack-global-firms-by-type/

Pew (2022). Broadband access still a challenge in rural affordable housing. https://www.pewtrusts.org/en/research-and-analysis/articles/2022/12/08/broadband-access-still-a-challenge-in-rural-affordable-housing

Pew Research Center. (2022). Public trust in government: 1958-2022. https://www.pewresearch.org/politics/2022/06/06/public-trust-in-government-1958-2022/

Piliang, Y. A. (2015). Redefining Ethics and Culture in the Virtual World. *Melintas, 31*(3), 236–251. DOI: 10.26593/mel.v31i3.1917.236-251

Pinto, B., Duarte, A., & Dias, P. (2022). The influence of YouTubers on children (8-12 years old): Current issues and brands. *Doxa Comunicación. Revista Interdisciplinar De Estudios De Comunicación Y. Ciencia y Sociedad*, (36), 321–340. DOI: 10.31921/doxacom.n36a1638

Polizzi, G., & Taylor, R. (2019). *Misinformation, digital literacy and the school curriculum. Media Policy Brief 22*. Media Policy Project, London School of Economics and Political Science.

Ponemon Institute & Accenture. (2019). *The Cost of Cybercrime: Ninth Annual Cost of Cybercrime Study*. Retrieved from https://iapp.org/media/pdf/resource_center/accenture_cost_of_cybercrime_study_2019.pdf

Ponemon Institute. (2016). *2016 Cost of Cyber Crime Study & the Risk of Business Innovation*. Retrieved from https://www.ponemon.org/local/upload/file/2016%20HPE%20CCC%20GLOBAL%20REPORT%20FINAL%203.pdf

Poole, E., & Williamson, M. (2021). Disrupting or reconfiguring racist narratives about Muslims? The representation of British Muslims during the Covid crisis. *Journalism: Theory. Practice & Criticism, 24*(2), 262–279. DOI: 10.1177/14648849211030129

Portes, A. (2014). Downsides of social capital. *Proceedings of the National Academy of Sciences of the United States of America, 111*(52), 18407–18408. DOI: 10.1073/pnas.1421888112 PMID: 25535346

Powell, P. A., & Roberts, J. (2017). Situational determinants of cognitive, affective, and compassionate empathy in naturalistic digital interactions. *Computers in Human Behavior, 68*, 137–148. DOI: 10.1016/j.chb.2016.11.024

Prastyaningtyas, E. W., Ausat, A. M. A., Muhamad, L. F., Wanof, M. I., & Suherlan, S. (2023). The Role of Information Technology in Improving Human Resources Career Development. *Jurnal Teknologi Dan Sistem Informasi Bisnis, 5*(3), 266–275. DOI: 10.47233/jteksis.v5i3.870

Pring, R. (2016). Preparing for citizenship: Bring back John Dewey. Citizenship, Social and Economics Education. 15(1), 6-14. https://journals.sagepub.com/doi/pdf/10.1177/2047173416646467

Prove, A. I. (2024). Blockchain & generative AI: Fueling innovation within digital economy. https://proveai.com/news/blockchain-and-generative-ai-fueling-innovation-within-the-digital-economy

Pueschel, A., Johnson, R. C., & Dhanani, L. Y. (2020). Putting Gen Z first: Educating with a generational mind-set. *Industrial and Organizational Psychology: Perspectives on Science and Practice, 13*(4), 594–598. DOI: 10.1017/iop.2020.103

Quan-Haase, A. (2016). *Technology and society: Social networks, power, and inequality* (2nd ed.). Oxford University Press.

Quan-Haase, A., & Ho, D. (2020). Online privacy concerns and privacy protection strategies among older adults in East York, Canada. *Journal of the Association for Information Science and Technology, 71*(9), 1089–1102. DOI: 10.1002/asi.24364

Radovanović, D., Holst, C., Belur, S. B., Srivastava, R., Houngbonon, G. V., Le Quentrec, E., Miliza, J., Winkler, A. S., & Noll, J. (2020). Digital literacy key performance indicators for sustainable development. *Social Inclusion (Lisboa), 8*(2), 151–167. DOI: 10.17645/si.v8i2.2587

Rahman, H. (2025). Blockchain-Driven Knowledge Ecosystems. In Blockchain Technology Applications in Knowledge Management (pp. 29-70). IGI Global Scientific Publishing. DOI: 10.4018/979-8-3693-3956-5.ch002

Rahman, H. (2009, November). Framework of e-governance at the local government level. In *Proceedings of the 3rd international conference on Theory and practice of electronic governance* (pp. 372-374). DOI: 10.1145/1693042.1693121

Rahman, H. (2021). Framework of E-governance at the grass roots level. In *Ubiquitous Technologies for Human Development and Knowledge Management* (pp. 78–114). IGI Global. DOI: 10.4018/978-1-7998-7844-5.ch005

Rahman, M. S., Liu, X., & Hossain, M. A. (2019). Anomaly detection in online social networks using machine learning algorithms. *Social Network Analysis and Mining*, *9*(1), 1–16. DOI: 10.1007/s13278-019-0566-6

Rahman, N. A. A., Sairi, I. H., Zizi, N. A. M., & Khalid, F. (2020). The importance of cybersecurity education in school. *International Journal of Information and Education Technology (IJIET)*, *10*(5), 378–382. DOI: 10.18178/ijiet.2020.10.5.1393

Rajput, A., & Nair, K. (2013). Significance of digital literacy in e-governance. *The SIJ transactions on industrial financial & business management, 1*(4).

Raman, R., Kumar Nair, V., Nedungadi, P., Kumar Sahu, A., Kowalski, R., Ramanathan, S., & Achuthan, K. (2024). Fake news research trends, linkages to generative artificial intelligence and sustainable development goals. *Heliyon, 10*(3), e24727. DOI: 10.1016/j.heliyon.2024.e24727 PMID: 38322879

Ranjan, A. (2011). Indian Performing Right Society Ltd. v. Sanjay Dalia & Anr. Delhi High Court.

Rao, N., & Lingam, L. (2020). Smartphones, youth and moral panics: Exploring print and online media narratives in India. *Mobile Media & Communication*, *9*(1), 128–148. DOI: 10.1177/2050157920922262

Read, A., & Fromm, J. (2018). *Marketing to Gen Z: The Rules for Reaching This Vast and Very Different Generation of Influencers.* AMACOM.

Recover Reputation. (n.d.). *5 online reputation management case studies.* Retrieved December 23, 2024 from https://www.recoverreputation.com/5-online-reputation-management-case-studies

Reddy, P., Sharma, B., & Chaudhary, K. (2020). Digital literacy: A review of literature. [IJT]. *International Journal of Technoethics*, *11*(2), 65–94. DOI: 10.4018/IJT.20200701.oa1

Reisman, D., Schultz, J., Crawford, K., & Whittaker, M. (2018). *ALGORITHMIC IMPACT ASSESSMENTS: a PRACTICAL FRAMEWORK FOR PUBLIC AGENCY ACCOUNTABILITY.* https://www.nist.gov/system/files/documents/2021/10/04/aiareport2018.pdf

Reiss, M. (2010). Ethical Thinking. In *Ethics in the Science and Technology Classroom.* Brill. DOI: 10.1163/9789460910715_003

Reputation, X. (n.d.). *Case studies.* Retrieved December 23, 2024 from https://blog.reputationx.com/case-studies

ReputationManagement.com. (n.d.). *Case studies.* Retrieved December 21, 2024 from https://www.reputationmanagement.com/case-studies

Reuters. (2024, June 7). *Yelp can sue reputation company for promising to suppress bad reviews.* Retrieved December 21, 2024 from https://www.reuters.com/legal/yelp-can-sue-reputation-company-promising-suppress-bad-reviews-2024-06-07

Rex, B. (2024). Breaking down barriers: building inclusive online communities. *Meeds DAO.* https://medium.com/meeds-dao/breaking-down-barriers-building-inclusive-online-communities-78b4e10ae379

Reynolds, T., & Cerutti, E. (2021). *Digital ethics: A framework for responsible technology use.* Palgrave Macmillan.

Rhee, H. S., Kim, C., & Ryu, Y. U. (2009). Self-efficacy in information security: Its influence on end users' information security practice behavior. *Computers & Security, 28*(8), 816–826. DOI: 10.1016/j.cose.2009.05.008

Ribble, M. (2008). *Passport to Digital Citizenship: Journey toward Appropriate Technology Use at School and at Home.* Retrieved September 21, 2024, from https://eric.ed.gov/?id=EJ904288

Ribble, M. S., Bailey, G. D., & Ross, T. W. (2004). Digital citizenship: Addressing appropriate technology behavior. *Learning & Leading with Technology*, 32/(1), 6-12.

Ribble, M. (2010). *Digital citizenship in schools: Nine elements all students should know.* International Society for Technology in Education.

Ribble, M., & Bailey, G. (2007). *Digital citizenship in schools.* ISTE.

Ribble, M., Bailey, G., & Ross, T. (2004). Digital citizenship: Addressing appropriate technology behavior. *Learning and Leading with Technology, 32*(1), 7–13.

Richards, N. M. (2013). The dangers of surveillance. *Harvard Law Review, 126*(7), 1934–1965.

Richardson, J., & Milovidov, E. (2019). *Digital citizenship education handbook: Being online, well-being online, and rights online.* Council of Europe.

Rifon, N. J., Quilliam, E., & LaRose, R. (2005, May 26–30). Consumer perceptions of online safety. Paper presented at the 2005 Annual Meeting of the International Communication Association, New York, NY.

Rijitha, R. (2021). The Impact of Social Media Marketing on Consumer Purchase Intention. In *International Conference on Contemporary Media Persuasive Technology and Psychology of Visual Communication* (pp. 100-119).

Riley, S. (2006). Password Security: What Users Know and What They Actually Do. *Usability News, 8*(1), 1–5.

Riso, B., Tupasela, A., Vears, D. F., Felzmann, H., Cockbain, J., Loi, M., Kongsholm, N. C. H., Zullo, S., & Rakic, V. (2017). Ethical sharing of health data in online platforms–which values should be considered? *Life Sciences, Society and Policy, 13*(1), 1–27. DOI: 10.1186/s40504-017-0060-z PMID: 28825221

Robbins, L. (1932). *An Essay on the Nature and Significance of Economic Science.* Macmillan.

Robertson, C. B. (2016). Online reputation management in attorney regulation. *The Georgetown Journal of Legal Ethics, 29*, 97.

Rogers, R. (1983). Cognitive and physiological processes in fear appeals and attitude change: A revised theory of protection motivation. In Cacioppo, J. T., & Petty, R. E. (Eds.), *Social Psychophysiology* (pp. 153–176). Guilford Press.

Rosário, A. T., Lopes, P. R., & Rosário, F. S. (2023). The Digital Marketing for Sustainable Development.

Rossomakha, A.. (2023). ChatGPT and conversational AI. *Journal of Natural Language Processing, 30*(1), 1–15.

Roy, S., Majumder, S., Bose, A., & Chowdhury, I. R. (2024). The rich-poor divide: Unravelling the spatial complexities and determinants of wealth inequality in India. *Applied Geography (Sevenoaks, England), 166*, 103267. DOI: 10.1016/j.apgeog.2024.103267

Ruo-Fei, S., Zeng, J. Y., & Jin, C. H. (2022). The role of consumer' social capital on ethical consumption and consumer happiness. *SAGE Open, 12*(2), 21582440221095026. Advance online publication. DOI: 10.1177/21582440221095026

Sabatini, F., & Sarracino, F. (2019). Online social networks and trust. https://ideas.repec.org/a/spr/soinre/v142y2019i1d10.1007_s11205-018-1887-2.html

Sachdeva, D. (2016). Gen Z – THE TECHILICIOUS GENERATION Case Study of a Digitally Hooked Gen Zee. TIJ's *Research Journal of Social Science & Management - RJSSM,* 6. https://api.semanticscholar.org/CorpusID:151938534

Safadi, H., Li, W., Rahmati, P., & Soleymani, S. (2020). Curtailing Fake News Propagation with Psychographics. SSRN Electronic Journal. https://doi.org/DOI: 10.2139/ssrn.3558236

Sage, A. Bagheri, Maryam & Bagheritaba, Mohsen & Alizadeh, Sohila & Salemizadeh Parizi, Sam & Matoufinia, Parisa & Luo, Yang. (2024). AI-Driven Decision-Making in Healthcare Information Systems: A Comprehensive Review. DOI: 10.20944/preprints202406.0790.v1

Sahoo, D. (2024). Chapter-5 Leveraging Digital HR and Marketing Focus for Sustainable Business Practices: A Radical Transformation. Business Management: A New Paradigm, 63.

Saini, R. (2023). Applications of AI in healthcare and finance. *Tech Innovations Journal, 15*(2), 45–60.

Salminen, J., Kaate, I., Kamel, A. M. S., Jung, S. G., & Jansen, B. J. (2021). How does personification impact ad performance and empathy? An experiment with online advertising. *International Journal of Human-Computer Interaction, 37*(2), 141–155. DOI: 10.1080/10447318.2020.1809246

Sambasivan, N., Arnesen, E., Hutchinson, B., Doshi, T., & Prabhakaran, V. (2021). Re-imagining Algorithmic Fairness in India and Beyond. In *FAccT '21: Proceedings of the 2021 ACM Conference on Fairness, Accountability, and Transparency* (pp. 315–328). https://doi.org/DOI: 10.1145/3442188.3445896

Samuelson, P. (2018). The Evolution of Copyright in the Digital Age: Sony Corp. and Its Legacy. *Harvard Law Review, 131*(5).

Samuelson, P. (2019). Copyright's Evolution in the Digital Age. *Harvard Law Review, 132*(5), 1530–1570.

Samuelson, P. (2019). Rethinking fair use in the digital age. *Columbia Journal of Law & the Arts, 42*(2), 145–176.

Samuelson, P. (2019). The Blurred Lines of Fair Use: Copyright, Creativity, and Digital Innovation. *Harvard Law Review, 132*(6).

Samuelson, P. (2019). The Complexities of Fair Use: The Role of Market Impact and Transformative Use. *Berkeley Technology Law Journal, 34*(2), 567–590.

Samuelson, P. (2019). The Evolution of Copyright in the Digital Age. *Harvard Law Review, 131*(5).

Sanders, C. K., & Scanlon, E. (2021). The digital divide is a human rights issue: advancing social inclusion through social work advocacy. *Journal of Human Rights and Social Work*. https://pmc.ncbi.nlm.nih.gov/articles/PMC7973804/

Sandrini, L., & Somogyi, R. (2023). Generative AI and deceptive news consumption. *Economics Letters, 232*, 111317–111317. DOI: 10.1016/j.econlet.2023.111317

Sangeen, M., Bhatti, N. A., Kifayat, K., Alsadhan, A. A., & Wang, H. (2023). Blind-trust: Raising awareness of the dangers of using unsecured public Wi-Fi networks. *Computer Communications, 209*, 359–367. DOI: 10.1016/j.comcom.2023.07.011

Santhosh, T., & Thiyagu, K. (2024). Fostering Responsible Behavior Online-Relevance of Cyber Ethics Education. *Malaysian Online Journal of Educational Technology, 12*(1), 32–38. DOI: 10.52380/mojet.2024.12.1.428

Santiago-Torner, C., Corral-Marfil, J. A., & Tarrats-Pons, E. (2024). Relationship Between E-Ethical Leadership and Emotional Exhaustion: A Moderated Mediation Model.

Schoenbrunner, A., Gosman, A., & Bajaj, A. K. (2019). Framework for the creation of ethical and professional social media content. *Plastic and Reconstructive Surgery, 144*(1), 118e–125e. DOI: 10.1097/PRS.0000000000005782 PMID: 31246836

Schouten, A., Janssen, L., & Verspaget, M. (2019). Celebrity vs. Influencer endorsements in advertising: The role of identification, credibility, and Product-Endorser fit. *International Journal of Advertising, 39*(2), 1–24. DOI: 10.1080/02650487.2019.1634898

Schwartz, S. H., & Bilsky, W. (1987). Toward a universal psychological structure of human values. *Journal of Personality and Social Psychology, 53*(3), 550–562. DOI: 10.1037/0022-3514.53.3.550

Seifert, C. F., Van Ness, R. K., Eddy, E. R., Buff, C., & D'Abate, C. P. (2023). Generational Work Ethic Differences: From Baby Boomers to Gen Z. *Journal of Managerial Issues, 35*(4), 401–422.

Selbst, A. D., Boyd, D., Friedler, S. A., Venkatasubramanian, S., & Vertesi, J. (2019). Fairness and Abstraction in Sociotechnical Systems. *FAT* '19: Proceedings of the Conference on Fairness, Accountability, and Transparency*, 59–68. https://doi.org/DOI: 10.1145/3287560.3287598

Selwyn, N. (2021). *Digital education and ethical challenges*. Routledge.

Selwyn, N., & Aagaard, J. (2021). Banning mobile phones from classrooms—An opportunity to advance understandings of technology addiction, distraction and cyberbullying. *British Journal of Educational Technology, 52*(1), 8–19.

Sengupta, R. (2011). *Super Cassettes Industries Ltd. v. MySpace Inc*. Delhi High Court.

Setiawan, B., Nugraha, D. P., Irawan, A., Nathan, R. J., & Zoltan, Z. (2021). User Innovativeness and Fintech Adoption in Indonesia. *Journal of Open Innovation*, *7*(3), 188. DOI: 10.3390/joitmc7030188

Shabahang, R., Shim, H., Aruguete, M. S., & Zsila, A. (2024). Oversharing on social media: Anxiety, attention-seeking, and social media addiction predict the breadth and depth of sharing. *Psychological Reports*, *127*(2), 513–530. DOI: 10.1177/00332941221122861 PMID: 35993372

Shafira, V. S., Ramadhani, G., & Rachman, I. F. (2024). How digital literacy can drive inclusive progress towards the 2030 SDGs. *Advances in Economics & Financial Studies*, *2*(2). Advance online publication. DOI: 10.60079/aefs.v2i2.260

Shams, R. A., Zowghi, D., & Bano, M. (2023). AI and the quest for diversity and inclusion: A systematic literature review. *AI and Ethics*. Advance online publication. DOI: 10.1007/s43681-023-00362-w

Sharma, D. (2020). Representing pride through hashtags. *Feminist Media Studies*, *21*(1), 162–164. DOI: 10.1080/14680777.2021.1864878

Sharma, R., Fantin, A. R., Prabhu, N., Guan, C., & Dattakumar, A. (2016). Digital literacy and knowledge societies: A grounded theory investigation of sustainable development. *Telecommunications Policy*, *40*(7), 628–643. DOI: 10.1016/j.telpol.2016.05.003

Shayakhmetova, Z. (2019, Sep 14). *Astana IT University opens in Kazakh capital*. Retrieved from: https://astanatimes.com/2019/09/astana-it-university-opens-in-kazakh-capital/

Shay, R., Komanduri, S., Kelley, P. G., Bauer, L., Leon, P. G., Christin, N., Mazurek, M. L., & Cranor, L. F. (2010). Encountering stronger password requirements: User attitudes and behaviors. *Symposium on Usable Privacy and Security (SOUPS)*. Redmond, WA, USA. DOI: 10.1145/1837110.1837113

Sherry, J. L., & Kozinets, R. V. (2008). Special issue: Consumer culture theory. *The Journal of Consumer Research*, *35*(1), 151–162.

Shillair, S., Boehmer, J., Cotten, S. R., LaRose, R., & Rifon, N. (2014, May 25). Communicating online safety: A critical content analysis of websites providing online safety information to consumers over time. Paper presented at the International Communication Association, Seattle, WA.

Shi, S., Liang, Y., Ang, E. L., & Zhao, H. (2019). Delta Integration CRISPR-Cas (Di-CRISPR) in Saccharomyces cerevisiae. *Methods in Molecular Biology (Clifton, N.J.)*, *1927*, 73–91. DOI: 10.1007/978-1-4939-9142-6_6 PMID: 30788786

Shockley, B., Lari, N., Elmaghraby, E., & Al-Ansari, M. (2020). Social media usage and support for women in community leadership: Evidence from Qatar. *Women's Studies International Forum*, *81*, 102374. DOI: 10.1016/j.wsif.2020.102374

Shopova, T. (2014). Digital literacy of students and its improvement at the university. *Journal on Efficiency and Responsibility in Education and Science*, *7*(2), 26–32. DOI: 10.7160/eriesj.2014.070201

Shubham, S. (2022). Caste and the Digital Sphere. *Research Gate*. https://www.researchgate.net/publication/381295637_Caste_and_the_Digital_Sphere

Shu, K., Mahudeswaran, D., & Liu, H. (2019). FakeNewsTracker: A tool for fake news collection, detection, and visualization. *Computational & Mathematical Organization Theory*, *25*(1), 60–71. Advance online publication. DOI: 10.1007/s10588-018-09280-3

Silvia, S. (2019). The Importance of Social Media and Digital Marketing to Attract Millennials' Behavior as a Consumer. Marketing, 4(2).

Simonds, T. A., & Brock, B. L. (2014). Relationship between age, experience, and student preference for types of learning activities in online courses. *The Journal of Educators Online*, *11*(1), n1. DOI: 10.9743/JEO.2014.1.3

Simsek, E., & Simsek, A. (2013). New Literacies for Digital Citizenship. *Contemporary Educational Technology*, *4*(2), 126–137. DOI: 10.30935/cedtech/6097

Singh, M. (2017). M/s. Ramesh D. Naik v. M/s. Rajhans. Delhi High Court.

Singh, R. (2013). Super Cassettes Industries Ltd. v. Chintamani Rao: A Case Study on Public Performance Rights. Journal of Indian Intellectual Property Law, 22(4).

Singh, A. (2014). Challenges and Issues of Generation Z. *IOSR Journal of Business and Management*, *16*(7), 59–63. DOI: 10.9790/487X-16715963

Singh, M. K., Ahmed, J., Raghuvanshi, K. K., & Alam, M. A.Manish Kumar Singh. (2023). BharatFakeNewsTracker: An Ensemble Learning Tool to Spot Fake News in India. *International Journal of Engineering Trends and Technology*, *71*(9), 76–91. DOI: 10.14445/22315381/IJETT-V71I9P208

Singh, N. (2004). The digital economy: An overview. In Singh, N. (Ed.), *The Handbook of Information Security*.

Singh, P., & Dhiman, D. B. (2023). Exploding AI-Generated Deepfakes and misinformation: A threat to global concern in the 21st century. Available at *SSRN* 4651093.

Singh, R. (2021). Fair Dealing and Digital Content in India: Challenges and Prospects. *Indian Journal of Intellectual Property Law*, *13*(1), 98–120.

Singh, R., & Sharma, N. (2021). Green marketing strategies for healthcare organizations: Opportunities and challenges. *Marketing and Sustainability*, *29*(3), 343–356.

Singh, T. (2017). Copyright and Fair Dealing in Indian Law: Analyzing Civic Chandran v. Ammini Amma (1996). *Economic and Political Weekly*, *52*(18), 34–40.

Sinha, S. (2017). Fragile Hegemony: Modi, Social Media and Competitive Electoral Populism in India. *International Journal of Communication*. https://eprints.soas.ac.uk/24533/

Şirzad, N. (2022). A review on online reputation management and online reputation components. *Doğuş Üniversitesi Dergisi*, *23*(1), 219–242.

Sivek, S. D. (2018). Both Facts and Feelings: Emotion and News Literacy. *The Journal of Media Literacy Education*, *10*(2), 123–138. DOI: 10.23860/JMLE-2018-10-2-7

Sjöberg, M., Chen, H. H., Floréen, P., Koskela, M., Kuikkaniemi, K., Lehtiniemi, T., & Peltonen, J. (2017). Digital me: Controlling and making sense of my digital footprint. In *Symbiotic Interaction: 5th International Workshop, Symbiotic 2016, Padua, Italy, September 29–30, 2016, Revised Selected Papers 5* (pp. 155-167). Springer International Publishing.

Smith, A. (2020). The impact of the California Consumer Privacy Act on data protection. *Technology in Society*, *42*(2), 113–128.

Smith, A. D., & Brower, T. R. (2012). Longitudinal examination of website sustainability practices: Insights for green marketing. *Sustainable Business Journal*, *7*(3), 155–169.

Smith, J. A., Spears, A. L., & Ellefsen, A. M. (2018). The impact of social media on youth: A review of the literature. *Developmental Review*, •••, 2018.

SOCAN v. Bell Canada, [2012] 2 S.C.R. 326 (Canada).

Social Media in India - 2023 Stats & Platform Trends - OOSGA. (n.d.). OOSGA. https://oosga.com/social-media/ind/#:~:text=With%20the%20latest%20data%20published,users%20account%20for%20around%2027.01%25

Sokolova, K., & Kefi, H. (2019). Instagram and YouTube bloggers promote it, why should I buy? How credibility and parasocial interaction influence purchase intentions. *Journal of Retailing and Consumer Services*, *53*, 101742. Advance online publication. DOI: 10.1016/j.jretconser.2019.01.011

Soler-Costa, R., Lafarga-Ostáriz, P., Mauri-Medrano, M., & Moreno-Guerrero, A. J. (2021). Netiquette: Ethic, education, and behavior on internet—a systematic literature review. *International Journal of Environmental Research and Public Health*, *18*(3), 1212–1236. DOI: 10.3390/ijerph18031212 PMID: 33572925

Solove, D. J. (2020). *Privacy law fundamentals*. IAPP.

Sombatruang, N., Kadobayashi, Y., Sasse, A. M., Baddeley, M., & Miyamoto, D. (2018). The continued risks of unsecured public WiFi and why users keep using it: Evidence from Japan. *2018 16th Annual Conference on Privacy, Security and Trust (PST)* (pp. 1-11). IEEE.

Sombatruang, N., Sasse, A. M., & Baddeley, M. (2016). Why Do People Use Unsecure Public WiFi: An Investigation of Behaviour and Factors Driving Decisions. *Proceedings of the 6th Workshop on Socio-Technical Aspects in Security and Trust* (pp. 61-72). Association for Computing Machinery. DOI: 10.1145/3046055.3046058

Sommestad, T., Karlzén, H., & Hallberg, J. (2015). A Meta-Analysis of Studies on Protection Motivation Theory and Information Security Behaviour. *International Journal of Information Security and Privacy*, *9*(1), 26–46. DOI: 10.4018/IJISP.2015010102

Spears, B. A. (2013). A review of initiatives using technology to promote cyber-safety and digital citizenship. *The Impact of Technology on Relationships in Educational Settings*, *15*(4), 188–203.

Spohr, D. (2017). Fake news and ideological polarisation: Filter bubbles and selective exposure on social media. *Business Information Review*, *34*(3), 150–160. DOI: 10.1177/0266382117722446

Stafford, T. F., & Poston, R. (2010). Online security threats and computer user intentions. *Computer, 43*(1), 58–64. DOI: 10.1109/MC.2010.18

Statista. (2024). Social media and politics in the United States. https://www.statista.com/study/40641/social-media-and-politics-in-the-united-states/

Statista. (2024, May 15). *Internet penetration rate in India 2014-2024*. https://www.statista.com/statistics/792074/india-internet-penetration-rate/

Stearns, S. A. (2017). Student Responsible Learning: Getting Students to Read Online Discussions. *College Teaching, 65*(2), 69–78. DOI: 10.1080/87567555.2016.1244654

Stefanoudis, P. V., Biancani, L. M., Cambronero-Solano, S., Clark, M. R., Copley, J. T., Easton, E., Elmer, F., Haddock, S. H. D., Herrera, S., Iglesias, I. S., Quattrini, A. M., Sigwart, J., Yesson, C., & Glover, A. G. (2022). Correction to 'moving conferences online: Lessons learned from an international virtual meeting'. *Proceedings. Biological Sciences, 289*(1971), 20211769. DOI: 10.1098/rspb.2022.0321 PMID: 35350862

Steinbrecher, S., Pingel, F., & Juschka, A. (2011). Reputation management as an extension of future identity management. In *Digital Privacy: PRIME-Privacy and Identity Management for Europe* (pp. 557–568). Springer Berlin Heidelberg. DOI: 10.1007/978-3-642-19050-6_19

Steiner, J. F., & Steiner, G. A. (1999). *Business, government, and society: A managerial perspective*. McGraw-Hill.

Stieglitz, S., & Dang-Xuan, L. (2013). Social media and political communication. *Journal of Computer-Mediated Communication, 18*(3), 379–395.

Stinson, C. (2022). Algorithms are not neutral. *AI and Ethics, 2*(4), 763–770. DOI: 10.1007/s43681-022-00136-w

Stobert, E., & Biddle, R. (2018). The Password Life Cycle. *ACM Transactions on Privacy and Security, 21*(3), 1–32. DOI: 10.1145/3183341

Subhashini, V. (2014). Ethical hacking and legal systems. [IJETCSE]. *Internal Journal of Emerging Technology in Computer & Electronics, 11*(4), 36–40.

Subramanian, S. (2015). From the streets to the web: Looking at feminist activism on social media. *Economic and Political Weekly, 50*(17). https://www.jstor.org/stable/24481828

Sujatha, R., Aarthy, S. L., & Vettriselvan, R. (Eds.). (2021). *Integrating Deep Learning Algorithms to Overcome Challenges in Big Data Analytics*. CRC Press. DOI: 10.1201/9781003038450

Suler, J. (2004). The online disinhibition effect. *Cyberpsychology & Behavior, 7*(3), 321–326. DOI: 10.1089/1094931041291295 PMID: 15257832

Sunder, M. (2018). From Goods to a Good Life: Intellectual Property and Global Justice. Yale University Press.Supreme Court of the United States, Google LLC v. Oracle America, Inc. (2021)

Sunstein, C. R. (2017). *Republic: Divided democracy in the age of social media*. Princeton University Press. DOI: 10.1515/9781400884711

Super Cassettes Industries Ltd. v. MySpace Inc., (2011) 174 DLT 279 (Del).

Supreme Court of the United States, Warhol Foundation v. Goldsmith (2023)

Suson, R. L. (2019). Appropriating digital citizenship in the context of basic education. *International Journal of Education, Learning and Development*, 7/(4), 44-66.

Sweeney, L., & Johnson, J. (2019). The impact of digital media on green marketing in the healthcare sector. *Journal of Environmental Marketing*, *22*(2), 115–132.

Tafesse, W., & Dayan, M. (2023). Content creators' participation in the creator economy: Examining the effect of creators' content sharing frequency on user engagement behavior on digital platforms. *Journal of Retailing and Consumer Services*, *73*, 103357. DOI: 10.1016/j.jretconser.2023.103357

Tambunan, R., Budimansyah, D., & Darmawan, C. (2024). THE IMPLEMENTATION OF DIGITAL LITERACY TO ENCOURAGE STUDENTS' DEMOCRATIC ENGAGEMENT IN CIVIC EDUCATION LEARNING. *JUPIIS JURNAL PENDIDIKAN ILMU-ILMU SOSIAL*, *16*(1), 109–118. DOI: 10.24114/jupiis.v16i1.53503

Tamò, A., & George, D. (2014). Oblivion, erasure and forgetting in the digital age. *J. Intell. Prop. Info. Tech. & Elec. Com. L.*, *5*, 71.

Tang, L. (2023). Loyalty marketing agency fined S$10,000 over data leak of Starbucks Singapore customers. *CNA*. Retrieved from https://www.channelnewsasia.com/singapore/starbucks-singapore-330000-customers-data-leak-ascentis-fined-3925281

Tang, L. (2023). ShopBack fined S$74,400 over leak of more than 1.4 million customers' personal data. *CNA*. Retrieved from https://www.channelnewsasia.com/singapore/shopback-fined-data-leak-million-customers-personal-3705006

Tavares, A. O., Areia, N. P., Mellett, S., James, J., Intrigliolo, D. S., Couldrick, L. B., & Berthoumieu, J. (2020). The European Media Portrayal of Climate Change: Implications for the Social Mobilization towards Climate Action. *Sustainability*, *12*(20), 8300. DOI: 10.3390/su12208300

Teixeira, R. L. P., & Teixeira, C. H. S. B. (2024). Exploring sustainable human manufacturing: industry 4.0, manufacturing, and green marketing for a balanced future.

Thakare, A., Sedani, S., Kriplani, S., Patel, A., & Umre, U. (2024). Chitosan: A Versatile Biomaterial Revolutionizing Endodontic Therapy. *Cureus*, *16*(6), e62506. DOI: 10.7759/cureus.62506 PMID: 39022517

Thakur, A. (2019). New media and the Dalit counter-public sphere. *Television & New Media*, *21*(4), 360–375. DOI: 10.1177/1527476419872133

The Committee on European Computing Education (CECE). (2017). Informatics Education in Europe: Are We All. In The Same Boat? Informatics Europe & ACM Europe. Retrieved from., Retrieved February 18, 2021, from, DOI: 10.1145/3106077

The Royal Society. (2012). *Shut down or restart? The way forward for computing in the UK schools.* [online] London: The Royal Society. Available at: https://royalsociety.org/topics-policy/projects/computing-in-schools/report/ [Accessed 28 May 2019].

The Royal Society. (2017). *After the Reboot: Computing Education in UK Schools*. London: Royal Society. Retrieved from http://royalsociety.org/computing-education [Accessed 18 February 2021].

The Shelf. (2023, May 15). *Gen Z Social Media: How They Use It and the Content They Want to See* [Website]. The Shelf Full-Service Influencer Marketing. https://www.theshelf.com/insights/gen-z-social-media/

Tiidenberg, K. (2018). Ethics in digital research. *The SAGE handbook of qualitative data collection*, 466-479.

Tiku, V. N., Jain, C., Raz, Y., Nakamura, S., Heestand, B., Liu, W., Späth, M., Suchiman, H. E. D., Müller, R. U., Slagboom, P. E., Partridge, L., & Antebi, A. (2017). Small nucleoli are a cellular hallmark of longevity. *Nature Communications*, *8*(1), 16083. DOI: 10.1038/ncomms16083 PMID: 28853436

Titus, D. (2018). Social media as a gateway for young feminists: Lessons from the #IWillGoOut campaign in India. *Gender and Development*, *26*(2), 231–248. DOI: 10.1080/13552074.2018.1473224

Tiwary, D. K., & Pradesh, U. (2011). Security and ethical issues in it: An organization's perspective. *International Journal of Enterprise Computing and Business Systems*, *1*(2), 2230–8849.

Toh, E. (2018). SingHealth cyber attack: PM Lee and other ministers among 1.5 million patients whose records were stolen. *The Straits Times.* Retrieved from https://www.straitstimes.com/singapore/pm-lee-and-other-ministers-among-15-million-patients-whose-records-were-stolen

Tokovska, M., Ferreira, V. N., Vallušova, A., & Seberíni, A. (2023). E-Government—The Inclusive Way for the Future of Digital Citizenship. *Societies (Basel, Switzerland)*, *13*(6), 141. DOI: 10.3390/soc13060141

Trandabă , D., & Gifu, D. (2023). Discriminating AI-generated Fake News. *Procedia Computer Science*, *225*, 3822–3831. DOI: 10.1016/j.procs.2023.10.378

Trier, M., Kundisch, D., Beverungen, D., Müller, O., Schryen, G., Mirbabaie, M., & Trang, S. (2023). Digital Responsibility: A Multilevel Framework for Responsible Digitalization. *Business & Information Systems Engineering*, *65*(4), 463–474. DOI: 10.1007/s12599-023-00822-x

Tsai, H. S., Jiang, M., Alhabash, S., LaRose, R., Rifon, N. J., & Cotten, S. R. (2016). Understanding online safety behaviors: A protection motivation theory perspective. *Computers & Security*, *59*, 138–150. DOI: 10.1016/j.cose.2016.02.009

Tsipursky, G., Votta, F., & Roose, K. M. (2018). Fighting fake news and post-truth politics with behavioral science: The Pro-Truth Pledge. *Behavior and Social Issues*, *27*(1), 47–70. DOI: 10.5210/bsi.v27i0.9127

Tsoy, D., Tirasawasdichai, T., & Kurpayanidi, K. I. (2021). Role of social media in shaping public risk perception during COVID-19 pandemic: A theoretical review. *International Journal of Management Science and Business Administration*, *7*(2), 35–41. DOI: 10.18775/ijmsba.1849-5664-5419.2014.72.1005

Tufekci, Z., & Wilson, C. (2012). Social media and the decision to participate in political protest: Observations from Tahrir square. *Journal of Communication*, *62*(2), 363–379. DOI: 10.1111/j.1460-2466.2012.01629.x

Turkle, S. (1995). *Life on the Screen: Identity in the Age of the Internet.* Simon & Schuster.

Twingate. (2020). Privacy for a premium. https://www.twingate.com/research/privacy-for-a-premium -exploring-peoples-sentiments-on-paying-for-social-media

U.S. Copyright Office. (2021). Copyright and Fair Use: Historical and Legal Perspectives. Washington, D.C.

U.S. Copyright Office. (2021). Fair use index. Retrieved from https://www.copyright.gov/fair-use

U.S. Copyright Office. (2021). Fair use index: A resource on fair use cases. Retrieved from https://www .copyright.gov/fair-use

U.S. Copyright Office. (2021). Fair Use. Retrieved from https://www.copyright.gov/fair-use

U.S. Court of Appeals (2015). Authors Guild v. Google, Inc., 804 F.3d 202 (2d Cir. 2015).

U.S. District Court (2022). McGucken v. Newsweek LLC, No. 1:20-cv-05063 (S.D.N.Y. 2022).

U.S. Supreme Court (1994). Campbell v. Acuff-Rose Music, Inc., 510 U.S. 569 (1994).

U.S. Supreme Court (2021). Google LLC v. Oracle America, Inc., 141 S. Ct. 1183 (2021).

U.S. Supreme Court (2023). Warhol Foundation v. Goldsmith, 598 U.S. 310 (2023).

UK Court of Appeal (2001). Ashdown v. Telegraph Group Ltd., (2001) EWCA Civ 1142.

UK High Court (2011). Sport Five GmbH v. Litman, (2011) EWHC 684 (Ch).

UNECSO. (2021). Recommendation on the Ethics of Artificial Intelligence. In *UNESCO Digital Library.* Retrieved February 6, 2025, from https://unesdoc.unesco.org/ark:/48223/pf0000381137

UNESCO. (2019). *Internet governance and ethical principles: The ROAM-X model.* UNESCO Publishing.

UNESCO. (2024). Global citizenship education in a digital age: teacher guidelines. *unesco.org.* https:// www.unesco.org/en/articles/global-citizenship-education-digital-age-teacher-guidelines

United Nations. (2019). *The right to privacy in the digital age.*

United Nations. 2021, January 14. SDG knowledge: Sustainable Development Goals. https://sdgs.un .org/goals

United States Bureau of the Census. (2001). *Measuring the digital economy.*

University of New South Wales v. Moorhouse, (1975) 133 CLR 1 (Australia).

Uršej, K. (2019). Digital literacy in the first three years of primary school: Case study in Slovenia. *International Journal of Management. Knowledge and Learning, 8*(1), 61–77.

Useful Websites, C. O. E. I. N. T. (2023). Democratic Schools for All - www.coe.int. Democratic Schools for All. https://www.coe.int/en/web/campaign-free-to-speak-safe-to-learn/useful-websites

Ushahidi (2023). Crowdsourcing solutions to empower communities. https://www.ushahidi.com/

Ushahidi (2024). Ushahidi's vision for AI and social impact. https://www.ushahidi.com/about/blog/empowering-communities-through-technology-ushahidi-s-vision-for-ai-and-social-impact/

USIDHR – US Institute of Diplomacy and Human Rights. (2022). What is digital citizenship and why is it important? https://usidhr.org/what-is-digital-citizenship-and-why-is-it-important

Vaccari, C., & Chadwick, A. (2020). Deepfakes and Disinformation: Exploring the Impact of Synthetic Political Video on Deception, Uncertainty, and Trust in News. *Social Media + Society*, *6*(1), 2056305120903408. Advance online publication. DOI: 10.1177/2056305120903408

Van der Sloot, B. (2020). *The general data protection regulation in plain language*. Amsterdam University Press. DOI: 10.5117/9789463726511

van Dijk, J. (2005) The deepening divide: inequality in the information society. https://www.researchgate.net/publication/292683691_The_Deepening_Divide_Inequality_in_the_Information_Society, DOI: DOI: 10.4135/9781452229812

van Dijk, J. (2020). The digital divide. *Polity*.

Van Zyl, A., Turpin, M., & Matthee, M. (2020). How Can Critical Thinking Be Used to Assess the Credibility of Online Information? In Hattingh, M., Matthee, M., Smuts, H., Pappas, I., Dwivedi, Y. K., & Mäntymäki, M. (Eds.), *Responsible Design, Implementation and Use of Information and Communication Technology* (Vol. 12067, pp. 199–210). Springer International Publishing., DOI: 10.1007/978-3-030-45002-1_17

Vanacker, B., & Heider, D. (2012). Ethical harm in virtual communities. *Convergence (London)*, *18*(1), 71–84. DOI: 10.1177/1354856511419916

Velmurugan, P. R., Arunkumar, S., Vettriselvan, R., Deepan, A., & Rajesh, D. (2025). Strategic Approaches to Corporate Social Responsibility and Sustainable Development: Integrating Leadership, Marketing, and Finance. In Navigating Corporate Social Responsibility Through Leadership and Sustainable Entrepreneurship (pp. 373-406). IGI Global Scientific Publishing.

Venkatesh, V., & Davis, F. D. (2000). A theoretical extension of the technology acceptance model: Four longitudinal field studies. *Management Science*, *46*(2), 186–204. DOI: 10.1287/mnsc.46.2.186.11926

Verma, A., & Misra, S. (2017). Dalits: In the era of Digital media. *Amity Journal of Media & Communication Studies, 7*(2). https://www.academia.edu/43894745/Dalits_In_the_Era_of_Digital_Media

Verma, V. (2019). A crisis of representation: Interests, identities and politics. *Journal of Social Inclusion Studies*, *5*(1), 1–18. DOI: 10.1177/2394481119849289

Vettriselvan, R. (2025). Harnessing Innovation and Digital Marketing in the Era of Industry 5.0: Resilient Healthcare SMEs. In The Future of Small Business in Industry 5.0 (pp. 163-186). IGI Global Scientific Publishing.

Viasat. (2023). Why is addressing the digital divide important? https://news.viasat.com/blog/corporate/why-is-addressing-the-digital-divide-important

Villegas-Simón, I., Anglada-Pujol, O., Lloveras, M. C., & Oliva, M. (2023). "I'm Not Just a Content Creator": Digital Cultural Communicators Dealing with Celebrity Capital and Online Communities. *International Journal of Communication, 17*, 19.

Vissenberg, J., d'Haenens, L., & Livingstone, S. (2022). Digital literacy and online resilience as facilitators of young people's well-being? *European Psychologist, 27*(2), 76–85. DOI: 10.1027/1016-9040/a000478

Voigt, P., & Von dem Bussche, H. (2017). *The EU General Data Protection Regulation (GDPR): A practical guide*. Springer. DOI: 10.1007/978-3-319-57959-7

Von Solms, R., & Van Niekerk, J. (2019). Information security management: Standards and best practices. *Computers & Security, 88*, 101–116.

Vrontis, D., Makrides, A., Christofi, M., & Thrassou, A. (2021). Social media influencer marketing: A systematic review, integrative framework and future research agenda. *International Journal of Consumer Studies, 45*(4), 617–644. DOI: 10.1111/ijcs.12647

Walden University. (n.d.). How is digital technology changing our interactions with each other? https://www.waldenu.edu/online-bachelors-programs/bs-in-psychology/resource/how-is-digital-technology-changing-our-interactions-with-each-other#:~:text=Digital%20technology%20has%20so%20changed,now%20a%20regularly%20used%20verb

Walorska, K. (2020). Deepfakes: Technology and implications. *Journal of Digital Forensics, 12*(3), 1–10.

Walther, J., & Jang, S. (2012). Social media likes and information exposure. *Journal of Computer-Mediated Communication, 17*(3), 249–265.

Walton, J., Turner, P., & Simpson, G. (2020). Sustainable healthcare practices: The role of green technologies. *Journal of Sustainable Health Care, 9*(1), 65–77.

Wang, P. A. (2013, June 23–28). Assessment of cybersecurity knowledge and behavior: An anti-phishing scenario. Paper presented at the eighth International Conference on Internet Monitoring and Protection (ICIMP 2013), Rome, Italy.

Wang, T., Guo, X., & Wu, T. (2021). Social capital and digital divide: implications for mobile health policy in developing countries. *Journal of Healthcare Engineering*. https://pmc.ncbi.nlm.nih.gov/articles/PMC7857903/

Wang, J., & Zhang, J. (2020). The impact of perceived credibility and trustworthiness on social media use: A cross-platform study. *Journal of Business Research, 2022*(112), 186–195.

Wang, W., Huang, H., Zhang, L., & Su, C. (2021). Secure and efficient mutual authentication protocol for smart grid under blockchain. *Peer-to-Peer Networking and Applications, 14*(5), 2681–2693. DOI: 10.1007/s12083-020-01020-2

Ward, C. L., Shaw, D., Sprumont, D., Sankoh, O., Tanner, M., & Elger, B. (2018). Good collaborative practice: Reforming capacity building governance of international health research partnerships. *Globalization and Health, 14*(1), 1–6. DOI: 10.1186/s12992-017-0319-4 PMID: 29310698

Wardle, C., & Derakhshan, H. (2017). *Information disorder: Toward an interdisciplinary framework for research and policy making*. Council of Europe.

Ward, S. J. (2014). Radical Media Ethics: Ethics for a global digital world. *Digital Journalism (Abingdon, England)*, *2*(4), 455–471. DOI: 10.1080/21670811.2014.952985

Warhol Foundation v. Goldsmith, 598 U.S. ___ (2023).

Węcel, K., Sawiński, M., Stróżyna, M., Lewoniewski, W., Księżniak, E., Stolarski, P., & Abramowicz, W. (2023). Artificial intelligence—Friend or foe in fake news campaigns. *Economics and Business Review*, *9*(2). Advance online publication. DOI: 10.18559/ebr.2023.2.736

Wei, C., Yu, Z., & Li, Y. (2021). Empathy impairs virtue: The influence of empathy and vulnerability on charitable giving. *Internet Research*, *31*(5), 1803–1822. DOI: 10.1108/INTR-07-2020-0407

Weigel, M., James, C., & Gardner, H. (2009). Learning: Peering backward and looking forward in the digital era. *International Journal of Learning and Media*, *1*(1), 1–18. DOI: 10.1162/ijlm.2009.0005

Wellness Program. (2024). Practicing Cyber Wellness. Ministry of Education, Singapore. https://www.moe.gov.sg/education-in-sg/our-programmes/cyber-wellness

White, K., MacDonnell, R., & Ellard, J. H. (2019). Belief in sustainability: The role of consumer attitudes and behavioral intention. *Journal of Marketing*, *83*(3), 78–91.

Wierzbicki, A. (2018). *Web content credibility*. Springer International Publishing. DOI: 10.1007/978-3-319-77794-8

Wikimedia (2024). Citizenship - Wikipedia. Wikimedia Foundation, Inc. https://en.wikipedia.org/wiki/Citizenship

Wilden, A. (1972). *System and structure: Essays on communications and exchange*. Tavistock.

Williams, B. A., Brooks, C. F., & Shmargad, Y. (2018). How algorithms discriminate based on data they lack: Challenges, solutions, and policy implications. *Journal of Information Policy*, *8*, 78–115. DOI: 10.5325/jinfopoli.8.2018.0078

Williams, M., Karim, W., Gelman, J., & Raza, M. (2024). Ethical data acquisition for LLMs and AI algorithms in healthcare. *NPJ Digital Medicine*, *7*(1). Advance online publication. DOI: 10.1038/s41746-024-01399-9

Wilson, B. (2019). Book Review: Marketing to Gen Z: The Rules for Reaching This Vast and Very Different Generation of Influencers. *Journal of Macromarketing*, *39*(2), 228–230. DOI: 10.1177/0276146719830156

Wilson, D. C. (2021). *Cybersecurity*. MIT Press. DOI: 10.7551/mitpress/11656.001.0001

Wood, N. (2019). *Virtue Rediscovered: Deontology*. Consequentialism, and Virtue Ethics in the Contemporary Moral Landscape.

World Intellectual Property Organization (WIPO). (2021). *Global Copyright and Fair Use Policies: A Comparative Analysis*. WIPO.

Wright, L. (2022). Automated platform governance through visibility and scale: On the transformational power of AutoModerator. *Social Media + Society*, *8*(1), 20563051221077020. DOI: 10.1177/20563051221077020

Yang, S. (2015). Understanding personal online reputation management: A grounded theory study.

Yang, L., Wang, H., & Li, J. (2023). Digital Literacy and Misinformation: A Global Perspective. *International Journal of Communication*, *17*, 1234–1256.

Yeke, S. (2023). Digital intelligence as a partner of emotional intelligence in business administration. *Asia Pacific Management Review*, *28*(4), 390–400. DOI: 10.1016/j.apmrv.2023.01.001

Yingxuan, Y. (2024). Deceptive information dissemination. *Media Studies Quarterly*, *46*(1), 56–70.

Yin, R. K. (2000). Case study evaluations: a decade of progress? In *Evaluation models: Viewpoints on educational and human services evaluation* (pp. 185–193). Springer Netherlands.

Younis, Y. A., & Alghamdi, M. Y. (2021). The use of computer games for teaching and learning cybersecurity in higher education institutions. *Journal of Engineering Research*, *9*(3), 16–29. DOI: 10.36909/jer.v9i3A.10943

Youn, S. (2008). Parental influence and teens' attitude toward online privacy protection. *The Journal of Consumer Affairs*, *42*(3), 362–388. DOI: 10.1111/j.1745-6606.2008.00113.x

You, S., Lee, J., & Lee, Y. (2024). Effect of young adults' political efficacy on digital citizenship: Mediating effect of online social capital. *Current Psychology (New Brunswick, N.J.)*, *43*(41), 32154–32161. Advance online publication. DOI: 10.1007/s12144-024-06766-z

YouTube. (2022). *Global Copyright Enforcement and Fair Use: A Digital Study of Cross-Border Disputes*. YouTube Legal Compliance Division.

Yuen, A. H., & Hew, T. K. (2018). Information and communication technology in educational policies in the Asian region. *Handbook of information technology in primary and secondary education*, 1-20.

Yun, J., & Ahn, J. (2018). Fake news and psychological effects. *Journal of Media Psychology*, *30*(2), 78–90.

Yustika, G. P., & Iswati, S. (2020). Digital Literacy in Formal Online Education: A Short review. *Dinamika Pendidikan*, *15*(1), 66–76. DOI: 10.15294/dp.v15i1.23779

Zafarani, R., Abbasi, M. A., & Liu, H. (2014). *Social media mining: an introduction*. Cambridge University Press.

Zahedi, F. M., Abbasi, A., & Chen, Y. (2015). Fake-website detection tools: Identifying elements that promote individuals' use and enhance their performance. *Journal of the Association for Information Systems*, *16*(6), 448–484. DOI: 10.17705/1jais.00399

Zannettou, S., Sirivianos, M., Blackburn, J., & Kourtellis, N. (2019). The Web of False Information: Rumors, Fake News, Hoaxes, Clickbait, and Various Other Shenanigans. *ACM Journal of Data and Information Quality*, *11*(3), 1–37. DOI: 10.1145/3309699

Zeng, J., & Kaye, D. B. V. (2022). From content moderation to visibility moderation: A case study of platform governance on TikTok. *Policy and Internet, 14*(1), 79–95. DOI: 10.1002/poi3.287

Zhang, Y., McAreavey, K., & Liu, W. (2022). Developing andExperimenting on Approaches to Explainability in AI Systems. In A. P.Rocha, L. Steels, & J. van den Herik (Eds.), Proceedings of the 14thInternational Conference on Agents and Artificial Intelligence (ICAART'22). Vol. 2, pp. 518-527. DOI: 10.5220/0010900300003116

Zhang, H., & Daim, T. U. (2023). *Cybersecurity: A Technology Landscape Analysis. Technology Assessment of Cybersecurity*. Springer Nature Switzerland.

Zhang, T., & McMullan, R. (2020). Social media and green healthcare: Consumer perceptions and influence. *Journal of Digital Healthcare Management, 12*(2), 67–81.

Zhang, X., Liu, Y., & Chen, Z. (2019). The Role of Social Media in Spreading Fake News. *Journal of Information Science, 45*(4), 567–582.

Zhong, J., Qiu, J., Sun, M., Jin, X., Zhang, J., Guo, Y., Qiu, X., Xu, Y., Huang, J., & Zheng, Y. (2022). To be ethical and responsible digital citizens or not: a linguistic analysis of cyberbullying on social media. *Frontiers in Psychology*. https://pmc.ncbi.nlm.nih.gov/articles/PMC9100568/

Zhou, D., Li, Y., & Tirasawasdichai, T. (2022, October 14). Do online social interactions cultivate social capital? Evidence from a longitudinal study. *Frontiers in Psychology, 13*, 989137. DOI: 10.3389/fpsyg.2022.989137 PMID: 36312069

Zhu, J. J., Tuo, L., You, Y., Fei, Q., & Thomson, M. (2024). A preemptive and curative solution to mitigate data breaches: Corporate social responsibility as a double layer of protection. *JMR, Journal of Marketing Research, 61*(4), 778–801. DOI: 10.1177/00222437231218969

Zuboff, S. (2019). *The age of surveillance capitalism: The fight for a human future at the new frontier of power.*

Zuboff, S. (2019). *The Age of Surveillance Capitalism: The Fight for a Human Future at the New Frontier of Power*. PublicAffairs.

Zulkarnain, Z., Heleni, S., & Thahir, M. (2020, October). Digital literacy skills of math students through e-learning in COVID-19 era: A case study in Universitas Riau. [). IOP Publishing.]. *Journal of Physics: Conference Series, 1663*(1), 012015. DOI: 10.1088/1742-6596/1663/1/012015

About the Contributors

Hakikur Rahman is an academic over 30 years has served leading education institutes in Bangladesh and abroad, and established various ICT4D projects funded by ADB, EU, UNDP and World Bank. He is a Professor and the Dean of School of Science and Engineering at the Presidency University. He is also currently serving as the Director at the Ansted University Sustainability Research Institute. Before joining Presidency University, he was a Professor, CSE department at the Asian University of Bangladesh. Before that he served the BRAC University as an Associate Professor (Chair, CSE Dept.) and a Post-Doctoral Researcher (Research Associate) at the University of Minho, Portugal under the Centro Algoritmi. He has written and edited over 25 books, more than 50 book chapters and contributed over 200 articles on computer education, ICTs, knowledge management, open innovation, data mining and e-government research in newspapers, journals and conference proceedings. Graduating from the Bangladesh University of Engineering and Technology in 1981, he has done his Master's of Engineering from the American University of Beirut in 1986 and completed his PhD in Computer Engineering from the Ansted University, BVI, UK in 2001; and PhD in ICT from the Empresarial University of Costa Rica in 2011.

***

A.K. Subramani completed his Ph.D. in Business Administration at Manonmaniam Sundaranar University, Tirunelveli, India. A teaching career spanning 18 years has been amassed, during which 74 articles have been published in international journals, including two papers in ABDC-A category journals, eight papers in the ABDC-B category, and two papers in the ABDC-C category. Additionally, 28 papers were published in Scopus-indexed journals, and eight in Web of Science-indexed journals. Numerous papers have been presented at international and national conferences. Over 31 courses have been completed from the Swayam Online platform, covering HR, Business Analytics, and Intellectual Property domains. NPTEL Star Certificates have been received, including those for NPTEL Evangelist, Motivated Learner, Enthusiast, and Discipline Star. Peer review has been conducted for several journals published by Inderscience and Emerald publishers. Research interests are focused on human resource management and organizational behavior. The 'Teaching and Research Excellence Award' from Grabs Charitable Trust, India, and the 'Distinguished Teacher Award-Management' from MTC Global Trust have been awarded. Resource person roles have been fulfilled in many training workshops on SPSS and AMOS software packages in reputed educational institutions across the country.

Alexandre Duarte is an Assistant Professor in Universidade NOVA de Lisboa, invited Professor in Universidade Católica Portuguesa and researcher of the R&D Unit ICNOVA (Universidade Nova de Lisboa). Throughout his career, Alexandre had the opportunity to work for several multinational advertising agencies such as TBWA, Saatchi & Saatchi, Ogilvy, W/Portugal, W/Brazil, Lowe&Partners, or

BrandiaCentral as a Senior Creative Copywriter. Author of many seminars on communication, regularly participates in different conferences and lectures in both business and academic contexts. Between 2012 and 2014, was the Tutor of the EUROBEST Young Marketers Academy and in the beginning of 2014, was appointed CEO of RESTART, an Institute of Creativity, Arts and New Technologies. In 2018 was invited to teach in the UNIVERSIDADE CATÓLICA PORTUGUESA, where now Coordinates the Post Graduations Courses in Advertising&Creativity and Service Design and, in 2019 was re-elected member of the Directors Board of EDCOM.

Ha Jin Hwang is the Director of the School of creative industries at Astana IT University, Kazakhstan. Prior to that, he served as Professor and Head of Department of Business Analytics at Sunway University, Malaysia (2016-2023), Professor at KIMEP University, Kazakhstan (2010-2016), Professor at Catholic University of Daegu, Korea (1991-2010), Associate Professor of MIS, Minnesota State University, Mankato, Minnesota, USA (1989-1991).

Jiow Hee Jhee has lectured and trained over 50,000 students, parents, government officials and leaders on new media topics, both locally and internationally. His research investigates the impact of media on the family, cyber wellness and cybercrime, and he has presented academic papers on parental mediation, video gaming and gamification. Prior to obtaining his PhD, Dr Jiow helped set up Singapore's first Cyber Wellness Centre in 2006 and was also involved in developing a body of cyber wellness values, ethics and best practices, to promote cyber wellness in schools. Dr Jiow currently serves as a member of the Films Consultative Panel (IMDA) and Media Literacy Council (IMDA), and he developed the BSc (Hons) Digital Communications and Integrated Media programme at Singapore Institute of Technology (SIT). He is Deputy Director at SIT Teaching and Learning Academy looking into nurturing the applied learning and competency-based education practice.

Youmei Liu earned her doctoral degree in Curriculum and Instruction focusing on instructional technology in 2003 from the University of Houston. She worked as an instructional designer and educational production specialist. Currently she is the Director of Assessment and Accreditation Services at the University of Houston. She also served as Vice President and Conference Chairs for the Society of International Chinese Educational Technology and Chinese American Educational Research & Development Association. Dr. Liu has rich research experiences and has numerous publications in the area of quality course design, innovative use of educational technology, distance education, mobile learning and student learning assessment. She has extensive accomplishments in collaborative research with faculty, in development of evaluation programs to assess teaching and learning quality, in conducting major innovative research projects on instructional technology, and in presenting research projects at important national and international conferences. She has also served as an evaluator for several NSF grants in higher educational institutions.

Zhanni Luo is an associate professor at Chongqing Normal University, China.Her research interests include e-learning, gamification, teacher education, behavior change, and learning engagement.

Vijayalakshmi M is an Associate Professor, AMET University, specialised in HRM and Marketing. He is acting as Mentor, Saraswathi Institute of Medical Sciences, Hapur. He served as Review Board member, National Council for Higher Education, Malawi, Head of the Department, School of

Commerce and Management Studies, St. Eugene University, Zambia, and Director, Research and Publication, St. John the Baptist University, Malawi. He has published 25 books and 93 articles in peer reviewed journals and edited book chapters. Under his guidance two have completed and five are pursuing their PhD.

Mohsen Mahmoudi-Dehaki is an Assistant Professor of TEFL at the Islamic Azad University of Najafabad, Iran. With over a decade of experience teaching undergraduate and postgraduate students, he has established himself as a respected educator and scholar. His research has been widely published in national and international journals, solidifying his reputation as a prominent voice in the field. Dr. Mahmoudi-Dehaki serves as a peer reviewer for several esteemed scientific journals, including RELP, ensuring the highest standards of academic rigor and integrity. His academic achievements include winning the prestigious first place in the English Writing Award in Iran, a distinction earned among all university students from state universities and institutes of higher education. His paper "COVID-19 Lingo or Covidipedia" has garnered significant attention, with over 70,000 visits, making it the most visited paper in Iranian scientific journals. This remarkable achievement underscores his significant impact on the academic community. Dr. Mahmoudi-Dehaki's research and teaching interests span a diverse range of novel branches in applied linguistics, including psycho/cognitive/neurocognitive/forensic/jury/sociocultural/ecological, and clinical linguistics. He is also well-versed in educational technology, educational psychology, and psychometrics, demonstrating his versatility and breadth of expertise in the interdisciplinary fields of education and linguistics.

Nasim Nasr-Esfahani, PhD, is an Assistant Professor of TEFL at the Islamic Azad University of Isfahan (Khorasgan), Iran. She earned her PhD in TEFL at the Islamic Azad University of Isfahan, Khorasgan. She has over two decades of experience teaching undergraduate and postgraduate students in Iran. Her publications have appeared in leading international journals, such as the Australian Journal of Teacher Education. She serves as a reviewer and article editor of RELP Journal. Her areas of research and teaching interests are dynamic and evolving and span language education, generative artificial intelligence (GenAI) in education, critical thinking skills, oral and written communication skills, student engagement, advancing equity in education, brain-based teaching and learning, neurolinguistics, psycholinguistics as well as teaching and research methodologies.

Vijayalakshmi R. is a PhD Research Scholar and an Assistant Professor at Saveetha School of Managment, SIMATS, Saveetha University, Chennai, Tamilnadu, India. She holds Master of Philosophy (M.Phil) from Sri Venkateshwara University, Tirupati and MBA from Anna University, Chennai. Her exceptional academic performance earned her first-class honors with distinction, along with the prestigious top rank. With a decade of experience in academics, she has made significant contributions to the students' community. Her dedication to scholarly research is evident through her publications in impact factor journals and springer nature. She has leveraged her experiences by receiving Best Paper Award at a prestigious international conference. She has served as a resource person for VIT University, extending her expertise to benefit the wider academic community.

Deepika Ramamoorthy is currently Faculty of Marketing at Rajalakshmi School of Business. She brings with her a wealth of corporate and academic experience that spans close to a decade, during which she has contributed significantly to teaching and research at prestigious institutions such as Apollo Hospitals, Loyola Institute of Business Administration (LIBA), and Anna University, where she earned her PhD from the esteemed CEG Campus. Notably, Dr. Deepika has excelled in her academic pursuits, achieving the remarkable feat of qualifying the UGC's NET (Management) thrice. She is also certified in Business Analytics from Great Lakes Institute of Management, reflecting her dedication to staying at the forefront of her field. Her scholarly achievements are evidenced by her extensive publications in national and international peer-reviewed ABDC/Scopus indexed journals. Recognized for her outstanding contributions to education, Dr. Deepika was honored with the "Best Teacher Award" by Rajalakshmi School of Business for the academic year 2022 - 2023. Dr. Deepika also serves as a Reviewer for esteemed ABDC and Scopus indexed journals such as the Journal of Retailing and Consumer Services and the Journal of Wealth Management. She is the author of the insightful book titled "The Game Changer – Smartphone Shopping Applications.". Dr. Deepika has addressed various business graduates across different domains, inspiring and guiding them in their professional journeys. Furthermore, she holds certifications as a Behaviour & Mentoring Analyst and an AIMA BizLab Certified Trainer, further underscoring her expertise and dedication to her profession.

Dheera C Sasidharan is currently a Teaching Assistant and PhD scholar at the Times School of Media, Bennett University, India. Her research revolves around Digital Media Literacy, with particular interests in development communication, civic participation, and social media engagement studies. With a background in journalism, Dheera has previously worked with The New Indian Express (Print), Jeevan TV (broadcast channel) and Smash Story (Digital marketing and advertising).

Catherin T.C. is an Assistant Professor at the Department of Management Studies, Saveetha Engineering College. With a strong research focus, she has published 8 papers in reputable journals. Dr. Catherin is dedicated to academic excellence and specializes in Operations, fostering innovative thinking and knowledge sharing among her students.

Adeline Tay is an undergraduate currently matriculating in Digital Communications and Integrated Media at Singapore Institute of Technology. Accredited with a diploma in Child Psychology and Early Education and having worked in a preschool for a year, she has extensive experience in teaching and interacting with young children. Having borne witness to how digital media affected children physiologically and psychologically, she is passionate about improving the well-being of young children. She has presented related papers at academic conferences, spoken to parents and preschool educators on digital device use in nurturing early learners, participated in radio interviews, and written an opinion piece discussing how broader familial ecosystems are at play in managing children's device use."

Bea Wohl's research focuses on digital innovation and computing education, having specifically studied the delivery of the 2014 English computing curriculum. They have worked in consultancy and communications for over a decade, particularly in the creative and cultural sectors. They have an MA in Media and Communications and an Advanced Certificate in PR. Their research interests are the intersection of the digital economy, media theory and digital literacy. As a teaching and supervising master's student at CCI, Bea provides research support across the Creative Computing Institute, shaping funded research that examines the social impact of technology. Bea is the Post-Doctoral Fellow on the Transforming the Gap: Inclusive Digital Arts and Humanities Research Skills (DAReS) CoLab.

Yuanlu Xiong is a postgraduate student at Chongqing Normal University, specializing in English Language Teaching. Her research focuses on technology-based English learning and English learning motivation. She has published papers and presented her work at national conferences, showcasing her enthusiasm for language education.

Index

A

B

C

D

E

F

G

H

I

K

L

M

N

O

www.ingramcontent.com/pod-product-compliance
Lightning Source LLC
LaVergne TN
LVHW062110300926
849161LV00010B/2417
* 9 7 9 8 3 6 9 3 6 6 7 5 2 *